Lecture Notes in Computer Science 2919
Edited by G. Goos, J. Hartmanis, and J. van Leeuwen

Springer
Berlin
Heidelberg
New York
Hong Kong
London
Milan
Paris
Tokyo

Enrico Giunchiglia   Armando Tacchella (Eds.)

Theory and Applications of Satisfiability Testing

6th International Conference, SAT 2003
Santa Margherita Ligure, Italy, May 5-8, 2003
Selected Revised Papers

 Springer

Series Editors

Gerhard Goos, Karlsruhe University, Germany
Juris Hartmanis, Cornell University, NY, USA
Jan van Leeuwen, Utrecht University, The Netherlands

Volume Editors

Enrico Giunchiglia
Armando Tacchella
Università di Genova
DIST
Viale Causa 13, 16145 Genova, Italy
E-mail: giunchiglia@unige.it, tac@dist.unige.it

Cataloging-in-Publication Data applied for

A catalog record for this book is available from the Library of Congress.

Bibliographic information published by Die Deutsche Bibliothek
Die Deutsche Bibliothek lists this publication in the Deutsche Nationalbibliografie;
detailed bibliographic data is available in the Internet at <http://dnb.ddb.de>.

CR Subject Classification (1998): F.4.1, I.2.3, I.2.8, I.2, F.2.2, G.1.6

ISSN 0302-9743
ISBN 3-540-20851-8 Springer-Verlag Berlin Heidelberg New York

Springer-Verlag is a part of Springer Science+Business Media

springeronline.com

© Springer-Verlag Berlin Heidelberg 2004
Printed in Germany

Typesetting: Camera-ready by author, data conversion by Boller Mediendesign
Printed on acid-free paper   SPIN: 10982143   06/3142   5 4 3 2 1 0

Preface

This book is devoted to the *6th International Conference on Theory and Applications of Satisfiability Testing (SAT 2003)* held in Santa Margherita Ligure (Genoa, Italy), during May 5–8, 2003. SAT 2003 followed the Workshops on Satisfiability held in Siena (1996), Paderborn (1998), and Renesse (2000), and the Workshop on Theory and Applications of Satisfiability Testing held in Boston (2001) and in Cincinnati (2002). As in the last edition, the SAT event hosted a SAT solvers competition, and, starting from the 2003 edition, also a Quantified Boolean Formulas (QBFs) solvers comparative evaluation.

There were 67 submissions of high quality, authored by researchers from all over the world. All the submissions were thoroughly evaluated, and as a result 42 were selected for oral presentations, and 16 for a poster presentation. The presentations covered the whole spectrum of research in propositional and QBF satisfiability testing, including proof systems, search techniques, probabilistic analysis of algorithms and their properties, problem encodings, industrial applications, specific tools, case studies and empirical results. Further, the program was enriched by three invited talks, given by Riccardo Zecchina (on *"Survey Propagation: from Analytic Results on Random k-SAT to a Message-Passing Algorithm for Satisfiability"*), Toby Walsh (on *"Challenges in SAT* (and QBF)*"*) and Wolfgang Kunz (on *"ATPG Versus SAT: Comparing Two Paradigms for Boolean Reasoning"*). SAT 2003 thus provided a unique forum for the presentation and discussion of research related to the theory and applications of propositional and QBF satisfiability testing.

The book includes 38 contributions. The first 33 are revised versions of some of the articles that were presented at the conference. The last 5 articles present the results of the SAT competition and of the QBF evaluation, solvers that won the SAT competition, and results on survey and belief propagation. All 38 papers were thoroughly reviewed.

We would like to thank the many people who contributed to the SAT 2003 organization (listed in the following pages), the SAT 2003 participants for the lively discussions, and the sponsors.

September 2003

Enrico Giunchiglia
Armando Tacchella

SAT 2003 Organization

SAT 2003 was organized by DIST (Dipartimento di Informatica, Sistemistica e Telematica), Università di Genova.

Chair

Enrico Giunchiglia, DIST, Università di Genova

Organizing Committee

John Franco, University of Cincinnati
Enrico Giunchiglia, Università di Genova
Henry Kautz, University of Washington
Hans Kleine Büning, Universität Paderborn
Hans van Maaren, University of Delft
Bart Selman, Cornell University
Ewald Speckenmayer, Universität Köln

SAT Competition Organizers

Daniel Le Berre, CRIL, Université d'Artois
Laurent Simon, LRI Laboratory, Université Paris-Sud

QBF Comparative Evaluation Organizers

Daniel Le Berre, CRIL, Université d'Artois
Laurent Simon, LRI Laboratory, Université Paris-Sud
Armando Tacchella, DIST, Università di Genova

Local Organization

Roberta Ferrara, Università di Genova
Marco Maratea, DIST, Università di Genova
Massimo Narizzano, DIST, Università di Genova
Adriano Ronzitti, DIST, Università di Genova
Armando Tacchella, DIST, Università di Genova (Chair)

Program Committee

Dimitris Achlioptas, Microsoft Research
Fadi Aloul, University of Michigan
Fahiem Bacchus, University of Toronto
Armin Biere, ETH Zurich
Nadia Creignou, Université de la Méditerranée, Marseille
Olivier Dubois, Université Paris 6
Uwe Egly, Technische Universität Wien
John Franco, University of Cincinnati
Ian Gent, St. Andrews University
Enrico Giunchiglia, DIST, Università di Genova
Carla Gomez, Cornell University
Edward A. Hirsch, Steklov Institute of Mathematics at St. Petersburg
Holger Hoos, University of British Columbia
Henry Kautz, University of Washington
Hans Kleine Büning, Universität Paderborn
Oliver Kullmann, University of Wales, Swansea
Daniel Le Berre, CRIL, Université d'Artois
Joo Marques-Silva, Instituto Superior Técnico, Univ. Técnica de Lisboa
Hans van Maaren, University of Delft
Remi Monasson, Laboratoire de Physique Théorique de l'ENS
Daniele Pretolani, Università di Camerino
Paul W. Purdom, Indiana University
Jussi Rintanen, Freiburg University
Bart Selman, Cornell University
Malik Sharad, Princeton University
Laurent Simon, LRI Laboratory, Université Paris-Sud
Ewald Speckenmeyer, Universität Köln
Armando Tacchella, DIST, Università di Genova
Allen Van Gelder, UC Santa Cruz
Miroslav N. Velev, Carnegie Mellon University
Toby Walsh, University of York

Additional Referees

G. Audemard	A. Kulikov	S. Porschen	D. Tang
F. Corradini	T. Lettmann	S. Prestwich	H. Tompits
S. Coste-Marquis	I. Lynce	B. Randerath	D. Tompkins
H. Daudé	M. Maratea	O. Roussel	H. Vollmer
L. Drake	M. Molloy	A. Rowley	S. Woltran
M. Fink	M. Narizzano	U. Schoening	Y. Yu
Z. Fu	S. Nikolenko	A. Shmygelska	
A. Kojevnikov	A. Polleres	K. Smyth	

Sponsoring Institutions

CoLogNet, Network of Excellence in Computational Logic
DIST, Università di Genova
IISI, Intelligent Information Systems Institute at Cornell University
Microsoft Research
MIUR, Ministero dell'Istruzione, dell'Università e della Ricerca

Table of Contents

Satisfiability and Computing van der Waerden Numbers

Michael R. Dransfield[1], Victor W. Marek[2], and Mirosław Truszczyński[2]

[1] National Security Agency, Information Assurance Directorate,
Ft. Meade, MD 20755
[2] Department of Computer Science, University of Kentucky, Lexington,
KY 40506-0046, USA

Abstract. In this paper we bring together the areas of combinatorics and propositional satisfiability. Many combinatorial theorems establish, often constructively, the existence of positive integer functions, without actually providing their closed algebraic form or tight lower and upper bounds. The area of Ramsey theory is especially rich in such results. Using the problem of computing van der Waerden numbers as an example, we show that these problems can be represented by parameterized propositional theories in such a way that decisions concerning their satisfiability determine the numbers (function) in question. We show that by using general-purpose complete and local-search techniques for testing propositional satisfiability, this approach becomes effective — competitive with specialized approaches. By following it, we were able to obtain several new results pertaining to the problem of computing van der Waerden numbers. We also note that due to their properties, especially their structural simplicity and computational hardness, propositional theories that arise in this research can be of use in development, testing and benchmarking of SAT solvers.

1   Introduction

In this paper we discuss how the areas of propositional satisfiability and combinatorics can help advance each other. On one hand, we show that recent dramatic improvements in the efficiency of SAT solvers and their extensions make it possible to obtain new results in combinatorics simply by encoding problems as propositional theories, and then computing their models (or deciding that none exist) using off-the-shelf general-purpose SAT solvers. On the other hand, we argue that combinatorics is a rich source of structured, parameterized families of hard propositional theories, and can provide useful sets of benchmarks for developing and testing new generations of SAT solvers.

In our paper we focus on the problem of computing van der Waerden numbers. The celebrated van der Waerden theorem [20] asserts that for every positive integers k and l there is a positive integer m such that every partition of $\{1, \ldots, m\}$ into k blocks (parts) has at least one block with an arithmetic progression of length l. The problem is to find the least such number m. This

E. Giunchiglia and A. Tacchella (Eds.): SAT 2003, LNCS 2919, pp. 1–13, 2004.
International Association for Cryptologic Research Springer-Verlag Berlin Heidelberg 2004

number is called the *van der Waerden number* $W(k,l)$. Exact values of $W(k,l)$ are known only for five pairs (k,l). For other combinations of k and l there are some general lower and upper bounds but they are very coarse and do not give any good idea about the actual value of $W(k,l)$. In the paper we show that SAT solvers such as POSIT [6], and SATO [21], as well as recently developed local-search solver *walkaspps* [13], designed to compute models for propositional theories extended by cardinality atoms [4], can improve lower bounds for van der Waerden numbers for several combinations of parameters k and l.

Theories that arise in these investigations are determined by the two parameters k and l. Therefore, they show a substantial degree of structure and similarity. Moreover, as k and l grow, these theories quickly become very hard. This hardness is only to some degree an effect of the growing size of the theories. For the most part, it is the result of the inherent difficulty of the combinatorial problem in question. All this suggests that theories resulting from hard combinatorial problems defined in terms of tuples of integers may serve as benchmark theories in experiments with SAT solvers.

There are other results similar in spirit to the van der Waerden theorem. The Schur theorem states that for every positive integer k there is an integer m such that every partition of $\{1, \ldots, m\}$ into k blocks contains a block that is not sum-free. Similarly, the Ramsey theorem (which gave name to this whole area in combinatorics) [16] concerns the existence of monochromatic cliques in edge-colored graphs, and the Hales-Jewett theorem [11] concerns the existence of monochromatic lines in colored cubes. Each of these results gives rise to a particular function defined on pairs or triples of integers and determining the values of these functions is a major challenge for combinatorialists. In all cases, only few exact values are known and lower and upper estimates are very far apart. Many of these results were obtained by means of specialized search algorithms highly depending on the combinatorial properties of the problem. Our paper shows that generic SAT solvers are maturing to the point where they are competitive and sometimes more effective than existing advanced specialized approaches.

2   Van der Waerden Numbers

In the paper we use the following terminology. By $\mathbb{Z}^+$ we denote the set of positive integers and, for $m \in \mathbb{Z}^+$, $[m]$ is the set $\{1, \ldots, m\}$. A *partition* of a set X is a collection $\mathcal{A}$ of nonempty and mutually disjoint subsets of X such that $\bigcup \mathcal{A} = X$. Elements of $\mathcal{A}$ are commonly called *blocks*.

Informally, the van der Waerden theorem [20] states that if a sufficiently long initial segment of positive integers is partitioned into a few blocks, then one of these blocks has to contain an arithmetic progression of a desired length. Formally, the theorem is usually stated as follows.

Theorem 1 (van der Waerden theorem). *For every* $k, l \in \mathbb{Z}^+$, *there is* $m \in \mathbb{Z}^+$ *such that for every partition* $\{A_1, \ldots, A_k\}$ *of* $[m]$, *there is* i, $1 \leq i \leq k$, *such that block* A_i *contains an arithmetic progression of length at least* l.

We define the *van der Waerden number* $W(k, l)$ to be the least number m for which the assertion of Theorem 1 holds. Theorem 1 states that van der Waerden numbers are well defined.

One can show that for every k and l, where $l \geq 2$, $W(k, l) > k$. In particular, it is easy to see that $W(k, 2) = k + 1$. From now on, we focus on the non-trivial case when $l \geq 3$.

Little is known about the numbers $W(k, l)$. In particular, no closed formula has been identified so far and only five exact values are known. They are shown in Table 1 [1,10].

l	3	4	5
k			
2	9	35	178
3	27		
4	76		

Table 1. Known non-trivial values of van der Waerden numbers

Since we know few exact values for van der Waerden numbers, it is important to establish good estimates. One can show that the Hales-Jewett theorem entails the van der Waerden theorem, and some upper bounds for the numbers $W(k, l)$ can be derived from the Shelah's proof of the former [18]. Recently, Gowers [9] presented stronger upper bounds, which he derived from his proof of the Szemerédi theorem [19] on arithmetic progressions.

In our work, we focus on lower bounds. Several general results are known. For instance, Erdős and Rado [5] provided a non-constructive proof for the inequality

$$W(k, l) > (2(l - 1)k^{l-1})^{1/2}.$$

For some special values of parameters k and l, Berlekamp obtained better bounds by using properties of finite fields [2]. These bounds are still rather weak. His strongest result concerns the case when $k = 2$ and $l - 1$ is a prime number. Namely, he proved that when $l - 1$ is a prime number,

$$W(2, l) > (l - 1)2^{l-1}.$$

In particular, $W(2, 6) > 160$ and $W(2, 8) > 896$.

Our goal in this paper is to employ propositional satisfiability solvers to find lower bounds for several small van der Waerden numbers. The bounds we find significantly improve on the ones implied by the results of Erdős and Rado, and Berlekamp.

We proceed as follows. For each triple of positive integers $\langle k, l, m \rangle$, we define a propositional CNF theory $\text{vdW}_{k,l,m}$ and then show that $\text{vdW}_{k,l,m}$ is satisfiable if and only if $W(k, l) > m$. With such encodings, one can use SAT solvers (at least in principle) to determine the satisfiability of $\text{vdW}_{k,l,m}$ and, consequently, find $W(k, l)$. Since $W(k, l) > k$, without loss of generality we can restrict our attention to $m > k$. We also show that more concise encodings are possible, leading ultimately to better bounds, if we use an extension of propositional logic by *cardinality atoms* and apply to them solvers capable of handling such atoms directly.

To describe $\text{vdW}_{k,l,m}$ we will use a standard first-order language, without function symbols, but containing a predicate symbol *in_block* and constants $1, \ldots, m$. An intuitive reading of a ground atom $in\_block(i, b)$ is that an integer i is in block b.

We now define the theory $\text{vdW}_{k,l,m}$ by including in it the following clauses:

vdW1: $\neg in\_block(i, b_1) \vee \neg in\_block(i, b_2)$, for every $i \in [m]$ and every $b_1, b_2 \in [k]$ such that $b_1 < b_2$,

vdW2: $in\_block(i, 1) \vee \ldots \vee in\_block(i, k)$, for every $i \in [m]$,

vdW3: $\neg in\_block(i, b) \vee \neg in\_block(i + d, b) \vee \ldots \vee \neg in\_block(i + (l-1)d, b)$, for every $i, d \in [m]$ such that $i + (l-1)d \leq m$, and for every b such that $1 \leq b \leq k$.

As an aside, we note that we could design $\text{vdW}_{k,l,m}$ strictly as a theory in propositional language using propositional atoms of the form $in\_block_{i,b}$ instead of ground atoms $in\_block(i, b)$. However, our approach opens a possibility to specify this theory as finite (and independent of data) collections of *propositional schemata*, that is, open clauses in the language of first-order logic without function symbols. Given a set of appropriate constants (to denote integers and blocks) such theory, after grounding, coincides with $\text{vdW}_{k,l,m}$. In fact, we have defined an appropriate syntax that allows us to specify both data and schemata and implemented a grounding program *psgrnd* [4] that generates their equivalent ground (propositional) representation. This grounder accepts arithmetic expressions as well as simple regular expressions, and evaluates and eliminates them according to their standard interpretation. Such approach significantly simplifies the task of developing propositional theories that encode problems, as well as the use of SAT solvers [4].

Propositional interpretations of the theory $\text{vdW}_{k,l,m}$ can be identified with subsets of the set of atoms $\{in\_block(i, b) : i \in [m], \ b \in [k]\}$. Namely, a set $M \subseteq \{in\_block(i, b) : i \in [m], \ b \in [k]\}$ determines an interpretation in which all atoms in M are true and all other atoms are false. In the paper we always assume that interpretations are represented as sets.

It is easy to see that clauses (vdW1) ensure that if M is a model of $\text{vdW}_{k,l,m}$ (that is, is an interpretation satisfying all clauses of $\text{vdW}_{k,l,m}$), then for every $i \in [m]$, M contains at most one atom of the form $in\_block(i, b)$. Clauses (vdW2) ensure that for every $i \in [m]$ there is at least one $b \in [k]$ such that $in\_block(i, b) \in M$. In other words, clauses (vdW1) and (vdW2) together ensure that if M is a model of $\text{vdW}_{k,l,m}$, then M determines a partition of $[m]$ into k blocks.

The last group of constraints, clauses (vdW3), guarantee that elements from $[m]$ forming an arithmetic progression of length l do not all belong to the same block. All these observations imply the following result.

Proposition 1. *There is a one-to-one correspondence between models of the formula* $vdW_{k,l,m}$ *and partitions of* $[m]$ *into* k *blocks so that no block contains an arithmetic progression of length* l. *Specifically, an interpretation* M *is a model of* $vdW_{k,l,m}$ *if and only if* $\{\{i \in [m]: in\_block(i, b) \in M\}: b \in [k]\}$ *is a partition of* $[m]$ *into* k *blocks such that no block contains an arithmetic progression of length* l.

Proposition 1 has the following direct corollary.

Corollary 1. *For every positive integers* $k, l,$ *and* $m,$ *with* $l \geq 2$ *and* $m > k,$ $m < W(k, l)$ *if and only if the formula* $vdW_{k,l,m}$ *is satisfiable.*

It is evident that if m has the property that $vdW_{k,l,m}$ is unsatisfiable then for every $m' > m$, $vdW_{k,l,m'}$ is also unsatisfiable. Thus, Corollary 1 suggests the following algorithm that, given k and l, computes the van der Waerden number $W(k, l)$: for consecutive integers $m = k+1, k+2, \ldots$ we test whether the theory $vdW_{k,l,m}$ is satisfiable. If so, we continue. If not, we return m and terminate the algorithm. By the van der Waerden theorem, this algorithm terminates.

It is also clear that there are simple symmetries involved in the van der Waerden problem. If a set M of atoms of the form $in\_block(i, b)$ is a model of the theory $vdW_{k,l,m}$, and π is a permutation of $[k]$, then the corresponding set of atoms $\{in\_block(i, \pi(b)): in\_block(i, b) \in M\}$ is also a model of $vdW_{k,l,m}$, and so is the set of atoms $\{in\_block(m + 1 - i, b): in\_block(i, b) \in M\}$.

Following the approach outlined above, adding clauses to break these symmetries, and applying POSIT [6] and SATO [21] as a SAT solvers we were able to establish that $W(4, 3) = 76$ and compute a "library" of counterexamples (partitions with no block containing arithmetic progressions of a specified length) for $m = 75$. We were also able to find several lower bounds on van der Waerden numbers for larger values of k and m.

However, a major limitation of our first approach is that the size of theories $vdW_{k,l,m}$ grows quickly and makes complete SAT solvers ineffective. Let us estimate the size of the theory $vdW_{k,l,m}$. The total size of clauses (vdW1) (measured as the number of atom occurrences) is $\Theta(mk^2)$. The size of clauses (vdW2) is $\Theta(mk)$. Finally, the size of clauses (vdW3) is $\Theta(m^2)$ (indeed, there are $\Theta(m^2/l)$ arithmetic progressions of length l in $[m]$)[3]. Thus, the total size of the theory $vdW_{k,l,m}$ is $\Theta(mk^2 + m^2)$.

To overcome this obstacle, we used a two-pronged approach. First, as a modeling language we used PS+ logic [4], which is an extension of propositional logic by cardinality atoms. Cardinality atoms support concise representations of constraints of the form "at least p and at most r elements in a set are true" and result in theories of smaller size. Second, we used a local-search algorithm, *walkaspps*, for finding models of theories in logic PS+ that we have designed and

---

[3] Goldstein [8] provided a precise formula. When $r = rm(m - 1, l - 1)$ and $q = q(m - 1, l - 1)$ then there are $q \cdot r + \binom{q-1}{2} \cdot (l - 1)$ arithmetic progressions of length l in $[m]$.

implemented recently [13]. Using encodings as theories in logic PS+ and *walka-spps* as a solver, we were able to obtain substantially stronger lower bounds for van der Waerden numbers than those know to date.

We will now describe this alternative approach. For a detailed treatment of the PS+ logic we refer the reader to [4]. In this paper, we will only review most basic ideas underlying the logic PS+ (in its propositional form). By a *propositional cardinality atom* (*c-atom* for short), we mean any expression of the form $m\{p_1,\ldots,p_k\}n$ (one of m and n, but not both, may be missing), where m and n are non-negative integers and $p_1,\ldots,p_k$ are propositional atoms from At. The notion of a clause generalizes in an obvious way to the language with cardinality atoms. Namely, a *c-clause* is an expression of the form

$$C = A_1 \vee \ldots \vee A_s \vee \neg B_1 \vee \ldots \vee \neg B_t, \tag{1}$$

where all A_i and B_i are (propositional) atoms or cardinality atoms.

Let $M \subseteq At$ be a set of atoms. We say that M *satisfies* a cardinality atom $m\{p_1,\ldots,p_k\}n$ if

$$m \leq |M \cap \{p_1,\ldots,p_k\}| \leq n.$$

If m is missing, we only require that $|M \cap \{p_1,\ldots,p_k\}| \leq n$. Similarly, when n is missing, we only require that $m \leq |M \cap \{p_1,\ldots,p_k\}|$. A set of atoms M *satisfies* a c-clause C of the form (1) if M satisfies at least one atom A_i or does not satisfy at least one atom B_j. W note that the expression $1\{p_1,\ldots,p_k\}1$ expresses the quantifier "There exists exactly one ..." - commonly used in mathematical statements.

It is now clear that all clauses (vdW1) and (vdW2) from vdW$_{k,l,m}$ can be represented in a more concise way by the following collection of c-clauses:

vdW′1: $1\{in\_block(i,1),\ldots,in\_block(i,k)\}1$, for every $i \in [m]$.

Indeed, c-clauses (vdW′1) enforce that their models, for every $i \in [m]$ contain exactly one atom of the form $in\_block(i,b)$ — precisely the same effect as that of clauses (vdW1) and (vdW2). Let vdW$'_{k,l,m}$ be a PS+ theory consisting of clauses (vdW′1) and (vdW3). It follows that Proposition 1 and Corollary 1 can be reformulated by replacing vdW$_{k,l,m}$ with vdW$'_{k,l,m}$ in their statements. Consequently, any algorithm for finding models of PS+ theories can be used to compute van der Waerden numbers (or, at least, some bounds for them) in the way we described above.

The adoption of cardinality atoms leads to a more concise representation of the problem. While, as we discussed above, the size of all clauses (vdW1) and (vdW2) is $\Theta(mk^2 + mk)$, the size of clauses (vdW′1) is $\Theta(mk)$.

In our experiments, for various lower bound results, we used the local-search algorithm *walkaspps* [13]. This algorithm is based on the same ideas as *walksat* [17]. A major difference is that due to the presence of c-atoms in c-clauses *walkaspps* uses different formulas to calculate the breakcount and proposes several other heuristics designed specifically to handle c-atoms.

3   Results

Our goal is to establish lower bounds for small van der Waerden numbers by exploiting propositional satisfiability solvers. Here is a summary of our results.

1. Using complete SAT solvers POSIT and SATO and the encoding of the problem as $vdW_{k,l,m}$, we found a "library" of all (up to obvious symmetries) counterexamples to the fact that $W(4,3) > 75$. There are 30 of them. We list two of them in the appendix. A complete list can be found at http://www. cs.uky.edu/ai/vdw/. Since there are 48 symmetries, of the types discussed above, the full library of counterexamples consists of 1440 partitions.
2. We found that the formula $vdW_{4,3,76}$ is unsatisfiable. Hence, we found that a "generic" SAT solver is capable of finding that $W(4,3) = 76$.
3. We established several new lower bounds for the numbers $W(k,l)$. They are presented in Table 3. Partitions demonstrating that $W(2,8) > 1295$, $W(3,5) > 650$, and $W(4,4) > 408$ are included in the appendix. Counterexample partitions for all other inequalities are available at http://www.cs. uky.edu/ai/vdw/. We note that our bounds for $W(2,6)$ and $W(2,8)$ are much stronger than those implied by the results of Berlekamp [2], which we stated earlier.

Table 2. Extended results on van der Waerden numbers

l \ k	3	4	5	6	7	8
2	9	35	178	> 341	> 604	> 1295
3	27	> 193	> 650			
4	76	> 408				
5	> 125					
6	> 180					

To provide some insight into the complexity of the satisfiability problems involved, in Table 3 we list the number of atoms and the number of clauses in the theories $vdW'_{k,l,m}$. Specifically, the entry k,l in this table contains the number of atoms and the number of clauses in the theories $vdW'_{k,l,m}$, where m is the value given in the entry k,l in Table 3.

4   Discussion

Recent progress in the development of SAT solvers provides an important tool for researchers looking for both the existence and non-existence of various combinatorial objects. We have demonstrated that several classical questions related

Table 3. Numbers of atoms and clauses in theories $\mathrm{vdW}'_{k,l,m}$, used to establish the results presented in Table 3.

l k	3	4	5	6	7	8
2	18, 41	70, 409	356, 7922	682, 23257	1208, 60804	2590, 239575
3	108, 534	579, 18529	1950, 158114			
4	304, 5700	1632, 110568				
5	625, 19345					
6	1080, 48240					

to van der Waerden numbers can be naturally cast as questions on the existence of satisfying valuations for some propositional CNF-formulas.

Computing combinatorial objects such as van der Waerden numbers is hard. They are structured but as we pointed out few values are known, and new results are hard to obtain. Thus, the computation of those numbers can serve as a benchmark ('can we find the configuration such that...') for complete and local-search methods, and as a challenge ('can we show that a configuration such that ...' does not exist) for complete SAT solvers. Moreover, with powerful SAT solvers it is likely that the bounds obtained by computation of counterexamples are "sharp" in the sense that when a configuration is not found then none exist. For instance it is likely that $W(5,3)$ is close to 126 (possibly, it is 126), because 125 was the last integer where we were able to find a counterexample despite significant computational effort. This claim is further supported by the fact that in all examples where exact values are known, our local-search algorithm was able to find counterexample partitions for the last possible value of m. The lower-bounds results of this sort may constitute an important clue for researchers looking for nonexistence arguments and, ultimately, for the closed form of van der Waerden numbers.

A major impetus for the recent progress of SAT solvers comes from applications in computer engineering. In fact, several leading SAT solvers such as zCHAFF [15] and *berkmin* [7] have been developed with the express goal of aiding engineers in correctly designing and implementing digital circuits. Yet, the fact that these solvers are able to deal with hard optimization problems in one area (hardware design and verification) carries the promise that they will be of use in another area — combinatorial optimization. Our results indicate that it is likely to be the case.

The current capabilities of SAT solvers has allowed us to handle large instances of these problems. Better heuristics and other techniques for pruning the search space will undoubtedly further expand the scope of applicability of generic SAT solvers to problems that, until recently, could only be solved using specialized software.

Acknowledgments

The authors thank Lengning Liu for developing software facilitating our experimental work. This research has been supported by the Center for Communication Research, La Jolla. During the research reported in this paper the second and third authors have been partially supported by an NSF grant IIS-0097278.

References

1. M.D. Beeler and P.E. O'Neil. Some new van der Waerden numbers, *Discrete Mathematics*, 28:135–146, 1979.
2. E. Berlekamp. A construction for partitions which avoid long arithmetic progressions. *Canadian Mathematical Bulletin* 11:409–414, 1968.
3. M. Davis and H. Putnam. A computing procedure for quantification theory, *Journal of the Association for Computing Machinery*, 7:201–215, 1960.
4. D. East and M. Truszczyński. Propositional satisfiability in answer-set programming. Proceedings of Joint German/Austrian Conference on Artificial Intelligence, KI'2001. Lecture Notes in Artificial Intelligence, Springer Verlag 2174, pages 138–153. (Full version available at http://xxx.lanl.gov/ps/cs.LO/0211033). 2001.
5. P. Erdös and R. Rado. Combinatorial theorems on classifications of subsets of a given set, *Proceedings of London Mathematical Society*, 2:417–439, 1952.
6. J.W. Freeman. *Improvements to propositional satisfiability search algorithms*, PhD thesis, Department of Computer Science, University of Pennsylvania, 1995.
7. E. Goldberg, Y. Novikov. BerkMin: a Fast and Robust SAT-Solver. DATE-2002, pages 142–149, 2002.
8. D. Goldstein. Personal communication, 2002.
9. T. Gowers. A new proof of Szemerédi theorem. *Geometric and Functional Analysis*, 11:465-588, 2001.
10. R.L. Graham, B.L. Rotschild, and J.H. Spencer. *Ramsey Theory*, Wiley, 1990.
11. A. Hales and R.I. Jewett. Regularity and positional games, *Transactions of American Mathematical Society*, 106:222–229, 1963.
12. R.E. Jeroslaw and J. Wang. solving propositional satisfiability problems, *Annals of Mathematics and Artificial Intelligence*, 1:167–187, 1990.
13. L. Liu and M. Truszczyński. Local-search techniques in propositional logic extended with cardinality atoms. In preparation.
14. J.P. Marques-Silva and K.A. Sakallah. GRASP: A new search algorithm for satisfiability, *IEEE Transactions on Computers*, 48:506–521, 1999.
15. M.W. Moskewicz, C.F. Magidan, Y. Zhao, L. Zhang, and S. Malik. Chaff: engineering an efficient SAT solver, in *SAT 2001*, 2001.
16. F.P. Ramsey. On a problem of formal logic, *Proceedings of London Mathematical Society*, 30:264–286, 1928.
17. B. Selman, H.A. Kautz, and B. Cohen. Noise Strategies for Improving Local Search. *Proceedings of AAAI'94*, pp. 337-343. MIT Press 1994.
18. S. Shelah. Primitive recursive bounds for van der Waerden numbers, *Journal of American Mathematical Society*, 1:683–697, 1988.
19. E. Szemerédi. On sets of integers containing no k elements in arithmetic progression, *Acta Arithmetica*, 27:199–243, 1975.
20. B.L. van der Waerden. Beweis einer Baudetschen Vermutung, *Nieuwe Archief voor Wiskunde*, 15:212–216, 1927.
21. H. Zhang. SATO: An efficient propositional prover, in *Proceedings of CADE-17*, pages 308–312, 1997. Springer Lecture Notes in Artificial Intelligence 1104.

Appendix

Using a complete SAT solver we computed the library of all partitions (up to isomorphism) of [75] showing that $75 < W(4, 3)$. Two of these 30 partitions are shown below:

Solution 1:
Block 1: 6 7 9 14 18 20 23 24 36 38 43 44 46 51 55 57 60 61 73 75
Block 2: 4 5 12 22 26 28 29 31 37 41 42 49 59 63 65 66 68 74
Block 3: 1 2 8 10 11 13 17 27 34 35 39 45 47 48 50 54 64 71 72
Block 4: 3 15 16 19 21 25 30 32 33 40 52 53 56 58 62 67 69 70

Solution 2:
Block 1: 6 7 9 14 18 20 23 24 36 38 43 44 46 51 55 57 60 61 73
Block 2: 4 5 12 22 26 28 29 31 37 41 42 49 59 63 65 66 68 74
Block 3: 1 2 8 10 11 13 17 27 34 35 39 45 47 48 50 54 64 71 72
Block 4: 3 15 16 19 21 25 30 32 33 40 52 53 56 58 62 67 69 70 75

These two and the remaining 28 partitions can be found at http://www.cs.uky.edu/ai/vdw/

Next, we exhibit a partition of [1295] into two blocks demonstrating that $W(2, 8) > 1295$.

Block 1:
1 3 4 5 7 8 10 11 13 14 15 16 17 18 21 26 27 29 31 35 38 40 42 43 45 46 51 53 56 62 63 64 67 68 69 71 73 74 75 77 79 80 83 85 86 88 90 94 96 97 98 101 102 103 104 107 110 112 114 116 118 120 123 124 125 130 131 132 135 138 139 142 145 149 152 153 155 157 159 160 161 163 165 166 169 170 171 174 178 179 181 187 188 189 190 192 193 195 198 200 202 205 207 208 209 210 211 212 213 215 216 221 222 224 225 226 228 229 231 232 236 241 247 249 252 253 254 255 259 260 261 262 264 267 268 269 270 272 274 277 278 279 286 288 290 292 293 294 295 296 297 298 301 306 308 309 311 312 313 317 319 320 321 322 323 326 327 328 334 335 336 338 342 346 349 356 358 359 360 367 368 369 370 373 374 377 378 379 382 383 384 385 386 388 395 396 398 399 400 401 402 404 405 408 410 413 414 416 417 420 423 424 426 429 430 433 434 436 437 443 445 446 447 448 449 451 452 453 456 459 463 464 467 469 470 473 475 476 477 478 479 481 485 486 487 488 490 491 494 495 497 499 502 503 504 505 507 508 510 513 515 518 521 522 528 529 530 533 534 539 540 542 546 547 550 555 558 559 560 561 564 571 577 578 579 580 581 583 584 587 589 590 591 594 595 596 597 601 609 611 612 613 614 615 616 618 619 623 624 625 626 627 628 632 634 636 637 639 640 642 643 647 648 651 652 653 660 661 662 663 665 666 668 670 674 675 676 677 678 680 681 683 684 687 688 690 694 695 696 697 698 700 701 702 703 704 706 709 710 715 717 718 722 725 726 727 728 734 739 742 743 744 746 748 752 753 755 756 757 759 763 766 768 770 771 774 775 776 779 781 788 792 795 796 799 801 802 806 807 809 812 816 817 818 819 821 825 826 832 833 835 836 840 841 843 844 845 846 847 848 852 853 855 856 859 862 863 864 867 868 871 872 874 875 876 877 879 881 882 885 886 893 897 898 899 901 902 903 904 905 906 908 909 910 913 915 917 922 923 925 927 928 929 930 931 932 936 937 939 940 941 944 946 947 948 951 952 954 957 960 961 963 964 965 966 967 974 977 982 983 984 986 989 990 993 994 1001 1003 1004 1008 1009 1010 1012 1013 1016 1017 1020 1022 1023 1025 1026 1028 1029 1033 1034 1036 1037 1038 1040 1045 1047 1050 1051 1052 1053 1058 1060 1065 1070

1073 1074 1075 1076 1077 1079 1083 1085 1087 1088 1089 1090 1091 1092 1094 1095
1096 1097 1098 1102 1103 1105 1106 1109 1111 1113 1116 1117 1118 1119 1121 1123
1124 1126 1129 1130 1133 1135 1139 1140 1141 1144 1150 1151 1152 1154 1155 1156
1157 1159 1161 1168 1170 1171 1174 1175 1179 1180 1184 1185 1186 1188 1189 1190
1191 1194 1196 1197 1200 1202 1205 1206 1213 1216 1217 1218 1219 1220 1221 1222
1224 1226 1227 1229 1234 1236 1237 1238 1239 1246 1247 1249 1251 1253 1257 1260
1261 1262 1263 1264 1268 1269 1272 1274 1275 1276 1278 1279 1283 1285 1286 1287
1288 1289 1290 1291 1294 1295
Block 2:
2 6 9 12 19 20 22 23 24 25 28 30 32 33 34 36 37 39 41 44 47 48 49 50 52 54 55 57 58
59 60 61 65 66 70 72 76 78 81 82 84 87 89 91 92 93 95 99 100 105 106 108 109 111 113
115 117 119 121 122 126 127 128 129 133 134 136 137 140 141 143 144 146 147 148 150
151 154 156 158 162 164 167 168 172 173 175 176 177 180 182 183 184 185 186 191 194
196 197 199 201 203 204 206 214 217 218 219 220 223 227 230 233 234 235 237 238 239
240 242 243 244 245 246 248 250 251 256 257 258 263 265 266 271 273 275 276 280 281
282 283 284 285 287 289 291 299 300 302 303 304 305 307 310 314 315 316 318 324 325
329 330 331 332 333 337 339 340 341 343 344 345 347 348 350 351 352 353 354 355 357
361 362 363 364 365 366 371 372 375 376 380 381 387 389 390 391 392 393 394 397 403
406 407 409 411 412 415 418 419 421 422 425 427 428 431 432 435 438 439 440 441 442
444 450 454 455 457 458 460 461 462 465 466 468 471 472 474 480 482 483 484 489 492
493 496 498 500 501 506 509 511 512 514 516 517 519 520 523 524 525 526 527 531 532
535 536 537 538 541 543 544 545 548 549 551 552 553 554 556 557 562 563 565 566 567
568 569 570 572 573 574 575 576 582 585 586 588 592 593 598 599 600 602 603 604 605
606 607 608 610 617 620 621 622 629 630 631 633 635 638 641 644 645 646 649 650 654
655 656 657 658 659 664 667 669 671 672 673 679 682 685 686 689 691 692 693 699 705
707 708 711 712 713 714 716 719 720 721 723 724 729 730 731 732 733 735 736 737 738
740 741 745 747 749 750 751 754 758 760 761 762 764 765 767 769 772 773 777 778 780
782 783 784 785 786 787 789 790 791 793 794 797 798 800 803 804 805 808 810 811 813
814 815 820 822 823 824 827 828 829 830 831 834 837 838 839 842 849 850 851 854 857
858 860 861 865 866 869 870 873 878 880 883 884 887 888 889 890 891 892 894 895 896
900 907 911 912 914 916 918 919 920 921 924 926 933 934 935 938 942 943 945 949 950
953 955 956 958 959 962 968 969 970 971 972 973 975 976 978 979 980 981 985 987
988 991 992 995 996 997 998 999 1000 1002 1005 1006 1007 1011 1014 1015 1018 1019
1021 1024 1027 1030 1031 1032 1035 1039 1041 1042 1043 1044 1046 1048 1049 1054
1055 1056 1057 1059 1061 1062 1063 1064 1066 1067 1068 1069 1071 1072 1078 1080
1081 1082 1084 1086 1093 1099 1100 1101 1104 1107 1108 1110 1112 1114 1115 1120
1122 1125 1127 1128 1131 1132 1134 1136 1137 1138 1142 1143 1145 1146 1147 1148
1149 1153 1158 1160 1162 1163 1164 1165 1166 1167 1169 1172 1173 1176 1177 1178
1181 1182 1183 1187 1192 1193 1195 1198 1199 1201 1203 1204 1207 1208 1209 1210
1211 1212 1214 1215 1223 1225 1228 1230 1231 1232 1233 1235 1240 1241 1242 1243
1244 1245 1248 1250 1252 1254 1255 1256 1258 1259 1265 1266 1267 1270 1271 1273
1277 1280 1281 1282 1284 1292 1293

Next, we exhibit a partition of [650] into three blocks demonstrating that $W(3,5)$ > 650.

Block 1:
1 2 5 6 10 16 18 21 22 23 27 28 31 35 40 44 45 46 55 56 58 59 67 69 73 75 81 82 84 85
86 95 96 97 100 102 103 105 107 110 111 117 121 122 127 130 131 132 133 136 138 141
142 147 148 152 155 156 157 158 163 165 168 171 175 180 181 183 185 186 189 203 207

210 211 212 215 216 218 221 223 225 227 236 238 240 241 242 247 250 252 254 256 259
260 261 262 266 271 277 280 282 287 288 290 291 292 296 300 302 306 310 328 330 331
334 340 345 346 347 348 350 355 362 365 366 367 371 374 375 378 380 383 384 386 390
392 393 395 396 397 399 400 405 407 408 411 412 413 422 433 435 436 439 443 448 449
453 455 456 457 460 463 472 481 485 486 491 493 500 503 505 506 508 509 511 515 517
521 524 525 528 530 532 535 543 548 550 551 552 560 561 565 566 568 569 571 575 583
585 587 596 597 598 607 608 610 616 620 624 625 626 629 630 640 641 642 646
Block 2:
3 4 7 8 9 12 15 24 26 29 32 34 37 39 42 43 49 51 60 61 63 65 68 70 71 74 76 78 79 80
83 87 89 90 91 94 109 112 113 115 118 120 129 134 135 139 140 143 145 149 153 159
160 162 164 167 172 173 176 177 178 179 188 190 195 197 200 205 209 213 214 217 219
220 222 224 230 232 233 234 235 239 244 245 248 249 253 255 270 273 275 279 281 284
285 286 297 299 301 305 308 315 318 323 324 325 327 332 333 335 336 338 339 342 343
344 349 354 356 357 358 360 361 364 368 369 370 377 379 382 385 387 389 394 398 410
415 418 424 425 426 430 432 437 440 445 446 450 452 458 461 465 468 471 474 475 476
480 482 483 487 488 490 492 495 496 499 504 514 519 520 523 526 527 529 534 537 539
540 545 549 555 558 567 570 572 574 577 579 580 581 582 584 588 590 593 599 600 602
604 605 611 612 614 615 618 619 633 636 637 639 644 645 648
Block 3:
11 13 14 17 19 20 25 30 33 36 38 41 47 48 50 52 53 54 57 62 64 66 72 77 88 92 93 98
99 101 104 106 108 114 116 119 123 124 125 126 128 137 144 146 150 151 154 161 166
169 170 174 182 184 187 191 192 193 194 196 198 199 201 202 204 206 208 226 228 229
231 237 243 246 251 257 258 263 264 265 267 268 269 272 274 276 278 283 289 293 294
295 298 303 304 307 309 311 312 313 314 316 317 319 320 321 322 326 329 337 341 351
352 353 359 363 372 373 376 381 388 391 401 402 403 404 406 409 414 416 417 419 420
421 423 427 428 429 431 434 438 441 442 444 447 451 454 459 462 464 466 467 469 470
473 477 478 479 484 489 494 497 498 501 502 507 510 512 513 516 518 522 531 533 536
538 541 542 544 546 547 553 554 556 557 559 562 563 564 573 576 578 586 589 591 592
594 595 601 603 606 609 613 617 621 622 623 627 628 631 632 634 635 638 643 647 649
650

Finally, we exhibit a partition of [408] into four blocks demonstrating that
$W(4,4) > 408$.

Block 1:
2 8 11 17 19 20 23 30 38 42 48 50 52 59 61 65 67 71 78 82 83 85 89 90 98 104 107 108
113 119 120 124 127 129 140 143 144 147 150 152 157 158 163 166 181 183 184 198 199
204 214 220 223 226 231 237 240 241 244 250 251 253 259 264 266 270 271 273 278 282
286 287 289 306 312 314 317 318 321 327 329 331 348 351 354 359 361 362 363 366 373
377 378 382 383 386 399 401 402 403 406
Block 2:
1 3 7 13 15 16 24 26 28 37 39 47 49 57 58 66 73 76 77 81 84 86 87 92 93 94 103 110
111 117 118 121 122 123 125 133 135 151 153 154 155 161 162 167 170 172 176 182 190
194 195 196 207 210 216 228 232 233 234 242 243 245 246 248 249 254 255 256 258 262
275 280 283 284 290 293 297 298 299 305 307 309 328 333 336 341 346 352 353 355 356
358 368 370 371 372 381 385 391 393 404
Block 3:
4 6 21 22 27 29 31 32 34 35 40 41 44 56 62 63 69 70 72 74 75 79 95 96 99 101 105 109
114 115 116 126 132 134 136 141 145 159 160 165 169 171 174 175 179 180 187 188 191
192 197 200 201 208 209 212 217 219 221 227 229 235 236 247 257 263 267 269 272 274

276 281 291 292 294 300 302 304 310 311 322 324 325 330 332 334 339 340 342 344 345
350 365 367 376 379 388 390 394 397 398 400 407
Block 4:
5 9 10 12 14 18 25 33 36 43 45 46 51 53 54 55 60 64 68 80 88 91 97 100 102 106 112
128 130 131 137 138 139 142 146 148 149 156 164 168 173 177 178 185 186 189 193 202
203 205 206 211 213 215 218 222 224 225 230 238 239 252 260 261 265 268 277 279 285
288 295 296 301 303 308 313 315 316 319 320 323 326 335 337 338 343 347 349 357 360
364 369 374 375 380 384 387 389 392 395 396 405 408

Configurations showing the validity of other lower bounds listed in Table 3 are
available at http://www.cs.uky.edu/ai/vdw/.

An Algorithm for SAT Above the Threshold

Hubie Chen

Department of Computer Science, Cornell University, Ithaca, NY 14853, USA
hubes@cs.cornell.edu

Abstract. We study algorithms for finding satisfying assignments of randomly generated 3-SAT formula. In particular, we consider distributions of highly constrained formulas (that is, "above the threshold" formulas) restricted to satisfiable instances. We obtain positive algorithmic results, showing that such formulas can be solved in low exponential time.

1 Introduction

Randomly generated instances of the boolean satisfiability problem have been used to study typical-case problem complexity. Consider the model for randomly generating a 3-SAT formula over a variable set of size n which includes each of the $2^3 \binom{n}{3}$ clauses independently with probability p. For very "low" p, such a formula tends not to have many clauses, and is usually satisfiable; for very "high" p, such a formula tends to have many clauses, and is likely to be unsatisfiable. A fundamental conjecture, the *satisfiability threshold conjecture*, states that there is a transition point at which such random formulas abruptly change from being almost always satisfiable to almost always unsatisfiable. Precisely, this conjecture posits the existence of a constant c such that a random formula is satisfiable almost always if $p < cn/2^3 \binom{n}{3}$, that is, the expected number of clauses is less than cn; and is unsatisfiable almost always if $p > cn/2^3 \binom{n}{3}$. (By *almost always*, we mean with probability tending to one as n approaches infinity.) There is ample empirical evidence for this conjecture [23], and theoretical work has provided rigorous lower and upper bounds on the value of the purported constant c [16,6,15,1].

There are a number of positive algorithmic results for finding satisfying assignments of randomly generated formulas *below* the threshold – namely, results giving polynomial time algorithms which find satisfying assignments *almost always* on below-threshold formula distributions. In contrast, satisfiability algorithms for formulas *above* the threshold have seen relatively little attention. The main result of this paper is the analysis of a novel random walk algorithm for finding satisfying assignments of 3-SAT formulas from above the threshold. Of course, deciding satisfiability of such formulas is rather uninteresting on the described distribution: the trivial algorithm which outputs "unsatisfiable" is correct almost always. Hence, we restrict the described distribution of formulas to satisfiable formulas, and seek algorithms which output satisfying assignments

E. Giunchiglia and A. Tacchella (Eds.): SAT 2003, LNCS 2919, pp. 14–24, 2004.

almost always. In essence, our result shows that almost all satisfiable random 3-SAT formulas with enough clauses have a *tractable structure* that can be exploited by an algorithm to obtain a satisfying assignment. This paper thus forms a contribution to the long-term research program aiming to fully understand the typical complexity of random satisfiability formulas.

1.1   Three Formula Distributions

Before formally stating our results and comparing them to previous work, it will be useful to identify three *distinct* 3-SAT formula distributions.

Standard distribution, $P_{n,p}[\cdot]$. Formulas over a variable set of size n are generated by including each of the $2^3\binom{n}{3}$ clauses independently with probability p. Let $\mathcal{F}$ be a formula over n variables with exactly m clauses; clearly, $P_{n,p}[\mathcal{F}] = p^m(1-p)^{2^3\binom{n}{3}-m}$.

Satisfiable distribution, $P_{n,p}^{\text{sat}}[\cdot]$. This is the standard distribution, but conditioned on the property of satisfiability. That is, unsatisfiable formulas have probability zero, and satisfiable formulas have probability proportional to their probability in the standard distribution. Formally, we define, for all formulas $\mathcal{F}$ over n variables, $P_{n,p}^{\text{sat}}[\mathcal{F}] = P_{n,p}[\mathcal{F}|S]$, where S is the event that a formula is satisfiable.

Planted distribution, $P_{n,p}^{\text{plant}}[\cdot]$. Formulas over a variable set of size n are generated by first selecting an assignment α, uniformly at random from the set of all 2^n assignments. Then, each of the $7\binom{n}{3}$ clauses under which α is true are included independently with probability p; intuitively, α is a "forced" or "planted" satisfying assignment of the resulting formula. Formally, we define, for all formulas $\mathcal{F}$ over n variables, $P_{n,p}^{\text{plant}}[\mathcal{F}] = \sum_\alpha \frac{1}{2^n} P_{n,p}[\mathcal{F}|S(\alpha)]$, where the summation is over all 2^n assignments to the variables of $\mathcal{F}$, and $S(\alpha)$ is the event that α is a satisfying assignment.

The standard distribution is clearly different from each of the other two, since in the standard distribution, every formula has a non-zero probability of occurring, whereas in the other two distributions, unsatisfiable formulas have zero probability of occurring. Moreover, the satisfiable and planted distributions are different, as the planted distribution is biased towards formulas with many satisfying assignments. For instance, the empty formula containing no clauses has a $(1-p)^{7\binom{n}{3}}$ probability of occurring in the planted distribution, whereas the empty formula has a lower probability of occurring in the satisfiable distribution, namely, probability $(1-p)^{8\binom{n}{3}}/P_{n,p}[S]$. (It is easy to see that $P_{n,p}[S] > (1-p)^{\binom{n}{3}}$.)

1.2 Our Results

Roughly speaking, we show that there are "low" exponential time algorithms for random formulas (from $P_{n,p}^{\text{sat}}[\cdot]$ and $P_{n,p}^{\text{plant}}[\cdot]$) with "high" constant clause-to-variable ratios. More precisely, our main theorem is that there is a monotonically increasing function $d : (0,1) \to (0,\infty)$ such that for any $\epsilon > 0$, our algorithm finds a satisfying assignment in time $2^{\epsilon n}$ for almost all formulas according to the distribution $P_{n,p}^{\text{sat}}[\cdot]$ with $p \geq d(\epsilon)n/2^3\binom{n}{3}$ – that is, p is set so that the expected number of clauses is $d(\epsilon)n$. Put differently, for any $b > 1$, we demonstrate that there is a constant d' such that an algorithm running in time b^n finds a satisfying assignment for almost all formulas from $P_{n,p}^{\text{sat}}[\cdot]$, with $p \geq d'(b)n/2^3\binom{n}{3}$. This rephrasing follows from the initial description by defining $d'(b)$ to be $d(\epsilon)$ for an $\epsilon \in (0,1)$ such that $2^\epsilon < b$. We also prove that this theorem holds for almost all formulas from $P_{n,p}^{\text{plant}}[\cdot]$.

An intriguing corollary we obtain is that if $p(n)$ is set so that the expected number of clauses is super-linear – formally, if $p(n)$ is $\omega(1/n^2)$ – then, for every $b > 1$, there is an algorithm finding a satisfying assignment for almost all formulas from $P_{n,p}^{\text{sat}}[\cdot]$ (as well as from $P_{n,p}^{\text{plant}}[\cdot]$) in time b^n. In other words, for such a setting of $p(n)$, we obtain a distribution of formulas which can be solved almost always in time b^n – for *all* $b > 1$! The obvious question suggested by this corollary is whether or not such distributions can be solved almost always in polynomial time, and we conjecture this to be the case.

1.3 Related Work

Our results are most directly comparable to the works of Koutsoupias and Papadimitriou [17], Gent [9], and Flaxman [7]. Koutsoupias and Papadimitriou proved that the "greedy" algorithm almost always finds a satisfying assignment when $p(n)$ is a sufficiently large constant times $1/n$ – that is, the expected number of clauses is a sufficiently large quadratic function of n.

Theorem 1. *[17] There exists a constant c_{KP} such that when $p(n) \geq c_{KP}/n$, the greedy algorithm finds a satisfying assignment in polynomial time, for almost all formulas drawn according to $P_{n,p}^{\text{sat}}[\cdot]$.*

Their theorem also holds for $P_{n,p}^{\text{plant}}[\cdot]$ in place of $P_{n,p}^{\text{sat}}[\cdot]$.

Using a similar analysis, Gent showed that a simple algorithm almost always finds a satisfying assignment when the expected number of clauses is larger than a particular constant times $n \log n$. However, his result is only shown to hold for the planted distribution.

Theorem 2. *[9] There exists a constant c_G and a polynomial time algorithm A such that when $p(n) \geq c_G \log n/n^2$, the algorithm A finds a satisfying assignment for almost all formulas drawn according to $P_{n,p}^{\text{plant}}[\cdot]$.*

Flaxman also studied the planted distribution, using spectral techniques to demonstrate polynomial time tractability at a sufficiently high linear clause density.

Theorem 3. *[7] There exists a constant c_F and a polynomial time algorithm A such that when $p(n) \geq c_F/n^2$, the algorithm A finds a satisfying assignment for almost all formulas drawn according to $P_{n,p}^{\text{plant}}[\cdot]$.*

In fact, Flaxman [7] studies a more general model of "planted" random formulas which includes our definition of planted distribution as a particular parameterization; we refer the reader to his paper for more details.

To review, our result is that for every $b > 1$, there exists a constant c' such that when $p(n) \geq c'/n^2$, an algorithm running in time b^n finds a satisfying assignment for almost all formulas drawn according to $P_{n,p}^{\text{sat}}[\cdot]$ (as well as $P_{n,p}^{\text{plant}}[\cdot]$). Thus, we basically lower by a factor of n the clause density required to give the result of Koutsoupias and Papadimitriou, but give an algorithm which requires exponential time. Although our algorithms require more time than those of Gent and Flaxman, our results on the satisfiable distribution are not directly comparable to their results, which concern only the planted distribution. We emphasize that Flaxman's Theorem 3 does not imply our results on the satisfiable distribution. This theorem does imply the results we give on the planted distribution; however, we still consider our proof technique of analyzing a simple random walk to be of independent interest.

We briefly discuss three other groups of results related to our study.

"Below the threshold" algorithms. For randomly generated formulas from below the threshold, polynomial time algorithms that work almost always have been given [2,8]. It is worth noting that at the times these results were initially presented, they gave the best threshold lower bounds to date.

Worst-case algorithms. The problem of finding a satisfying assignment for a 3-SAT formula has the trivial exponential time upper bound of 2^n; a sequence of results has improved this trivial upper bound [21,5,22]. For instance, [22] gives a probabilistic algorithm solving 3-SAT in time 1.3302^n. That is, they exhibit an algorithm which solves *any* given 3-SAT formula in time 1.3302^n with high probability. Hirsch [13] took the slightly different route of performing a worst-case analysis of particular local search algorithms that had been shown to work well in practice.

Our results do not by any means strictly improve these results, nor are they implied by these results. These upper bounds establish specific values of $b' > 1$ such that any formula from the entire class of all 3-SAT formulas can be solved in time b'^n. In contrast, we establish that for *any* $b > 1$, there is a *particular* class of distributions that can be solved *almost* always in time b^n.

Proof complexity. Formulas from above the threshold are almost always unsatisfiable. A natural question concerning such formulas is how difficult or easy it is to certify their unsatisfiability. The lengths of unsatisfiability proofs from above-threshold formula distributions is taken up in [3,11,10,12], for example; we refer the reader thither for further pointers into the current literature.

2 Preliminaries

We first note a lemma; both it and its proof are from [17].

Lemma 1. *Let M, M' be independent binomial random variables. If $m \geq \epsilon n$, then $P[M(p, \binom{n-1}{2}) - \binom{m}{2}) \geq M'(p, \binom{n-1}{2}))] \leq e^{-g(\epsilon)pn^2}$, where $g : (0,1) \to (0,\infty)$ is monotonically increasing.*

Proof. For any a, the probability is bounded above by $P[M(p, \binom{n-1}{2}) - \binom{m}{2}) \geq a] + P[M'(p, \binom{n-1}{2})) \leq a]$. Choosing $a = p\sqrt{[\binom{n-1}{2}) - \binom{m}{2}]\binom{n-1}{2})}$, this sum can be bounded above by routine computations using the Chernoff bounds $P[M(p, N) \leq (1 - \theta)pn] \leq e^{-\theta^2 np/2}$ and $P[M(p, N) \geq (1 + \theta)pn] \leq e^{-\theta^2 np/3}$. $\square$

We now present the notation and definitions used throughout the paper. Let $\mathcal{F}$ denote a 3-SAT formula, and let A and α denote true/false assignments to the variable set of $\mathcal{F}$.

Let S denote the event that a formula is satisfying, and let $S(\alpha)$ denote the event that a formula is satisfied by α. A property X holds for *almost all satisfiable formulas* if $\lim_{n\to\infty} P_{n,p}^{\text{sat}}[X] \to 1$. A property X holds for *almost all planted formulas* if $\lim_{n\to\infty} P_{n,p}^{\text{plant}}[X] \to 1$. Define the distribution $P_{n,p}^{\alpha}[\cdot]$ so that for all formulas $\mathcal{F}$, $P_{n,p}^{\alpha}[\mathcal{F}] = P_{n,p}[\mathcal{F}|S(\alpha)]$.

Let $T(A, \alpha)$ be the set of variables v such that $A(v) = \alpha(v)$, and $F(A, \alpha)$ be the set of variables v such that $A(v) \neq \alpha(v)$. (Notice that when α is the all-true assignment $T(A, \alpha)$ is the set of variables that are true in A, and $F(A, \alpha)$ is the set of variables that are false in A.) Let $\overline{A}$ denote the complement of assignment A, so that $\overline{A}$ assigns a variable to false if and only if A assigns the variable to true. Let $B(A, d)$ denote the set of assignments within Hamming distance d of A.

The following definitions are relative to a formula $\mathcal{F}$, but we do not explicitly include $\mathcal{F}$ in the notation, as the relevant formula will be clear from context. Let $L(A, v)$ denote the set of clauses that are "lost" at assignment A if variable v is flipped, that is, the set of clauses in $\mathcal{F}$ that are true under A but false under A with v flipped. For $\beta \in (0, 1)$, let $E_\beta(A)$ denote the set of variables v which have one of the βn lowest values of $|L(A, v)|$. We assume that ties are broken in an arbitrary but fixed manner, so that $E_\beta(A)$ is always of size βn. Define the set $F'_\beta(A, \alpha)$ to be $F(A, \alpha) \setminus E_\beta(A)$, and define the set $T'_\beta(A, \alpha)$ to be $T(A, \alpha) \cap E_\beta(A)$. The set $T'_\beta(A, \alpha)$ contains every variable which is in $E_\beta(A)$ but would move the assignment A further away from the assignment α, if flipped.

Let us say that a formula $\mathcal{F}$ is *β-bad relative to α* if $\beta \in (0, 1)$, α is a satisfying assignment, and there exists $A \notin B(\alpha, \beta n) \cup B(\overline{\alpha}, \beta n)$ such that $|T'_\beta(A, \alpha)| \geq \frac{\beta n}{2}$. Let us say that a formula $\mathcal{F}$ is *β-bad* if it is β-bad relative to every satisfying assignment α. Let us say that a formula $\mathcal{F}$ is *β-good* if it is not β-bad, that is, $\beta \in (0, 1)$ and there exists a satisfying assignment α such that for all $A \notin B(\alpha, \beta n) \cup B(\overline{\alpha}, \beta n)$, the inequality $|T'_\beta(A, \alpha)| < \frac{\beta n}{2}$ holds.

The next two sections constitute the technical portion of the paper. Roughly speaking, these sections will establish that the algorithm succeeds on any good formula, and that almost all satisfiable formulas are good – hence, the algorithm can handle almost all satisfiable formulas.

3   Algorithm

In this section, we present our algorithm, and show that it finds a satisfying assignment on all good formulas. The algorithm, which is parameterized by $\beta \in (0,1)$ and takes a 3-SAT formula $\mathcal{F}$ as input, is as follows:

– Pick a random assignment A
– While no assignment in $B(A, \beta n) \cup B(\overline{A}, \beta n)$ satisfies $\mathcal{F}$:
  Randomly pick a variable from $E_\beta(A)$, and flip it to obtain a new A
– Output satisfying assignment

The key property of this algorithm is given by the following theorem. Roughly, the theorem states that the property of β-goodness is sufficient to direct the algorithm to a satisfying assignment in a polynomial number of loop iterations.

Theorem 4. *Suppose that $\mathcal{F}$ is a β-good formula. Then, the expected number of flips required by the algorithm (parameterized with β) to find a satisfying assignment for $\mathcal{F}$ is $O(n^2)$.*

Proof. Consider a Markov chain with state set $\{0, \ldots, n\}$ such that when the chain is in state i (for $i \in \{1, \ldots, n-1\}$), it is only possible to make transitions to the two states $i \pm 1$. If the probability of transitioning from state i to $i - 1$ is $\geq 1/2$ (for all $i \in \{1, \ldots, n-1\}$), then, from any starting point, it is known that the expected time to hit either state 0 or state n is $O(n^2)$ [19, pp. 246-247].

Since $\mathcal{F}$ is β-good, there exists a satisfying assignment α such that for all $A \notin B(\alpha, \beta n) \cup B(\overline{\alpha}, \beta n)$, $|T'_\beta(A, \alpha)| < \frac{\beta n}{2}$.

We can consider the algorithm as taking a random walk on the Markov chain described above, where its state is the Hamming distance to the assignment α. Each time the loop test fails on an assignment A, we have $\alpha \notin B(A, \beta n) \cup B(\overline{A}, \beta n)$, which implies that $A \notin B(\alpha, \beta n) \cup B(\overline{\alpha}, \beta n)$. Thus, $|T'_\beta(A, \alpha)| < \frac{\beta n}{2}$, and randomly picking a variable from $E_\beta(A)$ will result in picking a variable from $F(A, \alpha)$ at least $1/2$ of the time, and thus the Hamming distance of A to α has a probability of at least $1/2$ of decreasing. If the Hamming distance of A to α is less than βn or greater than $n - \beta n$, the loop test is true, and the satisfying assignment α is found. By the initially stated fact on the type of Markov chain considered, this will occur in expected $O(n^2)$ flips. $\square$

4   Analysis: Almost All Satisfiable Formulas Are Good

We now prove a complement to the theorem of the last section: that almost all satisfiable formulas are β-good, when $p(n) \geq h(\beta)/n^2$, for a trade-off function h that we give. This will allow us to conclude that the algorithm "works" for almost all satisfiable formulas.

We begin by showing that formulas are almost never β-bad on the planted distribution $P_{n,p}^{\text{plant}}[\cdot]$.

Lemma 2. *If $\beta \in (0,1)$, and A, α are assignments such that $A \notin B(\alpha, \beta n) \cup B(\overline{\alpha}, \beta n)$, then $P_{n,p}^{\alpha}[|T_{\beta}'(A, \alpha)| \geq \frac{\beta n}{2}] \leq (\frac{4}{e^{g(\beta)\beta pn^2/2}})^n$, where $g : (0,1) \rightarrow (0, \infty)$ is the same function given by Lemma 1.*

Proof. We assume without loss of generality that α is the all-true assignment. Let $k = \beta n/2$ (which is equal to $|E_{\beta}(A)|/2$). Since $A \notin B(\alpha, \beta n) \cup B(\overline{\alpha}, \beta n)$, $|F(A, \alpha)| \geq \beta n = 2k$, and $|T_{\beta}'(A, \alpha)| \geq k$ implies that $|F'(A, \alpha)| \geq k$ (with respect to any formula). This in turn implies that there exist distinct variables $t_1, \ldots, t_k$ in $T_{\beta}'(A, \alpha)$ and distinct variables $f_1, \ldots, f_k$ in $F'(A, \alpha)$ such that for all $i = 1, \ldots, k$, $|L(A, t_i)| \leq |L(A, f_i)|$.

Thus, we have

$$P_{n,p}^{\alpha}[|T_{\beta}'(A, \alpha)| \geq k] = P_{n,p}^{\alpha}[[|T_{\beta}'(A, \alpha)| \geq k] \wedge [|F'(A, \alpha)| \geq k]]$$

which is bounded above by

$$\leq P_{n,p}^{\alpha}[\cup(|L(A, t_1)| \leq |L(A, f_1)| \wedge \ldots \wedge |L(A, t_k)| \leq |L(A, f_k)|)]$$

where the union is over all sequences $t_1, \ldots, t_k, f_1, \ldots, f_k$ such that the t_i are distinct elements of $T_{\beta}'(A, \alpha)$, and the f_i are distinct elements of $F'(A, \alpha)$. Applying the so-called union bound, we see that the previous expression is

$$\leq \sum P_{n,p}^{\alpha}[|L(A, t_1)| \leq |L(A, f_1)| \wedge \ldots \wedge |L(A, t_k)| \leq |L(A, f_k)|]$$

where the summation is over the same set of sequences as the union in the previous expression.

Observe that the clause set $L(A, t_i)$ is a subset of those clauses containing t_i positively and two other variables occurring positively if and only if they are false in A. Also, the clause set $L(A, f_i)$ is a subset of those clauses containing f_i negatively and two other variables occurring positively if and only if they are false in A, but without clauses containing three negative literals (as they are not satisfied by the forced all-true assignment).

Consequently, for a variable t_i, of all the $L(A, v)$, only $L(A, t_i)$ can contain a clause with t_i appearing positively, and all clauses of $L(A, t_i)$ do; likewise, for a variable f_i, only $L(A, f_i)$ can contain a clause with f_i appearing negatively, and all clauses of $L(A, f_i)$ do. Hence, the quantities $|L(A, v)|$ are independent random variables! It follows that the previous expression is

$$= \sum P_{n,p}^{\alpha}[|L(A, t_1)| \leq |L(A, f_1)|] \cdots P_{n,p}^{\alpha}[|L(A, t_k)| \leq |L(A, f_k)|].$$

As just described, $|L(A, t_j)|$ is the sum of $\binom{n-1}{2}$ Bernoulli trials, and $|L(A, f_j)|$ is the sum of $\binom{n-1}{2} - \binom{m}{2}$ Bernoulli trials, where m is the number of variables assigned to the value true in A – that is, $|T(A, \alpha)|$. Using Lemma 1 along with the developed chain of inequalities, we can upper bound the expression of interest, $P_{n,p}^{\alpha}[|T_{\beta}'(A, \alpha)| \geq k]$, by $4^n[e^{-g(\beta)pn^2}]^k = [\frac{4}{e^{g(\beta)\beta pn^2/2}}]^n$. $\square$

Lemma 3. *There exists a monotonically decreasing function $h : (0,1) \to (0,\infty)$ such that $p(n) \geq h(\beta)/n^2$ implies $\lim_{n\to\infty} 2^n P_{n,p}^\alpha[\beta\text{-bad relative to } \alpha] = 0$, for all assignments α on n variables.*

Proof. Let $\mathcal{A}$ denote the set of assignments A such that $A \notin B(\alpha, \beta n) \cup B(\overline{\alpha}, \beta n)$. We establish an upper bound on the probability of interest.

$$
\begin{aligned}
2^n P_{n,p}^\alpha[\beta - \text{bad relative to } \alpha] &\leq 2^n P_{n,p}^\alpha[\cup_{A\in\mathcal{A}} |T_\beta'(A,\alpha)| \geq \tfrac{\beta n}{2}] \\
&\leq 2^n \textstyle\sum_{A\in\mathcal{A}} P_{n,p}^\alpha[|T_\beta'(A,\alpha)| \geq \tfrac{\beta n}{2}] \\
&\leq 2^n \textstyle\sum_{A\in\mathcal{A}} [\tfrac{4}{e^{g(\beta)\beta pn^2/2}}]^n \\
&\leq 2^n 2^n [\tfrac{4}{e^{g(\beta)\beta pn^2/2}}]^n \\
&= [\tfrac{16}{e^{g(\beta)\beta pn^2/2}}]^n
\end{aligned}
$$

The third inequality holds by Lemma 2.

It suffices to choose h to be a sufficiently large monotonically decreasing function so that $[\tfrac{16}{e^{g(\beta)\beta h(\beta)/2}}] < 1$, for all $\beta \in (0,1)$. $\square$

The next lemma connects the $P_{n,p}^{\text{sat}}[\cdot]$ distribution to the $P_{n,p}^{\text{plant}}[\cdot]$ distribution, using the prior two lemmas to show the rareness of β-bad formulas in $P_{n,p}^{\text{sat}}[\cdot]$.

Lemma 4. *There exists a monotonically decreasing function $h : (0,1) \to (0,\infty)$ such that $p(n) \geq h(\beta)/n^2$ implies $\lim_{n\to\infty} P_{n,p}^{\text{sat}}[\beta\text{-bad}] = 0$.*

Proof. Let $\mathcal{A}$ denote the set containing all 2^n assignments on n variables, and let h be as in Lemma 3. Fix $\beta \in (0,1)$, and let D denote the event that a formula is β-bad. We have the following chain of inequalities.

$$
\begin{aligned}
P_{n,p}^{\text{sat}}[D] &= P_{n,p}[D \wedge S]/P_{n,p}[S] \\
&\leq \textstyle\sum_{\alpha\in\mathcal{A}} P_{n,p}[D \wedge S(\alpha)]/P_{n,p}[S] \\
&\leq \textstyle\sum_{\alpha\in\mathcal{A}} (P_{n,p}[D \wedge S(\alpha)]/P_{n,p}[S(\alpha)]) \\
&\leq \textstyle\sum_{\alpha\in\mathcal{A}} P_{n,p}[D|S(\alpha)] \\
&= \textstyle\sum_{\alpha\in\mathcal{A}} P_{n,p}^\alpha[D] \\
&\leq \textstyle\sum_{\alpha\in\mathcal{A}} P_{n,p}^\alpha[\beta - \text{bad relative to } \alpha]
\end{aligned}
$$

By Lemma 3, the last expression approaches zero as n approaches infinity. $\square$

Having established that almost all formulas are β-good, we are now in a position to combine this result with the theorem which showed that the algorithm "works" on β-good instances.

Theorem 5. *There exists a monotonically decreasing function $h' : (0,1) \to (0,\infty)$ such that when $p \geq h'(\epsilon)/n^2$, the given randomized algorithm finds a satisfying assignment for almost all satisfiable formulas in expected $O(2^{\epsilon n})$ time (where "expected time" is with respect to the random bits used by the algorithm).*

Proof. For all $\epsilon \in (0,1)$, define $h'(\epsilon) = h(\beta)$, where β is sufficiently small so that $\binom{n}{\beta n} \leq 2^{\frac{\epsilon n}{2}}$. (Such a value β exists, for $\binom{n}{\beta n} \leq (\frac{ne}{\beta n})^{\beta n} = (\frac{e}{\beta})^{\beta n}$; see e.g. [18, p. 434].) By Lemma 4, when $p \geq h'(\epsilon)/n^2 = h(\beta)/n^2$, $\lim_{n\to\infty} P_{n,p}^{\text{sat}}[\beta - \text{good}] \to 1$.

Thus, for such p, almost all satisfiable formulas are β-good, and by Theorem 4, the algorithm will find a satisfying assignment for almost all satisfiable formulas in $O(n^2)$ flips.

Before each flip, every assignment in $B(A, \beta n) \cup B(\overline{A}, \beta n)$ must be checked to see if it is satisfying; there are $2\binom{n}{\beta n} \leq 2 * 2^{\frac{\epsilon n}{2}}$ such assignments, and checking each takes time $t(n)$, for some polynomial t. Thus, $O(n^2)$ flips require time $2 * 2^{\frac{\epsilon n}{2}} t(n) O(n^2)$, which is $O(2^{\epsilon n})$. $\square$

By a similar proof, we obtain the same theorem for planted formulas.

Theorem 6. *There exists a monotonically decreasing function $h'' : (0,1) \to (0, \infty)$ such that when $p \geq h''(\epsilon)/n^2$, the given randomized algorithm finds a satisfying assignment for almost all planted formulas in expected $O(2^{\epsilon n})$ time (where "expected time" is with respect to the random bits used by the algorithm).*

Proof. Notice that Lemma 4 is true with $P_{n,p}^{\text{plant}}[\cdot]$ in place of $P_{n,p}^{\text{sat}}[\cdot]$; this is immediate from the definition of $P_{n,p}^{\text{plant}}[\cdot]$. (Notice that the extra 2^n factor of Lemma 3 is not needed.) Using this modified version of Lemma 4, the proof of the theorem is exactly the same as the proof of Theorem 5. $\square$

Finally, we observe in the next two corollaries that when the clause density is super-linear, then both the satisfiable and planted distributions are solvable almost always in time b^n – for *all* $b > 1$.

Corollary 1. *Suppose that $p(n)$ is $\omega(1/n^2)$. Then for every $b > 1$, there is a randomized algorithm finding a satisfying assignment for all satisfiable formulas in expected $O(b^n)$ time (where "expected time" is with respect to the random bits used by the algorithm).*

Proof. Let $\epsilon > 0$ be sufficiently small so that $2^\epsilon < b$. Observe that $p(n) \geq h'(\epsilon)/n^2$ for all but finitely many n. Modify the algorithm given by Theorem 5 so on formulas of variable set size n such that $p(n) < h'(\epsilon)/n^2$, the desired output is hard-coded. $\square$

Corollary 2. *Suppose that $p(n)$ is $\omega(1/n^2)$. Then for every $b > 1$, there is a randomized algorithm finding a satisfying assignment for all planted formulas in expected $O(b^n)$ time (where "expected time" is with respect to the random bits used by the algorithm).*

The proof of Corollary 2 is similar to that of Corollary 1.

Example 1. Let $p(n) = .0001(\log \log \log n)/n^2$, that is, p is set so that the expected number of clauses is a small constant times $n \log \log \log n$. Then there is an algorithm finding satisfying assignments for almost all satisfiable formulas in expected time $(1.000001)^n$.

5   Discussion

The work in this paper points to the question of whether or not there is a *polynomial time algorithm* that works almost always, for "above the threshold" satisfiable formulas. A number of SAT algorithms, for example, GSAT [24], can be viewed as performing random walks on a Markov chain with state set equal to the set of all assignments. From this perspective, each SAT formula induces a different Markov chain, and what we have really done here is shown that almost all such induced Markov chains (for the distribution of formulas of interest) have a desired convergence property. Perhaps one can perform a more detailed analysis of the Markov chains for some such random walk algorithm, without collapsing assignments with the same Hamming distance from some "target" assignment, together into one state – as is done here and in the analysis of a 2-SAT algorithm given in [20].

It may also be of interest to attempt to prove time lower bounds on restricted classes of random-walk algorithms. For instance, the algorithm in this paper picks a variable to flip based only on the *number* of clauses that would be lost by flipping each variable (that is, the quantity $|L(A, v)|$ for variable v at assignment A) and not based on, for instance, what the clauses of $L(A, v)$ look like, and at which variables they overlap. Can it be shown that all such algorithms with this property have an exponential expected time before converging on a satisfying assignment, say, for the distribution of formulas studied in this paper?

A broader but admittedly more speculative question we would like to pose is whether or not it is possible to develop a complexity theory for distributions of problem instances which takes *almost always polynomial time* as its base notion of tractability. Along these lines, we are curious whether or not distributions that are in almost always time b^n for *all* $b > 1$ – such as those identified by Corollaries 1 and 2 – are always in almost always polynomial time. Can this hypothesis, perhaps restricted to some class of "natural" or computable distributions, be related to any better-known complexity-theoretic hypotheses?

Acknowledgements. The author wishes to thank Amy Gale, Jon Kleinberg, Riccardo Pucella, and Bart Selman for useful discussions and comments. The author also thanks the anonymous referees for their many helpful comments.

References

1. D. Achlioptas, Y. Peres. The Threshold for Random k-SAT is 2^k (ln 2 + o(1)). 35th ACM Symposium on Theory of Computing (STOC 2003).
2. D. Achlioptas, G. B. Sorkin. Optimal myopic algorithms for random 3-SAT. Proc. 41st IEEE Symp. on Foundations of Comput. Sci., 2000, pp. 590-600.
3. P. Beame, R. Karp, T. Pitassi, M. Saks. On the complexity of unsatisfiability of random k-CNF formulas. In Proceedings of the 30th Annual ACM Symposium on Theory of Computing, pages 561-571, Dallas, TX, May 1998.
4. P. Beame, T. Pitassi. Propositional Proof Complexity: Past, Present and Future. Electronic Colloquium on Computational Complexity (ECCC) 5(067): (1998).
5. E. Dantsin, A. Goerdt, E. A. Hirsch, R. Kannan, J. Kleinberg, C. Papadimitriou, P. Raghavan, U. Schning. A Deterministic $(2 - 2/(k + 1))^n$ Algorithm for k-SAT Based on Local Search. Theoretical Computer Science, 289(2002).

6. O. Dubois, Y. Boufkhad, J. Mandler. Typical random 3 SAT formulae and the satisfiability threshold. Proc. 11th ACM SIAM Symp. on Discrete Algorithms, San Franscisco, CA, January 2000, pp. 124-126 and Electronic Colloquium on Computational Complexity, TR03-007 (2003).
7. A. Flaxman. A spectral technique for random satisfiable 3CNF formulas. SODA 2003.
8. A. M. Frieze, S. Suen. Analysis of two simple heuristics on a random instance of k-SAT. J. Comput. System Sci. 53 (1996) 312-355.
9. I. Gent. On the Stupid Algorithm for Satisfiability. APES Technical Report 03-1998.
10. A. Goerdt, T. Jurdzinski. Some Results On Random Unsatisfiable K-Sat Instances and Approximation Algorithms Applied To Random Structures. Combinatorics, Probability & Computing, Volume 12, 2003.
11. A. Goerdt, M. Krivelevich. Efficient recognition of random unsatisfiable k-SAT instances by spectral methods. Proceedings of the 18th International Symposium on Theoretical Aspects of Computer Science (STACS 2001). Lecture Notes in Computer Science, 2010:294-304, 2001.
12. A. Goerdt, A. Lanka. Recognizing more random unsatisfiable 3-SAT instances efficiently. LICS'03, Workshop on Typical case complexity and phase transitions, June 21, 2003, Ottawa, Canada.
13. E. Hirsch. SAT Local Search Algorithms: Worst-Case Study. Journal of Automated Reasoning 24(1/2):127-143, 2000.
14. A. Kamath, R. Motwani, K. Palem, P. Spirakis. Tail Bounds for Occupancy and the Satisfiability Threshold Conjecture. Random Structures and Algorithms 7 (1995) 59-80.
15. A. C. Kaporis, L. M. Kirousis, E. G. Lalas. The Probabilistic Analysis of a Greedy Satisfiability Algorithm. ESA 2002: 574-585.
16. L. M. Kirousis, E. Kranakis, D. Krizanc, Y.C. Stamatiou. Approximating the Unsatisfiablity Threshold of Random Formulas. Random Structures and Algorithms 12 (1998) 253-269.
17. E. Koutsoupias, C. H. Papadimitriou. On the greedy algorithm for satisfiability. Inform. Process. Lett. 43 (1992) 53-55.
18. R. Motwani, P. Raghavan. Randomized Algorithms. Cambridge University Press, Cambridge, England, 1995.
19. C. H. Papadimitriou. Computational Complexity. Addison-Wesley, New York, 1994.
20. C. H. Papadimitriou. On selecting a satisfying truth assignment (Extended Abstract). 32nd IEEE Symposium on Foundations of Comput. Sci., 1991, pp. 163-169.
21. U. Schöning. A probabilistic algorithm for k-SAT and constraint satisfaction problems. 40th IEEE Symposium on Foundations of Comput. Sci., 1999 pp. 410-414.
22. R. Schuler, U. Schöning, O. Watanabe, T. Hofmeister. A probabilistic 3-SAT algorithm further improved. Proceedings of 19th International Symposium on Theoretical Aspects of Computer Science, STACS 2002. Lecture Notes in Computer Science, Springer.
23. B. Selman, H. Levesque, D. Mitchell. Hard and Easy Distributions of SAT Problems. Proc. of the 10th National Conference on Artificial Intelligence, 1992, pp. 459-465.
24. B. Selman, H. Levesque, D. Mitchell. A New Method for Solving Hard Satisfiability Problems. Proc. of the 10th National Conference on Artificial Intelligence, 1992, pp. 440-446.

Watched Data Structures for QBF Solvers

Ian Gent[1], Enrico Giunchiglia[2], Massimo Narizzano[2], Andrew Rowley[1], and
Armando Tacchella[2]

[1] Dept. of Computer Science, University of St. Andrews
North Haugh, St. Andrews, Fife, KY16 9SS, Scotland
{ipg, agdr}@dcs.st-and.ac.uk
[2] DIST - Università di Genova
Viale Causa 13, 16145 Genova, Italy
{enrico, mox, tac}@mrg.dist.unige.it

Abstract. In the last few years, we have seen a tremendous boost in the
efficiency of SAT solvers, this boost being mostly due to CHAFF. CHAFF
owes some of its efficiency to its "two-literal watching" data structure.
In this paper we present watched data structures for Quantified Boolean
Formula (QBF) satisfiability solvers. In particular, we propose (*i*) two
CHAFF-like literal watching schemes for unit clause detection; and (*ii*)
two other watched data structures, one for detecting pure literals and
the other for detecting void quantifiers. We have conducted an experimental evaluation of the proposed data structures, using both randomly
generated and real-world benchmarks. Our results indicate that clause
watching is very effective, while the 2 and 3 literal watching data structures become more effective as the clause length increases. The quantifier
watching structure does not appear to be effective on the instances considered.

1 Introduction

In the last few years, we have seen a tremendous boost in the efficiency of SAT
solvers, this boost being mostly due to CHAFF. CHAFF is based on DPLL procedure [1, 2], and owes part of its efficiency to its data structures designed for
the specific look-ahead it implements, i.e., unit-propagation. The basic idea is
to detect unit clauses by watching two unassigned literals per clause. As soon
as one of the watched literals is assigned, another unassigned literal is looked
for in the clause: failure to find one implies that the clause is unit. The main
advantage of any such procedure is that, when a literal is given a truth value,
only its watched occurrences are assigned. This is to be contrasted to traditional
DPLL implementations where, when assigning a variable, all its occurrences are
considered. This simple idea can be realized in various ways, differing for the
specific operations done when assigning a watched literal or when backtracking (see, e.g., [3, 4, 5]). In CHAFF, backtracking requires a constant number of
operations. See [4] for more details.

In this paper we tackle the problem of designing, implementing and experimenting with watching data structures for DPLL-based QBF solvers. In particular, we propose (*i*) two CHAFF-like literal watching schemes for unit clause

E. Giunchiglia and A. Tacchella (Eds.): SAT 2003, LNCS 2919, pp. 25–36, 2004.
© Springer-Verlag Berlin Heidelberg 2004

detection; and (ii) two other watched data structures, one for detecting pure literals and the other for detecting void quantifiers. We have implemented such watching structures, and we conducted an experimental evaluation, using both randomly generated and real-world benchmarks. Our results indicate that clause watching is very effective, while the 2 and 3 literal watching data structures become more effective as the clause length increases. The quantifier watching structure does not appear to be effective on the instances considered.

The paper is structured as follows. We first introduce some basic terminology and notation (§2). In §3, we briefly present the standard data structures. The watched literal data structures that we propose are comprehensively described in §4 and the other watched data structures are described in §5. We end the paper with the experimental analysis (§6).

2 Basic Definitions

We take for granted the definitions of variable, literal, clause. Notationally, if l is a literal, we write $\bar{l}$ as an abbreviation for x if $l = \neg x$, and for $\neg l$ otherwise.

A *QBF* is an expression of the form

$$Q_1 \boldsymbol{x_1} \ldots Q_n \boldsymbol{x_n} \Phi \qquad (n \geq 0) \tag{1}$$

where every Q_i ($1 \leq i \leq n$) is a quantifier (either existential $\exists$ or universal $\forall$); $\boldsymbol{x_1}, \ldots, \boldsymbol{x_n}$ are sets of variables; and Φ is a set of clauses in $\boldsymbol{x_1} \cup \ldots \cup \boldsymbol{x_n}$. We assume that no variable occurs twice in a clause; that $\boldsymbol{x_1}, \ldots, \boldsymbol{x_n}$ are pairwise disjoint; and that $Q_i \neq Q_{i+1}$ ($1 \leq i < n$). In (1), $Q_1 \boldsymbol{x_1} \ldots Q_n \boldsymbol{x_n}$ is the *prefix*, Φ is the *matrix*, and Q_i is the *bounding quantifier* of each variable in $\boldsymbol{x_i}$.

The semantics of a QBF φ can be defined recursively as follows:

1. If the matrix of φ contains an empty clause then φ is FALSE.
2. If the matrix of φ is the empty set of clauses then φ is TRUE.
3. If φ is $\exists \boldsymbol{x} \psi$ and $x \in \boldsymbol{x}$, φ is TRUE if and only if φ_x or $\varphi_{\neg x}$ are TRUE.
4. If φ is $\forall \boldsymbol{x} \psi$ and $x \in \boldsymbol{x}$, φ is TRUE if and only if φ_x and $\varphi_{\neg x}$ are TRUE.

If φ is a QBF and l is a literal, φ_l is the QBF

1. whose matrix Φ is obtained from the matrix of φ by deleting the clauses C such that $l \in C$, and removing $\bar{l}$ from the others, and
2. whose prefix is obtained from the prefix of φ by deleting the variables not occurring in Φ. Void quantifiers (i.e., quantifiers not binding any variable) are also eliminated.

As usual, we say that a QBF φ is *satisfiable* iff φ is TRUE.

On the basis of the semantics, a simple recursive procedure for determining the satisfiability of a QBF φ, simplifies φ to φ_x and/or $\varphi_{\neg x}$ if x is in the leftmost set of variables in the prefix, until either an empty clause or the empty set of clauses is produced: On the basis of the satisfiability of φ_x and $\varphi_{\neg x}$, the satisfiability of φ can be determined according to the semantics of QBFs.

Most of the current QBF solvers are based on such simple procedure. However, in order to prune the search tree, they introduce some improvements.

The first improvement is that it is possible to directly conclude that a QBF is unsatisfiable if the matrix contains a *contradictory clause*, i.e., a clause with no existential literals. (Notice that the empty clause is also contradictory).

Then, if a literal l is unit or pure in a QBF φ, then φ can be simplified to φ_l. We say that a literal l is

- *Unit* if the matrix contains a *unit clause in l*, i.e., a clause of the form $\{l, l_1, \ldots, l_m\}$ ($m \geq 0$) with (i) l existential; and (ii) each literal l_i ($1 \leq i \leq m$) universally quantified inside the quantifier binding l. For example, both x_1 and x_2 are unit in any QBF of the form:

$$\ldots \exists x_1 \forall y_1 \exists x_2 \ldots \{\{x_1, y_1\}, \{x_2\}, \ldots\}.$$

- *Pure* if either l is existential and $\bar{l}$ does not belong to any clause in Φ; or l is universal and l does not belong to any clause in Φ. For example, in the following QBF, the pure literals are y_1 and x_1:

$$\forall y_1 \exists x_1 \forall y_2 \exists x_2 \{\{\neg y_1, y_2, x_2\}, \{x_1, \neg y_2 \neg x_2\}\}.$$

In the above example, notice that after y_1 and x_1 are assigned, $\neg y_2$ can be assigned because it is pure, and then x_2 can be assigned because it is unit. This simple example shows the importance of implementing pure literal fixing in QBFs: The assignment of a pure existential literal may cause the detection of a pure universal literal, and the assignment of a pure universal literal may cause the detection of unit literals.

Finally, all QBF solvers implement some heuristic in order to decide the best (among those admissible) literal to be used for branching.

3   Unwatched Data Structures

We need to compare our new, watched, data structures with an implementation which is identical in terms of propagation, heuristics, etc, but in which standard data structures are used. To do this, we provided an alternative implementation of CSBJ identical except for the use of unwatched data structures. Thus, all our results in terms of run time compare executions with identical number of backtracks. For a fair comparison, we tried our best to implement 'state-of-the-art', but unwatched, data structures. In the rest of this section we describe these.

The main requirements of any data structure in a QBF solver is to detect key events. The key events that we want to detect are

1. The occurrence of unit or pure literals.
2. The presence of contradictory clauses in the matrix.
3. The presence of void quantifiers in the prefix: This allows the removal of the quantifier from the prefix.

4. The presence of the empty set of clauses: This allows us to immediately backtrack to the last universal variable whose right branch has not yet been explored.

All such events are to be detected while descending the search tree assigning variables. When a variable is assigned, data structures get updated and each condition checked. Of course, changes are stored so that they can be undone while backtracking. Here we briefly describe how such events are detected in our standard procedure.

Unit literals and contradictory clauses, assuming a literal l is assigned true, are detected while removing $\bar{l}$ from any clauses it occurs in. To perform this operation more efficiently, each clause is first sorted into existential and universal literals. These sets are then sorted into the order in which the variables occur in the prefix. Further, since a literal can be removed from any point in the clause, it is assumed that a linked list data structure is used to hold the literals.

For pure literals, we store which clauses a variable's literals are contained in. In the same way that a clause contains literals, a variable can be thought to contain c-literals. Each of these c-literals consists of a clause and a sign. The sign of the c-literal is the same as the sign of the literal of the variable in the clause. The c-literals are then stored in the variable, split into negative and positive c-literals. Again, a linked list data structure allows removal of any c-literal efficiently. When a clause is removed, the c-literals of the clause can be removed from the variables left in the clause. Pure literals have no positive or no negative c-literals.

For void quantifiers, the procedure is the same since we can think of a quantifier in a similar way to a clause. A quantifier contains q-variables, which consist of a variable and a quantification. As with literals in clauses, a linked list data structure is required here to allow removal from any part of the quantifier. When a q-variable is assigned, it is removed from the quantifier in which it occurs.

For detecting the empty matrix, we keep a count of the number of clauses. When a clause is marked as removed, this count is decremented and when a clause is restored, the count is incremented; clauses are never actually removed.

4   Literal Watching

As it has been shown in SAT solvers such as CHAFF, lazy data structures can improve efficiency. This is attributed to the fact that cache memory is used more efficiently and less operations are performed. One of the requirements of these data structures that make this true is that no work should be done on the data structure during backtracking. To allow this to happen, no literals are ever removed from clauses. This allows the data structure to use arrays in place of linked lists. Here we outline two data structures for watching literals.

4.1   Two Literal Watching

In SAT solvers, two literal watching is used for the removal of clauses in addition to the removal of literals from clauses. In a SAT solver, we are only interested in finding a solution; once this has been done, no backtracking is required. This means that we do not care how many variable assignments it takes to get to the solution, or if these variable assignments are superfluous. In QBF solvers this is no longer the case. We are likely to need to backtrack upon finding a solution and so it is important that the empty set of clauses is detected as soon as possible, and that no variable assignments are made that are not absolutely necessary. To facilitate this, when assigning a literal, l, true, we only deal with watched literals from clauses containing $\bar{l}$, but remove all clauses containing l.

The invariant that we wish to uphold in a clause is that one of the following holds:

1. The clause contains a true literal and is therefore removed; or
2. The clause contains no true existential literals and is therefore false; or
3. The clause contains one unassigned existential literal and all unassigned universals are quantified inside the existential and is therefore unit; or
4. The clause contains two unassigned watched existential literals; or
5. The clause contains one unassigned watched existential literal and one unassigned watched universal literal quantified outside the existential.

These should hold in such a way that nothing has to be done upon backtracking. As before, we assume the literals of the clause are sorted. When removing a literal from a clause, if ever we find a literal that satisfies the clause, the operation is stopped.

If the initial literal is an existential, e_{old}, the rules are as follows:

1. If we find an unassigned, unwatched existential, e_{new} to the right of the current one, watch e_{new}. Due to sorting, e_{new} must be inside e_{old}, and so invariant 5 can still hold.
2. Scan left to find an unassigned, unwatched existential, e_{new}.
3. If we found the other pointer, and e_{new}, watch e_{new}. There must still be two existentials watched.
4. If we didn't find a new pointer or the other pointer, the clause is now contradictory.
5. If we found the other pointer e_{other}, but not e_{new}, we must scan the universals from the left to find an unassigned, unmarked universal, u_{new}, quantified outside e_{other}.
   (a) If we find u_{new}, watch it.
   (b) If we don't, we have a unit clause in e_{other}.
6. If we didn't find the other pointer, but found e_{new}, we must carry on scanning to the left to find the other pointer. If we encounter another unassigned unwatched existential, call it e_{new2}.
   (a) If we find the other pointer, watch the new existential. There must still be two existentials watched.
   (b) If we don't, we must scan the universals to find the watched universal, u_{other}.
      i. If we found e_{new} and e_{new2}, watch e_{new} in place of e_{old} and e_{new2} in place of u_{other}.

    ii. If u_{other} is quantified outside e_{new}, watch e_{new}.

    iii. If u_{other} is quantified inside e_{new}, we must scan to the left to find a new universal, u_{new}, that is quantified outside the existential.

      A. If this is not possible, the clause is unit in e_{new}.

      B. If it is found, watch e_{new} and move the u_{other} pointer to u_{new}.

If the initial literal is a universal, u_{old}, the rules are as follows:

1. Scan to the left and try to find an unwatched existential, e_{new}.
2. If we find e_{new}, watch it.
3. If we do not find e_{new}, we must have found e_{other} and then we must scan left and right over the universals to find one that is quantified outside e_{other}.
   (a) If we find it, we watch it.
   (b) If we don't, the clause must be unit in e_{other}.

4.2 Three Literal Watching

In the above, we can be watching an existential and a universal as in case 5 but there might be two unassigned existentials in the clause. To reference this problem, we suggest a method where by three literals are watched in a clause: two existentials, and one universal. The invariant for this is that one of the following should hold:

1. The clause contains a true literal and is therefore removed; or
2. The clause contains no true existential literals and is therefore false; or
3. The clause contains one unassigned existential literal and all unassigned universals are quantified inside the existential and is therefore unit; or
4. The watched existentials are both unassigned; or
5. One of the two watched existentials is assigned, and the watched universal literal is unassigned and is quantified outside the watched unassigned existential literal.

In order to determine the other watched literals in the clause as quickly as possible, each clause contains a set of watched literals. These point to the actual watched literals in the clause. It is now less important that the existential literals in the clause are sorted, but universal sorting is still important, since we still need to scan for universals with a proper position in the prefix. As before, search is stopped if a literal that satisfies the clause is found.

If the initial literal is an existential, e_{old}, the rules are as follows:

1. Determine the other existential watched literal, e_{other}, and the universal watched literal u.
2. If e_{other} is assigned false, find a universal literal, u_{sat} that satisfies the clause.
   (a) If u_{sat} exists, stop.
   (b) If u_{sat} does not exist, the clause is contradictory.
3. If e_{other} is unassigned find another unwatched existential literal, e_{new}.
   (a) If e_{new} exists, watch it.

(b) If e_{new} does not exist, scan the universals to the right until an unassigned universal u_{new} is found that is quantified outside e_{other}.

    i. If u_{new} exists, watch it.

    ii. If u_{new} does not exist, the clause is unit in e_{other}.

If the initial literal is universal, the rules are as follows:

1. Determine the existential watched literals, e_1 and e_2.
2. If e_1 and e_2 are both unassigned, stop.
3. If only one of e_1 and e_2 are assigned, scan the universals until an unassigned universal, u_{new}, is found that is quantified outside the unassigned existential watched literal.

    (a) If u_{new} exists, watch it.

    (b) If u_{new} does not exist, the clause is unit.

5 Other Watched Data Structures

The basic idea of watching literals is in fact a single instance of a more general idea. We now show that it is possible to extend the concept of watching to other data structures, namely clause watching and quantifier watching. So far, literal watching is the only lazy data structure to have been used in SAT solvers. We will see that our new technique of clause watching could be used in SAT as well as QBF.

5.1 Clause Watching

In clause watching, we need to detect if either or both of the signs of the c-literals become empty. For this, we require two watched c-literals per variable, one of positive sign, and the other of negative sign.

The invariants for c-literal watching are:

1. The variable is pure in one or other of the signs.
2. The variable is removed.
3. There are two watched c-literals in the variable, one of each sign.

When a c-literal is removed, the rules are as follows:

1. Search for a new c-literal of the same sign, c_{new}.

    (a) If c_{new} exists, watch it.

    (b) If c_{new} does not exist, search for an unassigned c-literal of the opposite sign, c_o.

        i. If c_o exists, the variable is pure in the sign of c_o.

        ii. If c_o does not exist, the variable is removed.

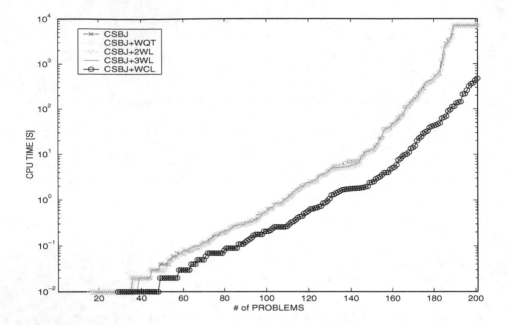

Fig. 1. Performances of CSBJ augmented with watched data structures on real-world instances.

5.2   Quantifier Watching

In two literal watching in SAT solvers, the two literals allow us to detect when a clause only contains one item, as well as when it is empty. In quantifier watching, we only need to know when the quantifier is empty, and for this, only one watched q-variable is needed per quantifier.

The invariants for q-variable watching are:

1. The quantifier is empty and so removed.
2. There is one watched unassigned q-variable in the quantifier.

When we remove the watched q-variable, q_{old}, the rules are as follows:

1. Search left and right for an unassigned q-variable, q_{new}.
   (a) If q_{new} exists, watch it.
   (b) If q_{new} does not exist, remove the quantifier.

6   Experimental Analysis

We implemented the above ideas in a QBF solver featuring both conflict and solution directed backjumping [6]. In order to test the effectiveness of the watched data structures, we run the 5 different versions of the solver:

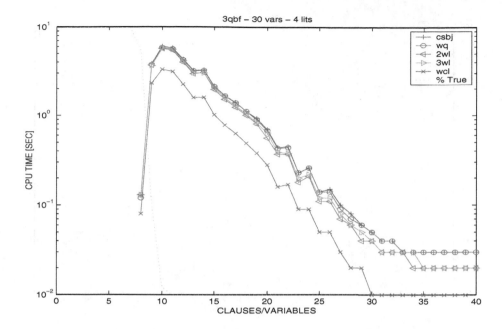

Fig. 2. Performance of CSBJ augmented with watched data structures on random problems with 4 literals per clause

1. CSBJ represents the basic solver with the standard data structures,
2. CSBJ+2WL is CSBJ plus two literal watching, as described in § 4.1,
3. CSBJ+3WL is CSBJ plus three literal watching, as described in § 4.2,
4. CSBJ+WCL is CSBJ plus watching clauses, as described in § 5.1, and
5. CSBJ+WQT is CSBJ plus watching quantifiers, as described in § 5.2.

All the versions implement the same heuristic. In particular, the literal occurring in most clauses is chosen, and ties are broken by selecting the literal with the smallest index. We considered all of the real world problems available at www.qbflib.org as of 1st May 2003. The problems considered are the Ayari family, the Castellini family, the Letz family and the Rintanen family, giving a total of 322 problems. Of these 322 problems, all the solvers (*i*) timed out on 93, and (*ii*) were able to solve 28 in 0 seconds. The results on the remaining 201 problems are shown in Figure 1. In all the figures, on the *y*-axis there is the time taken by each procedure to solve the number of instances specified on the *x*-axis. The time out is 7200s. We run the exeriments on a farm of eight identical PCs equipped with a Pentium IV 2.4Ghz processor, and 1024MB of RAM (DDR 333Mhz), running Linux RedHat 7.2(kernel version 2.4.7-10).

As it can be seen, neither the two nor the three literal watching structures cause a speed-up in the performances. One reason could be that the average length of clauses on these problems is 3.8. On the positive side, we see that watching clauses provides a significant boost; for example, to solve 167 instances

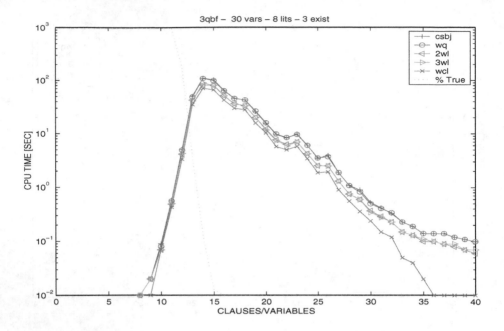

Fig. 3. Performance of CSBJ augmented with watched data structures on random problems with 8 literals per clause

out of the 201 considered, CSBJ+WCL takes 10s, while the other solvers take 100s.

We have considered also instances randomly generated according to model B of [7], with 3 alternations, 30 variables per alternation, 3 existentials per clause, with a total clause length varying between 4 and 12 and a clause to variable ratio from 1 to 40. The results of these experiments can be seen in figures 2, 3 and 4. We see that the clause watching is consistently good in all these graphs. Literal watching does not appear to affect the performance with shorter clauses, but starts to show a good increase in performance as the clause length increases. Watching Quantifiers never shows an increase in performance on these experiments. There appears to be no difference in performance between watching 2 and watching 3 Literals. Indeed, when assigning a watched literal in a clause, in the worst case, both the 3 and the 2 watching literals algorithms have to scan the whole clause. On the other hand, when all but one of the existential literals are negatively assigned, the 3 watching literal algorithm does not need to scan the list of existential literals. This advantage does not produce any significative speed-up on this test set, in which each clause contains only three existential literals. The results for clause watching are not too surprising. It is well known that in QBF the pure literal heuristic plays an important role. Detection of pure literals is difficult however, since we normally need to check every literal in a clause when the clause is removed. Clause watching removes

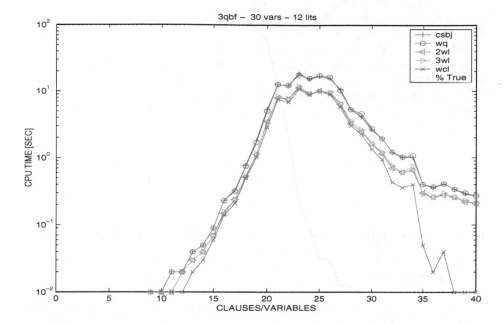

Fig. 4. Performance of CSBJ augmented with watched data structures on random problems with 12 literals per clause

this need. Additionally, most variables have many occurrences in large enough formula, giving rise to large c-literal lists. This means that many clauses will be removed with no work being done.

The results for literal watching are what we would expect; with smaller clauses, the watched literal pointers are likely to be chosen with a high probability. As the size of the clauses increases, this probability becomes smaller and so the algorithm operates more efficiently. This is due to the fact that if a literal is not watched, then no work will be done when it is removed. This is similar to the results seen in SAT, where watching literals in larger clauses (i.e. learned clauses) is more effective than watching literals in small clauses.

Finally, the results for quantifier watching are also not surprising. In QBF solvers, there is not a lot of work done on the prefix. Additionally, quantifier watching performs the same operations as the standard data structure.

7 Conclusions

We have presented four new watched data structures for QBF solvers. We have implemented and evaluated these implementations on both structured problems and randomly generated problems. The experimental analysis shows that watched data structures do not slow down the solvers significantly. Moreover, watching clauses gives consistently good results on all problems, and watching both 2 and 3 literals gives better results as the size of the clauses increase, as would be expected from previous results in SAT. Watching quantifiers almost never helps to increase the performance of search in the problems we have observed.

Finally, it is well known that in QBF, pure literal detection plays an important role. In SAT, it is a common belief that detecting and assigning pure literals does not pay off. Nevertheless, we think that clause watching could be used to increase the efficiency of pure literal detection, and so prove ultimately that pure literals may be of some importance also in SAT search.

Acknowledgements

The work of the authors in Genova is partially supported by MIUR and ASI.

References

[1] Davis, M., Putnam, H.: A computing procedure for quantification theory. Journal of the ACM **7** (1960) 201–215
[2] Davis, M., Logemann, G., Loveland, D.: A machine program for theorem proving. Journal of the ACM **5** (1962)
[3] Zhang, H., Stickel, M.E.: An efficient algorithm for unit propagation. In: Proceedings of the Fourth International Symposium on Artificial Intelligence and Mathematics (AI-MATH'96), Fort Lauderdale (Florida USA) (1996)
[4] Moskewicz, M.W., Madigan, C.F., Zhao, Y., Zhang, L., Malik, S.: Chaff: Engineering an Efficient SAT Solver. In: Proc. DAC. (2001)
[5] Lynce, I., Marques-Silva, J.: Efficient data structures for fast sat solvers. In: Proceedings of the 5th International Symposium on the Theory and Applications of Satisfiability Testing (SAT'02). (2002)
[6] Giunchiglia, E., Narizzano, M., Tacchella, A.: Backjumping for Quantified Boolean Logic Satisfiability. Artificial Intelligence **145** (2003) 99–120
[7] Gent, I., Walsh, T.: Beyond NP: the QSAT phase transition. In: Proc. AAAI. (1999) 648–653

How Good Can a Resolution Based SAT-solver Be?

Eugene Goldberg[1] and Yakov Novikov[2]

[1] Cadence Berkeley Labs, USA,
egold@cadence.com
[2] The United Institute of Informatics Problems, National Academy of Sciences,
Belarus,
nov@newman.bas-net.by

Abstract. We introduce a parameterized class $M(p)$ of unsatisfiable formulas that specify equivalence checking of Boolean circuits. If the parameter p is fixed, a formula of $M(p)$ can be solved in general resolution in a linear number of resolutions. On the other hand, even though there is a polynomial time deterministic algorithm that solves formulas from $M(p)$, the order of the polynomial is a monotone increasing function of parameter p. We give reasons why resolution based SAT-algorithms should have poor performance on this very "easy" class of formulas and provide experimental evidence that this is indeed the case.

1 Introduction

In the long run, studying the complexity of proofs in various proof systems of propositional logic is aimed at establishing lower bounds on performance of deterministic algorithms for solving SAT. The first result on complexity of resolution proofs was obtained by Tseitin in 1966 [12]. He proved exponential lower bounds for proofs in a restricted version of general resolution called regular resolution. Two decades later Haken established exponential lower bounds for general resolution [7]. Some proof systems (like extended resolution) are so powerful that all attempts to find a class of formulas that admit only proofs of exponential length have failed so far. For that reason, along with trying to break non-deterministic proof systems, a new line of research has started that studies the automatizability of proof systems [4]. A proof system P is automatizable if there is a deterministic algorithm that for any unsatisfiable CNF formula F finds a proof whose length is bounded by a polynomial of the length of the shortest proof in P (that establishes the unsatisfiability of F). In [9] it was shown that general resolution is non-automatizable under an assumption.

In this paper, we introduce a parameterized class $M(p)$ of CNF formulas. This class should be very helpful in studying the complexity of finding short resolution proofs by a deterministic algorithm, that is in studying the automatizability of resolution. A formula F of $M(p)$ represents an instance of checking the equivalence of two Boolean circuits N_1,N_2 obtained from a common "specification" S. Here, by specification we mean a circuit consisting of multi-valued

E. Giunchiglia and A. Tacchella (Eds.): SAT 2003, LNCS 2919, pp. 37–52, 2004.

gates and so computing a multi-valued function of multi-valued arguments. (A multi-valued gate will be further referred to as a block). Boolean circuits N_1,N_2 are obtained from S by encoding values of variables of S with binary codes. After encoding the values of the input and output variables of a block G, the obtained multi-output Boolean function can be implemented by a Boolean sub-circuit. Circuits N_1,N_2 are obtained from S by replacing each block G of S with Boolean sub-circuits $I_1(G)$ and $I_2(G)$ respectively. The formula F is satisfiable if and only if N_1 and N_2 are not equivalent. The parameter p bounds the number of gates in implementations of a block G of S. (I.e. each implementation of a block has at most p gates.)

We show that the unsatisfiability of a CNF formula F of $M(p)$ describing equivalence checking of circuits N_1 and N_2 can be proven in general resolution in no more than $d * n * 3^{6p}$ resolution steps. Here, d is a constant and n is the number of blocks in a common specification S of N_1 and N_2. So, if the value of p is fixed, the formulas of $M(p)$ have linear size proofs in the formula length. (We will further refer to these proofs as specification guided ones.) On the other hand, the maximal length (number of literals) of resolvents one has to deduce in those proofs is bounded by $6 * p$. So if p is fixed, the length of resolvents is bounded by a constant and there is a trivial deterministic algorithm that solves all the formulas from $M(p)$ in polynomial time. This algorithm just derives all the possible resolvents whose length does not exceed a threshold. If no empty clause is derived the algorithm just increases the threshold by 1 and repeats derivation of resolvents. We show that given a value r, there is always p' such that in a specification guided proof of the unsatisfiability of a formula from $M(p')$ one has to produce a clause of length r. This means (see Section 4) that the order of the polynomial bounding the runtime of this trivial deterministic algorithm is a monotonic increasing function of p. So the gap between the complexity of nondeterministic proofs and that of this trivial deterministic algorithm widens as the value of p grows.

Since formulas of $M(p)$ have linear complexity in general resolution and appear to be hard for deterministic algorithms they can be used to gauge the performance of resolution based SAT-solvers. Of course, in this paper we just show that formulas of $M(p)$ are hard (from a practical point of view) for a trivial deterministic algorithm. However, there is a reason to believe that these formulas are hard for any deterministic algorithm. Let us assume that specification guided resolution proofs are "significantly" shorter (at least for some classes of specifications) than any other resolution proofs. On the one hand, a specification guided proof of the unsatisfiability of a formula F from $M(p)$ describing equivalence checking of N_1 and N_2 closely follows the topology of a common specification S of N_1 and N_2. So knowing a short proof of unsatisfiability for F one could recover S from N_1 and N_2. On the other hand, the problem of finding a common specification of N_1 and N_2 appears to be hard (most likely NP-hard).

In this paper we give some experimental evidence that the class $M(p)$ is indeed hard for existing SAT-solvers. Namely, we test the performance of state-of-the-art SAT-solvers 2clseq [1], Zchaff [8], BerkMin [6] and show that their performance quickly degrades as the size of parameter p grows.

2 Formal Definition of Class $M(p)$

In this section, we specify the class of formulas we consider in this paper.

2.1 Definition of Specification and Implementation

In this subsection, we introduce the notion of a specification and its implementation that play a key role in the following exposition.

Let S be a combinational circuit of multi-valued gates specified by a directed acyclic graph H. The circuit S will be further referred to as a **specification**. The sources and sinks of H correspond to **primary inputs and outputs** of S. Each internal node of H corresponds to a multi-valued gate further referred to as **a block**. A block is a device computing a multi-valued function of multi-valued arguments (like a "regular" gate computes a Boolean function of Boolean arguments.) Let n_1 and n_2 be nodes of H connected by an edge directed from n_1 to n_2. Then in the specification S the output of the block G_1 corresponding to n_1 (or the primary input corresponding to n_1 if n_1 is a source of H) is connected to an input of the block G_2 corresponding to n_2. Each node n of H is associated with **a multi-valued variable V**. If n is a source (respectively a sink) of H then the corresponding variable specifies values taken by the corresponding primary input (respectively primary output) of S. If n is an internal node of H (i.e. it is neither a source nor a sink) than the corresponding variable (also called an **internal variable**) specifies values taken by the output of the block specified by n. Let G be the block of S specified by a node n of H. We will use the notation $C = G(A, B)$ to indicate that a) the output of the block G is associated with a variable C; b) the function computed by the block G is $G(A, B)$; c) only two nodes of H are connected to the node n in H by edges directed to n and these nodes are associated with variables A and B.

Denote by $D(A)$ the **domain** of the variable A associated with a node of H. The number of values taken by A i.e. the value $|D(A)|$ is called the **multiplicity** of A. If the multiplicity of each variable A associated with a node of H is equal to 2 then S is a **Boolean circuit**.

Henceforth, by a variable A of a specification S we will mean the variable associated with a node of the graph H. Let $D(A) = \{a_1, \ldots, a_t\}$ be the domain of a variable A of S. Denote by $q(A)$ **a Boolean encoding** of the values of $D(A)$ that is a mapping $q : D(A) \to \{0, 1\}^m$. Denote by $length(q(A))$ the number of bits in the encoding q (that is the value of m). The value $q(a_i)$, $a_i \in D(A)$ is called the **code** of a_i. Given an encoding q of length m of a variable A, denote by $v(A)$ the set of all m **coding Boolean variables** that is the variables specifying the Boolean space of codes for the values of A.

In the following exposition we make the assumptions below.

Assumption 1 *Each block of a multi-valued and each gate of a Boolean circuit has two inputs and one output.*

Assumption 2 *For each variable A of any specification, $|D(A)|$ is a power of 2.*

Assumption 3 *For each variable A of any specification S, a minimal length encoding is used. That is $length(q(A)) = log_2(|D(A)|)$.*

Assumption 4 *Any two different values of a variable A of S have different codes. That is if $a_i, a_j \in D(A)$ and $a_i \neq a_j$ then $q(a_i) \neq q(a_j)$.*

Remark 1. From Assumption 2 and Assumption 3 it follows that for each variable A of any specification, any encoding $q : D(A) \rightarrow \{0,1\}^m$ is bijective. In particular, any assignment to the variables of $v(A)$ is a code of some $a_i \in D(A)$.

Now we describe how a Boolean circuit N is obtained from a specification S.

Definition 1. *Denote by $Inp(I)$ (respectively $Out(I)$) the set of primary input (respectively primary output) variables of a Boolean circuit I.*

Definition 2. *Let X_1 and X_2 be sets of Boolean variables and $X_2 \subseteq X_1$. Let y be an assignment to the variables of X_1. Denote by $proj(y, X_2)$ **the projection** of y on X_2 i.e. the part of y that consists of assignments to the variables of X_2.*

Definition 3. *Let $C = G(A, B)$ be a block of specification S. Let $q(A), q(B), q(C)$ be encodings of variables A,B, and C respectively. A Boolean circuit I is said to **implement the block G** if the following three conditions hold:*

1. *The set $Inp(I)$ is a subset of $v(A) \cup v(B)$.*
2. *The set $Out(I)$ is equal to $v(C)$.*
3. *If the set of values assigned to $v(A)$ and $v(B)$ form codes $q(a)$ and $q(b)$ respectively where $a \in D(A)$, $b \in D(B)$, then $I(z') = q(c)$ where $c = G(a,b)$. Here, z' is the projection of assignment $z = q(a),q(b)$ on $Inp(I)$ and $I(z')$ is the value taken by I at z'.*

Remark 2. If $Inp(I) = v(A) \cup v(B)$, then the third condition above just says $I(q(a), q(b)) = q(c)$. However, in general, one can implement a block G of S with a circuit having fewer input variables than $|v(A)| + |v(B)|$ (because the output of the block G may take the same value for different assignments to variables A,B). This is why we used the notion of projection to define block implementation.

Definition 4. *Let S be a specification. A Boolean circuit N is said to **implement a specification S**, if it is built according to the following two rules.*

1. *Each block G of S is replaced with an implementation I of G.*
2. *Let the output of block $C = G_1(A, B)$ of S be connected to the input C of a block $G_2(C, D)$. Then the outputs of the circuit I_1 implementing G_1 are properly connected to inputs of circuit I_2 implementing G_2. Namely, the primary output of I_1 specified by a Boolean variable $v_i \in v(C)$ is connected to the input of I_2 specified by the same variable, if $v_i \in Inp(I_2)$.*

Let N be an implementation of a specification S. From Remark 1 it follows that for any value assignment y to the primary input variables of N there is a unique set of values $(x_1, \ldots, x_k)$, where $x_i \in D(X_i)$ such that $y = (q(x_1), \ldots, q(x_k))$. In other words there is one-to-one correspondence between assignments to primary inputs of S and N.

Definition 5. *Let N be an implementation of S. Given a Boolean vector y of assignments to the primary inputs of N, the corresponding vector $Y = (x_1, \ldots, x_k)$ such that $y = (q(x_1), \ldots, q(x_k))$ is called the **pre-image** of y.*

Definition 6. *The **topological level** of a block G in a circuit S is the length of the longest path going from a primary input of S to G. (The length of a path is measured in the number of blocks on it. So, the topological level of a primary input is 0.) Denote by **level(G)** the topological level of G in S.*

Proposition 1. *Let N be a circuit implementing specification S. Let y be a value assignment to the primary input variables of N and Y be the pre-image of y. Then, for any variable C of S, the values assigned to variables $v(C)$ in N form the code $q(c)$ where c is the value taken by variable C when the inputs of S take the values specified by Y.*

Proof. The proposition can be proven by induction in topological levels of variables of the specification S.

Base step. The correctness of the theorem for the variables with topological level 0 (that is primary input variables of S) follows from Remark 1.

Induction step. Let C be a variable of S and $level(C) = n$, $n > 1$. The induction hypothesis is that the proposition holds for all the variables of S with topological level less than n. Let $G(A, B)$ be the block of S whose output is associated with C. Let $I(G)$ be the implementation of G in N. Obviously, the topological level of A and B is less than n. Then under the assignment y to the primary inputs of N the variables $v(A)$ and $v(B)$ take values $q(a)$ and $q(b)$. Here, a and b are the values taken by A and B under the assignment Y to primary inputs of S. Then from Definition 3 it follows that the value taken by the outputs of $I(G)$ is $q(c)$ where $c = G(a, b)$. □

Proposition 2. *Let N_1, N_2 be circuits implementing a specification S. Let each primary input (or output) variable Q have the same encoding in N_1 and N_2. Then Boolean circuits N_1 and N_2 are functionally equivalent.*

Proof. Let y be an arbitrary assignment of values to the variables of N_1 and N_2. Let us prove that all the corresponding primary outputs of N_1 and N_2 take identical values. Denote by Y_1 and Y_2 the pre-images of y with respect to encodings of variables of S used when obtaining implementations N_1 and N_2. Since N_1 and N_2 have identical encodings of corresponding primary input variables, then $Y_1 = Y_2$. Let C be the primary output variable of S associated with a block G. Denote by c the value taken by the output of G under the input assignment Y_1 (or Y_2). From Definition 3 it follows that the value assignment taken by the outputs of the implementation I_1 (respectively I_2) of the block G in the circuit N_1 (respectively N_2) is $q_1(c)$ (respectively $q_2(c)$). Since all the primary output variables have the same encoding, then $q_1(c)=q_2(c)$.   □

Definition 7. *Let N_1, N_2 be two functionally equivalent Boolean circuits implementing a specification S. Let for each primary input and output variable X of S, encodings $q_1(X)$ and $q_2(X)$ (used when producing N_1 and N_2 respectively) be identical. Then S is called **a common specification (CS)** of N_1 and N_2.*

Definition 8. *Let S be a CS of N_1 and N_2. Let the maximal number of gates in an implementation of a block from S in N_1 and N_2 be equal to p. Then we will say that S is a CS of N_1 and N_2 of **granularity** p.*

2.2   Equivalence Checking as Satisfiability Problem

In this subsection, we introduce the class $M(p)$ that is considered in this paper

Definition 9. *A disjunction of literals of Boolean variables not containing more than one literal of the same variable is called a **clause**. A conjunction of clauses is called a **conjunctive normal form (CNF)**.*

Definition 10. *Given a CNF F, **the satisfiability problem (SAT)** is to find a value assignment to the variables of F for which F evaluates to 1 (also called a **satisfying assignment**) or to prove that such an assignment does not exist. A clause C of F is said to be **satisfied** by a value assignment if a literal of C is set to 1 by this assignment.*

The standard conversion of an equivalence checking problem into an instance of SAT is performed in two steps. Let N_1 and N_2 be two Boolean circuits to be checked for equivalence. In the first step, a circuit M called **a miter** [5] is formed from N_1 and N_2. The miter M is obtained by 1) identifying the corresponding primary inputs of N_1 and N_2; 2) adding d two-input XOR gates (where d is the number of primary outputs in N_1 and N_2), each XOR gate being fed by the j-th primary outputs of N_1 and N_2, $1 \leq j \leq d$; 3) adding a d-input OR gate, inputs of which are connected to the outputs of the d XOR gates. So the miter of N_1 and N_2 evaluates to 1 if and only if for some input assignment a primary output of N_1 and the corresponding output of N_2 take different values. Therefore, the problem of checking the equivalence of N_1 and N_2 is equivalent to testing the satisfiability of the miter of N_1 and N_2.

In the second step of conversion, the satisfiability of the miter is **reduced** to that of a CNF formula F. This formula is a conjunction of CNF formulas $F_1, \ldots, F_n$ specifying the functionality of the gates of M and a one-literal clause that is satisfied only if the output of M is set to 1. A CNF F_i specifies a gate g_i of M if and only if each assignment satisfying F_i is consistent with the functionality of g_i and each assignment falsifying a clause of F_i is inconsistent with the functionality of g_i. For instance, the AND gate $y=x_1 x_2$ is specified by the following three clauses $\overline{x_1} \vee \overline{x_2} \vee y,\ x_1 \vee \overline{y},\ x_2 \vee \overline{y}$.

Definition 11. *Given a constant p, a CNF formula F is a member of the **class M(p)** if and only if it satisfies the following two conditions.*

1. *F is the CNF formula (obtained by the procedure described above) specifying the miter of a pair of funcionally equivalent circuits N_1,N_2.*
2. *N_1,N_2 have a CS of granularity p', where $p' \leq p$.*

3   Solving Formulas of $M(p)$ in General Resolution

In this section, we show that the unsatisfiability of formulas from $M(p)$ can be proven in general resolution in a linear number of resolution steps.

Definition 12. *Let K and K' be clauses having opposite literals of a variable x and there is only one such variable. The **resolvent of K , K' in x** is the clause that contains all the literals of K and K' but the positive and negative literals of x. The operation of producing the resolvent of K and K' is called **resolution**.*

Definition 13. General resolution *is a proof system of propositional logic that has only one inference rule. This rule is to resolve two existing clauses to produce a new one. Given a CNF formula F, the proof of unsatisfiability of F in the general resolution system consists of a sequence of resolution steps resulting in the derivation of an empty clause (i.e. a clause without literals).*

Definition 14. *Let F be a set of clauses. Denote by **supp(F)** the support of F i.e. the set of variables whose literals occur in F.*

The following three propositions are used in the proof of Proposition 6.

Proposition 3. *Let A,B,C be Boolean functions and $A \wedge B \to C$. (The symbol $\to$ means implication.) Then for any function A' such that $A' \to A$, it is true that $A' \wedge B \to C$.*

Proof. The proof follows from the definition of implication.

Proposition 4. *Let X_1 and X_2 be sets of Boolean variables, $F(X_1, X_2)$ and $H(X_2)$ be CNF formulas, and F imply H. Then one can derive a CNF formula H' implying H, where $supp(H') \subseteq supp(H)$, from clauses of F in at most $3^{|supp(F)|}$ resolution steps.*

Proof. Let $H = C_1 \wedge \ldots \wedge C_k$. We show that for any $C_i, 1 \le i \le k$ one can derive a clause C_i' (by resolving clauses of F) such that C_i' implies C_i. Then the CNF formula $H' = C_1' \wedge \ldots \wedge C_k'$ implies H.

Since C_i is a clause of H and F implies H, then F implies C_i. Then from the completeness of resolution it follows that by resolving clauses of F one can derive a clause C_i' implying C_i. Indeed, by making the assignments setting the literals of C_i to 0, we turn F into an unsatisfiable CNF formula F'. Then an empty clause can be derived from clauses of F'. From a derivation of an empty clause from F', one can easily construct the derivation of a clause C_i' (from clauses of F) implying C_i. The total number of resolvents that can be obtained from clauses of F is bounded by $3^{|supp(F)|}$. So all the clauses of H' can be obtained in no more than in $3^{|supp(F)|}$ resolution steps.

Definition 15. *Let G be a block of S, I_1 and I_2 be the implementations of G in N_1 and N_2 and C be the multi-valued variable associated with the output of G. A set of clauses $R(v_1(C), v_2(C))$ is called **a restricting set** for the block G if the following two conditions hold:*

1. *The support of $R(v_1(C), v_2(C))$ is a subset of $v_1(C) \cup v_2(C)$ (here, $v_1(C)$, $v_2(C)$ are the coding variables of encodings q_1 and q_2 for the values of C).*
2. *Let z_1, z_2 be assignments to the variables of $v_1(C)$ and $v_2(C)$ respectively. An assignment (z_1, z_2) satisfies $R(v_1(C), v_2(C))$ if and only if $z_1 = q_1(c)$ and $z_2 = q_2(c)$ where $c \in D(C)$.*

Proposition 5. *Let S be a CS of circuits N_1, N_2. Let $C = G(A, B)$ be a block of S. Let F be the CNF formula specifying the miter of N_1, N_2. Let $F(I_1(G))$ and $F(I_2(G))$ be the parts of F specifying the implementation $I_1(G)$ and $I_2(G)$ respectively. Then the CNF $P = Restriction \wedge Implementation$ implies $R(v_1(C), v_2(C))$. Here, $Restriction = R(v_1(A), v_2(A)) \wedge R(v_1(B), v_2(B))$ and $Implementation = F(I_1(G)) \wedge F(I_2(G))$.*

Proof. To prove that P implies $R(v_1(C), v_2(C))$ one needs to show that any assignment satisfying P satisfies $R(v_1(C), v_2(C))$ as well. It is not hard to see that the support of all CNFs of the expression $P \rightarrow R(v_1(C), v_2(C))$ is a subset of $supp(F(I_1(G))) \cup supp(F(I_2(G)))$. Let $h = (x_1, x_2, y_1, y_2, z_1, z_2)$ be an assignment satisfying P where $x_1, x_2, y_1, y_2, z_1, z_2$ are assignments to $v_1(A)$, $v_2(A), v_1(B), v_2(B), v_1(C), v_2(C)$ respectively. Since h satisfies P it has to satisfy *Restriction* and *Implementation*. Since h satisfies *Restriction*, then $x_1 = q_1(a)$, $x_2 = q_2(a)$ and $y_1 = q_1(b)$, $y_2 = q_2(b)$ where $a \in D(A)$ and $b \in D(B)$. So h can be represented as $(q_1(a), q_2(a), q_1(b), q_2(b), z_1, z_2)$. Since h satisfies *Implementation* then z_1 has to be equal to $q_1(c)$, $c = G(a, b)$ and z_2 has to be equal to $q_2(c)$. Hence h satisfies $R(v_1(C), v_2(C))$. $\square$

Proposition 6. *Let F be a formula of $M(p)$ specifying the miter of circuits N_1, N_2 obtained from a CS S of granularity p' where $p' \leq p$. The unsatisfiability of F can be proven by a resolution proof of no more than $d * n * 3^{6p}$ resolution steps where n is the number of blocks in S and d is a constant. (These proofs will be further referred to as **specification guided proofs**).*

Proof. From Proposition 5 it follows that one can deduce a restricting set of clauses for each variable of S in topological order starting with blocks of topological level 1 and proceeding to outputs. Indeed, if a variable A is a primary input variable of S, then $R(v_1(A), v_2(A))$ is empty. (The latter is true because, since S is a CS of N_1 and N_2, then $q_1(A)$ and $q_2(A)$ are identical and, besides, variables $v_1(A)$ and $v_2(A)$ in a miter are identified.) Let $C = G(A, B)$ be a block of topological level 1. Then A and B are primary input variables and restricting sets of clauses for them are known (namely, they are empty). From Proposition 5 it follows that $R(v_1(C), v_2(C))$ is implied by $F(I_1(G)) \cup F(I_2(G))$. From Proposition 4 it follows that a CNF R' formula implying $R(v_1(C), v_2(C))$ such that $supp(R') \subseteq supp(R)$ can be derived by resolving clauses of $F(I_1(G)) \cup F(I_2(G))$. From Proposition 3 it follows that replacing CNF formula $R(v_1(C), v_2(C))$ with a CNF R' that implies R does not break the inductive scheme of deducing restricting sets in topological order. After restricting CNFs are computed for the variables of topological level 1, the same procedure can be applied to variables of topological level 2 and so on.

The complexity of the derivation of a restricting set $R(v_1(A), v_2(A))$ is 3^{6p}. Indeed, the support of all CNFs forming the expression of Proposition 5 is a subset of $supp(F(I_1(G))) \cup supp(F(I_2(G)))$ where G is the block whose output is associated with the variable A. The number of gates in each implementation is bounded by p. Besides, each gate has two inputs and one output. So, the total number of variables in the clauses used when deriving $R(v_1(A), v_2(A))$ is at most $6 * p$. Hence the complexity of derivation is bounded by 3^{6p}. Since the total number of blocks in S in n, then the complexity of deriving restricting sets for all the variables of S is bounded by $n * 3^{6p}$.

Now we show that after restricting sets are deduced for all the primary output variables of S, an empty clause can be derived in the number of resolution steps that is linear in $n * p$. Let G be a block of S whose output (associated with variable C) is a primary output of S. From the definition of the class $M(p)$ it follows that the encodings $q_1(C)$ and $q_2(C)$ are identical and have minimal length. According to Remark 1 any assignment to the variables of $v_1(C)$ (or $v_2(C)$) is a code of a value $c \in D(C)$. Then any assignment h satisfying $R(v_1(C), v_2(C))$ has to assign the same values to corresponding variables of $v_1(C)$ and $v_2(C)$. This means that $R(v_1(C), v_2(C))$ implies 2-literal clauses specifying the equivalence of corresponding variables of $v_1(C)$ and $v_2(C)$. Then these 2-literal equivalence clauses can be deduced from $R(v_1(C), v_2(C))$ in the number of resolution steps bounded by 3^{6p}. (These steps are already counted in the expression $n * 3^{6p}$). Having deduced these equivalence clauses one can derive $|v_1(C))|$ single literal clauses (each variable representing the output of a gate XORing a pair of corresponding outputs of I_1 and I_2) that can be satisfied only by setting outputs of

XOR gates to 0. The number of such clauses is bounded by $n*p$ and the complexity of the derivation of such a clause is bounded by a constant. Finally, in the number of resolution steps bounded by $n*p$ we can derive the empty clause. It is deduced from the single-literal clause that requires setting the output of the miter to 1, the clauses specifying the last OR gate, and the single-literal clauses that require setting the outputs of the XOR gates to 0. □

Remark 3. If the value of p is fixed, then the expression 3^{6p} is just a constant and so the formulas of $M(p)$ can be proven to be unsatisfiable in a linear number of resolution steps.

Proposition 7. *Given a number r, there is always p such that in a specification driven resolution proof of unsatisfiability of a formula $F \in M(p)$ one has to produce a clause of at least r literals.*

Proof. Let $C = G(A, B)$ be a block of a specification S and N_1 and N_2 be circuits with a CS S. Let $q_1(C)$, $q_2(C)$ be the binary encodings of the variable C used when obtaining N_1 and N_2. Let F be the formula specifying the miter of N_1 and N_2. The key feature of a specification guided resolution proof is that it deduces restricting sets of clauses. To prove the required proposition it suffices to show that for any r there are always binary encodings $q_1(C)$, $q_2(C)$ such that any restricting set of clauses $R(v_1(C), v_2(C))$ has to contain a clause of at least r literals. Then in any specification guided resolution proof of the unsatisfiability of F one has to produce a clause of at least r literals. Since S is finite, there exists p such that F is in $M(p)$.

Let $C=G(A,B)$ be a block of a specification S and $level(C) = 1$. Assume that the multiplicity of variables A,B and C is 2^r. Let $I_1(G)$ and $I_2(G)$ be two implementations of G where $q_1(A) = q_2(A)$ and $q_1(B) = q_2(B)$ and the variable C is encoded with two different r-bit encodings. Now we show that there is a pair of encodings $q_1(C)$ and $q_2(C)$ such that any restricting set of clauses $R(v_1(C), v_2(C))$ has to include a clause of at least r literals

Denote by Z the vector whose all r components are equal to 0. Denote by B_i the vector of r components i-th component of which is equal to 1 and the rest of the components are equal to 0. Let $c_0,\ldots,c_m$ be the values of C where $m = 2^r - 1$. Let $q_1(C)$ and $q_2(C)$ be picked in such a way that the following two conditions hold:

1) The codes for $c_1,\ldots,c_r$ are chosen as follows: $q_1(c_1)=q_2(c_1)=B_1,\ \ldots,$ $q_1(c_i)=q_2(c_i)=B_i,\ \ldots,\ q_1(c_r)=q_2(c_r)=B_r.$

2) Encodings q_1 and q_2 assign the code Z to different values of C. In other words, there are $c_j, c_p \in D(C)$, where $j, p > r$ such that $q_1(c_j) = Z$, $q_2(c_j) \neq Z$ and $q_2(c_p) = Z$, $q_1(c_p) \neq Z$ and $j \neq p$.

Let us show that among the clauses of $R(v_1(C), v_2(C))$ there has to be a clause of at least r literals. When the output of $I_1(G)$ is equal to Z the output of $I_2(G)$ cannot be equal to Z. So in any restricting set $R(v_1(C), v_2(C))$ there must be a clause K falsified by the vector Z,Z (obtained by the concatenation of two vectors Z.) Let us show that any clause of $R(v_1(C), v_2(C))$ that is falsified by Z,Z has to contain at least r literals. The idea is that if a clause K contains less than r literals, then it falsifies one of the allowed combinations B_i, B_i where $i = 1, \ldots, k$. Indeed, to be falsified by Z,Z the clause should contain only positive literals of variables $q_1(C)$, $q_2(C)$. Let x be a positive literal contained in K. Any two allowed combinations B_i, B_i and B_j, B_j where $i \neq j$ have different components that are equal to 1. Hence only one of the allowed combinations B_i, B_i can set the literal x to 1. So if the clause C contains less than r literals than there exists an allowed combination B_i, B_i that falsifies K. $\square$

4   Class $M(p)$ and Non-automatizability of General Resolution

It is not hard to show that if p is fixed, the unsatisfiability of a formula F from $M(p)$ can be proven by a polynomial time deterministic algorithm. Indeed, from the proof of Proposition 6 it follows that the maximal length of a resolvent one needs to derive when producing an empty clause is bounded by $6*p$. So an empty clause can be obtained by the following trivial algorithm. This algorithm derives all the resolvents whose length does not exceed a threshold value. (Initially, this value is equal to 1.) If an empty clause is not derived, the algorithm increases the threshold value by 1 and derives all the resolvents whose length does not exceed the new threshold value. As soon as the threshold value reaches $6 * p$, the algorithm deduces an empty clause, and since the value of p is fixed, the algorithm completes in polynomial time (in formula length). From Proposition 7 it follows that the order of the polynomial bounding the performance of the described algorithm is a monotonic increasing function of p. (Indeed, according to Definition 11 if $p' < p''$, then $M(p') \subset M(p'')$. Denote by $L(p)$ the length of the longest clause one has to deduce in a specification guided proof of a formula from $M(p)$. Then $L(p') \leq L(p'')$. From Proposition 7 it follows that the value of $L(p)$ actually increases as p grows.) So even though the class $M(p)$ is polynomially tractable for every fixed p, the complexity of the described algorithm increases as p grows, which makes it impractical.

Due to polynomial tractability, formulas from $M(p)$ cannot be used for proving general resolution to be non-automatizable. (Recall that a proof system is non-automatizable if there is no algorithm for finding proofs whose length is a polynomial of the length of the shortest proof.) However, one can form harder classes of formulas by letting the parameter p be a function of formula length L. Denote by $M(\infty)$ the union of classes $M(p)$ for all p. Let $M(p \leq t * log_2(L))$ be a class of formulas from $M(\infty)$ for which the value of p is bounded by the value of $t * log_2(L)$ where t is a constant. (So the value of p "slowly" grows as the length of the formula increases.) Since for a formula from $M(p)$ there is a resolution proof whose length is bounded by $d*n*3^{6p}$, then the size of resolution proofs for the formulas from $M(p \leq t*log_2(L))$ is bounded by L^{12t}. That is these formulas have polynomial size resolution proofs of unsatisfiability. On the other hand, since the value of p is not fixed, the length of resolvents one has to derive for the formulas of $M(p \leq t * log_2(L))$ is not bounded any more. So there is no trivial polynomial time deterministic algorithm like the one described above for solving formulas from $M(p \leq t * log_2(L))$.

A natural question is whether there is a better deterministic algorithm for solving formulas from a class $M(p)$ or harder classes that can be derived from $M(\infty)$. We believe that the answer is negative unless at least one of the following two assumptions does not hold. The first assumption is that, given circuits N_1, N_2 having a CS S, finding S is hard. (In particular, the problem of testing if a pair of circuits has a CS of granularity less or equal to p is most probably NP-complete.) The second assumption is that specification guided proofs are the only short resolution proofs for a class of formulas from $M(p)$. Then, for a formula of this class, the only way to find a short resolution proof is to construct a specification guided proof. But knowing such a proof one could recover the underlying specification. On the other hand, according to our first assumption finding a CS should be hard. So, for this class of formulas there should not be an efficient algorithm for finding short resolution proofs.

5   Experimental Results

The goal of experiments was to prove that formulas of $M(p)$ are hard for existing SAT-solvers and that the hardness of formulas increases as p grows. Namely, we would like to test for satisfiability formulas from classes $M(p')$ and $M(p'')$ where $p' < p''$ and show that formulas from $M(p'')$ are much harder. The main problem here is to guarantee that the formulas of $M(p'')$ we build are not in $M(p')$. (That is one needs to make sure that circuits with a CS of granularity $p \leq p''$ whose equivalence checking a formula from $M(p'')$ describes, do not have a "finer" CS whose granularity is less or equal to p'.) We solve this problem by constructing formulas of classes $M(1)$ and $M(p), p > 1$ and using the fact that proving that a formula of $M(p), p > 1$ is not in $M(1)$ is easy. (Of course, generating formulas with a greater variety of values of p will make experimental results more convincing. We hope that such experiments will be carried out in the near future.)

To construct formulas of $M(1)$ we just form the miter of two identical copies N_1, N_2 of a Boolean circuit N. These two copies can be considered as obtained from the "specification" N where each two-valued variable X was encoded using the trivial encoding $q(0) = 0$, $q(1) = 1$. Since each block G of specification is replaced with a "sub-circuit" consisting of one gate (identical to G), then the formula F specifying the miter is in $M(1)$.

To construct formulas from a class $M(p')$, $p' > 1$ that are not in $M(1)$ we used the following observation. The fact that a formula F is in $M(1)$ means that F specifies equivalence checking of circuits N_1, N_2 that are identical modulo negation of some gate pins. That is if g is a gate of N_1 then its counterpart in N_2 is identical to g modulo negation of an input (or both inputs) and/or the output. This means that each internal point y_1 of N_1 has a corresponding point y_2 of N_2 that is functionally equivalent to y_1 (modulo negation) in terms of primary input variables. So one can obtain a formula F that is in $M(p')$, $p' > 1$ and not in $M(1)$ by producing two circuits from a specification S that have no functionally equivalent internal points (modulo negation). This can be achieved by taking a specification S and picking two sets of "substantially different" encodings of the internal multi-valued variables of S

In the experiments we tested the performance of 2clseq (downloaded from [15], Zchaff (downloaded from [13]), BerkMin (version 561 that can be downloaded from [3]). BerkMin and Zchaff are the best representatives of the conflict driven learning approach. The program 2clseq was used in experiments because it employs preprocessing to learn short clauses and uses a special form of resolution called hyperresolution. The experiments were run on a SUNW Ultra-80 system with clock frequency 450MHz. In all the experiments the time limit was set to 60,000 sec. (16.6 hours). The best result is shown in bold.

Table 1. Equivalence checking of circuits from $M(1)$

Name of specifica-tion	Number of var.	Number of clauses	2clseq (sec.)	Zchaff (sec.)	BerkMin (sec.)
9symml	480	1,410	0.2	0.1	**0.04**
C880	806	2,237	**0.1**	5.4	0.5
ttt2	1,385	4,069	1.3	0.3	**0.1**
frg1	1,615	4,759	1.3	0.6	**0.3**
term1	1,752	5,157	2.0	2.0	**0.6**
x4	2,083	5,929	1.5	1.1	**0.3**
alu4	2,368	7,057	29.9	2.0	**0.5**
i9	2,477	7,105	14.3	1.0	**0.3**
c3540	2,624	7,723	**3.0**	71.0	13.4
rot	2,990	8,497	2.1	6.1	**0.9**
x1	4,380	12,955	27.8	3.8	**0.9**
dalu	4,713	13,899	36.8	5.7	**1.9**
c6288	4,770	14,245	**1.2**	*	*
frg2	5,158	14,967	17.5	6.5	**1.3**
c7552	5,781	16,621	**4.6**	327.5	83.6
k2	5,848	17,369	406.8	2.1	**1.1**
i10	6,499	18,653	31.2	237.4	**24.4**
i8	7,262	21,307	207.7	8.1	**2.7**
t481	9,521	28,512	3,250.4	13.1	**8.7**
des	14,451	42,343	164.2	271.7	**15.1**
too_large	29,027	86,965	8,838.9	384.47	**162.6**

'*' means that the program was aborted after 60,000 seconds

In Table 1 we test the performance of the three SAT-solvers on formulas from $M(1)$. Each formula describes equivalence checking of two copies of an industrial circuit from the LGSynth-91 benchmark suite [16]. First, a circuit N of LGSynth-91 was transformed by a logic optimization system SIS [10] into a circuit N' consisting only of two-input AND gates and invertors. Then the formula specifying the miter of two copies of N' was tested for satisfiability. BerkMin has the best performance on almost all the instances except four. On the other hand, 2clseq is the only program to have finished all the instances within the time limit. In particular, it is able to solve the C6288 instance (a 16-bit multiplier) in about one second while BerkMin and Zchaff fail to solve it in 60,000 seconds. This should be attributed to successful formula preprocessing that results in deducing many short clauses describing equivalences of variables.

In Table 2 we test the performance of 2clseq, Zchaff, BerkMin on formulas that are in a class $M(p'), p' > 1$ and not in $M(1)$. These circuits were obtained by the following technique. First, we formed multi-valued specifications. Each specification S was obtained from a circuit N of the LGSynth-91 benchmark suite by replacing each binary gate of N with a four-valued gate. (In other words, the obtained specification S had different functionality while inheriting the "topology" of N.) Then two functionally equivalent Boolean circuits N_1, N_2 were produced from S using two "substantially different" sets of two-bit encodings of four-valued values. This guaranteed that the internal points of N_1 had no (or had very few) functionally equivalent (modulo negation) counterparts in N_2 and vice versa. So each produced formula was not in $M(1)$.

It is not hard to see that the performance of all three SAT-solvers is much worse regardless of the fact that each formula of Table 2 is only a few times larger than its counterpart from Table 1. Namely, the number of variables in each formula of Table 2 is only two times and the number of clauses is only four times that of its counterpart from Table 1. Such kind of performance degradation is not unusual for formulas having exponential complexity in general resolution. However, all the formulas we used in experiments are in $M(p')$ (class $M(1)$ is a subset of any class $M(p')$, $p' > 1$) that has linear complexity in general resolution.

2clseq is able to solve only three instances, which indicates that preprocessing that helped for the instances of $M(1)$ does not work for $M(p'), p' > 1$. The probable cause is that "useful" clauses are longer than for formulas of $M(1)$. BerkMin shows best performance on the instances of Table 2. Nevertheless it fails to solve five instances and is much slower on instances that it is able to complete. For example, BerkMin solves the instance dalu from $M(1)$ in 1.6 seconds while testing the satisfiability of its counterpart from $M(p'), p' > 1$ takes more than 20,000 seconds. These results suggest that indeed formulas $M(p)$ are hard for existing SAT-solvers and the "hardness" of formulas increases as p grows.

6   Conclusions

We introduced a class $M(p)$ of formulas specifying equivalence checking of Boolean circuits obtained from the same specification. If the value of p is fixed, the formulas of $M(p)$ can be proven to be unsatisfiable in a linear number of

Table 2. Equivalence checking of circuits from $M(p'), p' > 1$

Name of specification	Number of var.	Number of clauses	2clseq (sec.)	Zchaff (sec.)	BerkMin (sec.)
9symml	960	6,105	206.3	210.6	**58.2**
C880	1,612	9,373	*	*	**200.1**
ttt2	2,770	17,337	*	799.4	**77.2**
frg1	3,230	20,575	*	*	**3,602.4**
term1	3,504	22,229	*	*	**1,183.6**
x4	4,166	24,733	*	769.5	**139.0**
alu4	4,736	30,465	45,653.5	25,503.6	**2,372.6**
i9	4,954	29,861	2,215.5	23.5	**11.5**
c3540	5,248	33,199	*	*	**4,172.0**
rot	5,980	35,229	*	*	**1,346.9**
x1	8,760	55,571	*	*	*
dalu	9,426	59,991	*	*	**20,234.4**
c6288	9,540	61,421	*	*	*
frg2	10,316	62,943	*	7,711.0	**1,552.9**
c7552	11,282	69,529	*	803.5	**74.5**
k2	11,680	74,581	*	*	*
i10	12,998	77,941	*	*	**38,244.6**
i8	14,524	91,139	*	35,721.5	**4,039.7**
t481	19,042	123,547	*	*	*
des	28,902	179,895	*	1,390.3	**331.1**
too_large	58,054	376,801	out of memory	*	*

'*' means that the program was aborted after 60,000 seconds

resolution steps. On the other hand, the formulas of $M(p)$ can be solved in a polynomial time by a trivial deterministic algorithm but the order of the polynomial increases as the value of p grows. We give reasons why formulas from $M(p)$ should be hard for any deterministic algorithm. We show that formulas of $M(p)$ are indeed hard for the state-of-the-art SAT-solvers we used in experiments.

References

1. F.Bacchus. *Exploring the computational tradeoff of more reasoning and less searching.* Fifth International Symposium on Theory and Applications of Satisfiability Testing, pages 7-16, 2002.
2. E.Ben-Sasson, R. Impagliazzo, A.Wigderson. *Near optimal separation of Treelike and General resolution* SAT-2000: Third Workshop on the Satisfiability Problem. - May 2000.
3. BerkMin web page. http://eigold.tripod.com/BerkMin.html
4. M.Bonet, T.Pitassi, R.Raz. On interpolation and automatization for Frege Systems. SIAM Journal on Computing, 29(6),pp 1939-1967,2000.

5. D. Brand. *Verification of large synthesized designs.* Proceedings of ICCAD-1993,pp 534-537.
6. E. Goldberg, Y. Novikov. *BerkMin: A fast and robust SAT-solver.* Design, Automation, and Test in Europe (DATE '02), pages 142-149, March 2002.
7. A. Haken. The intractability of resolution. Theor. Comput. Sci. 39 (1985),297-308.
8. M. Moskewicz, C. Madigan, Y. Zhao, L.Zhang, and S.Malik.*Chaff: Engineering an efficient SAT-solver.* Proceedings of DAC-2001.
9. M. Alekhnovich, A.Razborov. Resolution is not automatizable unless W[p] is tractable. Proceedings of FOCS,2001.
10. Sentovich, E. e.a. *Sequential circuit design using synthesis and optimization.* Proceedings of ICCAD, pp 328-333, October 1992.
11. J.P.M.Silva, K.A.Sakallah. GRASP: A Search *Algorithm for Propositional Satisfiability* . IEEE Transactions of Computers. -1999. -V. 48. -P. 506-521.
12. G.S.Tseitin. On the complexity of derivations in propositional calculus. Studies in Mathematics and Mathematical Logic. Part II (Consultants Bureau, New York/London, 1970) 115-125.
13. Zchaff web page. http://ee.princeton.edu/~chaff/zchaff.php
14. H.Zhang. SATO: *An efficient propositional prover.* Proceedings of the International Conference on Automated Deduction. -July 1997. -P.272-275.
15. 2clseq web page. http://www.cs.toronto.edu/~fbacchus/ 2clseq.html.
16. http://www.cbl.ncsu.edu/CBL_Docs/lgs91.html

A Local Search SAT Solver
Using an Effective Switching Strategy
and an Efficient Unit Propagation

Xiao Yu Li, Matthias F. Stallmann, and Franc Brglez

Dept. of Computer Science, NC State Univ., Raleigh, NC 27695, USA
{xyli,mfms,brglez}@unity.ncsu.edu

Abstract. Advances in local-search SAT solvers have traditionally been presented in the context of local search solvers only. The most recent and rather comprehensive comparisons between UnitWalk and several versions of WalkSAT demonstrate that neither solver dominates on all benchmarks. QingTing2 (a 'dragonfly' in Mandarin) is a SAT solver script that relies on a novel switching strategy to invoke one of the two local search solvers: WalkSAT or QingTing1. The local search solver Qing-Ting1 implements the UnitWalk algorithm with a new unit-propagation technique. The experimental methodology we use not only demonstrates the effectiveness of the switching strategy and the efficiency of the new unit-propagation implementation – it also supports, on the very same instances, statistically significant performance evaluation between local search and other state-of-the-art DPLL-based SAT solvers. The resulting comparisons show a surprising pattern of solver dominance, completely unanticipated when we began this work.

1 Introduction

SAT problems of increasing complexity arise in many domains. The performance concerns about SAT solvers continue to stimulate the development of new SAT algorithms and their implementations. The number of SAT-solver entries entered into the 2003 SAT-solver competition [1] exceeds 30. The Web has become the universal resource to access large and diverse directories of SAT benchmarks[2], SAT discussion forums [3], and SAT experiments [4], each with links to SAT-solvers that can be readily down-loaded and installed.

For the most part, this paper addresses the algorithms and performance issues related to the (stochastic) local search SAT solvers. To date, relative improvements of such solvers have been presented in the context of local search solvers only. Up to seven local search SAT algorithms have been evaluated (under benchmarks-specific 'noise levels') in [5], including four different versions of WalkSAT [6, 7, 8]. The most recent comparisons between several versions of WalkSAT and UnitWalk demonstrate that neither solver dominates on all problem instances [9].

The performance analysis of local search SAT solvers as reported in [5] and [9] is based on statistical analysis of multiple runs with the same inputs, using

E. Giunchiglia and A. Tacchella (Eds.): SAT 2003, LNCS 2919, pp. 53–68, 2004.

different seeds. On the other hand, DPLL-based (deterministic) solvers such as chaff [10] and sato [11], are compared either on basis of a single run or multiple runs with incomparable inputs. This explains the lack of statistically significant performance comparisons between the two types of SAT solvers – until now. By introducing syntactical transformations of a problem instance we can now generate, for each reference cnf formula, an *equivalence class* of as many instances as we find necessary for an experiment of statistical significance involving both deterministic and stochastic solvers [12, 13]. For the stochastic local search solvers, results are (statistically) the same whether we do multiple runs with different seeds on identical inputs or with the same seed on a class of inputs that differ only via syntactical transformations.

Our research, as reported in this paper, proceeded by asking and pursuing answers to the three questions in the following order:

1. Can we accelerate UnitWalk by improving the implementation of its unit-propagation?
2. Can we analyze each problem instance to devise a switching strategy to correctly select the dominating local search algorithm?
3. How much would state-of-the-art DPLL-based solvers change the pattern of solver dominance on the very same problem instances?

Answers to these questions are supported by a large number of experiments, making extensive use of the SATbed features as reported in [12]. Raw data from these experiments as well as formatted tables and statistics are openly accessible on the Web, along with instance classes, each of minimum size of 32. All experiments can be readily replicated, verified, and extended by other researchers.

Following a suggestion of Johnson [14], we treat each algorithm as a fixed entity with no "hand-tuned" parameters. For two of the algorithms, UnitWalk2 and QingTing2 , some tuning takes place during execution and is charged as part of the execution time. Extensive studies [5] have been done on how to tune WalkSAT for some of the benchmarks discussed in this paper. Given that the tuning is not part of the execution, we had to make a choice to be fair to all SAT solvers reported in this paper.[a] We used the default heuristics ("best") and noise level (0.5) for WalkSAT. We ran all our experiments on a P-II@266 MHz Linux machine with 196 MBytes of physical memory and set the timeout value for each experiment to be 1800 seconds.

This paper is organized as follows. In Section 2, we provide the motivation and background information. We describe how unit propagation is implemented in

---

[a] See the extended discussion about the "PET PIEVE 10. *Hand-tuned algorithm parameters*" that concludes with a recommended rule as follows [14]:

> If different parameter settings are to be used for different instances, the adjustment process must be well-defined and algorithmic, the adjustment must be described in the paper, and the time for the adjustment must be included in all reported running times.

QingTing1 and its effect on reducing clause and literal visits in Section 3. We then present the switching strategy and QingTing2 in Section 4. Comparisons between deterministic and stochastic solvers are presented in Section 5. We conclude the paper and describe some future research directions in Section 6.

2   Motivation and Background

In this section, we first give the motivation of our work. We then introduce some basic concepts and briefly discuss the key differences between UnitWalk and WalkSAT.

2.1   Motivation

Recently, Hirsch and Kojevnikov [9] proposed and implemented a new local search algorithm called UnitWalk. Despite its simple implementation, the Unit-Walk solver outperforms WalkSAT on structured benchmarks such as planning and parity learning. Unlike GSAT and WalkSAT, UnitWalk relies heavily on unit propagation and any speedup gained on this operation will directly boost the solver's performance. We demonstrate that such speedup is in fact possible using Zhang's unit propagation algorithm [15] with chaff's lazy data structure [10].

However, UnitWalk is not competitive with WalkSAT on randomly generated 3-SAT benchmarks and some benchmarks from problem domains such as microprocessor verification. We show the large performance discrepancies between the two solvers in Figure 1. For a given benchmark, it would be highly desirable to have a strategy that can find the faster algorithm before solving it. Using such a strategy a solver can simply solve the benchmark with the more efficient algorithm of the two.

Below is a summary of the two versions of UnitWalk and two versions of our solver we refer to in the rest of the paper:

- UnitWalk1 refers to version 0.944 of the UnitWalk solver proposed in [9]. It implements the UnitWalk algorithm in a straightforward way.
- UnitWalk2 refers to version 0.981 of the UnitWalk solver [16]. UnitWalk2 incorporates both WalkSAT and a 2-SAT solver into its search strategy and it also uses Zhang's unit propagation algorithm.
- QingTing1[b] refers to version 1.0 of our solver. It improves upon UnitWalk1 by implementing unit propagation more efficiently.
- QingTing2 refers to version 2.0 of our solver. It is a SAT solver script that relies on a novel switching strategy to invoke one of the two local search solvers: WalkSAT or QingTing1.

---

[*] QingTing1 is available at http://pluto.cbl.ncsu.edu/EDS/QingTing.

The examples in the two following tables shows that UnitWalk1 (version 0.944) and WalkSAT can outperform each other significantly on different benchmarks. In the upper table, UnitWalk1 is shown to dominate WalkSAT on the blocks-world benchmarks (`bw_large_a` and `bw_large_b`) and the parity learning benchmarks (`par16-4-c` and `par16-5-c`). However, the lower table shows that Walk-SAT runs as many as 62 times faster than UnitWalk1 on the random 3-SAT benchmarks (`uf250-1065-027` and `uf250-1065-087`) and microprocessor verification benchmarks (`dlx2_cc_a_bug17` and `dlx2_cc_a_bug39`). Therefore, for the eight benchmarks, UnitWalk1 and WalkSAT can complement each other's performance, and this can be made possible if there is an efficient strategy that can choose the faster solver from the two.

Solvers	bw_large_a	bw_large_b	par16-4-c	par16-5-c
UnitWalk1	0.13/0.17	3.90/3.58	75.3/68.6	69.0/55.8
WalkSAT	0.27/0.26	17.4/17.1	timeout	timeout

Solvers	uf250-1065-027	uf250-1065-087	dlx2_cc_a_bug17	dlx2_cc_a_bug39
UnitWalk1	6.84/6.46	12.4/12.7	61.6/61.9	86.4/85.4
WalkSAT	0.11/0.09	0.15/0.15	2.44/1.82	4.18/3.09

Fig. 1. Performance comparisons for UnitWalk1 and WalkSAT on 32 PC-class instances for each of the respective benchmarks (see details in Section 5). Table entries report the mean/standard-deviation of runtime (in seconds). The timeout is set to be 1800 seconds.

2.2 Basic Concepts

We consider algorithms for SAT formulated in conjunctive normal form (CNF). A CNF formula F is the conjunction of its clauses where each clause is a disjunction of literals. A literal x in F appears in the form of either v or its complement $\bar{v}$, for some variable v. A clause is *satisfied* if at least one literal in the clause is true. F is satisfiable if there exists an assignment for variables that satisfies each clause in F; otherwise, F is unsatisfiable. A clause containing only one literal is called a *unit clause*.

Consider a CNF formula F and let x be a literal in the formula. Then F can be divided into three sets:

- $A = \{x \vee A_1, \cdots, x \vee A_m\}$: the clauses that contain x.
- $B = \{\bar{x} \vee B_1, \cdots, \bar{x} \vee B_n\}$: the clauses that contain $\bar{x}$.
- $R = \{R_1, \cdots, R_l\}$: the clauses that contain neither x nor $\bar{x}$.

When x is set true, the *unit propagation* operation, $F := F[x \leftarrow true]$, will delete $\bar{x}$ from B and remove A from F. New unit clauses may arise and unit propagation continues as long as there are unit clauses in the formula. Next, we present the UnitWalk algorithm and the WalkSAT algorithm.

2.3   The Algorithms

Both UnitWalk and WalkSAT generate initial assignments for the variables uniformly at random. These assignments are modified by complementing (flipping) the value of some variable in each step. In WalkSAT the variable to be flipped is chosen from a random unsatisfied clause. Variations of WalkSAT vary in how the variable is chosen. If a solution is not reached after a specified number of flips, WalkSAT tries again with a new random assignment. The number of such tries is not limited in our experiments, but the algorithm is terminated after a *timeout* of 1800 seconds.

UnitWalk takes advantage of unit propagation whenever possible, delaying variable assignment until unit clauses are resolved. An iteration of the outer loop of UnitWalk is called a *period*. At the beginning of a period, a permutation of the variables and their assignments are randomly chosen. The algorithm will start doing unit propagation using the assignment for the first variable in the permutation. The unit propagation process modifies the current assignment and will continue as long as unit clauses exist. When there are no unit clauses left and some variables remain unassigned, the first unassigned variable in the permutation along with its current assignment is chosen to continue the unit propagation process. At least one variable is flipped during each period, thus ensuring progress. If at the end of a period, the formula becomes empty, then the current assignment is returned as the satisfying solution. The parameter MAX_PERIODS determines how long the program will run. In our experiments with UnitWalk1 and QingTing1, MAX_PERIODS is set to infinity (the same setting used by Hirsch and Kojevnikov [9]), but, as with WalkSAT, the program is terminated if it runs for longer than our timeout.

3   Efficient Unit Propagation Using a Lazy Data Structure

As we have seen, UnitWalk relies heavily on unit propagation. Therefore, the performance of any UnitWalk-based solver depends on how fast it can execute this operation, which, in turn, is largely determined by the underlying data structure. In this section, we briefly review counter-based adjacency lists as a data structure for unit propagation. This appears to be the structure used in UnitWalk1. We then present the data structure used in QingTing1. Last, we show empirically that QingTing1's data structure reduces the number of memory accesses in terms of clause and literal visits.

3.1   Counter-Based Adjacency Lists

Variations of *adjacency lists* have traditionally been used as the data structure for unit propagation. One example is the counter-based approach. Each variable in the formula has a list of pointers to the clauses in which it appears. Each clause has a counter that keeps track of the number of unassigned literals in the clause. When a variable is assigned *true*, the following actions take place.

1. The clauses that contain its positive occurrences are declared *satisfied.*
2. The counters for the clauses that contain its negative occurrences are decremented by 1 and all its negative occurrences are marked *assigned.*
3. If a counter becomes 1 and the clause is not yet satisfied, then the clause is a unit clause and the surviving literal is found using linear search.

The case when the variable is assigned *false* works analogously. It is important to realize that when a variable is assigned, every clause containing that variable must be visited. Moreover, the average number of literal visits to find the surviving literal in a unit clause is half of the clause length.

3.2   A Lazy Data Structure

Recently, more efficient structures (often referred to as "lazy" data structures) have been used in complete SAT solvers to facilitate unit propagation [17]. Some examples include sato's Head/Tail Lists [18] and chaff's Watched Literals [10]. Head/Tail Lists and Watched Literals are very similar. In our solver QingTing1, we use the Watched Literals approach. Every clause has two watched literals, each a pointer to an unassigned literal in the clause. Each variable has a list of pointers to clauses in which it is one of the watched literals. Initially, the two watched literals are the first and last literal in the clause. When assigning variables, a clause is visited *only* when one of the two watched literals becomes false. Then the next unassigned literal will be searched for, which leads to three cases:

1. If a satisfied literal is found, the clause is declared *satisfied.*
2. If a new unassigned literal is found and it is not the same as the other watched literal, it becomes a new watched literal.
3. If the only unassigned literal is the other watched literal, the clause is declared *unit.*

A unit clause is declared *conflicted* when it is unsatisfied by the current assignment of the variable in the unit clause. As new unassigned literals are discovered, the list of pointers associated with each variable is dynamically maintained. The unit propagation method we just described is based on Zhang's algorithm [15].

3.3   Reducing Clause and Literal Visits

Using the Watched Literals data structure, we have reduced the number of clause visits during unit propagation:

- Unlike the adjacency list approach, assignments to clause literals other than the two watched literals do not cause a clause visit: a clause does not become unit as long as the two variables of the watched literals are unassigned.
- In addition, the process of declaring a clause satisfied is implicit instead of explicit. An assignment satisfying a watched literal doesn't result in a clause visit. However, there is a trade-off. Since the clause is not explicitly

UnitWalk1 and QingTing_AL use counter-based adjacency lists; QingTing1 uses Watched Literals. We run each solver 128 times on each benchmark with different seeds and report the mean and standard deviation (in the form mean/stdev) of the runtime, number of periods and flips, and number of clause and literal visits. There are two exceptions: (1) the number of clause and literal visits is not reported in the program output by UnitWalk1, and (2) runtime is not reported for QingTing_AL because it is implemented inefficiently — its execution time is not comparable with the other two solvers.

We observe that, though the reported number of periods and flips is virtually the same for UnitWalk1, QingTing_AL, and QingTing1, QingTing1 has fewer clause and literal visits, especially for queen19 and bw_large_b. For uf250-1065_087, QingTing1 has fewer clause and literal visits, but runs slower than UnitWalk1. This slow-down is the cost of dynamically maintaining the Watched-Literals data structure when an assignment does not result in a unit propagation.

128 experiments on the queen19 benchmark

Solver	Periods	Flips	Visits	runtime
UnitWalk1	29/34	363/228	N/A	0.27/0.30
QingTing_AL	29/30	358/207	6.30e5/6.67e5	–/–
QingTing1	22/23	316/160	8.50e4/8.61e4	0.05/0.05

128 experiments on the bw_large_b benchmark

Solver	Periods	Flips	Visits	runtime
UnitWalk1	204/207	1.20e4/1.19e4	N/A	3.00/3.03
QingTing_AL	176/173	1.03e4/9.64e3	5.71e6/5.59e6	–/–
QingTing1	200/234	1.19e4/1.35e4	3.46e6/4.03e6	1.71/2.00

128 experiments on the uf250-1065_087 benchmark

Solver	Periods	Flips	Visits	runtime
UnitWalk1	9.59e3/8.60e3	5.64e5/4.61e5	N/A	10.9/9.77
QingTing_AL	1.29e4/1.08e4	6.89e5/5.81e5	4.42e7/3.72e7	–/–
QingTing1	1.10e4/1.15e4	6.15e5/6.41e5	3.27e7/3.40e7	15.8/16.5

Fig. 2. Performance comparisons of UnitWalk1, QingTing_AL and QingTing1 on three benchmark. Table entries report mean/stdev of each metric.

declared satisfied, a new unassigned literal in the clause will still instigate a search when the other watched literal is made false. With the traditional counter-based structure, these extra searches are avoided.

We implemented the UnitWalk algorithm with each of the data structures described above and then compared their performances with the UnitWalk1 solver. In local search solvers such as UnitWalk1 and QingTing1, significant variability of performance metrics can be observed by simply repeating experiments on the same instance, randomly choosing the seed. To ensure statistical significance, we run each solver 128 times on each benchmark with different seeds. In Figure 2 we report for each solver the sample mean and standard deviation of periods,

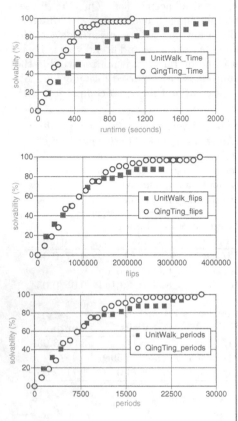

Experiments with a SAT-solver A on N instances from a well-defined equivalence class of CNF formulas [19] show that parameters such as *flips* and *runtime* are random variables X with a cumulative distribution $F_X^A(x)$. For runs that *do not time-out*, we define the estimate of $F_X^A(x)$ as the *solvability function* $\mathcal{S}^A(x)$ [13]:

$$\mathcal{S}^A(x) = \frac{\cdot}{N}(\text{num of observations} \leq x)$$

The solvability functions on the left are based on experiments with Unit-Walk1 and QingTing1 on 32 PC-class instances of **bw_large_c**. We observe that for both solvers, *runtime, flips,* and *periods* all have *exponential distribution*, also confirmed at 5% level of significance by the $\chi^{\cdot}$-test. The mean values of *flips* and *periods* are approximately the same while the mean values of *runtime* differ significantly for the two solvers. Sample means and standard deviations are summarized below.

	UnitWalk1 mean/stdev	QingTing1 mean/stdev
runtime	591/554	267/219
flips	986510/923651	924915/758691
periods	7613/7121	6988/5746

Fig. 3. Solvability functions of QingTing1 and UnitWalk1 induced by solving 32 PC-class instances of **bw_large_c**.

flips and runtime (the time to find the first solution). All three solvers have approximately the same number of periods and flips for the three benchmarks from different problem domains. However, the sample mean and standard deviation of the number of clause and literal visits are consistently less with the Watched-Literals approach. The ratio of improvement varies on different benchmarks.

To further demonstrate the effectiveness of the lazy data structures, we took one of the most challenging benchmarks for UnitWalk1, **bw_large_c**, and looked at the distribution of three random variables on 32 runs each of UnitWalk1 and QingTing1 — see Figure 3. As expected, the number of periods and flips exhibits exponential distribution with roughly the same mean and standard deviation for both solvers. However, the mean and standard deviation of runtime for Qing-Ting1 are less than half those of UnitWalk1.

Algorithm QingTing2
Input: A formula F in CNF containing n variables $v., \cdots, v_n$
Output: A satisfying assignment or "No solution found"
Method:
variable_immunity = 0
random_assignments = 0
for $i := 1$ to MAX_TRIALS **do**
    while G is not empty **do**
        assign a random value to an unassigned variable chosen at random
        random_assignments = random_assignments + 1
        do unit propagation until no unit clauses exist
    end do
end do
variable_immunity = random_assignments / n / MAX_TRIALS
if variable_immunity ≤ 0.07, **then return** QingTing1(F)
else return WalkSAT(F)

Fig. 4. The QingTing2 algorithm.

4 A Switching Strategy

The experimental work in [9] shows that variants of WalkSAT have better performance than UnitWalk1 on randomly generated 3-SAT benchmarks as well as some from the more realistic problem domains; however, for some highly structured benchmarks, UnitWalk1 clearly dominates. The latest release of UnitWalk has incorporated WalkSAT and it switches periodically between its internal Unit-Walk and WalkSAT engines. As we will show later, UnitWalk2 is able to improve significantly upon UnitWalk1 on some benchmarks but clearly slows down the search process on many others. In this section, we propose a clearly delineated switching strategy that chooses which engine to use right at the beginning. Using this switching strategy, QingTing2 is also able to improve upon QingTing1 on many benchmarks but doesn't suffer the slowdown on the others.

4.1 Heuristic for the Switching Strategy

The fact that UnitWalk1 works well on highly structured benchmarks suggests that it is able to take advantage of the dependencies among the variables via unit propagation. However, when such dependencies are weak, e.g., in randomly generated 3-SAT benchmarks, WalkSAT appears to run faster. Intuitively, if we can measure such dependencies in an effective way, then we can predict which solver is faster for a given benchmark. QingTing2 shown in Figure 4, implements the switching strategy based on this idea.

Before solving a benchmark with either QingTing1 or WalkSAT, QingTing2 samples a benchmark for a fixed number of times specified by MAX_TRIALS, which we set to 128 in our experiments. During each trial, it starts by assigning a random value to an unassigned variable chosen at random. Such a step is called

Benchmark	Time	Variable Immunity	Distribution	QingTing1/WalkSAT
sched06s_v00828	0.21	0.01/0.01	normal	< 0.1[*a*]
sched07s_v01386	0.94	0.01/0.01	normal	< 0.1[*a*]
bw_large_a	0.49	0.01/0.01	normal	0.48
bw_large_b	1.43	0.01/0.00	near-normal	0.17
logistics_d	4.86	0.02/0.00	normal	0.05
sched04z07s_v00655	0.57	0.04/0.01	near-normal	0.75
sched04z08s_v01094	1.14	0.04/0.01	near-normal	0.46
sched04z06s_v00354	0.26	0.06/0.02	near-normal	0.44
flat200-87	0.38	0.06/0.01	normal	3.12
flat200-33	0.38	0.07/0.00	near-normal	2.98
flat200-79	0.38	0.07/0.01	normal	1.75
par16-4-c	0.27	0.07/0.02	near-normal	0.21
par16-5-c	0.28	0.07/0.01	normal	0.22
logistics_c	1.00	0.09/0.01	normal	12.15
logistics_a	0.68	0.10/0.02	normal	83.33
logistics_b	0.67	0.11/0.02	near-normal	1.80
uf250-1065_027	0.24	0.15/0.02	normal	92.73
uf250-1065_034	0.24	0.15/0.03	normal	151.52
uf250-1065_087	0.24	0.16/0.02	normal	111.33
hgen2_v250_s26609_093	0.21	0.16/0.02	near-normal	1024.14
hgen2_v250_s53446_089	0.22	0.16/0.02	normal	17.71
dlx2_cc_a_bug39	2.59	0.18/0.05	near-normal	11.12
dlx2_cc_a_bug40	3.10	0.19/0.04	near-normal	13.67
dlx2_cc_a_bug17	7.08	0.21/0.05	near-normal	8.69

[a] WalkSAT times out at 1800 seconds for the majority of instances of these classes.

Fig. 5. Time to measure variable immunity, its mean/stdev and distribution, and the runtime performance ratio of QingTing1 and WalkSAT.

a random assignment. It then propagates its value through unit propagation. When the unit propagation stops, it does another random assignment. Such a process repeats and doesn't stop until all the clauses in the formula are either conflicted or satisfied (the formula is now empty). We define *variable immunity* as the ratio between the number of random assignments in a trial and the number of variables. Intuitively, the higher the variable immunity, the less chance that variables will be assigned during unit propagation.

We considered a wide range of benchmarks and measured their variable immunities. These benchmarks include instances from the domains of blocks-world (the bw_large series [20]), planning (the logistics series [2]), scheduling (the sched series [12]), graph 3-coloring (the flat series [2]), parity learning (the par series [2]), randomly generated 3-SAT (the uf250 series [2] and hgen2 series by the hgen2 generator [21]) and microprocessor verification (the dlx2 series [22]). In Figure 5, the column data show the time it takes to measure the variable

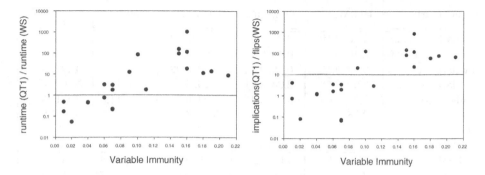

Fig. 6. The left plot shows the correlation between *variable immunity* and Qing-Ting1(QT1)/WalkSAT(WS) runtime ratio. The right plot shows the correlation between *variable immunity* and the ratio of QingTing1(QT1) implications and WalkSAT(WS) flips.

immunity, the sample mean, standard deviation and distribution of the variable immunity, and the QingTing1/WalkSAT runtime ratio for each benchmark (based the experiments we show in section 5). The rows are sorted based on the sample mean of variable immunity. We make the following observations.

The sample mean of variable immunity is correlated with the QingTing1/ WalkSAT runtime ratio: (a) For the top section of the table where variable immunity ≤ 0.06, the runtime for QingTing1 is a fraction of that for WalkSAT. (b) For the middle section of the table where variable immunity is between 0.06 and 0.07, either solver can dominate. (c) For the bottom section of the table where variable immunity > 0.07, WalkSAT clearly outperforms QingTing1.

The correlation is depicted in the left scatter plot in Figure 6. It is easy to see that when variable immunity is below 0.06, the QingTing1/WalkSAT runtime ratio is less than 1, which means QingTing1 is the faster solver; and when variable immunity is above 0.07, the QingTing1/WalkSAT runtime ratio is greater than 1, which means WalkSAT is the faster solver. The same trend can be observed in the right scatter plot. Here, instead of runtime, we consider two machine independent parameters that are closely related to the solvers' runtime: the number of implications in QingTing1 and the number of flips in WalkSAT. Their ratio exhibits the same pattern as the runtime ratio.

The distributions of variable immunity are mostly normal or near-normal. Thus the mean values we report for 128 trials are relatively stable. Variable immunity is similar for benchmarks from the same problem domain (the only exception we observe is `logistics_d`). This is an indication that variable immunity has the ability to capture the intrinsic structure of the benchmarks.

In QingTing2, we choose 0.07 as the threshold for our switching strategy: if the variable immunity is less than or equal to 0.07, QingTing1 is invoked; otherwise, WalkSAT is invoked.

Benchmark	UnitWalk1	QingTing1	UnitWalk2	WalkSAT	Dominating Solver
sched06s_v00828	0.20/0.19	0.04/0.03	0.29/0.29	1800*^{a}*	QT1
sched07s_v01386	0.40/0.42	0.10/0.09	0.67/0.70	1800*^{a}*	QT1
bw_large_a	0.13/0.17	0.13/0.12	0.22/0.22	0.27/0.26	QT1/UW1*^{c}*
bw_large_b	3.90/3.58	2.93/3.11	6.23/5.44	17.4/17.1	QT1
logistics_d	0.96/0.63	0.75/0.45	1.34/0.71	13.7/11.5	QT1/UW1*^{c}*
sched04z07s	2.58/3.23	0.73/0.98	3.36/3.76	1.59/1.53	QT1
sched04z08s	6.47/8.12	1.79/1.98	10.2/10.8	4.08/4.06	QT1
sched04z06s	0.07/0.11	0.03/0.03	0.14/0.16	0.04/0.04	QT1
flat200-87	15.9/13.9	14.9/12.1	23.9/21.2	4.77/5.08	WS*^{b}*
flat200-33	3.43/3.78	4.26/3.79	5.18/5.66	1.43/1.27	WS*^{b}*
flat200-79	11.1/10.1	8.80/10.9	16.9/15.7	5.04/4.88	WS*^{b}*
par16-4-c	75.3/68.6	104/81.0	137/116	1800*^{a}*	UW1/QT1*^{c}*
par16-5-c	69.0/55.8	110/144	114/97.1	1800*^{a}*	UW1/QT1*^{c}*
logistics_c	281/230	107/105	307/326	8.81/8.00	WS
logistics_a	375/327	130/177	576/531	1.56/1.32	WS
logistics_b	15.4/15.6	5.25/4.41	20.0/18.9	2.91/2.47	WS
uf250-1065_027	6.84/6.46	10.2/10.1	0.48/0.44	0.11/0.09	WS
uf250-1065_034	126/118	150/112	3.80/3.78	0.99/0.95	WS
uf250-1065_087	12.4/12.7	16.7/17.2	0.45/0.36	0.15/0.15	WS
hgen2_v250..._093	448/375	594/458	2.01/2.28	0.58/0.59	WS
hgen2_v250..._089	835/506	487/765	59.3/61.5	27.5/24.9	WS
dlx2_cc_a_bug39	86.4/85.4	46.5/41.9	129/133	4.18/3.09	WS
dlx2_cc_a_bug40	42.2/41.7	28.7/25.4	65.9/65.3	2.10/2.15	WS
dlx2_cc_a_bug17	61.6/61.9	21.2/23.1	94.2/102	2.44/1.82	WS

^{a} WalkSAT times out at 1800 seconds for the majority of instances of these classes.

^{b} QingTing2 chooses the slower QingTing1 solver according to the switching strategy.

^{c} A t-test (at 5% level of significance) shows that QingTing1 and UnitWalk1 have the same performance on these instances.

Fig. 7. Performance comparisons of QingTing1, UnitWalk1, UnitWalk2 and WalkSAT. Numerical table entries represent mean/stdev of runtime (in seconds).

The overhead of the switching strategy is shown in the second column of Figure 5. Even though the overhead tends to grow as the size of the problem increases, it is still small relative to the runtime of the slower solver.

4.2   Comparisons of Incomplete Solvers

Figure 7 shows our experiments with four local search SAT solvers on the benchmarks we considered in Figure 5. For each reference benchmark, we performed experiments on the reference as well as 32 instances from the associated PC-class (see Section 5).

First, we compare the performance of UnitWalk1 and QingTing1. QingTing1 outperforms UnitWalk1 on all benchmarks except the uf250 and dlx2 series. For these two series, we observe that even though we save modestly on the number of clause and literal visits, significant amount of time (around 27%) is spent on updating the dynamic data-structure in QingTing1.

For the benchmarks in the top section of Figure 7, QingTing1 is the dominating solver. For the benchmarks in the bottom section, WalkSAT is the dominating solver. QingTing2 is successful in choosing the dominating solver for all these benchmarks. For the benchmarks in the middle section, no solver clearly dominates. According to the switching strategy, QingTing2 chooses QingTing1 instead of the faster WalkSAT for the flat series.

UnitWalk2 improves significantly upon UnitWalk1 on the uf250 and hgen2 series by utilizing its WalkSAT component. However, UnitWalk2 suffers a 10% (logistics_c) to 100% (sched04z06s) slowdown on all other benchmarks (for most of them, the slowdown is about 50%). This shows that the switching strategy implemented in UnitWalk2 is not as effective as the one introduced in this paper. Notably, QingTing2 improves upon QingTing1 for all the benchmarks in the bottom section of Figure 7 by using WalkSAT. The only slowdown it suffers is from the overhead of the switching strategy. As shown in Figure 5, this extra cost becomes a very small fraction of the runtime for harder benchmarks.

5   Comparisons with DPLL-based Solvers

As noted earlier and in the companion paper [12], our use of syntactic equivalence classes gives us an unprecedented opportunity to compare solvers across categories. The key question we are able to ask at this point is how the local-search solvers we have evaluated compare with DPLL-based SAT solvers, the former being stochastic and the latter regarded as deterministic.

We consider three DPLL-based SAT solvers that include chaff [10] and two versions of the sato [11] solver, where *sato* uses the trie data structure and *satoL* uses the linked-list data structure. In Figure 8, we summarize the runtime data for the best solvers in both categories. It is clear from the table that

1. WalkSAT is the best solver for the randomly generated benchmarks: the uf250 series and the hgen2 series.
2. For the scheduling benchmarks [12], QingTing1 outperforms chaff.
3. For all other benchmarks, chaff dominates. Moreover, since chaff dominates for some benchmarks in each of the three benchmark categories (originally introduced in Figure 5), our switching criterion is not an effective decider when chaff is included.

Thus the unitwalk-based strategy excels only for the scheduling instances, which are also shown to exhibit atypical behavior with other solvers [12]. While we have not included enough different DPLL-based solvers in this study, it appears that in almost all circumstances where a local-search solver is competitive, Walk-SAT is the solver of choice. It remains to be seen whether this is an inherent

Benchmark	Best Local Search Solver	Best DPLL Solver	Dominating Solver
sched06s_v00828	QingTing1 (0.03/0.04)	sato (4.52/18.3)	QingTing1
sched07s_v01386	QingTing1 (0.10/0.09)	chaff (13.4/11.2)	QingTing1
bw_large_a	QingTing1 (0.12/0.12)	chaff (0.01/0.01)	chaff
bw_large_b	QingTing1 (2.93/3.11)	chaff (0.09/0.04)	chaff
logistics_d	QingTing1 (0.75/0.45)	chaff (0.34/0.13)	chaff
sched04z07s_v00655	QingTing1 (0.73/0.98)	chaff (28.9/23.4)	QingTing1
sched04z08s_v01094	QingTing1 (1.79/1.98)	chaff (165/192)	QingTing1
sched04z06s_v00354	QingTing1 (0.03/0.03)	chaff (24.2/37.1)	QingTing1
flat200-87	WalkSAT (4.77/5.08)	chaff (1.04/1.45)	chaff
flat200-33	WalkSAT (1.43/1.27)	chaff (0.74/0.72)	chaff
flat200-79	WalkSAT (5.04/4.88)	chaff (0.85/1.01)	chaff
par16-4-c	UnitWalk1 (75.3/68.6)	chaff (1.33/0.82)	chaff
par16-5-c	UnitWalk1 (69.0/55.8)	chaff (2.19/2.07)	chaff
logistics_c	WalkSAT (8.81/8.00)	chaff (0.37/0.08)	chaff
logistics_a	WalkSAT (1.56/1.32)	chaff (0.14/0.03)	chaff
logistics_b	WalkSAT (2.91/2.47)	chaff (0.14/0.04)	chaff
uf250-1065_027	WalkSAT (0.11/0.09)	satoL (26.5/19.6)	WalkSAT
uf250-1065_034	WalkSAT (0.99/0.95)	satoL (27.5/16.0)	WalkSAT
uf250-1065_087	WalkSAT (0.15/0.15)	satoL (30.9/19.5)	WalkSAT
hgen2_v250..._093	WalkSAT (0.58/0.59)	Unknown[*a*]	WalkSAT
hgen2_v250..._089	WalkSAT (27.5/24.9)	Unknown[*a*]	WalkSAT
dlx2_cc_a_bug39	WalkSAT (4.18/3.09)	chaff (1.49/1.42)	chaff
dlx2_cc_a_bug40	WalkSAT (2.10/2.15)	chaff (3.59/1.12)	WalkSAT
dlx2_cc_a_bug17	WalkSAT (2.44/1.82)	chaff (0.68/1.25)	chaff

[a] All DPLL solvers time out at 1800 seconds for almost all instances of these classes.

Fig. 8. Performance comparisons between local search solvers (*QingTing1, Unit-Walk1, UnitWalk2, WalkSAT*) and DPLL solvers (*chaff, sato, satoL*). The time-out value is set to 1800 seconds.

limitation of doing unit propagation in the context of local search (all of the DPLL-based solvers do unit propagation as an integral part of their algorithms) or whether the current selection of benchmarks is not sufficiently rich. In other words, are the scheduling instances (unintentionally) contrived to be unlike those encountered in any other application, or are they representative of a large class of instances that naturally arise in applications heretofore not considered?

6   Conclusions and Future Research

We have introduced a new local-search-based SAT solver QingTing. With an efficient implementation using the Watched Literals data structure, QingTing1 is able to improve upon UnitWalk1. We have also experimentally measured the effect data structures have on the number of clause and literal visits required for unit propagation. Combining WalkSAT with QingTing1 using a lower-overhead switching strategy, QingTing2 has a better overall performance than either solver used exclusively.

More important, however, are the lessons learned in this process. We set out to improve UnitWalk without first considering the bigger picture — in this case: are there circumstances under which an improved version of UnitWalk would be the undisputed solver of choice? The answer in retrospect is no; the reader should regard with suspicion the fact that the only benchmarks on which Qing-Ting dominates the performance of other solvers are those of our own creation (however unintended this may have been).

The switching strategy, an important concept presented here, also has less impact on the bigger picture than we had hoped. What is really needed is a switching strategy that also takes DPLL-based solvers into account, but the basis for such a strategy is not immediately clear. It should also be instructive to see if QingTing can be significantly improved with the use of learning techniques, such as clause recording, that are behind the effectiveness of state-of-the-art DPLL solvers.

Acknowledgments. The experiments, as reported in this paper, could not have taken place without the SAT solvers *chaff*, *sato*, *satoL*, *UnitWalk* and *WalkSAT*. We thank authors for the ready access and the exceptional ease of installation of these software packages.

References

[1] Daniel Le Berre and Laurent Simon. SAT Solver Competition, in conjunction with 2003 SAT Conference, May 2003. For more information, see www.satlive.org/SATCompetition/2003/comp03report/index.html.
[2] H. Hoos and T. Stuetzle. SATLIB: An online resource for research on SAT, 2000. For more information, see www.satlib.org.
[3] Daniel Le Berre. SAT Live! Up-to-date links for the SATisfiability problem, 2003. For more information, see www.satlive.org.
[4] Laurent Simon. Sat-Ex: The experimentation web site around the satisfiability, 2003. For more information, see www.lri.fr/~simon/satex/satex.php3.
[5] Holger H. Hoos and Thomas Stützle. Local Search Algorithms for SAT: An Empirical Evaluation. *Journal Of Automated Reasoning*, 24, 2000.
[6] B. Selman and H. Kautz. WalkSAT Homepage: Stochastic Local Search for Satisfiability, 2002. The source code is available at www.cs.washington.edu/homes/kautz/walksat/.
[7] David A. McAllester, Bart Selman, and Henry A. Kautz. Evidence for invariants in local search. In *AAAI/IAAI*, pages 321–326, 1997.
[8] B. Selman, H. Kautz, and B.Cohen. Noise Strategies for Improving Local Search. In *Proceedings of AAAI'94*, pages 46–51. MIT Press, 1994.
[9] E. Hirsch and A. Kojevnikov. UnitWalk: A new SAT solver that uses local search guided by unit clause elimination. In *Electronic Proceedings of Fifth International Symposium on the Theory and Applications of Satisfiability Testing*, 2002.
[10] Matthew Moskewicz, Conor Madigan, Ying Zhao, Lintao Zhang, and Sharad Malik. Chaff: Engineering an efficient SAT solver. In *IEEE/ACM Design Automation Conference (DAC)*, 2001. Version 1.0 of Chaff is available at www.ee.princeton.edu/~chaff/zchaff/zchaff.2001.2.17.src.tar.gz.

[11] Hantao Zhang and Mark E. Stickel. Implementing the Davis-Putnam method. *Kluwer Academic Publisher*, 2000.

[12] F. Brglez, M. F. Stallmann, and X. Y. Li. SATbed: An Environment For Reliable Performance Experiments with SAT Instance Classes and Algorithms. In *Proceedings of SAT 2003, Sixth International Symposium on the Theory and Applications of Satisfiability Testing, May 5-8 2003, S. Margherita Ligure - Portofino, Italy*, 2003. For a revised version, see www.cbl.ncsu.edu/publications/ .

[13] F. Brglez, X. Y. Li, and M. Stallmann. On SAT Instance Classes and a Method for Reliable Performance Experiments with SAT Solvers. *Annals of Mathematics and Artificial Intelligence, Special Issue on Satisfiability Testing*, 2003. Submitted to AMAI as the revision of the paper published at the Fifth International Symposium on the Theory and Applications of Satisfiability Testing, Cincinnati, Ohio, USA, May 2002. Available at www.cbl.ncsu.edu/publications/ .

[14] David Johnson. A Theoretician's Guide to the Experimental Analysis of Algorithms. pages 215–250, 2002.

[15] H. Zhang and M. E. Stickel. An efficient algorithm for unit propagation. In *Proceedings of the Fourth International Symposium on Artificial Intelligence and Mathematics (AI-MATH'96)*, Fort Lauderdale (Florida USA), 1996.

[16] E. Hirsch and A. Kojevnikov. UnitWalk Home Page. See http://logic.pdmi.ras.ru/~arist/UnitWalk/.

[17] Ines Lynce and Joao Marques-Silva. Efficient data structure for backtrack search sat solvers. In *Fifth International Symposium on the Theory and Applications of Satisfiability Testing*, 2002.

[18] Hantao Zhang. SATO: An efficient propositional prover. In *Conference on Automated Deduction*, pages 272–275, 1997. Version 3.2 of SATO is available at ftp://cs.uiowa.edu/pub/hzhang/sato/sato.tar.gz.

[19] F. Brglez, X. Y. Li, and M. Stallmann. The Role of a Skeptic Agent in Testing and Benchmarking of SAT Algorithms. In *Fifth International Symposium on the Theory and Applications of Satisfiability Testing*, May 2002. Available at www.cbl.ncsu.edu/publications/ .

[20] H. Hoos. SATLIB - The Satisfiability Library, 2003. See http://www.satlib.org.

[21] E. Hirsch. Random generator hgen2 of satisfiable formulas in 3-CNF. See http://logic.pdmi.ras.ru/~hirsch/benchmarks/hgen2.html.

[22] M.N. Velev. Benchmark suite SSS1.0. See http://www.ece.cmu.edu/~mvelev.

Density Condensation of Boolean Formulas

Youichi Hanatani, Takashi Horiyama, and Kazuo Iwama

Graduate School of Informatics, Kyoto University,
Kyoto 606-8501, Japan
{hanatani,horiyama,iwama}@i.kyoto-u.ac.jp

Abstract. The following problem is considered: Given a Boolean formula f, generate another formula g such that: (i) If f is unsatisfiable then g is also unsatisfiable. (ii) If f is satisfiable then g is also satisfiable and furthermore g is "easier" than f. For the measure of this easiness, we use the *density* of a formula f which is defined as (the number of satisfying assignments) / 2^n, where n is the number of Boolean variables of f. In this paper, we mainly consider the case that the input formula f is given as a 3-CNF formula and the output formula g may be any formula using Boolean AND, OR and negation. Two different approaches to this problem are presented: One is to obtain g by reducing the number of variables and the other by increasing the number of variables, both of which are based on existing SAT algorithms. Our performance evaluation shows that, a little surprisingly, better SAT algorithms do not always give us better density-condensation algorithms. This is a preliminary report of the on-going research.

1 Introduction

There are many situations that for given Boolean formula f we wish to obtain another formula g which keeps the satisfiability of f and has some "nice property." For example, suppose that one has a beautiful benchmark formula derived from a real-world problem but this is a single, specific formula of, say, 129 variables. It is apparently desirable if he/she can generate several formulas of the same type with different number of variables. It might be also nice if we can change the formula so that it satisfies some shape-property such as including no exclusive-or operations. Thus, there can be many possibilities for the nice property of the formula f. However, a most natural one seems to be the easiness of obtaining satisfying assignments of f (if it is satisfiable).

As one can see easily, it is not always true that formulas of fewer variables are easier. However, it is quite reasonable to claim that formulas of larger *density* of satisfying assignments are easier. Here the density is defined as (the number of satisfying assignments) / 2^n, where n is the number of variables. While higher density may not always give better situations (e.g., we may loose a chance for applying unit propagations), we emphasize here that it is still desirable to control the density arbitrarily. For example, the time complexity of the most basic SAT algorithm, i.e., testing random assignments, is completely governed by this

E. Giunchiglia and A. Tacchella (Eds.): SAT 2003, LNCS 2919, pp. 69–77, 2004.

measure. Now our problem, called the *density condensation problem* (DCP), is to obtain, given a Boolean formula f, a formula g such that g is satisfiable iff f is satisfiable and moreover if f is satisfiable then g's density is larger than f's. It should be noted that deciding the satisfiability of f can be regarded as a special DCP, namely outputting formula "True" or "False" for a given f. In this case, therefore, we are required to increase the density up to 1.0. Since this "complete condensation" is well-known to be hard, DCP should play important roles as its generalization.

Our Contribution. In this paper, we present several algorithms for DCP and compare their performances. It is probably reasonable to assume that the input formula is 3-CNF, but there are at least three possibilities for the restriction of output formulas: (i) Only CNFs are allowed. (ii) The shape of the output formulas is not restricted, but we can only use logical AND, OR and negation operations. (iii) Functional substitution is allowed. As for the case (i), DCP seems to be hard since, if we could do that then by repeating it, we may obtain a completely condensed formula, i.e., True or False. As will be mentioned in Section 2.2, the case (iii) is relatively straightforward, namely, there is a polynomial-time algorithm which can increase the density exponentially. The remaining one, case (ii), appears most interesting.

In this paper, two different approaches to this problem are presented: One is to obtain g by increasing the number of variables and the other by reducing the number of variables, both of which are based on existing SAT algorithms and generate formulas that simulate the behavior of the algorithms. The former approach is applied to two probabilistic algorithms; random flipping of variables, and the probabilistic local search by Schöning [12]. In this approach, we introduce new variables to represent random selections. In the latter approach, we use a search space splitting algorithm that repeats assigning constants to some of the variables in f.

We evaluate the proposed algorithms by two measures; the density d and the ratio $|g|/d$. According to the measure d, the density-condensation algorithm based on random flipping does not work well since the density increases only in a constant factor. Surprisingly, the condensation algorithm based on the search space splitting method achieves better results than the one based on probabilistic local search in both measures d and $|g|/d$.

Related Work. Our general goal is to modify a given formula f so that it will satisfy some desirable property. One of the most interesting results in this category is given by Papadimitriou and Wolfe [10]. They showed that from a given CNF formula f, one can construct g in polynomial time such that if f is satisfiable then so is g and if f is not satisfiable then so is g but it becomes satisfiable if any one clause is removed from g. Iwama [8] investigated how to relate the satisfiability of the original formula f and the length of resolution proofs for the target formula g. If we wish to keep not only the satisfiability but the functionality, there has been a large literature, for example, to convert a CNF positive formula to an equivalent DNF one in pseudo-polynomial time [6].

Also there is a huge literature for SAT algorithms; a very partial list is as follows: The history of 3-SAT bounds begins with a 1.839^n time deterministic algorithm [9] based on the search space splitting technique by Monien and Speckenmeyer. In [9], they slightly improved the bound to 1.769^n by looking closer clauses. Recently, Schöning showed a probabilistic local search algorithm [12], which works in 1.334^n time. The algorithm starts with a random initial assignment, repeat flipping an assignment to a variable randomly selected from unsatisfied clauses. Its improved version [7], is known as the fastest probabilistic algorithm with time complexity 1.331^n. (Very recently 1.329^n has been claimed [2].) Dantsin et al. has proposed a derandomized variant of Schöning's local search algorithm [3]. By simulating Schöning's random flip deterministically and by selecting a set of initial assignments for traversing the search space $\{0, 1\}^n$ efficiently, this algorithm achieves time complexity 1.481^n.

2   Density Condensation Algorithms

2.1   Condensation Based on Random Flipping

We consider the following naive random flipping algorithm:

Procedure RFlip(f).

1. Select an initial assignment $a \in \{0, 1\}^n$ uniformly at random.
2. Flip the value of a randomly selected variable.
3. Repeat Step 2 for t times.

If a satisfying assignment is found in the repetition, output "True", otherwise "False".

We construct a formula g that simulates the above algorithm. Let $x^0 = (x_1^0, \ldots, x_n^0)$ denote the initial assignment, and $x^k = (x_1^k, \ldots, x_n^k)$ denote the assignment after the k-th flip. Let $v^k = (v_1^k, \ldots, v_{\lceil \log n \rceil}^k)$ denote the random choice of a variable in the k-th flip, where v^k is considered as a binary representation to denote the selected variable x_{v^k}. The following formula simulates algorithm RFlip;

$$\bigvee_{k=0}^{t} f(x_1^k, \ldots, x_n^k). \tag{1}$$

The assignment $(x_1^k, \ldots, x_n^k)$ after the k-th flip can be represented as a combination of variables $x^0, v^1, v^2, \ldots, v^k$;

$$x_i^k = x_i^0 \oplus (v^1 = i) \oplus (v^2 = i) \oplus \cdots \oplus (v^k = i), \tag{2}$$

where $v^k = i$ is a concatenation of exactly $\lceil \log n \rceil$ literals. By substituting variables x_i^k in formula (1), we obtain the resulting formula g with variables $x^0, v^1, v^2, \ldots, v^t$.

Now, we consider the length of the resulting formula g and its density d of the satisfying assignments. Since Eq. (2) says x_i^k has length

$$|x_i^k| = 1 + k\lceil \log n \rceil,$$

the length of g becomes

$$|g| = \sum_{k=0}^{t} |f| \cdot (1 + k\lceil \log n \rceil)$$

$$\sim |f| \, t^2 \log n.$$

Since g describes the behavior of the random flipping algorithm, the probability of finding a satisfying assignment a^* of f by random flip corresponds to the density d of the satisfying assignments for g. Given an initial assignment a whose Hamming distance from a^* is k, the probability that k flips transfer the assignment to a^* is $k!/n^k$. Thus, we have

$$d = \left(\frac{1}{2}\right)^n \sum_{k=0}^{t} \binom{n}{k} (k!/n^k) \sim \left(\frac{1}{2}\right)^n$$

for the formula g that simulates random flipping algorithm with t flips. Let T denote the time for constructing g. It is obvious that $T \sim |g|$ holds.

2.2 Condensation Based on Probabilistic Local Search

Let us recall Schöning's probabilistic local search algorithm.

   Procedure PSearch(f).

   1. Select an initial assignment $a \in \{0,1\}^n$ uniformly at random.
   2. Select an arbitrary unsatisfied clause and flip one of the variables in it.
   3. Repeat Step 2 for $3n$ times.

We modify above algorithm as follows. In the flipping step, we do not select an unsatisfied clause at random, but the first unsatisfied clause C_i. Namely, C_i is selected when clauses $C_1, C_2, \ldots, C_{i-1}$ are satisfied. This modification gives no affect on the probability of finding a satisfying assignment a^*, since the probability analysis in [12] depends on the fact that every selected clause has at least one literal (out of three) whose value needs to be flipped so that the Hamming distance between a^* and the flipped assignment a decreases by 1.

We construct a formula g that simulates the modified Schöning's algorithm, where the flipping step is repeated t times. Let $x^0 = (x_1^0, \ldots, x_n^0)$ denote the initial assignment, and (y^k, z^k) denote the random choice of a literal in the k-th flip, where $(y^k, z^k) = (0,1)$ (respectively, $(1,0)$ and $(1,1)$) represents the case that the first (respectively, the second and the third) literal in the selected clause is flipped. Encoding $(y^k, z^k) = (0,0)$ has several interpretations; no literal is flipped, or we may flip a literal according to another rule (e.g., the first literal in the selected clause is flipped). Since the results on their performance analysis have no distinct difference, we adopt the no flipping rule. Then, the assignment $(x_1^{k+1}, \ldots, x_n^{k+1})$ after the k-th flip is obtained as

$$x_i^{k+1} = x_i^k \oplus \sum_{\ell \text{ s.t. } x_i \in C_\ell \text{ or } \overline{x_i} \in C_\ell} \left(\prod_{1 \leq j < \ell} C_j(x^k) \right) \overline{C_\ell(x^k)} \, pos_{i,\ell}(y^k, z^k), \quad (3)$$

where $pos_{i,\ell}(y^k, z^k)$ denotes $\overline{y^k} \cdot z^k$ (respectively, $y^k \cdot \overline{z^k}$ and $y^k \cdot z^k$) if variable x_i appears in the first (respectively, the second and the third) position of clause C_ℓ. In Eq. (3), clause C_i is also used as a Boolean formula, i.e., $C_i(x^k)$ denotes the substitutions of variables in clause C_i. For example, for a clause $C_i = x_{i_1} \vee x_{i_2} \vee x_{i_3}$, $C_i(x^k)$ denotes $x_{i_1}^k \vee x_{i_2}^k \vee x_{i_3}^k$. Let us assume that x_i appears in clauses $C_{j_1}, C_{j_2}, \ldots, C_{j_{\ell_i}}$. Then, Eq. (3) can be rewritten as

$$x_i^{k+1} = x_i^k \oplus \left(\begin{array}{l} C_1(x^k) \cdots C_{j_1-1}(x^k)(\overline{C_{j_1}(x^k)}\, pos_{i,j_1}(y^k, z^k) \\ + C_{j_1}(x^k) \cdots C_{j_2-1}(x^k)(\overline{C_{j_2}(x^k)}\, pos_{i,j_2}(y^k, z^k) + \cdots \\ + C_{j_{\ell_i}-1}(x^k) \cdots C_{j_{\ell_i}-1}(x^k)(\overline{C_{j_{\ell_i}}(x^k)}\, pos_{i,j_{\ell_i}}(y^k, z^k)))) \end{array} \right). \quad (4)$$

By iterative substitutions of Eq. (4), we can represent x_i^{k+1} as a function of variables $(x_1^0, \ldots, x_n^0), (y^0, z^0), (y^1, z^1), \ldots, (y^k, z^k)$. Thus, we can rewrite $f(x_1^t, \ldots, x_n^t)$ as a function g of $n + 2t$ variables $(x_1^0, \ldots, x_n^0)$ and (y^k, z^k) for $k = 0, 1, \ldots, t-1$.

Now, we consider the length of the resulting formula g and its density d of the satisfying assignments. By iterative substitutions of Eq. (4), x_i^{k+1} has length $|x_i^{k+1}| < |x_i^k| + (3 \max|x_i^k| + 2)m$, where m denotes the number of clauses in f. This means that

$$|x_i^t| \sim (3m)^t \max|x_i^0|$$

which implies g has length

$$|g| \sim (3m)^t |f|.$$

Since we use variables (y^k, z^k) for the random choice of a literal in each flipping step, the Hamming distance between a^* and the flipped assignment a decreases by 1 with probability at least $1/4$. The sequence of $(y^0, z^0), (y^1, z^1), \ldots, (y^k, z^k)$ gives the computation tree of Schöning's algorithm, whose path of finding a^* corresponds to a satisfying assignment for g. Thus, the density of the satisfying assignments for g is equal to the probability of finding a satisfying assignment of f by Schöning's algorithm, where the Hamming distance from a^* decreases by 1 with probability at least $1/4$. According to the analysis in [12], we have

$$d = \left(\frac{1}{2}\right)^n \sum_{j=0}^{t} \binom{n}{j} \left(\frac{1}{4}\right)^j > \frac{1}{2^n} \cdot \frac{n^t}{t! \cdot 4^t} \sim \frac{1}{2^n} \cdot \left(\frac{en}{4t}\right)^t$$

for the formula g that simulates Schöning's algorithm with t flips. Let T denote the time for constructing formula g, i.e., $T \sim |g|$. Then, we obtain d as a function of T (not t);

$$d \sim T^{\frac{1}{\log 3n^3}} \log \frac{4n \log 3n^3}{e \log T}.$$

Note that, if functional substitutions are allowed then the iterative substitutions of Eq. (4) are not needed. Namely, we can easily obtain a formula of polynomial size with exponential increase of the density.

2.3   Condensation Based on Search Space Splitting

The last condensation algorithm is based on the search space splitting method. We use the following Procedure Split(f) as the splitting method [9].

> Procedure Split(f).
>
> 1. If f is a 2-CNF, then solve f by the polynomial time 2-SAT algorithm [1].
> 2. Select a clause $(x_i \vee x_j \vee x_k)$ in f. (We can obtain the form by renaming the variables.)
> 3. Return $(\mathrm{Split}(f|_{x_i=1}) \vee \mathrm{Split}(f|_{x_i=0,x_j=1}) \vee \mathrm{Split}(f|_{x_i=0,x_j=0,x_k=1}))$.

We can easily modify Procedure Split(f) so that at least t variables are eliminated, i.e., we stop the recursion at the t-th depth. Intermediate formulas are stored, and their logical OR is outputted as the resulting formula g. For example, when $t = 1$, we obtain intermediate formulas $f|_{x_i=1}$, $f|_{x_i=0,x_j=1}$ and $f|_{x_i=0,x_j=0,x_k=1}$. Note that, if f is satisfiable, at least one of the intermediate formulas is also satisfiable. In the case of $t > 1$, different intermediate formulas may have different elimination of variables, e.g., one has variables (x_1, x_2, x_3) and another may have variables (x_4, x_5, x_6). In such case, we rename the variables so that the resulting formula contains at most $n - t$ variables.

Now, we consider the length of the resulting formula g and its density d. The recursive application of the original Procedure Split(f) (i.e., the recursion until the n-th depth) can be viewed as a search tree, which has 1.839^n leaf nodes [9]. In case recursion is stopped at the t-th depth, we have 1.839^t leaf nodes. Since each node has its corresponding intermediate formula of size at most $|f|$, the length $|g|$ is evaluated as

$$|g| = |f| \cdot 1.839^t.$$

The density d is

$$d = \left(\frac{1}{2}\right)^{n-t},$$

since g preserves the satisfiability of f and the search space is 2^{n-t}. Let T denote the time for constructing formula g, i.e., $T \sim |g|$. Since $t \sim \log_2 T \cdot \log_{1.839} 2$ holds, we can obtain d as a function of T (not t);

$$d \sim \left(\frac{1}{2}\right)^{-t} \sim T^{1.137}.$$

Here, note that our approach is applicable to any algorithm based on the search space splitting method (in other words, DPLL-algorithms [4,5]). For example, consider the pure literal lookahead technique [11] that gives time complexity $O(1.497^n)$. By a similar argument as above, it leads better condensation $|g| = |f| \cdot 1.497^t$ and $d = (\frac{1}{2})^{n-t}$, which implies $|g|/d = |f| 2^n 0.749^t$. As for the efficiency with respect to the consumed time T, we have $|d| \sim T^{1.717}$ and $|g|/d \sim T^{-0.717}$.

Table 1. Comparison among the proposed algorithms.

	Length $|g|$	Density d	$|g|/d$
Random flipping	$|f|\, t^{\cdot}\, \log n$	$1/2^n$	$|f|\, 2^n t^{\cdot}\, \log n$
Probabilistic local search	$|f|\, (3n^{\cdot})^t$	$(en/4t)^t/2^n$	$|f|\, 2^n (12tn^{\cdot}/e)^t$
Search space splitting	$|f|\, 1.839^t$	$1/2^{n-t}$	$|f|\, 2^n 0.920^t$

Table 2. Comparison among the proposed algorithms with respect to T.

	Density d	$|g|/d$
Random flipping	Constant factor	T
Probabilistic local search	$T^{\varphi \cdot n \cdot}$	$T^{\chi \cdot n \cdot}$
Search space splitting	$T^{\cdot . \cdots}$	$T^{- \cdot . \cdots}$

$$\varphi(n) = \frac{\cdot}{\cdots \cdot n^3} \log \frac{\cdot n \cdots \cdot n^3}{e \cdots T}, \quad \chi(n) = \frac{\cdot}{\cdots \cdot n^3} \log \frac{\cdot \cdot n^2 \cdots T}{e \cdots \cdot n^3}.$$

3   Evaluation of the Proposed Algorithms

We first propose two evaluation measures for density condensation algorithms. One is the density of the satisfying assignments for the resulting formula. This is quite natural since our concern is density condensation. The second measure is the ratio $|g|/d$. This is based on the observations that we can find a satisfying assignment of g by the random selection of expectedly $1/d$ assignments if the given CNF f is satisfiable, and that it requires $|g|$ time to determine the satisfiability of g for an assignment. A desirable algorithm is the one that achieves a large d and a small $|g|/d$ at the same time.

Tables 1 and 2 show the comparisons among the proposed three algorithms, where constant factors are ignored. Figures 1 and 2 respectively illustrate the behaviors of the algorithms concerning d and $|g|/d$ with respect to the computation time T. The thick solid, thin solid, and dotted lines respectively correspond to the condensation algorithms based on a search space splitting, a probabilistic local search, and a random flipping. Our three algorithms can arbitrary condense a given formula (i.e., obtain arbitrary high density) according to the time T spent on the computation, while the lengths $|g|$ of the resulting formulas become longer and longer.

The condensation algorithm based on random flipping increase the density with a small constant factor, while the length of the output formula grows proportionally to the computation time. Thus, random flip without strategy does not work well. Contrary to the random flip, Schöning's flip achieves an exponential increase of the density. On the other hand, the length of the output formula is $(3n^3)^t$ whose growth is quite faster than that of the density d. Figures 1 and 2 say that the condensation algorithm based on search space splitting works well on both measures d and $|g|/d$, since their efficiency improves according to the computation time.

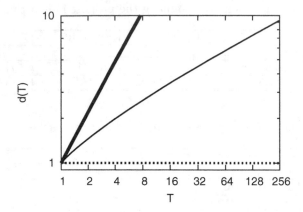

Fig. 1. Behaviors of the algorithms concerning d.

4   Concluding Remarks

Our performance evaluation shows that better SAT algorithms do not always give us better density-condensation algorithms. Therefore, we need to construct an algorithm that is independent from existing SAT algorithms, and specific to DCP.

In every proposed algorithm, the computation time T for generating a formula g is proportional to the length $|g|$. This implies that a good algorithm from the view point of a measure d always works well for the other measure $|g|/d$, and vice versa. However, other density-condensation algorithms may spend almost all computation time to reduce the size of the output formula. In such case, $|g|$ may be exponentially smaller than T. Thus, it may happen that one algorithm wins in measure d and another wins in measure $|g|/d$. To find such algorithm should be addressed in the future research.

References

1. B. Aspvall, M. F. Plass, and R. E. Tarjan, "A Linear-Time Algorithm for Testing the Truth of Certain Quantified Boolean Formulas", Information Processing Letters, 8(3), pp.121–123, 1979.
2. S. Baumer and R. Schuler, "Improving a probabilistic 3-SAT Algorithm by Dynamic Search and Independent Clause Pairs", Proc. 6th International Coference on Theory and Applications of Satisfiability Testing, pp.36–44, 2003.
3. E. Dantsin, A. Goerdt, E. A. Hirsch, R. Kannan, J. Kleinberg, C. Papadimitriou, P. Raghavan, and U. Schöning, "A Deterministic $(2 - 2/(k + 1))^n$ Algorithm for k-SAT Based on Local Search", Theoretical Computer Science, 289(1), pp.69–83, 2002.
4. M. Davis, G. Logemann, D. Loveland, "A Machine Program for Theorem-Proving", Comm. ACM, 5, pp.394–397, 1962.
5. M. Davis, H. Putnam, "A Computing Procedure for Quantification Theory", J. ACM, 7, pp.201–215, 1960.

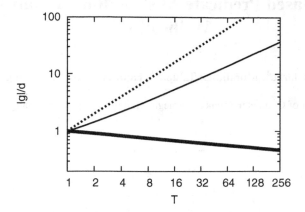

Fig. 2. Behaviors of the algorithms concerning $|g|/d$.

6. M.L. Fredman and L. Khachiyan, "On the Complexity of Dualization of Monotone Disjunctive Normal Forms", J. Algorithms, 21 (3), pp.618–628, 1996.
7. T. Hormeister, U. Schöning, R. Schuler, and O. Watanabe, "Probabilistic 3-SAT Algorithm Further Improved", Proc. 10th Symposium on Theoretical Aspects of Computer Science, LNCS 2285, pp.193–202, 2002.
8. K. Iwama, "Complexity of Finding Short Resolution Proofs", Proc. 22nd Symposium on Mathematical Foundation of Computer Science, LNCS 1295, pp.309–318, 1997.
9. B. Monien and E. Speckenmeyer, "3-Satisfiability is Testable in $O(1.62^r)$ Steps", Technical Report Bericht Nr. 3/1979, Universität-Gesamthochschule-Paderborn, 1979. Reihe Theoretische Informatik.
10. C. H. Papadimitriou and D. Wolfe, "The Complexity of Facets Resolved", J. Computers and System Sciences, 37, pp.2–13, 1988.
11. I. Schiermeyer, "Pure Literal Look Ahead: An $O(1.497^n)$ 3-Satisability Algorithm", Proc. the Workshop on the Satisability Problem, pp.63–72, 1996.
12. U. Schöning, "A Probabilistic Algorithm for k-SAT and Constraint Satisfaction Problems", Proc. 40th Annual Symposium on Foundations of Computer Science, pp.410–414, 1999.

SAT Based Predicate Abstraction for Hardware Verification[*]

Edmund Clarke[1], Muralidhar Talupur[1], Helmut Veith[2], and Dong Wang[1]

[*] School of Computer Science, Carnegie Mellon University, Pittsburgh, USA
{emc,tmurali,dongw}@cs.cmu.edu
[*] Institut für Informationsysteme, Technische Universität Wien, Vienna, Austria
veith@dbai.tuwien.ac.at

Abstract. Predicate abstraction is an important technique for extracting compact finite state models from large or infinite state systems. Predicate abstraction uses decision procedures to compute a model which is amenable to model checking, and has been used successfully for software verification. Little work however has been done on applying predicate abstraction to large scale finite state systems, most notably, hardware, where the decision procedures are SAT solvers. We consider predicate abstraction for hardware in the framework of *Counterexample-Guided Abstraction Refinement* where in the course of verification, the abstract model has to be repeatedly refined. The goal of the refinement is to eliminate *spurious behavior* in the abstract model which is not present in the original model, and gives rise to false negatives (spurious counterexamples).

In this paper, we present two efficient SAT-based algorithms to refine abstract hardware models which deal with spurious transitions and spurious counterexamples respectively. Both algorithms make use of the conflict graphs generated by SAT solvers. The first algorithm extracts constraints from the conflict graphs which are used to make the abstract model more accurate. Once an abstract transition is determined to be spurious, our algorithm does not need to make any additional calls to SAT solver. Our second algorithm generates a compact predicate which eliminates a spurious counterexample. This algorithm uses the conflict graphs to identify the important concrete variables that render the counterexample spurious, creates an additional predicate over these concrete variables, and adds it to the abstract model. Experiments over hardware designs with several thousands of registers demonstrate the effectiveness of our methods.

---

[*] This research was sponsored by the Semiconductor Research Corporation (SRC) under contract no. 99-TJ-684, the National Science Foundation (NSF) under grant no. CCR-9803774, the Office of Naval Research (ONR), the Naval Research Laboratory (NRL) under contract no. N00014-01-1-0796, and by the Defense Advanced Research Projects Agency, and the Army Research Office (ARO) under contract no. DAAD19-01-1-0485, the General Motors Collaborative Research Lab at CMU, the Austrian Science Fund Project N Z29-N04, and the EU Research and Training Network GAMES. The views and conclusions contained in this document are those of the authors and should not be interpreted as representing the official policies, either expressed or implied, of SRC, NSF, ONR, NRL, DOD, ARO, or the U.S. government.

E. Giunchiglia and A. Tacchella (Eds.): SAT 2003, LNCS 2919, pp. 78–92, 2004.
© Springer-Verlag Berlin Heidelberg 2004

1   Introduction

Counterexample-Guided Abstraction Refinement.   Model checking [11, 10] is a widely
used automatic formal verification technique. Despite the recent advancements in model
checking technology, practical applications are still limited by the state explosion problem, i.e., by the combinatorial explosion of system states. For model checking large real
world systems, symbolic methods such as BDDs and SAT-sovlers need to be complemented by abstraction methods [6, 5]. By a conservative abstraction we understand a
(typically small) finite state system which preserves the behavior of the original system, i.e., the abstract system allows more behavior than the original (concrete) system.
If the abstraction is conservative then we have a preservation theorem to the effect that
the correctness of universal temporal properties (in particular properties specified in
universal temporal logics such as $ACTL^*$) on the abstract model implies the correctness of the properties on the concrete model. In this paper, we are only concerned with
universal safety properties whose violation can be demonstrated on a finite counterexample trace. The simplest and most important examples of such properties are system
invariants, i.e., ACTL* specifications of the form $\mathbf{AG}p$.

A preservation theorem only ensures that universal properties which hold on the
abstract model are indeed true for the concrete model. If, however, a property is violated
on the abstract model, then the counterexample on the abstract model may possibly not
correspond to any real counterexample path. False negatives of this kind are called
spurious counterexamples, and are consequences of the information loss incurred by
reducing a large system to a small abstract one. Since the spurious behavior is due to
the approximate nature of the abstraction, we speak of *overapproximation*.

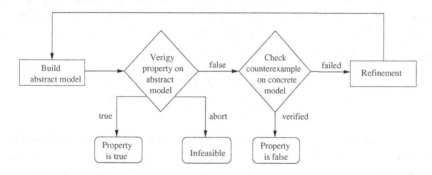

Fig. 1. Counterexample-Guided Abstraction Refinement.

In order to remedy the effect of abstraction and overapproximation, the abstraction needs to be refined in such a way that the spurious counterexample is eliminated.
Counterexample guided abstraction refinement (CEGAR) [7] automates this procedure.
CEGAR analyzes the spurious abstract counterexample to find a refinement of the current abstraction, such that the spurious counterexample is eliminated from the refined
abstract model. This procedure is repeated until the property is either confirmed or

refuted, cf. Figure 1. During the last years, CEGAR Methods have found many applications in verification. See [9] for an overview.

Predicate Abstraction. Predicate abstraction [17, 1, 13, 12, 16, 18] is a specific construction to obtain conservative abstractions. In predicate abstraction, we identify a set of predicates $P_1, \ldots, P_m$ which describe important properties of the concrete system. The predicates are given by formulas which depend on the variables of the concrete system. The crucial idea is now to construct 2^m abstract states such that each abstract state corresponds to one valuation of the predicates, and thus to a conjunction of literals over $B_1, \ldots, B_m$.

More formally, we use the predicates P_i as the atomic propositions that label the states in the concrete and abstract transition systems, that is, the set of atomic propositions is $A = \{P_1, P_2, .., P_m\}$. A state in the concrete system is labeled with all the predicates it satisfies. The abstract state space contains one boolean variable B_j for each predicate P_j. An abstract state is labeled with predicate P_j if the corresponding Boolean variable B_j has value 1 in this state.

The predicates are also used to define a relation ρ between the concrete and the abstract state spaces. A concrete state s will be related to an abstract state $\hat{s}$ through ρ if and only if the truth value of each predicate on s equals the value of the corresponding boolean variable in the abstract state $\hat{s}$, i.e.,

$$\rho(s, \hat{s}) \quad = \quad \bigwedge_{1 \leq j \leq m} P_j(s) \Leftrightarrow B_j(\hat{s}).$$

We now define the *concretization function* γ, which maps a set of abstract states to the corresponding set of concrete states. Each set of abstract states can be described by a Boolean formula over the variables $B_1, \ldots, B_m$. For a propositional formula $\hat{f}$ over the abstract state variables, we define $\gamma(\hat{f}) = \hat{f}[B_j \leftarrow P_j]$ to be the formula obtained by replacing each occurrence of some B_j by the corresponding P_j. The abstract initial states $\hat{S}_0$ and the abstract transition relation $\hat{R}$ are defined as

$$\hat{S}_0 \quad = \quad \bigwedge\{\hat{Y}_1 \mid S_0 \rightarrow \gamma(\hat{Y}_1)\} \tag{1}$$

$$\hat{R} \quad = \quad \bigwedge\{\hat{Y} \rightarrow \hat{Y}' \mid (R \wedge \gamma(\hat{Y})) \rightarrow \gamma(\hat{Y}')\} \tag{2}$$

where $\hat{Y}$ ($\hat{Y}_1$) denotes conjunctions (disjunctions resp.) of the literals over the current abstract state variables $\{B_1, \ldots, B_m\}$ and $\hat{Y}'$ denotes disjunctions of literals over the next state variables $\{B'_1, \ldots, B'_m\}$. The abstract model built according to equations (1) and (2) is called the *most accurate abstract model*. In the most accurate abstract model, every abstract initial state has at least one corresponding concrete initial state, and every abstract transition has at least one corresponding concrete transition.

Building the most accurate abstract model is expensive because the number of implications that need to be checked in the worst case is exponential in the number of predicates. Thus, to reduce the abstraction time, in practice an *approximate abstract model* is constructed by intentionally excluding certain implications from consideration. Therefore, there will be more behaviors in this approximate model than in the most accurate abstract model. We call those abstract transitions that do not have any corresponding concrete transitions *spurious transitions*. (Precise definitions are given in Section 3.1). Since an approximate abstract model preserves all behaviors of the original concrete system, the preservation theorem still holds.

Contribution. For software model checking, the use of predicate abstraction or similar abstraction techniques is essential because most software systems are infinite state and the existing model checking algorithms either cannot handle infinite state systems, or only very specific ones. Predicate abstraction can extract finite state abstract models which are amenable to model checking from infinite state systems. Since hardware systems are finite state, model checking, possibly combined with simple forms of abstraction such as the localization reduction [14]) has been traditionally used to verify them. Existing predicate abstraction techniques for verifying software however are not efficient when applied to the verification of large scale hardware systems.

There are many proof obligations involved in predicate abstraction that require the use of decision procedures. Proof obligations can arise from equations (1) and (2) and also from determining whether an abstract counterexample is spurious or not. For software verification, these proof obligations are solved using decision procedures or general theorem provers. For the verification of hardware systems, which usually have compact representation in conjunctive normal form (CNF), we can use SAT solvers instead of general theorem provers. With the advancements in SAT technology, discharging the proof obligations using SAT solvers becomes much faster than using general theorem provers.

When refining the abstract model, we distinguish two cases of spurious behavior:

1. **Spurious Transitions** are abstract transitions which do not have any corresponding concrete transitions. By definition, spurious transitions cannot appear in the most accurate abstract model.
2. **Spurious Prefixes** are prefixes of abstract counterexample paths which do not have corresponding concrete paths. These are the typical spurious counterexamples described in the literature.

Our first SAT based algorithm deals with the first case, i.e., spurious transitions. As argued above, it is time consuming to build the most accurate abstract model when the number of predicates is large. We use a heuristic similar to the one given in [1] to build an approximate abstract model. Instead of considering all possible implications of the form $\hat{Y} \rightarrow \hat{Y}'$ we impose restriction on the lengths of $\hat{Y}$ and $\hat{Y}'$ in equation (2), and similarly for equation (1)). If the resulting abstract model is too coarse, an abstract counterexample with a spurious transition might be generated. This spurious transition can be removed by adding an appropriate constraint to the abstract model. This constraint however should be made as general as possible so that many related spurious transitions are removed simultaneously. An algorithm for this has been proposed in [12] which in the worst case requires $2m$ number of calls to a theorem prover, where m is the number of predicates. In this paper, we propose a new algorithm based on SAT conflict dependency analysis (presented in Section 2) to generate a general constraint without any additional calls to the SAT solver. Our algorithm works by analyzing the conflict graphs generated when detecting the spurious transition. Thus our algorithm can be much more efficient than the algorithm in [12]. We give a detailed description in Section 3.1.

Even after removing spurious transitions the abstract counterexample can have a spurious prefix. This happens when the set of predicates is not rich enough to capture the relevant behaviors of the concrete system, even for the most accurate abstract model. In this case, a new predicate is identified and added to the current abstract model to invalidate the spurious abstract counterexample. To make the abstraction refinement process efficient, it is desirable to compute a predicate that can be compactly represented. Large predicates are difficult to compute and discharging proof obligations involving them will be slow. We propose an algorithm, again based on SAT conflict dependency analysis, to reduce the number of concrete state variables that the new predicate depends on. The new predicate is then calculated by a projection-based SAT enumeration algorithm. Our experiments show that this algorithm can efficiently compute the required predicates for designs with thousands of registers.

Related work. SAT based localization reduction has been investigated in [2]. To identify important registers for refinement, SAT conflict dependency analysis is used. Their method is similar to our algorithm for reducing the support of the predicates. However, there are several important differences: First, we have generalized SAT conflict dependency analysis to find the set of predicates which disables a spurious transition, while the algorithm in [2] only finds important registers. Second, in this paper, we present a projection-based SAT enumeration algorithm to determine a new predicate that can be used to refine the abstract model. Third, we approximate the most accurate abstract model by intentionally excluding certain implications, while in [2], approximation is achieved through pre-quantifying invisible variables during image computation. Finally, our experimental results show significant improvement over the method in [2].

An algorithm to make the abstract model more accurate given a fixed set of predicates is presented in [12]. Given a spurious transition, their algorithm requires $2m$ number of calls to a theorem prover, where m is the number of predicates. Our algorithm is more efficient in that no additional calls to a SAT solver are required. Note that, in general, their algorithm can come up with a more general constraint than ours. However, we can obtain the same constraints, probably using much less time, by combining the two algorithms together. Furthermore, the work in [12] does not consider the problem of introducing new predicates to refine the abstract model.

Other refinement algorithms in the literature compute new predicates using techniques such as syntactical transformations [16] or pre-image calculation [7, 18]. In contrast to this, our algorithm is based on SAT. They also neglect the problem of making the representation of the predicates compact. This could result in large predicates which affects the efficiency of abstraction and refinement.

More recently, [8] shows that predicate abstraction can benefit from localization reduction when verifying control intensive systems. By selectively incorporating state machines that control the behavior of the system under verification, more compact abstract models can be computed than those only based on predicate abstraction.

Outline of the paper. The rest of the paper is organized as follows. In Section 2 we

Section 3. Section 4 contains the experimental results, and Section 5 concludes the paper.

2 SAT Conflict Dependency Analysis

In this section, we give a brief review of *SAT conflict dependency analysis* [2]. Modern SAT solvers rely on conflict driven learning to prune the search space; we assume that the reader is familiar with the basic concepts of SAT solvers such as CHAFF [19]. As presented in [19], a conflict graph is an implication graph whose sink is the conflict vertex, and a conflict clause is obtained from a vertex cut of the conflict graph that separates the decision vertices from the conflict vertex.

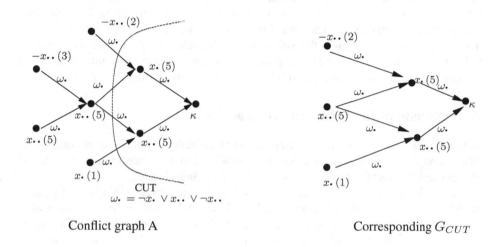

Conflict graph A   Corresponding G_{CUT}

Fig. 2. Part of an implication graph and its corresponding G_{CUT}.

Let G be a conflict graph, κ be the conflict vertex in G, and CUT be a vertex cut which corresponds to the conflict clause $cl(CUT)$. Then G_{CUT} is the subgraph of G whose sources are the vertices in CUT and whose sink is κ. An example of a conflict graph and a vertex cut is given in Figure 2. For a subgraph G' of a conflict graph G, let $\Omega(G')$ be the set of clauses that occur as labels on the edges in the graph G'. Since G_{CUT} includes the conflict vertex κ, it is easy to see that $\neg cl(CUT) \wedge \Omega(G_{CUT}) \Rightarrow$ false. Therefore we obtain

$$\Omega(G_{CUT}) \Rightarrow cl(CUT). \tag{3}$$

Given a CNF formula f, a SAT solver concludes that f is unsatisfiable if and only if the SAT solver derives a conflict graph without decision vertices. We associate the empty conflict clause, denoted by θ, with this last conflict graph. Note that since θ is an empty clause, it is logically equivalent to false.

A conflict clause $cl(CUT)$ *directly depends* on a clause b iff b is one of the clauses in $\Omega(G_{CUT})$. We say the conflict clause a *depends* on clause b iff there exist $a = c_1, c_2, \ldots, b = c_n$, such that for $1 \leq i < n$, c_i directly depends on c_{i+1}. Given a CNF formula f, the set of clauses in f that a given set of conflict clauses cls depend on is called the *dependent set* and the set is denoted by $dep(cls)$. Based on equation (3), it is easy to see that $dep(cls) \Rightarrow cls$. If f is an unsatisfiable CNF formula, let $SUB(f) = dep(\theta)$ denote the clauses actually used for showing that f is unsatisfiable. Since $dep(\theta) \Rightarrow \theta$, $SUB(f) \subseteq f$ is also unsatisfiable, i.e.,

$$f \equiv \text{false} \quad \Rightarrow \quad SUB(f) \equiv \text{false} \tag{4}$$

During SAT search, our conflict dependency analysis algorithm keeps track of the set of clauses on which a conflict clause directly depends. After the SAT solver concludes that f is unsatisfiable, our algorithm identifies the unsatisfiable subset $SUB(f)$ based on these dependencies. Note that the dependencies and the unsatisfiable subset that our algorithm computes are determined by the conflict graphs and the conflict clauses generated by a concrete run of the SAT solver during SAT search. Thus, for an unsatisfiable CNF formula f, $SUB(f)$ is in general not the minimal unsatisfiable subset of f; in many practical cases, however, $SUB(f)$ is significantly smaller than f.

3   Refinement for Predicate Abstraction

In this section, we describe the two SAT-based refinement algorithms for spurious transitions and spurious prefixes. To this end, we first need to introduce some more notation to represent the unrolling of a transition system from initial states.

Let V be the set of system variables, and V' be the corresponding set of primed variables used to describe the next state. Both V and V' are called *untimed variables*. For every variable v in V we maintain a version v^i of that variable for each time point $i \geq 0$. V^i, is the set of timed versions of variables in V at time $i \geq 0$. The variables V^i are called *timed variables* at time i.

Let $f(V)$ be a boolean function describing a property of concrete states; thus, it maps the set of states over variables V to $\{0, 1\}$. The *timed* version of f at time i, denoted by $f^i(V^i)$, is the same function as f except that it is defined over the timed variables V^i. We define an operator, called utf ("untimed function") which for a given timed function $f^i(V^i)$ returns the untimed function $f(V)$, i.e., $f(V) = utf(f^i(V^i))$.

Similarly, given a relation $r(V, V')$, which relates a current state over variables V to the next state over variables V', $r^i(V^i, V^{i+1})$ is the *timed* version of r at time i. We define an operator, called utr (for untimed relation), which for a given timed relation $r^i(V^i, V^{i+1})$, returns the untimed relation $r(V, V')$, i.e., $utr(r^i(V^i, V^{i+1})) = r(V, V')$.

Let $B = \{B_1, \ldots, B_m\}$ be the set of abstract state variables, and B^i be the corresponding timed variables. Given a timed abstract expression f ranging over variables B^i at time i, its concretization is a timed concrete expression $\gamma(f)$ in terms of V^i obtained by replacing each B_j^i in f with P_j^i, the timed version of the corresponding predicate.

An *abstract counterexample* $ce(B^0, \ldots, B^n)$ is a sequence of abstract states

$$\langle ce_0(B^0), ce_1(B^1), \ldots, ce_n(B^n) \rangle$$

where each $ce_i(B^i)$ is a cube (i.e., a conjunction of literals) over the set of abstract variables B^i at time i. When it is clear from the context, we sometimes represent a counterexample without explicitly mentioning timed variables.

Let $ce = \langle ce_0, ce_1, \ldots, ce_n \rangle$ be an abstract counterexample, and i a natural number, such that $0 < i \leq n$. The set of pairs of concrete states corresponding to the abstract transition from ce_{i-1} to ce_i is

$$trans(i - 1, i) = \gamma(ce_{i-1}) \wedge R^{i-1} \wedge \gamma(ce_i) \tag{5}$$

The set of concrete paths which correspond to the prefix of the abstract counterexample up to time i is given by

$$prf(i) = S_0 \wedge \gamma(ce_0) \wedge R^0 \wedge \cdots \wedge \gamma(ce_{i-1}) \wedge R^{i-1} \wedge \gamma(ce_i). \tag{6}$$

Let BV be a set of boolean variables and let $BV_1 \subseteq BV$. If c is a conjunction of literals over BV, the projection of c to BV_1, denoted by $\text{proj}[BV_1](c)$, is a conjunction of literals over BV_1 that agrees with c over the literals in BV_1, i.e., the conjunction obtained from c by removing all literals based on atoms in $BV \setminus BV_1$.

If f is a CNF formula over BV, the *satisfiable set* of f over BV_1, denoted by $SA[BV_1](f)$, is the set of all satisfying assignments of f projected on to BV_1. Thus, $SA[BV_1](f) = \text{proj}[BV_1](SA[BV](f))$. For a SAT solver with conflict based learning, there is a well known algorithm to compute $SA[BV_1](f)$ without first computing $SA[BV](f)$, cf. [15]. This SAT enumeration technique works as follows: Once a satisfiable solution is found, a blocking clause over BV_1 is created to avoid regenerating the same projected solution. After this blocking clause is added, the SAT search continues, pretending that no solution was found. This process repeats until the SAT solver concludes that the set of clauses is unsatisfiable, i.e., there are no further solutions. The set of all satisfying assignments over BV_1 is the required result, which can be represented as a DNF formula.

Given a set of variables SV that are not necessarily boolean, let BSV be the set of boolean variables in the boolean encoding of the variables in SV. Let f be a CNF formula over BSV. The *scalar support of the CNF formula* f, denoted by $ssuppt[SV](f)$, is a subset of SV that includes a variable $v \in SV$ iff at least one of v's corresponding boolean variables is in f.

Recall from the above discussion that an abstract counterexample $ce = \langle ce_0, ce_1, \ldots, ce_n \rangle$ corresponds to a real counterexample if and only if the set $prf(n)$ is not empty. If the abstract counterexample is a real counterexample, then by virtue of the preservation theorem, the property to be verified is also false for the concrete model. Otherwise, the counterexample is spurious and we need to refine the current abstract model. As argued above, there are two possible reasons for the existence of a spurious counterexample: One is that the computed abstract model is a too coarse over-approximation of the most accurate abstract model. The other possibility is that the set of predicates we used is not rich enough to model the relevant behaviors of the system, and thus we have a spurious prefix. In Section 3.1, we describe how our algorithm deals with the first case. In Section 3.2 we deal with the case where the set of predicates has to be extended.

3.1   Refinement for Spurious Transitions

Let $ce = \langle ce_0, ce_1, \dots, ce_n \rangle$ be an abstract counterexample as defined above. If there exists an index i, $0 < i \leq n$, such that the set $trans(i-1, i) = R^{i-1} \wedge \gamma(ce_{i-1}) \wedge \gamma(ce_i)$ is empty, then we call the transition from ce_{i-1} to ce_i a *spurious transition*. This means that there are no concrete transitions which correspond to the abstract transition from ce_{i-1} to ce_i. It is evident that in this case the abstract counterexample does not have a corresponding real counterexample.

Recall that in the most accurate abstract model, there is at least one concrete transition corresponding to every abstract transition; consequently, the spurious transitions exist only for approximate abstract transition relations. Since spurious transitions are not due to the lack of predicates but due to an approximate abstract transition relation, our algorithm removes spurious transitions by adding appropriate constraints to $\hat{R}$.

To determine whether $trans(i-1, i)$ is empty or not, we convert it into a SAT unsatisfiability problem. For the spurious transition from ce_{i-1} to ce_i, we have

$$R^{i-1} \wedge \gamma(ce_{i-1}) \wedge \gamma(ce_i) \quad \Leftrightarrow \quad \text{false},$$

and therefore,

$$R^{i-1} \Rightarrow (\gamma(ce_{i-1}) \to \gamma(\neg ce_i)).$$

Note that ce_{i-1} is a conjunction over the abstract state variables at time $i-1$, and $\neg ce_i$ is a disjunction over the abstract state variables at time i. Since the concrete transition relation does not allow any transition from $\gamma(ce_{i-1})$ to $\gamma(ce_i)$, a natural (but naive) approach is to add the constraint

$$utr(ce_{i-1} \to \neg ce_i)$$

to $\hat{R}$. It is evident that the resulting transition relation is correct and eliminates the spurious transition.

The constraint $ce_{i-1} \to \neg ce_i$ however can potentially involve a much larger number of the abstract state variables than necessary; it will therefore be very specific and not very useful in practice. It is thus reasonable to make the constraint as general as possible as long as the cost of achieving this is not too large. In the rest of this subsection, we describe an efficient algorithm which removes some of the literals from ce_{i-1} and ce_i from the constraint $ce_{i-1} \to \neg ce_i$ in order to obtain a more general constraint.

Computing a General Constraint. Let m be the number of predicates. The problem of finding a general constraint which eliminates a spurious transition can be viewed as follows: Given propositional formulas f and f_j, $1 \leq j \leq 2m$, which make

$$f \wedge \bigwedge_{1 \leq j \leq 2m} f_j$$

unsatisfiable, we need to find a small subset $care \subseteq \{1, \dots, 2m\}$, such that $f \wedge \bigwedge_{j \in care} f_j$ is unsatisfiable.

Returning to our problem of computing a general constraint, it is easy to see that if we set $f = R^{i-1}$ and let each f_j correspond to the concretization of a literal in

ce_{i-1} or ce_i, then we can drop those literals that are not in *care* from $ce_{i-1} \rightarrow \neg ce_i$. Consequently, the resulting constraint will be made more general.

The set *care* can be efficiently calculated using the conflict dependency analysis algorithm described in Section 2. Before we run the SAT solver we need to convert $f \wedge f_1 \wedge f_2 \wedge \cdots \wedge f_{2m}$ to CNF, and in this process some of the f_j's might be split into smaller formulas. Hence it may not be possible to keep track of all f_j's. To overcome this difficulty, we introduce a new boolean variable t_j for each f_j in the formula and convert the formula into

$$F = \exists t_1, t_2 \ldots, t_{2m}. \, f \wedge \bigwedge_{j \in \{1,\ldots,2m\}} (t_j \wedge (t_j \equiv f_j)). \qquad (7)$$

It is easy to see that this formula is unsatisfiable iff the original formula is unsatisfiable. Once (7) is translated to a CNF formula, for each t_j there will be a clause T_j containing only the literal t_j. So, instead of keeping track of the f_j's directly we keep track of the clauses T_j's. Since the CNF formula F corresponding to (7) is unsatisfiable, we know that $SUB(F) \subseteq F$ is unsatisfiable, where $SUB(F)$ is defined as in Section 2. Since $SUB(F)$ denotes the clauses used in the refutation of F, it is easy to see that we can set $care = \{j \mid T_j \in SUB(F)\}$ to obtain the indices for the relevant f_j's as desired. Using only the f_j's given by *care*, we can now add a more general constraint to $\hat{R}$.

It is easy to see that our algorithm only analyzes the search process of the SAT problem during which the spurious transition was identified. Using the approach in [12], a potentially more general constraint than the one computed by our algorithm can be found. It works by testing for each f_j whether removing it keeps the resulting formula unsatisfiable. Their algorithm however requires $2m$ calls to a theorem prover, which is time consuming when the number m of predicates is large. As presented in Section 2, the unsatisfiable subset $SUB(F)$ may not be a minimal unsatisfiable subset of F. Consequently, in general, the set *care* our algorithm computes is not minimal. In practice, however, its size is comparable to a minimal set. Note that it is easy to modify our algorithm in such a way as to make *care* minimal: After the set *care* is computed, we can try to eliminate the remaining literals one by one as in [12], which requires $|care|$ additional calls to the SAT solver. Since the size of *care* is already small, this is not very expensive.

3.2 Refinement for Spurious Prefixes

Even after we have ensured that there are no spurious transitions in the counterexample ce, the counterexample itself can still be spurious. Such a situation is possible because even in the absence of spurious transitions we only have the most accurate abstract model; even in the most accurate abstract model it is not necessarily the case that there exists a concrete path which corresponds to the abstract counterexample, cf. [7]. In the current section we deal with this case.

Let n be the length of the given abstract counterexample. We are interested in finding a k such that $1 < k \leq n$ and the prefix $p_{k-1} = \langle ce_0, ce_1, \ldots, ce_{k-1} \rangle$ of the counterexample corresponds to a valid path but $p_k = \langle ce_0, ce_1, \ldots, ce_k \rangle$ does not. Formally, we call p_k a *spurious prefix* if and only if $prf(k-1) \neq \emptyset \wedge prf(k) = \emptyset$. If there is no such k then the counterexample is real.

Otherwise, let us consider the states in V^{k-1} in more detail, adopting the terminology of [7]:

- The set of states $SA[V^{k-1}](prf(k-1))$ is called the set of *deadend states*, denoted by *deadend*. Deadend states are those states in $\gamma(ce_{k-1})$ that can be reached but do not have any transition to $\gamma(ce_k)$.
- The set of states $SA[V^{k-1}](trans(k-1,k))$ is called the set of *bad states*, denoted by *bad*. The states in *bad* are those states in $\gamma(ce_{k-1})$ that have a transition to some state in $\gamma(ce_k)$.

For a spurious abstract counterexample ce without spurious transitions, let k be the length of the spurious prefix of ce. By construction we know that

$$deadend \neq \emptyset, \qquad bad \neq \emptyset, \qquad (deadend \cap bad) = \emptyset.$$

As pointed out in [7], it is impossible to distinguish between the states in *deadend* and *bad* using the existing set of predicates, because all involved states correspond to the same abstract state ce_{k-1}. Therefore, our refinement algorithm aims to find a *new* separating predicate *sep*, such that

$$deadend \subseteq sep \qquad \text{and} \qquad sep \cap bad = \emptyset.$$

After introducing *sep* as a new predicate, the abstract model will be able to distinguish between the deadend and bad states. Note that we can also use an alternative symmetric definition for *sep* which satisfies $bad \subseteq sep$ and $deadend \cap sep = \emptyset$.

We call the set of concrete state variables over which a predicate is defined the *support* of the predicate. In order to compute a predicate *sep* with minimal support, we will describe an algorithm which first identifies a minimal set of concrete state variables. Then a predicate over these variables that can separate the deadend and bad states is computed. The details of this procedure are described in the rest of this section.

Minimizing the Support of the Separating Predicate. An important goal of our refinement algorithm is to compute a *compact predicate*, i.e., a predicate that can be represented compactly. For large scale hardware designs, existing refinement algorithms such as weakest precondition calculation, preimage computation, syntactical transformations etc., may fail because the predicates they aim to compute become too big. Our algorithm avoids this problem by first identifying a minimal set of concrete state variables that are responsible for the failure of the spurious prefix. Our algorithm guarantees that there exists a separating predicate over this minimal set that can separate the deadend and bad states. It is reasonable to assume that a predicate with a small support has a compact representation.

Our algorithm for computing the support of *sep* is similar to the one used in finding the important registers for the localization reduction in [2]. By assumption, the CNF formula for $prf(k)$ is unsatisfiable. Thus, we can use the conflict dependency analysis from Section 2 to identify an unsatisfiable subset $SUB(prf(k))$ of the clauses in $prf(k)$. Let $\mu(ce, k-1)$ denote the concrete state variables at time $k-1$ whose CNF variables are in $SUB(prf(k))$, i.e.,

$$\mu(ce, k-1) = ssuppt[V^{k-1}](SUB(prf(k))).$$

When the context is clear we will for simplicity refer to $\mu(ce, k - 1)$ as μ. Let $deadend_\mu = \texttt{proj}[\mu](deadend)$ be the projection of the deadend states on μ, and $bad_\mu = \texttt{proj}[\mu](bad)$ be the projection of the deadend states on μ. By the definition of SUB and μ it follows that

$$\mu \neq \emptyset \qquad \text{and} \qquad deadend_\mu \cap bad_\mu = \emptyset. \tag{8}$$

Thus any concrete set of states S_1 that satisfies

$$(S_1 \supseteq deadend_\mu) \wedge (S_1 \cap bad_\mu = \emptyset)$$

is a candidate separating predicate.

To further reduce the size of μ and to make it minimal we use the refinement minimization algorithm in [2], which eliminates any unnecessary variables in μ while ensuring that equation (8) still holds. In most of our experiments, the size of μ was less than 20, which is several orders of magnitude less than the total number of concrete state variables.

Computing Separating Predicates using SAT. As argued above, any set of concrete states that separates $deadend_\mu$ and bad_μ is a separating predicate. We propose a new *projection based SAT enumeration algorithm* to compute such a separating set, which can be represented efficiently as a CNF formula or a conjunction of DNF formulas. Our algorithm proceeds in several steps.

– First, we try to compute bad_μ using a SAT enumeration algorithm, which avoids computing bad first. To this end, the set of bad states bad is converted to a CNF formula. Once a satisfying assignment is found, we project the satisfying assignment to μ and add it into the set of solutions. A blocking clause over μ is also added to the set of clauses and the SAT search is continued. This procedure repeats until there are no more solutions. The collected set of solutions over μ is naturally represented in DNF. Since the size of μ is pretty small, this procedure can often terminate quickly. If that is the case, our algorithm terminates and

$$\neg bad_\mu$$

is the required separating predicate. Note that $\neg bad_\mu$ is represented as a CNF formula.

– Otherwise, we try to compute $deadend_\mu$ using a similar method. If this procedure finishes in a reasonably short amount of time, our algorithm terminates and

$$deadend_\mu$$

is the desired separating predicate, which is represented as a DNF formula.

– In the third case when both $deadend_\mu$ and bad_μ can not be computed within a given time limit, we compute an over-approximation of $deadend_\mu$ denoted by ODE. Note that it is possible that the set ODE overlaps with bad_μ. We define $SODE = \texttt{proj}[\mu](ODE \wedge bad)$ as the intersection of the two sets. Then the desired separating predicate is given by

$$ODE \wedge \neg SODE,$$

which is represented as a conjunction of DNF formulas. In most cases, $SODE$ is much smaller than bad_μ, so it can often be enumerated using SAT.

The over-approximation of $deadend_\mu$ is computed by a projection-based method: We partition the variables in μ into smaller sets $\mu_1, \ldots, \mu_l$ based on the closeness of the variables, whereby the criterion for closeness is based on circuit structure [3]. Since each set μ_i is small, we can compute each $deadend_{\mu_i}$ easily. The over-approximation is then given by

$$ODE = \bigwedge_{1 \le i \le l} deadend_{\mu_i}.$$

- If in a rare case, even $SODE$ can not be efficiently enumerated using SAT we identify important registers using the algorithms in [2], and add them as a new predicate to make sure the abstract model is refined. We did not encounter this case in any of our experiments.

After the calculated separating predicate sep is added as a new predicate, we introduce a new abstract Boolean variable B_{m+1} for sep. Then we add the constraint $B_{m+1} \to utr(ce_{k-1} \to \neg ce_k)$ to the abstract transition relation. Thus, the spurious counterexample is eliminated in the refined abstract model.

4   Experimental Results

We have implemented our predicate abstraction refinement framework on top of the NuSMV model checker [4]. We modified the SAT checker zChaff [19] to support conflict dependency analysis. We also developed a Verilog parser to extract useful predicates from Verilog designs directly; due to lack of space, we omit a detailed description of this parser. All experiments were performed on a dual 1.5GHz Athlon machine with 3GB of RAM running Linux. We have two verification benchmarks: one is the integer unit (IU) of the picoJava microprocessor from Sun Microsystems; the other is a programmable FIR filter (PFIR) which is a component of a system-on-chip design. All

circuit	# regs	# gates	ctrex length	Localization			Predicate Abstraction		
				time	iters	# regs	time	iters	# predicates
IUscr2	4855	149143	20	29115.0	69	115	13515.0	22	14
IUscr3	4855	149143	true	4794.1	9	31	2003.0	10	6
IUscr7	4855	149143	12	7332.1	17	73	3869.8	10	8
IUprop4	4855	149143	8	5603.7	36	61	3495.9	13	9
PFIRprop8	244	2304	true	> 24 hours	>37	>91	288.5	68	35
PFIRprop9	244	2304	true	>24 hours	>33	>85	2448.7	146	46
PFIRprop10	244	2304	true	>24 hours	>46	>94	6229.3	161	55
PFIRprop12	247	2317	true	>24 hours	>46	>91	707.0	111	45

Table 1. Comparison between localization reduction [2] and predicate abstraction.

properties verified were simple **AG** p properties where p specifies a certain combination of values for several control registers. For all the properties shown in the first column of Table 1, we have performed the cone-of-influence reduction before the verification. The resulting number of registers and gates are shown in the second and third columns. We compare three abstraction refinement systems, including the BDD based aSMV [7], the SAT based localization reduction [2] (SLOCAL), and the SAT based predicate abstraction (SPRED) described in this paper. The detailed results obtained using aSMV are not listed in Table 1 because aSMV can not solve any of the properties within the 24hr time limit. This is not surprising because aSMV uses BDD based image computation and it can handle only circuits with several hundred state variables, provided that good initial variable orderings are given. Since the time to generate good BDD variable orderings can be substantial, we did not pre-generate them for any of the properties. For the first four properties from IU, SLOCAL takes about twice the time taken by SPRED. Furthermore, the numbers of registers in the final abstract models from SLOCAL are much larger than the corresponding numbers of predicates in the final abstract models from SPRED. For the rest of the four properties from PFIR, SLOCAL can not solve any of them in 24 hours because all the abstract models had around 100 registers. SPRED could solve each of them easily using about 50 predicates.

To create the abstract transition relation in our experiments, we only considered implications in equation (2) where the left hand side has at most 2 literals, and the right hand side has 1 literal, and relied on the refinement algorithm to make the abstract transition relation as accurate as necessary. In all our experiments, our projection based SAT enumeration algorithm can successfully create the new predicates when necessary. Thus we never have to resort to other methods for creating new separating predicates.

5 Conclusion

We have presented two SAT-based counterexample guided refinement algorithms to enable efficient predicate abstraction of large hardware designs. In order to reduce the abstraction time, we initially construct an approximate abstract model which may lead not only to spurious counterexamples, but also to spurious transitions, i.e., abstract transitions which do not have a corresponding concrete transition. Our first SAT based refinement algorithm is used to eliminate spurious transitions. Our second SAT based refinement algorithm eliminates spurious counterexample prefixes. It extends the predicate abstraction by a new predicate ranging over a minimal number of state variables. The predicates computed by our algorithm can be represented compactly as CNF formulas or conjunctions of DNF formulas. Our experimental results demonstrate significant improvement of our predicate abstraction algorithms over popular abstraction algorithms for hardware verification.

References

[1] Thomas Ball, Rupak Majumdar, Todd Millstein, and Sriram K. Rajamani. Automatic Predicate Abstraction of C Programs. In *PLDI 2001*.

[2] Pankaj Chauhan, Edmund M. Clarke, Samir Sapra, James Kukula, Helmut Veith, and Dong Wang. Automated abstraction refinement for model checking large state spaces using SAT' based conflict analysis. In *FMCAD'02*, 2002.

[3] H. Cho, G. Hachtel, E. Macii, M. Poncino, and F. Somenzi. Automatic state space decomposition for approximate fsm traversal based on circuit analysis. *IEEE TCAD*, 15(12):1451–1464, December 1996.

[4] A. Cimatti, E. M. Clarke, F. Giunchiglia, and M. Roveri. NuSMV: A New Symbolic Model Verifier. In *CAV'99*, pages 495–499, 1999.

[5] E. Clarke, O. Grumberg, S. Jha, Y. Lu, and H. Veith. Progress on the state explosion problem in model checking. In *Informatics, 10 Years Back, 10 Years Ahead*, volume 2000 of *LNCS*, pages 176–194, 2001.

[6] E. Clarke, O. Grumberg, and D. Long. Model checking and abstraction. In *POPL*, pages 343–354, 1992.

[7] Edmund Clarke, Orna Grumberg, Somesh Jha, Yuan Lu, and Helmut Veith. Counterexample-guided Abstraction Refinement. In *CAV'00*, 2000. Extended Version to appear in J.ACM.

[8] Edmund Clarke, Orna Grumberg, Muralidhar Talupur, and Dong Wang. High level verification of control intensive systems using predicate abstraction. In *MEMOCODE*, 2003.

[9] Edmund Clarke and Helmut Veith. Counterexamples revisited: Principles, algorithms, applications. In *Proc. International Symposium on Verification in Honor of Zohar Manna*, volume 2772 of *LNCS*, 2003.

[10] Edmund M. Clarke, Orna Grumberg, and Doron Peled. *Model Checking*. MIT Press, 1999.

[11] E.M. Clarke and E.A. Emerson. Design and synthesis of synchronization skeletons using branching time temporal logic. In *Proc. Workshop on Logic of Programs, volume 131 of Lect. Notes in Comp. Sci.*, pages 52–71, 1981.

[12] S. Das and D. Dill. Successive approximation of abstract transition relations. In *LICS'01*, 2001.

[13] S. Das, D. Dill, and S. Park. Experience with predicate abstraction. In *CAV'99*, pages 160–171, 1999.

[14] R. P. Kurshan. *Computer-Aided Verification*. Princeton Univ. Press, Princeton, New Jersey, 1994.

[15] K. McMillan. Applying SAT methods in unbounded symbolic model checking. In *CAV'02*, pages 250–264, 2002.

[16] K. Namjoshi and R. Kurshan. Syntactic program transformations for automatic abstraction. In *CAV'00*, 2000.

[17] S. Graf and H. Saidi. Construction of abstract state graphs with PVS. In *CAV'97*, pages 72–83, 1997.

[18] H. Saidi and N. Shankar. Abstract and model check while you prove. In *CAV'99*, pages 443–454, 1999.

[19] Lintao Zhang, Conor F. Madigan, Matthew W. Moskewicz, and Sharad Malik. Efficient conflict driven learning in a Boolean satisfiability solver. In *ICCAD'01*, 2001.

On Boolean Models for Quantified Boolean Horn Formulas

Hans Kleine Büning[1], K. Subramani [*2], and Xishun Zhao[3]

* Department of Computer Science,
Universität Paderborn
33095 Paderborn (Germany)
kbcsl@upb.de
* LDCSEE,
West Virginia University,
Morgantown, WV
ksmani@csee.wvu.edu
* Institute of Logic and Cognition,
Zhongshan University
510275, Guangzhou, (P.R. China)
hsdp08@zsu.edu.cn

Abstract. For a Quantified Boolean Formula (QBF) $\Phi = Q\phi$, an assignment is a function $\mathcal{M}$ that maps each existentially quantified variable of Φ to a Boolean function, where ϕ is a propositional formula and Q is a linear ordering of quantifiers on the variables of Φ. An assignment $\mathcal{M}$ is said to be proper, if for each existentially quantified variable y_i, the associated Boolean function f_i does not depend upon the universally quantified variables whose quantifiers in Q succeed the quantifier of y_i. An assignment $\mathcal{M}$ is said to be a model for Φ, if it is proper and the formula $\phi^{\mathcal{M}}$ is a tautology, where $\phi^{\mathcal{M}}$ is the formula obtained from ϕ by substituting f_i for each existentially quantified variable y_i. We show that any true quantified Horn formula has a Boolean model consisting of monotone monomials and constant functions only; conversely, if a QBF has such a model then it contains a clause–subformula in $QHORN \cap SAT$.

Keywords: quantified Boolean formula, Boolean model, Horn formulas, complexity

1 Introduction

The language of quantified Boolean formulas (QBF) is becoming increasingly important, since QBF act as a natural tool for modelling problems in diverse areas such as planning, abductive reasoning, non-monotonic reasoning and modal logics. On the theoretical front, they represent one of the natural paradigms

* This research was conducted in part at Aalborg Universitet, where the author was supported by a CISS Faculty Fellowship.

E. Giunchiglia and A. Tacchella (Eds.): SAT 2003, LNCS 2919, pp. 93–104, 2004.

for representing the infinite sequence of complexity classes in the polynomial hierarchy [9]. Although the decision problem of finding a model for an arbitrary QBF is PSPACE-complete, there is tremendous interest in designing efficient algorithms for the restricted classes of $QBFs$ that arise in practice.

The naive decision procedure to determine the truth of a QBF formulas is based on generating all valid 0/1 assignments to the variables and testing the truth value of the resultant formula. While this procedure takes exponential time, it is unlikely that a more efficient procedure exists for the general case, although there exist backtracking strategies that are empirically superior to exhaustive testing, over specific application domains [2,6,10,14,16]. In [8], a generalization of simple resolution called Q-resolution was introduced; Q-resolution serves as a refutation procedure for $QBFs$ in exactly the same way as simple resolution does for propositional formulas. Q-resolution also takes exponential time and space in the worst-case; in fact, so does simple resolution over propositional formulas [11].

It is well known that the satisfiability problem is a special case of the evaluating problem for QBF formulas. More precisely, checking whether a propositional formula ϕ over variables $y_1, \cdots, y_m$ is satisfiable is equivalent to checking whether the quantified formula $\exists y_1 \cdots \exists y_m \phi$ is true. Because of the hardness of the satisfiability problem, some tractable subclasses of propositional formulas (such as Horn, 2CNF, q-Horn formulas) has been obtained by syntactically restricting the structure of formulas. Recently, however, some people consider the problem of determining if a propositional formula can be satisfied by some special models (truth assignments). For example, weighted satisfiability problem is to decide whether a CNF formulas has a satisfying truth assignment which sets exactly k (where $k \geq 0$ is a fixed natural number) propositional variables to be true (see [5]). More recently, the matching satisfiability and linearly satisfiability have been investigated and shown to be tractable, which are to determining whether a formula can be satisfied by a matching truth assignment or a linear truth assignment (for details see [13,18]).

For general QBF formulas, several researchers also studied some restrictions of the syntactical structure of the matrices hoping to obtain relatively easier subclasses. However, only very restricted subclasses (e.g., $QHORN$, $Q2$-CNF [7]) are tractable. Inspired by the research of satisfiability problems mentioned in the above paragraph, we would like to consider the problem of determining whether a QBF formula can be evaluated to be true by some specific models. At first we have to introduce Boolean models for QBF formulas. We briefly discuss possible structures for Boolean models of Φ. Let $\Phi = Q(x, y) \phi$ denote an arbitrary QBF, with x denoting the vector of universally quantified variables and y denoting the vector of existentially quantified variables, ϕ denoting a propositional formula and $Q(x, y)$ specifying an arbitrary linear ordering on the variables of $\{x \cup y\}$; When Φ contains no universal variables, that is, Φ is of the form $\exists y_1 \cdots \exists y_m \phi$, a model for Φ is nothing but a truth assignment assigning a truth value to each existential variables y_i. However, the presence of universally quantified variables in Φ, may, in general, forces the value of an existentially quantified variable y_i

to depend on the values of universally quantified variables that precede it in the quantifier string $Q(x, y)$. That is, a model for QBF should assign generally a Boolean function to each existential variable. Please note that every Boolean function can be represented by a propositional formula. We assume in this paper that the Boolean functions are given as propositional formulas. Observe that the existentially quantified variables in a $QBFs$ can be eliminated by Skolemization. We can see that a QBF is true if and only if there is an assignment which maps each existential variable to a Boolean function and for which the generated formula after Skolemization is a tautology.

Consider an assignment $F : Y \to \mathcal{M}$, where Y is the set of existential variables in Φ and $\mathcal{M}$ is a set of functions. We say that F is a proper assignment, if for each existentially quantified variable y_i, $F(y_i)$ is a Boolean function that does not depend upon the universally quantified variables whose quantifiers succeed $\exists y_i$ in $Q(x, y)$. F is said to be a model for Φ, if F is proper and the formula ϕ^F is a tautology, where ϕ^F is the formula obtained from ϕ by substituting $F(y_i) \in \mathcal{M}$ for each y_i. When F is an onto function, we do not distinguish between F and $\mathcal{M}$. Let K be a specific class of Boolean functions. If each function in $\mathcal{M}$ is in the class K, then we say that $\mathcal{M}$ is a K-model for Φ.

$QCNF$ is the class of closed quantified Boolean formulas in prenex form with matrices in CNF. The satisfiability problem or evaluation problem is the question whether an arbitrarily given $QCNF$ formula is true. We denote this class as $QCNF \cap SAT$ or $Eval(QCNF)$.

Let $\mathcal{R}$ be a class of closed QBFs, with $\mathcal{R} \subseteq QCNF$ and let K be a class of Boolean functions. As mentioned before, we are interested in $QCNF$ formulas which can be satisfied by some specific models. More accurately, this paper is concerned with the following issues:

- The K-model problem for $\mathcal{R}$: deciding whether an arbitrary $\mathcal{R}$-formula has a model in K.
- The K-model checking problem for $\mathcal{R}$: deciding whether $\mathcal{M} \subseteq K$ is a K-model for Φ.
- Characterizing the class of quantified Boolean formulas having a K-model.

Observe that when there are no restrictions on K and $\mathcal{R}$, the K-model problem is **PSPACE-complete**. We shall show that the corresponding model-checking problem is **coNP-complete**. We will focus on two types of models: (a) K_1—the class consisting of constant functions and functions which can be characterized by a literal; (b) K_2—the class consisting of constant functions and functions which can be characterized by conjunctions of some variables (literals, respectively). The K_1-model checking problem is solvable in linear time, but the K_1 model problem is **NP-complete**. The class $Q2\text{-}CNF$ can be characterized by K_1-models in the following sense: Every true $Q2\text{-}CNF$ formula has a K_1-model and conversely every formula with a K_1-model contains a true $Q2\text{-}CNF$ subformula with respect to clause-inclusion. The K_2-model checking problem is also tractable, further, K_2 models characterize the class $QHORN$.

2   Preliminaries

2.1   Quantified Boolean Formulas and Models

A literal is a propositional variable or a negated variable. Clauses are disjunctions of literals. A *CNF* formula is a conjunction of clauses. A *QBF* formula Φ is of the form

$$Q_1 x_1 \cdots Q_n x_n \phi,$$

where $Q_i \in \{\forall, \exists\}$ and where ϕ is a propositional formula over variables $x_1, \cdots, x_n$. $Q_1 x_1 \cdots Q_n x_n$ is called the prefix and ϕ the matrix of Φ. Usually, we simply write Φ as $Q\phi$.

A literal x or $\neg x$ is called a universal (respectively existential) literal, if the variable x is bounded by a universal quantifier (respectively existential quantifier). Universally (respectively existentially) quantified variables are also denoted as $\forall$-variables (respectively $\exists$-variables). A clause containing only $\forall$-literals is called universal clause or $\forall$-disjunct. We also consider the empty clause as a $\forall$-disjunct. Observe that if QBF Φ has a $\forall$-disjunct, it is unsatisfiable, i.e., it does not have a model.

QCNF denotes the class of *QBF* formulas with matrices in *CNF*.

Let $\Phi = Q\phi$, and $\Phi' = Q\phi'$ be two formulas in *QCNF* with $\phi = \alpha_1 \wedge \cdots \wedge \alpha_n$ and $\phi' = \alpha_1' \wedge \cdots \wedge \alpha_n'$. Suppose α_i' is a non-empty subclause of α_i for every i. Then we write $\Phi' \subseteq_{cl} \Phi$.

QHORN (*Q2-CNF*, respectively) is the class of formulas in *QCNF* such that every clause is a Horn clause (2-clause, respectively). For a subclass $\mathcal{R}$ of *QBF*, $\mathcal{R} \cap SAT$ is the class of all true formulas in $\mathcal{R}$.

For a *QCNF* formula $\Phi = Q\phi$, a clause α in ϕ is called a $\exists$-unit clause if α contains exactly one $\exists$-literal (besides some $\forall$-literals). If the $\exists$-literal is positive, then we call α a positive $\exists$-unit clause.

We consider a Boolean function f as a propositional formula over variables $x_1, \cdots, x_m$.

Definition 1. *(Ordering of Literals)*
Let $\Phi = Q\phi$ be a quantified Boolean formula and let l_1 and l_2 be literals in Φ. Then l_1 is less than l_2 ($l_1 < l_2$), if the variable of l_1 occurs in Q before the variable of l_2.

Remark: For a *QCNF* formula Φ, clauses are written in the form $(l_1 \vee \cdots \vee l_m)$ such that l_i is not less than l_{i-1}. Suppose in a clause α there is a $\forall$-literal which is not less than any $\exists$-literal in the clause, then removing this $\forall$-literal does not affect the truth of Φ.

Let $\Phi = Q\phi$ be a formula in *QCNF* with existential variables $y_1, \cdots, y_n$ and universal variables $x_1, \cdots, x_m$. Further, let $\mathcal{M} = (f_1, \cdots, f_n)$ be a sequence of functions such that each f_i is a propositional formula over universal variables which are less than y_i according to the order given in Definition 1. Then $\phi^{\mathcal{M}}$ is the propositional formula obtained from ϕ by replacing each occurrence of the existential variable y_i ($i = 1, \cdots, n$) by the function f_i. $\Phi^{\mathcal{M}}$ denotes the formula

$\forall x_1 \cdots \forall x_m \phi^{\mathcal{M}}$. We say $\mathcal{M}$ is a *model* for $Q\phi$ if and only if $\Phi^{\mathcal{M}}$ is true (that is, $\phi^{\mathcal{M}}$ is a tautology).

Obviously, a formula $\Phi \in QCNF$ is true if and only if Φ has a model.

Let K be a class of Boolean functions. Suppose $\mathcal{M} = (f_1, \cdots, f_n)$ is a model for a $\Phi \in QCNF$. We say $\mathcal{M}$ is a K-*model* for Φ if $f_i \in K$ for $1 \le i \le n$.

K-Model Checking Problem: *Instance:* A formula Φ in $QCNF$ and a sequence $\mathcal{M} = (f_1, \cdots, f_n)$ of Boolean functions.
   Query: Is $\mathcal{M}$ a K-model for Φ?
K-Model Problem: *Instance:* A formula Φ in $QCNF$.
   Query: Does Φ have a K-model?

If there is no restriction on the class of Boolean models K, the model problem is easily seen to be **PSPACE-complete** [15], whereas Lemma 1 shows that the model–checking problem is **coNP-complete**.

Lemma 1. *The model checking problem in QBF is* **coNP-complete** *for unrestricted Boolean models K.*

Proof. Let Φ be a formula in $QCNF$ and $\mathcal{M} = (f_1, \cdots, f_n)$ be a sequence of proper functions. Then $\mathcal{M}$ is a model for Φ if and only if $\phi^{\mathcal{M}}$ is a tautology. Since the tautology problem is in **coNP** and $|\phi^{\mathcal{M}}| \le |\phi||\mathcal{M}|$ (here $|X|$ denotes the size of X), the problem lies in **coNP**. The **coNP-hardness** follows from the following reduction of the tautology problem for propositional formulas in 3-*DNF*. We associate $\alpha = \bigvee_{1 \le i \le n}(l_{i1} \wedge l_{i2} \wedge l_{i3})$, where the l_{ij} are literals over the variables $x_1, \cdots, x_m$ with the formula

$$\Phi = \forall x_1 \cdots \forall x_m \exists y_1 \cdots \exists y_n \phi \text{ with } \phi = (y_1 \vee \cdots \vee y_n).$$

For each y_i, define $f_{y_i}(x_1, \cdots, x_m) = (l_{i1} \wedge l_{i2} \wedge l_{i3})$ and $\mathcal{M} = (f_{y_1}, \cdots, f_{y_n})$. Clearly, $\phi^{\mathcal{M}}$ is α. Thus, α is a tautology if and only if $\mathcal{M}$ is a model for Φ. $\square$

2.2   Q–Pos–Unit–Resolution

For some classes of Boolean functions and subclasses of $QCNF$ we are interested in, the so called Q-pos-unit-resolution plays an important role. The Q-pos-unit-resolution is a Q-resolution, for which one of the parent clauses is a positive $\exists$-unit-clause. We recall the definition and some known facts [12]. At first we introduce an ordering of literals which will be used in order to simplify the representations. Subsequently, we denote universal variables by $x_1 \cdots, x_m$ and existential variables by $y_1, \cdots, y_n$.

Definition 2. *(Q–pos–unit–resolution)*
Let $\Phi = Q\phi$ be a formula in $QCNF$ with matrix ϕ and prefix Q. Let α_1 and α_2 be non–tautological clauses of Φ, where α_1 is a positive $\exists$–unit clause with $\exists$-literal y, and α_2 an arbitrary clause containing the literal $\neg y$.

   We obtain a Q–pos–unit–resolution resolvent σ from α_1 and α_2 by applying steps 1-4:

1. *Eliminate the occurrence of y from α_1 and the occurrence of $\neg y$ from α_2. We get α_1' and α_2'.*
2. *$\sigma' = \alpha_1' \vee \alpha_2'$.*
3. *If σ' is a tautological clause then no resolvent exists.*
4. *Remove from σ' each $\forall$-literal l such that there is in σ' no $\exists$-literal l' with $l < l'$. The resulting clause is called σ (clearly, if σ' is a $\forall$-clause then σ is the empty clause).*

We write $\Phi \vdash_{\overline{Q\text{-}pos\text{-}unit\text{-}res}}^{1} \sigma$. For convenience, we shall consider each clause α in ϕ as derivable by zero resolution steps.

Usually we add the resolvent σ to $\Phi = Q\phi$. Then $\Phi' = Q(\phi \wedge \sigma)$ is true if and only if Φ is true. For a clause σ, we write $\Phi \vdash_{\overline{Q\text{-}pos\text{-}unit\text{-}res}} \sigma$, if there is a sequence (maybe empty) of Q–pos–unit–resolution steps leading to the clause σ. We demonstrate the behavior by an example.

$$\Phi = \forall x_0 \forall x_1 \exists y_1 \forall x_2 \forall x_3 \exists y_2 (x_1 \vee y_1) \wedge (\neg x_1 \vee \neg y_1 \vee x_3 \vee \neg y_2) \wedge (x_0 \vee \neg y_1) \wedge (x_2 \vee y_2)$$

Then resolve the second clause and the fourth clause we get

$$\Phi \vdash_{\overline{Q\text{-}pos\text{-}unit\text{-}res}}^{1} (\neg x_1 \vee \neg y_1).$$

Resolve the first clause and the third clause we get

$$\Phi \vdash_{\overline{Q\text{-}pos\text{-}unit\text{-}res}}^{1} \sqcup.$$

Please note according to Definition 2 that the first and the second clause cannot be resolved, since there is the complementary pair of literals $x_1, \neg x_1$ in the intermediate clauses.

The following theorem can be found in [12]

Theorem 1. *[12]*
1) For $\Phi \in QHORN : \Phi$ is unsatisfiable $\Longleftrightarrow \Phi \vdash_{\overline{Q\text{-}pos\text{-}unit\text{-}res}} \sqcup$.
2) Let $\mathcal{M}$ be a model for $\Phi = Q\phi \in QCNF$. If $\Phi \vdash_{\overline{Q\text{-}pos\text{-}unit\text{-}res}} \alpha$, then $\mathcal{M}$ is a model for $\Phi' = Q(\phi \wedge \alpha)$.

3   QHORN and K_2-models

QHORN is a proper subclass of QCNF for which the satisfiability problem is solvable in polynomial time [7,8,12]. A simple example of a true QHORN formula is the formula $\Phi = \forall x_1 \forall x_2 \exists y (\neg x_1 \vee \neg x_2 \vee y) \wedge (x_1 \vee \neg y) \wedge (x_2 \vee \neg y)$. A model for Φ is the function $f(x_1, x_2) = x_1 \wedge x_2$. Obviously, no function depending on at most one variable can be a model for Φ (This can be verified by exhaustive testing.). More generally, we will show that QHORN formulas have models consisting of monotone monomials and the constant functions 0 or 1 only.

Definition 3.
$$K_2 := \{ f \mid \exists I \subseteq \{1, \cdots, n\}, f(x_1, ..., x_n) = \bigwedge_{i \in I} x_i, \ n \in \mathbf{N} \} \cup \{0, 1\}$$

Theorem 2. *1. The K_2–model checking problem is decidable in quadratic time.*

2. The K_2–model problem is NP-complete.

3. Any formula $\Phi \in QHORN \cap SAT$ has a K_2–model and a K_2–model can be computed in polynomial time.

4. A formula $\Phi \in QCNF$ has a K_2–model if and only if there exists some $\Phi' \subseteq_{cl} \Phi : \Phi' \in QHORN \cap SAT$.

We prove Theorem 2, through the following 4 lemmas.

Lemma 2. *The K_2-model checking problem is decidable in quadratic time.*

Proof. Let $\Phi = Q\phi \in QCNF$ be a formula with $\exists$-variables $y_1, \cdots, y_n$ and $\forall$-variables $x_1, \cdots, x_m$. Let $\mathcal{M}$ be a sequence of K_2-functions $(f_{y_1}, \cdots, f_{y_n})$. To decide whether $\mathcal{M}$ is a model for Φ, it suffices to check whether the formula $\alpha^{\mathcal{M}}$ is a tautology, for every clause $\alpha \in \phi$. Rather than checking whether the formula $\alpha^{\mathcal{M}}$ is a tautology, we focus on checking whether its complement is unsatisfiable. Let

$$\alpha = (w_1 \vee y_1^{\epsilon_1} \vee \cdots \vee w_r \vee y_r^{\epsilon_r})$$

be a clause in Φ, where w_j is a disjunction of universal literals and $\epsilon_j \in \{0,1\}$. ($y^1 := y$ and $y^0 := \neg y$) Then we get

$$\alpha^{\mathcal{M}} = (w_1 \vee f_{y_1}^{\epsilon_1} \vee \cdots \vee w_r \vee f_{y_r}^{\epsilon_r})$$

and the complement is

$$\neg\alpha^{\mathcal{M}} = (\neg w_1 \wedge f_{y_1}^{1-\epsilon_1} \wedge \cdots \wedge \neg w_r \wedge f_{y_r}^{1-\epsilon_r}).$$

If f_{y_i} is not a constant then $f_{y_i} = \bigwedge x_j$ and therefore $f_{y_i}^0 = \bigvee \neg x_j$. That means, $\neg\alpha^{\mathcal{M}}$ is a *CNF* formula consisting of unit-clauses and some clauses containing only negative literals. For such formulas the satisfiability problem can be solved in linear time by means of the propositional unit resolution. The length of $\alpha^{\mathcal{M}}$ is bound by $|\alpha||\Phi|$. In other words, we have an algorithm that decides the K_2-model checking problem in quadratic time.   □

Lemma 3. *The K_2-model problem is* NP-complete *for an arbitrary QBF Φ.*

Proof. The problem belongs to NP, because the size of any model function is bound by the length of the prefix and because the K_2-model checking problem is solvable in quadratic time. We can choose randomly some $\mathcal{M} = (f_{y_1}, \cdots, f_{y_n})$ in K_2, and decide whether $\mathcal{M}$ is a model of Φ. The NP-hardness follows from the following observation: Let ϕ be a propositional *CNF*-formula over the variables $y_1, \cdots, y_n$. Then ϕ is satisfiable if and only if $\exists y_1 \cdots \exists y_n : \phi$ has a K_2-model. The model functions must be the constants 1 or 0.   □

Lemma 4. *Any formula $\Phi \in QHORN \cap SAT$ has a K_2–model and a K_2–model can be computed in polynomial time.*

Proof. Let $\Phi = Q\phi \in QHORN \cap SAT$ be a formula with universal variables $x_1, ..., x_m$ and existential variables $y_1, ..., y_n$. Now we construct a K_2-model $\mathcal{M}$ for Φ. We define

$$U := \{y \in \{y_1, \cdots, y_n\} : \Phi \vdash_{\overline{Q\text{-}pos\text{-}unit\text{-}res}} (v \vee y) \text{ for some } \forall\text{-disjunction } v\}$$

and for each $y \in U$ we define

$$W(y) := \{w \mid w \text{ is a } \forall\text{-disjunction}, \Phi \vdash_{\overline{Q\text{-}pos\text{-}unit\text{-}res}} (w \vee y)\}.$$

Further, we define $V(y) := \bigcap_{w \in W(y)} w$.

The v's and w's can be empty. If not, then they consists of negative literals only.

Let $\mathcal{M} = (f_{y_1}, \cdots, f_{y_n})$ be defined as follows:
For each $y \in \{y_1, \cdots, y_n\}$ (suppose $x_1, \cdots, x_r$ are the universal variables less than y), define

$$f_y(x_1, ..., x_r) := \begin{cases} 0, & \text{if } y \notin U \\ \neg V(y), & \text{if } y \in U, V(y) \text{ is not empty} \\ 1, & \text{if } y \in U, V(y) \text{ is empty} \end{cases}$$

We now show that $\mathcal{M} = (f_{y_1}, ..., f_{y_n})$ is a model for Φ by proving that for any clause α of ϕ the formula $\alpha^{\mathcal{M}}$ is a tautology.

Suppose there is a non-tautological formula $\alpha^{\mathcal{M}}$. We proceed by enumerating the individual cases.

Case 1 (negative clause): $\alpha = (w_1 \vee \neg y_1 \vee \cdots \vee w_r \vee \neg y_r)$, where w_j is a $\forall$–disjunction of negative literals.

Since $\alpha^{\mathcal{M}}$ is not a tautology, f_{y_j} is not the constant 0 for every j ($1 \leq j \leq r$). Then we have $f_{y_j} = 1$ or $f_{y_j} = \neg V(y_j)$, and therefore $y_j \in U$ for every j. That implies that there are v_j with $\Phi \vdash_{\overline{Q\text{-}pos\text{-}unit\text{-}res}} (v_j \vee y_j)$. Note that w_j and v_j consists of negative literals. Applying Q-pos-unit-resolution to α and the $\exists$-unit-clauses $(v_j \vee y_j)$ leads to the empty clause, which contradicts our assumption that Φ is true. It follows that $\alpha^{\mathcal{M}}$ is a tautology.

Case 2 (positive $\exists$-literal y_r):
$\alpha = (\bigvee_{1 \leq j \leq r-1}(w_j \vee \neg y_j) \vee w_r \vee y_r \vee \bigvee_{r+1 \leq j \leq s}(w_j \vee \neg y_j))$.

Since by our assumption $\alpha^{\mathcal{M}}$ is not a tautology, we get for all $j \neq r$ that f_{y_j} is not the constant 0 and f_{y_r} is not the constant 1. Then for all negative literals $\neg y_j$ we have $y_j \in U$. That means there exists some v_j with $\Phi \vdash_{\overline{Q\text{-}pos\text{-}unit\text{-}res}} (v_j \vee y_j)$. By means of the Q-pos-unit-resolution we get $\Phi \vdash_{\overline{Q\text{-}pos\text{-}unit\text{-}res}} (w \vee y_r)$ for some w and therefore $y_r \in U$.

Then we know that $f_{y_r} = \neg V(y_r)$, since $f_{y_r} \neq 1$. W.l.o.g. we suppose $V(y_r) = (\neg x_1 \vee \cdots \vee \neg x_t)$. Then we have

$$\alpha^{\mathcal{M}} = \left(\bigvee_{1 \leq j \leq r-1}(w_j \vee \neg f_{y_j}) \vee w_r \vee \bigwedge_{1 \leq i \leq t} x_i \vee \bigvee_{r+1 \leq j \leq s}(w_j \vee \neg f_{y_j}) \right)$$

where for $j \neq r$ we have $f_{y_j} = 1$ ($W(y_j)$ is not empty, but $\bigcap_{w \in W(y_j)} w$ is empty) or $f_{y_j} = \neg V(y_j)$.

We recall that $\alpha^{\mathcal{M}}$ is not a tautology by our assumption. Therefore, w.l.o.g we suppose that for some i ($1 \leq i \leq t$) $\neg x_i \notin V(y_j)$ for all $j \neq r$ and $\neg x_i \notin w_j$ for all j.

This implies, that for any $j \neq r$, $\Phi \vdash_{\overline{Q\text{-}pos\text{-}unit\text{-}res}} (v_j \vee y_j)$ for some $\forall$-disjunction v_j such that $\neg x_i \notin v_j$.

Now we can resolve α with $(v_j \vee y_j)$ for all $j \neq r$ and obtain a clause $(w_0 \vee y_r)$, for which $\neg x_i \notin w_0$. That is a contradiction to $\neg x_i \in V(y_r)$. Therefore, for any i we have $\neg x_i \in V(y_j)$ or $\neg x_i \in w_j$ for some $j \neq r$.

Then in $\alpha^{\mathcal{M}}$ for all i ($1 \leq i \leq t$) the literal $\neg x_i$ occurs in some w_j or in $\neg f_{y_j}$ for some $j \neq r$. Thus, $\alpha^{\mathcal{M}}$ is a tautology because of $f_{y_r} = \bigwedge_{1 \leq i \leq t} x_i$.

Case 3 (positive $\forall$-literal x):

$$\alpha = \left(\bigvee_{1 \leq j \leq r-1} (w_j \vee \neg y_j) \vee w_r \vee x \vee \neg y_r \vee \bigvee_{r+1 \leq j \leq s} (w_j \vee \neg y_j) \right)$$

By our assumption f_{y_j} is not the constant 0 for every ($1 \leq j \leq s$). Then we have $f_{y_j} = 1$ or $f_{y_j} = \neg V(y_j)$ and therefore $y_j \in U$ for every j.

Suppose $f_{y_j} = 1$. Then $V(y_j)$ is empty. Therefore there exists some v_j with $\neg x \notin v_j$ and $\Phi \vdash_{\overline{Q\text{-}pos\text{-}unit\text{-}res}} (v_j \vee y_j)$. Now we can resolve $(v_j \vee y_j)$ with α removing the negative literal $\neg y_j$. The resolution may add some new negative universal literals different from $\neg x$ to the clause, but that has no effect to whether the formula $\alpha^{\mathcal{M}}$ is a tautology. Therefore, w.l.o.g. we assume that every function f_{y_j} is not a constant.

Suppose for every j we have $\neg x \notin V(y_j)$. Then there are w_j^* for which $\Phi \vdash_{\overline{Q\text{-}pos\text{-}unit\text{-}res}} (w_j^* \vee y_j)$ and $\neg x \notin w_j^*$. Then we resolve the clause α with these unit clauses. We get the empty clause, in contradiction to the fact that Φ is true.

Hence there is some j for which $\neg x$ is in $V(y_j)$. Since $f_{y_j} = \neg V(y_j)$ (therefore $\neg f_{y_j} = V(y_j)$), we see that $\alpha^{\mathcal{M}}$ is a tautology, in contradiction to our assumption. That completes the proof. $\square$

A brute force method to compute a K_2–model for $\Phi \in QHORN \cap SAT$ is to determine $V(y) := \bigcap_{w \in W(y)} w$, where $W(y) := \{w \mid \Phi \vdash_{\overline{Q\text{-}pos\text{-}unit\text{-}res}} (w \vee y)\}$. It is known that $QHORN \cap SAT$ can be solved in time $O(n|\Phi|)$, where n is the number of $\forall$–variables occurring positively in Φ. The problem of constructing a K_2–model for $\Phi \in QHORN \cap SAT$ in quadratic time is currently open.

Corollary 1. *Any formula $\Phi \in Q2\text{-}HORN \cap SAT$ has a model consisting of constant functions and/or functions of the form $f(x_1, \cdots, x_n) = x_i$ for some i.*

Proof. We suppose $\Phi \in Q2\text{-}HORN \cap SAT$. Since Φ is true and Q-pos-unit-resolution applied with Φ generates only 2-clauses, no clause with more than one $\forall$-literal can be derived. In case of two $\forall$-literals the formula would be false. In our construction of the functions f_y in the proof of Lemma 4, the function f_y is not a constant, if $V(y)$ is not empty. Since $V(y) = \bigcap_{w \in W(y)} w$, where $w \in W(y)$ if $\Phi \vdash_{Q\text{-}pos\text{-}unit\text{-}res} (w \vee y)$, the set $V(y)$ consists of at most one literal. Hence, we have for non–empty $V(y)$ that $f_y(x_1, \cdots, x_n) = x_i$ for some variable x_i. $\square$

Lemma 5. *A formula $\Phi \in QCNF$ has a K_2-model if and only if there is some $\Phi' \subseteq_{cl} \Phi : \Phi' \in QHORN \cap SAT$.*

Proof. The direction from right to left follows from Lemma 4. For the other direction let $\Phi \in QCNF$ be a formula with $\exists$-variables $y_1, ..., y_n$.
We suppose that $\mathcal{M} = (f_1, \cdots, f_m)$ is a K_2-model for Φ. Let α be a clause in Φ.

At first we simplify the formula by investigating the constant functions of $\mathcal{M}$. For constant functions f_i we define $\alpha' = y_i$ resp. $\neg y_i$, if $f_i = 1$ and $y_i \in \alpha$ resp. $f_i = 0$ and $\neg y_i \in \alpha$. Clearly, α' is a Horn clause and $\alpha'^{\mathcal{M}}$ is a tautology. Furthermore, we remove all occurrences of the literal $\neg \alpha'$ from all clauses. Obviously, $\mathcal{M}$ is also a K_2-model for the resulting formula. Based on this simplification, we assume w.l.o.g. that $\mathcal{M}$ contains no constant functions. Then we have $f_i = \bigwedge_{j \in I_i} x_j$ for some $I_i \subseteq \{1, \cdots, n\}$.
W.l.o.g. let $\alpha = (w_1 \vee y_1 \vee \cdots \vee w_r \vee y_r \vee w_{r+1} \vee \neg y_{r+1} \vee \cdots \vee w_{r+s} \vee \neg y_{r+s})$ be a clause in Φ. Then we get
$$\alpha^{\mathcal{M}} = (w_1 \vee f_1 \vee \cdots \vee w_r \vee f_r \vee w_{r+1} \vee \neg f_{r+1} \vee \cdots \vee w_{r+s} \vee \neg f_{r+s}) =$$
$$(w_1 \vee \bigwedge_{j \in I_1} x_j \vee \cdots \vee w_r \vee \bigwedge_{j \in I_r} x_j \vee w_{r+1} \vee \bigvee_{j \in I_{r+1}} \neg x_j \vee \cdots \vee w_{r+s}$$
$$\vee \bigvee_{j \in I_{r+s}} \neg x_j).$$

The formula $\alpha^{\mathcal{M}}$ is a tautology if and only if

(1) one positive $\forall$-literal x in $w = w_1 \vee \cdots \vee w_{r+s}$ is the complement of a literal in some $\neg f_q = \bigvee_{j \in I_q} \neg x_j$. Then $\alpha' = (x \vee \neg y_q)$ is the desired Horn clause. Or,

(2) for one $f_q = \bigwedge_{j \in I_q} x_j$ the complementary literals $\neg x_j$ occur in w or in one or more $\neg f_k$. In that case let α' be the clause containing the literal y_q and all negative literals of α. Again α' is a Horn clause.

Now any clause α' is a Horn clause and $\alpha'^{\mathcal{M}}$ is a tautology by the construction of α'. That proves our desired result. $\square$

4   Q2-CNF and K_1-models

In this section we investigate models for $Q2\text{-}CNF \cap SAT$. Please note that $Q2\text{-}CNF \cap SAT$ is solvable in linear time [1,17]. For every $\Phi \in Q2\text{-}CNF \cap SAT$ there is a renaming, such that the resulting formula is in $Q2\text{-}HORN \cap SAT$. Since formulas in this class have models consisting of constant functions or functions of the form $f(x_1, \cdots, x_n) = x_i$, we introduce the following class of formulas.

Definition 4. $K_1 := \{f \mid \exists i \leq n \exists \epsilon \in \{0, 1\} : f(x_1, ..., x_n) = x_i^\epsilon, n \in \mathbf{N}\} \cup \{0, 1\}$

Theorem 3. *1. The K_1-model checking problem is solvable in linear time and the K_1-model problem is* NP-complete.
2. *Any formula $\Phi \in Q2$-$CNF \cap SAT$ has a K_1-model.*
3. *A formula $\Phi \in QCNF$ has a K_1-model if and only if there is some $\Phi' \subseteq_{cl} \Phi : \Phi' \in Q2$-$CNF \cap SAT$.*

Proof. **Ad 1**: $\mathcal{M}$ is a model for $\Phi = Q\phi$ if and only if for each clause α in Φ the propositional formula $\alpha^{\mathcal{M}}$ is a tautology. Please notice that $\alpha^{\mathcal{M}}$ is a clause for any clause α, if $\mathcal{M}$ consists of K_1-functions. Thus, $\alpha^{\mathcal{M}}$ is a tautology if and only if $\phi^{\mathcal{M}}$ contains the constant 1 or a complementary pair of literals. Hence, whether $\mathcal{M}$ is a K_1-model for Φ is decidable in linear time.

The K_1-model problem lies in NP, because the K_1-model checking problem is solvable in linear time. The NP-hardness will be shown by a reduction from SAT . Let ϕ be a CNF-formula with variables $y_1, \cdots, y_n$. Then ϕ is satisfiable if and only if $\exists y_1 \cdots \exists y_n \phi$ is true for some constant functions $f_{y_i} = 1$ and $f_{y_j} = 0$.

Ad 2 , 3: Part 2 and part 3 follow from the decision procedure established in [1] for $Q2$-$CNF \cap SAT$ or from the transformation into $Q2$-$HORN \cap SAT$.
□

5   Conclusions and Future Work

In this paper, we have introduced the notion of Boolean models. A Boolean model is a mapping (or, a sequence) $\mathcal{M}$ which associate each existential variable to a Boolean function over some universal variables. $\mathcal{M}$ is a model for a $QCNF$ formula $\Phi = Q\phi$ if the formula $\phi^{\mathcal{M}}$ is a tautology, where $\phi^{\mathcal{M}}$ is obtained from ϕ by replacing each existential variable by its corresponding function. Φ is true if and only if Φ has a Boolean model. This result enables us to investigate quantified Boolean formulas via Boolean models theory. From the implementation of view, we are interested in the model checking problem and the existence of models. Since $QCNF$ is PSPACE-complete, both problems are intractable. Then we restrict the two problems to some simple models. Let K be a class of some Boolean functions, a model $\mathcal{M}$ is a K-model if each function of $\mathcal{M}$ is in K. We have studied two classes (K_1 and K_2) of models. And we have shown that in some cases the above two decision problems becomes tractable or has lower complexity. In addition, we have constructed interconnection between some subclasses of $QCNF$ formulas and some classes of models. For example, every true $QHORN$ formula has a K_2-model, and conversely, every formula with a K_2-model contains a true subformula in $QHORN$ with respect to clause inclusion.

This paper is just the first step to explore the Boolean formulas by studying Boolean model theory. Several issues remains for further work. We will make a concentrated effort to find new natural classes of models with richer expressive capability but the complexity does not increase (with respect to polynomial reduction). For example, we are interested in the class K_3 of Boolean functions which can be represented by conjunctions of literals. Another important issue is to investigate the relationship between different classes of models. For example, we plan to establish dual theorems and renaming theorems about Boolean models.

References

1. B. Aspvall, B., M. F. Plass, M. F., and Tarjan, R. E.: A Linear-Time Algorithm for Testing the Truth of Certain Quantified Boolean Formulas, *Information Processing Letters*, **8** (1979), pp. 121-123
2. Cadoli, M., Schaerf, M., Giovanardi, A., and Giovanardi, M.: An Algorithm to Evaluate Quantified Boolean Formulas and its Evaluation, In: *highlights of Satisfiability Research in the Year 2000*, IOS Press, 2000.
3. Cook, S., Soltys, M.: Boolean Programs and Quantified Propositional Proof Systems, *Bulletin of the Section of Logic*, **28** (1999), 119-129.
4. Davis, M., Logemann, G., and Loveland, D.: A Machine Program for Theorem Proving, *Communications of the ACM*, **5** (1962), 394-397.
5. Downey, R., Fellows, M.: *Parameterized Complexity*, Springer, New York, 1999.
6. Feldmann, R., Monien, B., and Schambergers, S.: A Distributed Algorithm to Evaluate Quantified Boolean Formulas, In: *proceedings of AAAI*, 2000.
7. Flögel, A.: *Resolution für quantifizierte Bool'sche Formeln*, Ph. D thesis, 1993, Universität Paderborn.
8. Flögel, A., Karpinski, M., and Kleine Büning, H.: Resolution for Quantified Boolean Formulas, *Information and Computation* **117** (1995), pp. 12-18
9. Garey, M.R., Johnson, D.S.: *Computers and Intractability: A Guide to the Theory of NP-Completeness*. W.H. Freeman Company, San Francisco, 1979.
10. Giunchiglia, E., Narizzano, M., and Tacchella, A.: QuBE: A System for Deciding Quantified Boolean Formulas, In: *Proceedings of IJCAR*, Siena, 2001.
11. A. Haken, The Intractability of Resolution, *Theoretical Computer Science*, **39** (1985), 297-308.
12. Kleine Büning, H., Lettmann, T.: *Propositional Logic: Deduction and Algorithms*, Cambridge University Press, 1999
13. Kullmann, O.: Lean Clause-Sets: Generalizations of Minimal Unsatisfiable Clause-Sets, to appear in *Discrete Applied Mathematics*, 2003.
14. R. Letz, Advances in Decision Procedure for Quantified Boolean Formulas, In: *Proceedings of IJCAR*, Siena, 2001.
15. Papadimitriou, C. H.: *Computational Complexity*, Addison-Wesley, New York, 1994.
16. Rintanen, J.T.: Improvements to the Evaluation of Quantified Boolean Formulae, In: *Proceedings of IJCAI*, 1999.
17. Schaefer, T.J.: The Complexity of Satisfiability Problem, In: *Proceedings of the 10th Annual ACM Symposium on Theory of Computing (ed. A. Aho)*, 216-226, New York City, ACM Press, 1978.
18. Szeider, S.: Generalization of Matched CNF Formulas, In: *5th International Symposium on Theory and Applications of Satisfiability Testing*, 2002, Cincinnati, Ohio, USA.

Local Search on SAT-encoded Colouring Problems

Steven Prestwich

Cork Constraint Computation Centre
Department of Computer Science
University College, Cork, Ireland
s.prestwich@cs.ucc.ie

Abstract. Constraint satisfaction problems can be SAT-encoded in more than one way, and the choice of encoding can be as important as the choice of search algorithm. Theoretical results are few but experimental comparisons have been made between encodings, using both local and backtrack search algorithms. This paper compares local search performance on seven encodings of graph colouring benchmarks. Two of the encodings are new and one of them gives generally better results than known encodings. We also find better results than expected for two variants of the log encoding, and surprisingly poor results for the support encoding.

1 Introduction

Encoding problems as Boolean satisfiability (SAT) and solving them with very efficient SAT algorithms has recently caused considerable interest. In particular, local search algorithms have given impressive results on many problems. Unfortunately there are several ways of SAT-encoding constraint satisfaction (CSP) and other problems, and few guidelines on how to choose among them. This aspect of problem modelling, like most others, is currently more of an art than a science, yet the choice of encoding can be as important as the choice of search algorithm. It would be very useful to have some guidelines on how to choose an encoding. Experimental comparisons hint at such guidelines; for example the log encoding is regarded as inferior to the direct encoding, and a variant of the direct encoding is regarded as superior for local search (these encodings are defined below). But such comparisons are few and some encodings have not been compared.

This paper compares local search performance on seven encodings of graph colouring, in order to explore further what makes a good encoding. The remainder of this section reviews known encodings, discusses previous results, and makes some conjectures on the effects of different encodings on local search cost. Section 2 introduces a family of new encodings in order to test a conjecture: that encodings with greater solution density are easier to solve by local search. Section 3 compares local search costs on a set of graph colouring benchmarks under seven encodings. Section 4 discusses the results and future work.

E. Giunchiglia and A. Tacchella (Eds.): SAT 2003, LNCS 2919, pp. 105–119, 2004.

1.1 A Survey of SAT Encodings

We first review the best-known encodings. For simplicity we describe encodings of binary CSPs only, but they all generalize to non-binary problems. To illustrate them we use a simple graph colouring problem as a running example, with two adjacent vertices v and w and three available colours $\{0, 1, 2\}$. We also use graph colouring for our experiments below. Though arguably not a typical CSP it has the advantage of being well-studied as a SAT problem, and many benchmarks are available. The problem is defined as follows. A graph $G = (V, E)$ consists of a set V of vertices and a set E of edges between vertices. Two vertices connected by an edge are said to be *adjacent*. The aim is to assign a colour to each vertex in such a way that no two adjacent vertices have the same colour. The task of finding a k-colouring for a given graph can be modeled as a CSP in which each variable represents a vertex, each value a colour, and variables representing adjacent vertices are constrained by disequalities ($\neq$).

The Direct Encoding The most natural and widely-used encoding is called the *direct encoding* by Walsh [23]. A SAT variable x_{vi} is defined as true if and only if the CSP variable v has the domain value i assigned to it (the symbols i and j denote values in $\mathrm{dom}(v) = \{0, \ldots, d - 1\}$). The direct encoding consists of three sets of clauses. Each CSP variable must take at least one domain value, expressed by *at-least-one* clauses:

$$x_{v0} \vee x_{v2} \vee \ldots \vee x_{v\,d-1}$$

No CSP variable can take more than one domain value, expressed by *at-most-one* clauses:

$$\bar{x}_{vi} \vee \bar{x}_{vj}$$

Conflicts are enumerated by *conflict* clauses:

$$\bar{x}_{vi} \vee \bar{x}_{wj}$$

Unit propagation on a direct encoding is equivalent to forward checking on the CSP [7,23]. The direct encoding of the running example is shown in Figure 1.

$$x_{v\cdot} \vee x_{v\cdot} \vee x_{v\cdot} \qquad\qquad x_{w\cdot} \vee x_{w\cdot} \vee x_{w\cdot}$$

$$\bar{x}_{v\cdot} \vee \bar{x}_{v\cdot} \qquad \bar{x}_{v\cdot} \vee \bar{x}_{v\cdot} \qquad \bar{x}_{v\cdot} \vee \bar{x}_{v\cdot}$$
$$\bar{x}_{w\cdot} \vee \bar{x}_{w\cdot} \qquad \bar{x}_{w\cdot} \vee \bar{x}_{w\cdot} \qquad \bar{x}_{w\cdot} \vee \bar{x}_{w\cdot}$$

$$\bar{x}_{v\cdot} \vee \bar{x}_{w\cdot} \qquad \bar{x}_{v\cdot} \vee \bar{x}_{w\cdot} \qquad \bar{x}_{v\cdot} \vee \bar{x}_{w\cdot}$$

Fig. 1. Direct encoding example

The Multivalued Encoding The direct encoding has a variant in which the at-most-one clauses are omitted, allowing each CSP variable to be assigned more than one value simultaneously. A solution to the CSP can be recovered from a SAT solution by selecting any one of the assigned values for each CSP variable. This variant is well-known, but it does not appear to have a unique name and is often referred to as the direct encoding. To distinguish between the two variants we shall call it the *multivalued encoding*. The multivalued encoding of the running example is shown in Figure 2.

$$x_{v^\bullet} \vee x_{v^\bullet} \vee x_{v^\bullet} \qquad x_{w^\bullet} \vee x_{w^\bullet} \vee x_{w^\bullet}$$

$$\bar{x}_{v^\bullet} \vee \bar{x}_{w^\bullet} \qquad \bar{x}_{v^\bullet} \vee \bar{x}_{w^\bullet} \qquad \bar{x}_{v^\bullet} \vee \bar{x}_{w^\bullet}$$

Fig. 2. Multivalued encoding example

The Log Encoding Iwama & Miyazaki [12] define an encoding for Hamiltonian circuit, clique and colouring problems. Walsh [23] terms this the *log encoding* (the name used in this paper) and shows that unit propagation on the log encoding is less powerful than on the direct encoding. Hoos [10] calls it the *compact encoding* and applies it to Hamiltonian circuit problems. Ernst, Millstein & Weld [4] apply it to planning problems and call it the *bitwise representation*. van Gelder [6] uses it for graph colouring, calls it a *circuit-based encoding*, and describes a variant with auxiliary variables.

The aim of the log encoding is to generate models with fewer SAT variables (logarithmic instead of linear in the CSP domain sizes). Each CSP variable/domain value bit has a corresponding SAT variable, and each conflict has a corresponding clause prohibiting that combination of bits in the specified CSP variables. We define variables x_v^b where $x_v^b = 1$ if and only if bit b of the domain value assigned to v is 1. There are no at-least-one or at-most-one clauses. If $|\text{dom}(v)|$ is a power of 2 then we need only modified conflict clauses. For domains that are not a power of 2 we add clauses to prohibit combinations of bits representing non-domain values, which we shall call *prohibited-value* clauses. The log encoding of the running example is shown in Figure 3, including two prohibited-value clauses to prevent the use of colour 3. (In this example it is unimportant which bit is the most significant.)

$$x_v^\bullet \vee x_v^\bullet \vee x_w^\bullet \vee x_w^\bullet \qquad \bar{x}_v^\bullet \vee x_v^\bullet \vee \bar{x}_w^\bullet \vee x_w^\bullet \qquad x_w^\bullet \vee \bar{x}_w^\bullet \vee x_w^\bullet \vee \bar{x}_w^\bullet$$

$$\bar{x}_v^\bullet \vee \bar{x}_v^\bullet \qquad \bar{x}_w^\bullet \vee \bar{x}_w^\bullet$$

Fig. 3. Log encoding example

The Binary Encoding Frisch & Peugniez [5] describe a variant of the log encoding called the *binary transform* (and also a hybrid of the binary transform and the direct encoding), which for the sake of uniformity we shall call the *binary encoding*. It generates no surplus bit combinations when the domain size is not a power of 2. Instead all bit combinations are allowed, each value not in the domain is interpreted as a value that is, and extra conflict clauses are added for the extra values. For example if B bits are needed for a domain then we could interpret any domain value $i \geq 2^{B-1}$ as the value $i - |\text{dom}(v)|$. No prohibited-value clauses are required. The binary encoding of the running example is shown in Figure 4; the value 3 is allowed but interpreted as the real domain value 0. (As with the log encoding, in this example it is unimportant which bit is the most significant.)

$$x_v^{\bullet} \vee x_v^{\bullet} \vee x_w^{\bullet} \vee x_w^{\bullet} \qquad \bar{x}_v^{\bullet} \vee \bar{x}_v^{\bullet} \vee \bar{x}_w^{\bullet} \vee \bar{x}_w^{\bullet}$$
$$\bar{x}_v^{\bullet} \vee x_v^{\bullet} \vee \bar{x}_w^{\bullet} \vee x_w^{\bullet} \qquad x_v^{\bullet} \vee x_v^{\bullet} \vee \bar{x}_w^{\bullet} \vee \bar{x}_w^{\bullet}$$
$$x_v^{\bullet} \vee \bar{x}_v^{\bullet} \vee x_w^{\bullet} \vee \bar{x}_w^{\bullet} \qquad \bar{x}_v^{\bullet} \vee \bar{x}_v^{\bullet} \vee x_w^{\bullet} \vee x_w^{\bullet}$$

Fig. 4. Binary encoding example

A drawback pointed out by Frisch & Peugniez is that the binary encoding may generate exponentially more clauses than the log encoding. However, it has the interesting property that CSP solutions may now have more than one SAT representation, and it can be seen as a bitwise analogue of the multivalued encoding. More on this point below.

The Support Encoding Gent [8] revives and studies the *support encoding* of Kasif [13] which encodes support instead of conflict. The support encoding consists of the at-least-one and at-most-one clauses from the direct encoding, plus the following *support* clauses. If $i_1 \ldots i_k$ are the supporting values in the domain of CSP variable v for value j in the domain of CSP variable w, then add a clause

$$x_{vi_1} \vee \ldots \vee x_{vi_k} \vee \bar{x}_{wj}$$

Unit propagation on a support encoding maintains arc-consistency in the CSP [8], and causes DPLL [3] (the standard SAT backtracker) to perform an equivalent search to the MAC algorithm [20]. Gent also shows that the support clauses can be derived from the direct (or multivalued) encoding by binary resolution steps, thus the support encoding can be derived from the direct encoding by preprocessing or dynamically during search. It can also be generalized to non-binary CSPs and other forms of consistency [1]. The support encoding of the running example is shown in Figure 5.

1.2   Previous Experiments

Several experimental comparisons of encodings have been made. The log encoding gave inferior results to the direct encoding with backtrack [6] and local search [4,10]

$$x_{v\bullet} \vee x_{v\bullet} \vee x_{v\bullet} \qquad x_{w\bullet} \vee x_{w\bullet} \vee x_{w\bullet}$$

$$\bar{x}_{v\bullet} \vee \bar{x}_{v\bullet} \qquad \bar{x}_{v\bullet} \vee \bar{x}_{v\bullet}$$
$$\bar{x}_{v\bullet} \vee \bar{x}_{v\bullet} \qquad \bar{x}_{w\bullet} \vee \bar{x}_{w\bullet}$$
$$\bar{x}_{w\bullet} \vee \bar{x}_{w\bullet} \qquad \bar{x}_{w\bullet} \vee \bar{x}_{w\bullet}$$

$$x_{v\bullet} \vee x_{v\bullet} \vee \bar{x}_{w\bullet} \qquad x_{v\bullet} \vee x_{v\bullet} \vee \bar{x}_{w\bullet}$$
$$x_{v\bullet} \vee x_{v\bullet} \vee \bar{x}_{w\bullet} \qquad x_{w\bullet} \vee x_{w\bullet} \vee \bar{x}_{v\bullet}$$
$$x_{w\bullet} \vee x_{w\bullet} \vee \bar{x}_{v\bullet} \qquad x_{w\bullet} \vee x_{w\bullet} \vee \bar{x}_{v\bullet}$$

Fig. 5. Support encoding example

algorithms, though it was sometimes better than the direct encoding for one backtracker [6]. The binary encoding gave worse results than the direct encoding and sometimes generated impractically large models [5]. The support encoding and its generalizations gave better results than the direct encoding with both backtrack [1,8] and local search algorithms [8,17]. The multivalued encoding gave better results than the direct encoding [22]. (Note that these comparisons were done by different researchers using different methodologies.)

A possible explanation for the superiority of the multivalued over the direct encoding for local search is that the former has more solutions. More precisely, it has a higher *solution density* which can be defined as the number of solutions divided by 2^n, where n is the number of SAT variables. The idea that greater solution density makes a problem easier to solve seems so natural that it was taken as axiomatic in [9]. It was also pointed out by Clark et al. [2] that one might expect a search algorithm to solve a (satisfiable) problem more quickly if it has higher solution density. In detailed experiments on random problems across solvability phase transitions they did find a correlation between local and backtrack search performance and the number of solutions, though not a simple one. Problem features other than solution density also seem to affect search cost. For example Yokoo [24] showed that adding more constraints to take a random problem beyond the phase transition removes local minima, making them easier for local search algorithms to solve despite removing solutions. Other research has shown that adding symmetry breaking clauses can make a problem easier to solve by backtrack search, yet harder to solve by local search [16,19]. Symmetry breaking removes solutions, which may explain why it can make a problem harder to solve by local search. Conversely, adding symmetry to a problem can make it easier to solve by local search [16]. Thus there is considerable evidence that local search performs better on problems with higher solution density.

1.3 Conjectures and Aims

We aim to test several conjectures and to re-examine some previous results. All the points below refer to local search.

1. The binary encoding has more solutions than the log encoding when the CSP domain size is not a power of two. This suggests a conjecture not tested by [5]: *the binary encoding will give better results than the log encoding, at least in terms of flips.*
2. More speculatively: *new encodings with more solutions than the multivalued encoding will give even better results.*
3. In graph colouring each domain value supports all or all but one value in every other domain. This suggests that *the support encoding will have no advantage on graph colouring problems.* [1] (It is also noted in [1] that enforcing arc-consistency on disequality constraints is pointless.) Though this prediction concerns backtrack search, local search performance is also of interest, especially given the fact that the support encoding improved local search performance in [8,17].
4. We test a result found by [4,10]: *the log encoding is inferior to the direct encoding.*
5. Also a result found by [5]: *the binary encoding is inferior to the direct encoding.*
6. Finally, a result found by [22]: *the direct encoding is inferior to the multivalued encoding.*

2   Weakened Encodings

To test conjecture (2) in Section 1.3 on new encodings, we define a new family of encodings for (binary and non-binary) CSPs called *weakened encodings*. These weaken the multivalued encoding further by allowing even more values to be assigned to the CSP variables, giving a high ratio of SAT solutions to CSP solutions.

2.1   Definition

Weakened encodings are parameterized by an acyclic binary relation $\leadsto_v$ on the values of $\mathrm{dom}(v)$ for each CSP variable v. They are multivalued encodings in which each conflict clause

$$\bar{x}_{vi} \vee \bar{x}_{wj}$$

is replaced by

$$\bar{x}_{vi} \vee \bar{x}_{wj} \vee \bigvee_{a \leadsto_v i} x_{va} \vee \bigvee_{a \leadsto_w j} x_{wa}$$

Thus the multivalued encoding is the weakened encoding with the empty relation, that is $i \not\leadsto j$ for all i,j.

2.2   Extracting a CSP Solution

Recall that under the multivalued encoding we can obtain a CSP solution by taking any assigned value for each CSP variable. An analogous result holds for weakened encodings:

---

[1] Ian Gent, personal communication.

Theorem. *A CSP solution can always be extracted from a SAT solution under a weakened encoding, by choosing for each CSP variable v any assigned value i such that a $\not\leadsto_v$ i for all values a assigned to v.*

Proof. Because the $\leadsto_v$ are acyclic such a value can always be found for each CSP variable v. We prove by contradiction that this is guaranteed to be a CSP solution. Suppose that we have a SAT solution under a weakened encoding, and that taking such values does not yield a CSP solution. Then we can find two CSP variables v and w and two such values i and j such that the assignments $[v=i, w=j]$ are a CSP conflict. Each CSP conflict has a corresponding conflict clause

$$\bar{x}_{vi} \vee \bar{x}_{wj} \vee \bigvee_{a \leadsto_v i} x_{va} \vee \bigvee_{a \leadsto_w j} x_{wa}$$

But by construction $a \not\leadsto_v i$ and $a \not\leadsto_w j$ for all a, so the conflict clause is simply

$$\bar{x}_{vi} \vee \bar{x}_{wj}$$

We know that these two assignments are a CSP conflict, so this clause is violated. Therefore this is not a SAT solution, contradicting our initial assumption. (This proof extends easily to non-binary constraints.) □

2.3 Total Weakening

We aim to test the conjecture that greater solution density aids local search, so we test a weakened encoding based on the total ordering $<$ for each vertex v. We call this the *totally weakened encoding*. For graph colouring it has at-least-one clauses and modified conflict clauses of the form

$$\bar{x}_{vi} \vee \bar{x}_{wi} \vee \bigvee_{a<i} x_{va} \vee \bigvee_{a<i} x_{wa}$$

The totally weakened encoding of the running example is shown in Figure 6. To demonstrate the increase in the number of solutions, Figure 7 shows that the colouring [$v=0$, $w=1$] has 1 representation under the direct, log and support encodings, 2 under the binary encoding, 3 under the multivalued encoding, and 8 under the totally weakened encoding. Any direct solution is also a multivalued solution, and any multivalued solution is also a weakened solution. To extract a colouring we simply take the smallest colour assigned to each vertex.

$$x_{v\cdot} \vee x_{v\cdot} \vee x_{v\cdot} \qquad\qquad x_{w\cdot} \vee x_{w\cdot} \vee x_{w\cdot}$$

$$\bar{x}_{v\cdot} \vee \bar{x}_{w\cdot} \qquad \bar{x}_{v\cdot} \vee \bar{x}_{w\cdot} \vee x_{v\cdot} \vee x_{w\cdot} \qquad \bar{x}_{v\cdot} \vee \bar{x}_{w\cdot} \vee x_{v\cdot} \vee x_{w\cdot} \vee x_{v\cdot} \vee x_{w\cdot}$$

Fig. 6. Totally weakened encoding example

direct/log/support	binary	multivalued		totally weakened	
v=0 w=1	v=0 w=1	v={0}	w={1}	v={0}	w={1}
	v=3 w=1	v={0,2}	w={1}	v={0,2}	w={1}
		v={0}	w={1,2}	v={0}	w={1,2}
				v={0,1}	w={1}
				v={0,1,2}	w={1}
				v={0,1}	w={1,2}
				v={0,2}	w={1,2}
				v={0,1,2}	w={1,2}

Fig. 7. Representations of the colouring [v=0, w=1]

For graph colouring the multivalued encoding has $O(|V|)$ at-least-one clauses of size k and $O(k|E|)$ conflict clauses of size 2, so the encoding has space complexity $O(k(|V|+|E|))$. The totally weakened encoding has $O(|V|)$ at-least-one clauses of size k, and $O(k|E|)$ modified conflict clauses of size $O(k)$, giving the greater space complexity $O(k(|V|+k|E|))$. Using partial instead of total domain orderings this can be reduced, but it is worth testing this maximally-weakened encoding.

2.4   The Reduced Encoding

We also test a less-weakened variant using a weakening relation $\leadsto_v$ with $i \leadsto k$ for all $i \in \{1, \ldots, k-1\}$ on each vertex v. Then we obtain a partially-weakened encoding containing at-least-one clauses plus the conflict clauses:

$$\bar{x}_{vi} \vee \bar{x}_{wi}$$

for $i < k$ and the modified conflict clauses

$$\bar{x}_{vk} \vee \bar{x}_{wk} \vee \bigvee_{i=1}^{k-1} x_{vi} \vee \bigvee_{i=1}^{k-1} x_{wi}$$

where $v < w$ are adjacent. So far this is just an example of a weakened encoding, but it is possible to eliminate some SAT variables. Resolve the modified conflict clauses with the at-least-one clauses containing literals x_{vk} and x_{wk}, giving

$$\bigvee_{i=1}^{k-1} x_{vi} \vee \bigvee_{i=1}^{k-1} x_{wi}$$

We shall call these *weak-conflict* clauses. They subsume the modified conflict clauses which can be deleted. Now each x_{vk} occurs only in one place: an at-least-one clause. They are therefore pure (positive) literals and we can assign them to T. The at-least-one clauses then become tautologous and can be deleted. This leaves the conflict clauses for $i < k$ plus the weak-conflict clauses, which we shall call the *reduced encoding*. The reduced encoding of the running example is shown in Figure 8.

$$x_{v\cdot} \lor x_{v\cdot} \lor x_{w\cdot} \lor x_{w\cdot}$$

$$\bar{x}_{v\cdot} \lor \bar{x}_{w\cdot} \qquad \bar{x}_{v\cdot} \lor \bar{x}_{w\cdot}$$

Fig. 8. Reduced encoding example

Note that by eliminating the variables we have lost some multiple solutions. In the running example the reduced encoding has only one SAT representation of the colouring [v=0, w=1] corresponding to the assignments [v={0, 2}, w={1, 2}]. But it has fewer variables than the direct, multivalued and totally weakened encodings, effectively eliminating a colour from the problem. By reducing the number of variables we increase the solution density. To extract a solution take the colour assigned to each vertex if one exists; otherwise assign the colour k, which is implicitly assigned to each vertex. (The reduced encoding can be generalized to CSPs but the details are omitted here.)

The reduced encoding has $O(k\,|E|)$ conflict clauses of size 2 and $O(|E|)$ weak-conflict clauses of size $O(k)$, giving a space complexity of $O(k\,|E|)$ which is smaller than the multivalued encoding. However, weak-conflict clauses are almost twice as long as at-least-one clauses, so in practice the reduced encoding may not be smaller than the multivalued encoding.

3   Experiments

Tables 1, 2 and 3 compare the direct (dir), multivalued (mul), support (sup), log (log), binary (bin), reduced (red) and totally weakened (tot) encodings. The tables show (i) the number of flips and (ii) the number of CPU seconds for Walksat to find a solution, and (iii) the encoding sizes. For clauses $i = 1 \ldots c$ containing l_i literals then the encoding size is taken to be $\sum_{i=1}^{c} l_i$. The graphs are taken from the Colouring Symposium at Ithaca, NY, USA, 2002. [2] We selected instances that are not very large or difficult as SAT problems, so that Walksat can be executed a reasonable number of times. The optimal or best known k (number of colours) was used in each case. The algorithm used is Walksat with the default (SKC) heuristic [21]. This algorithm has the advantages of being fast, well-known, available from at least one web page, and believed to be *probabilistically approximately complete* [11], avoiding the need to tune a restart parameter. Each table entry shows the mean number of Walksat flips (search steps) or seconds over 10000 runs, or 100 for the harder problems, taking the best result over noise levels 10%, 20%, ... 90%. Entries "—" denote either problems that took unreasonably long to solve (for some totally weakened encodings) or that were too large for Walksat to read in (for some support encodings). Problems in which k is a power of two have identical log and binary encodings so the same figures appear in both columns. The winner in each case is shown in **bold**. The experiments were performed on a 733 MHz Pentium III. We summarize the results in terms of the aims stated in Section 1.3:

---

[2] http://mat.gsia.cmu.edu/COLORING02/

graph	k	sup	dir	mul	log	bin	red	tot
mug88_1	4	145	140	120	**31**	**31**	76	53
mug88_25	4	146	140	121	**31**	**31**	77	54
mug100_1	4	166	159	137	**35**	**35**	87	61
mug100_25	4	166	159	137	**36**	**36**	87	61
anna	11	1001	1070	768	203	**95**	489	806
david	11	732	712	677	222	**112**	438	1117
homer	13	3870	4102	2945	643	**300**	1988	3677
huck	11	396	378	320	67	**33**	265	623
jean	10	409	391	292	111	**55**	236	438
games120	9	575	500	485	174	**71**	419	552
miles250	8	671	618	531	**125**	**125**	409	455
miles500	20	2259	1860	1966	1848	**699**	1841	136305
myciel3	4	19	18	15	**4**	**4**	10	7
myciel4	5	61	54	48	24	**12**	34	23
myciel5	6	185	155	143	63	**45**	111	88
myciel6	7	533	452	421	163	**132**	349	338
myciel7	8	1506	1382	1280	**424**	**424**	1063	1229
queen5_5	5	521	256	257	440	**253**	284	338
queen6_6	7	8891	7932	**7232**	10672	10865	10617	24272
queen7_7	7	15429	**7378**	7911	20087	13792	12180	33421
school1	14	—	**8024**	8337	104908	33619	10128	—
zeroin.i.1	49	—	11578	8862	16069	7486	**4836**	—
zeroin.i.2	30	42381	55867	55181	15790	9886	**5393**	—
zeroin.i.3	30	42083	52451	54431	14559	10348	**5352**	—
mulsol.i.1	49	—	5832	4189	5451	**3228**	3978	—
mulsol.i.2	31	14263	9194	7131	5396	4916	**4421**	—
mulsol.i.3	31	14892	9318	7137	5943	5619	**4532**	—
mulsol.i.4	31	15484	9263	7974	5685	5631	**4678**	—
mulsol.i.5	31	16251	9764	7110	5274	4971	**4497**	—
DSJR500.1	12	5614	4654	4806	2719	**1923**	4083	19742

Table 1. Walksat flips on encodings of selected graphs

1. The binary encoding almost always beats the log encoding in the number of flips, which supports our conjecture. But it is usually slower in time so the result is mainly of academic interest. Nevertheless, on several graphs the binary encoding used the fewest flips of all encodings, so it seems worth further attention. It suffers from two drawbacks: (i) it has a low flip rate on some graphs, and (ii) it can generate large models (as pointed out by [5]). Both problems might be cured by using a *lifted* SAT model in which clauses are represented intensionally (as described in [15] for example). Locating violated clauses can sometimes be done more quickly on a lifted model than on the more usual extensional models. The question of what makes a good encoding for lifted solvers may be an interesting research direction.
2. Does increasing the solution density of the multivalued encoding help local search? The totally weakened encoding results are erratic in the number of flips but generally much slower in time. But the reduced encoding gives good results. In terms

graph	k	sup	dir	mul	log	bin	red	tot
mug88_1	4	0.0017	0.0011	0.00052	**0.00042**	**0.00042**	0.00045	0.00092
mug88_25	4	0.0016	0.0011	0.00052	**0.00043**	**0.00043**	0.00052	0.00060
mug100_1	4	0.0020	0.0010	0.00064	**0.00035**	**0.00035**	0.00060	0.00058
mug100_25	4	0.0020	0.0010	0.0060	**0.00037**	**0.00037**	0.00044	0.00079
anna	11	0.059	0.018	**0.0070**	0.011	0.021	**0.0070**	0.019
david	11	0.043	0.011	**0.0056**	0.010	0.018	0.0058	0.022
homer	13	0.301	0.108	0.049	0.052	0.067	**0.047**	0.118
huck	11	0.022	0.0062	**0.0028**	0.006	0.011	0.0039	0.011
jean	10	0.017	0.0052	**0.0023**	0.0045	0.013	0.0028	0.0077
games120	9	0.033	0.0093	**0.0050**	0.011	0.029	0.0061	0.012
miles250	8	0.020	0.0064	**0.0033**	0.0043	0.0040	0.0036	0.0077
miles500	20	0.376	0.072	**0.040**	0.100	0.180	0.070	4.90
myciel3	4	0.00020	0.00013	0.00011	**0.000068**	**0.000068**	0.000113	0.000064
myciel4	5	0.00094	0.00032	0.00022	0.00043	0.00095	**0.000033**	0.00024
myciel5	6	0.0046	0.0015	0.0010	0.0017	0.0033	**0.00098**	0.0021
myciel6	7	0.028	0.0064	**0.0048**	0.0080	0.011	0.0057	0.010
myciel7	8	0.16	0.042	**0.031**	0.034	0.034	0.044	0.068
queen5_5	5	0.0063	0.0014	**0.0011**	0.0038	0.0059	0.0019	0.0029
queen6_6	7	0.197	0.035	**0.026**	0.110	0.164	0.092	0.230
queen7_7	7	0.480	0.040	**0.035**	0.245	0.256	0.129	0.360
school1	14	—	1.11	**1.07**	25.8	12.9	2.59	—
zeroin.i.1	49	—	1.28	**0.771**	5.64	5.67	0.937	—
zeroin.i.2	30	8.52	4.54	3.73	3.27	2.70	**0.895**	—
zeroin.i.3	30	25.2	4.49	3.70	3.27	2.74	**0.96**	—
mulsol.i.1	49	—	0.676	**0.361**	1.70	2.78	0.570	—
mulsol.i.2	31	8.45	0.732	**0.432**	1.09	1.09	0.550	—
mulsol.i.3	31	8.71	0.745	**0.424**	1.16	1.29	0.609	—
mulsol.i.4	31	9.10	0.741	**0.482**	1.09	1.31	0.596	—
mulsol.i.5	31	9.72	0.78	**0.44**	1.09	1.11	0.58	—
DSJR500.1	12	0.477	0.122	**0.081**	0.104	0.175	0.114	0.505

Table 2. Walksat CPU times on encodings of selected graphs

of flips it is second only to the binary encoding; in terms of time it is the best encoding for some of the graphs (for example *zeroin* and *myciel*) and is rarely much worse than the multivalued encoding. Solution density is clearly not the only factor affecting local search performance, but it seems worth experimenting with other weakened encodings based on different weakening relations. There was another surprising aspect of these experiments, for which we currently have no explanation. The reduced encoding removes a colour so we expected it to pay off for problems with small k, and this was true of the smaller *myciel* problems. However, it also did well on the high-k *zeroin* and *mulsol* problems.

3. It was predicted that the support encoding would have no advantage over the direct encoding for graph colouring. In fact it turns out to be much worse, using more flips than most of the other encodings and with a lower flip rate. It gives very good results on generalized (bandwidth) colouring problems [17] as well as other problems [8]

graph	k	sup	dir	mul	log	bin	red	tot
mug88_1	4	6080	2576	1520	2336	2336	1752	3272
mug88_25	4	6080	2576	1520	2336	2336	1752	3272
mug100_1	4	6912	2928	1728	2656	2656	1992	3720
mug100_25	4	6912	2928	1728	2656	2656	1992	3720
anna	11	136004	27544	12364	46144	102544	19720	66594
david	11	108779	19459	9889	37468	84448	16240	54549
homer	13	645073	137137	49621	176044	286528	78144	303589
huck	11	81796	15576	7436	27968	62608	12040	40546
jean	10	58800	13080	5880	22240	56896	9144	28740
games120	9	113076	21204	12564	49296	153120	20416	58500
miles250	8	57728	14384	7216	18576	18576	10836	28888
miles500	20	987200	98000	49360	241680	655200	88920	493960
myciel3	4	816	336	204	320	320	240	444
myciel4	5	4125	1285	825	2337	5964	1136	2245
myciel5	6	18684	4524	3114	8778	16992	4720	10194
myciel6	7	78645	15225	11235	31995	45300	18120	42945
myciel7	8	314304	49984	39288	113280	113280	66080	171448
queen5_5	5	8625	2225	1725	5025	13440	2560	4925
queen6_6	7	30184	5824	4312	12288	17400	6960	16492
queen7_7	7	49049	9065	7007	20139	28560	11424	26999
school1	14	7560700	610120	540050	2141720	3055200	992940	4015340
zeroin.i.1	49	20194811	908411	412139	2429790	4624800	787200	10055339
zeroin.i.2	30	6563700	402360	218790	1064410	1274760	410756	3299460
zeroin.i.3	30	6557400	397800	218580	1064060	1274400	410640	3298380
mulsol.i.1	49	19320847	857647	394303	2325630	4427400	753600	9625903
mulsol.i.2	31	7647638	421538	246698	1205290	1320900	466200	3859748
mulsol.i.3	31	7703376	419616	248496	1214880	1331440	469920	3890376
mulsol.i.4	31	7761997	422437	250387	1224185	1341640	473520	3920167
mulsol.i.5	31	7814852	425072	252092	1232560	1350820	476760	3946982
DSJR500.1	12	1095840	157320	91320	349280	682560	156420	560580

Table 3. Encoding sizes of selected graphs

so these colouring results are probably atypical; but we believe that these are the first reported negative results for the support encoding.

4. The log encoding does surprisingly well compared to the direct encoding, and is not dominated by it. It does less well compared to the multivalued encoding but gives comparable results on some graphs. This shows that the log encoding may not be as bad as expected from previous results. In fact we have based a generalized SAT (linear pseudo-Boolean) model of multiple sequence alignment on a log encoding, with promising results [18].

5. Similarly the binary encoding is not dominated by the direct encoding. It does less well compared to the multivalued encoding but is not dominated by it. This roughly agrees with previous results, but as noted above the binary encoding seems worth investigating further.

6. The multivalued encoding is uniformly better than the direct encoding, with higher flip rates and fewer flips, in agreement with previous results.

The multivalued encoding is always smallest, followed by either the direct or reduced encodings. The support encoding is generally the largest, followed by the totally weakened encoding: on some graphs they are more than an order of magnitude larger than the other encodings. The binary encoding is less than three times larger than the log encoding, but because of their larger clauses both are several times larger than the smallest encodings. Not shown in the table is the number of SAT variables: the support, direct, multivalued and totally weakened encodings all have the same number of variables; the reduced encoding has $(k - 1)/k$ as many; and the log and binary encodings have the fewest with $(\log k)/k$ as many.

A note on encoding times. The problems were encoded several times faster than Walksat could solve them, but a fast graph colouring algorithm solves them several times faster again. We do not advocate a SAT approach to colouring problems in general: we use them only as a testbed for a comparison of various encodings.

4   Conclusion

Our results confirm some previous experimental results on local search and CSP encodings, but also contain interesting surprises. Firstly, the log and binary encodings perform much better than expected, and in terms of local moves the binary encoding gives the best results overall. Secondly, while we expected the support encoding to be no better than the direct encoding for local search on graph colouring, it turns out to be surprisingly poor. Thirdly, a particular aim of this paper was to explore whether encodings with higher solution densities are easier to solve by local search. This is confirmed in the cases of the multivalued vs direct, the reduced vs multivalued, and (in terms of local moves) the binary vs log encodings. But taken to an extreme (with the totally weakened encoding) this idea gave very erratic results, showing that solution density is not the only important property of an encoding. Testing other local search algorithms may help to decide whether the negative results are caused by a feature of the search space, or an incompatibility with Walksat heuristics in particular.

Experimental comparisons such as this one are necessarily limited. In future work we will experiment with different weakenings and CSPs other than graph colouring. The work can be extended to linear pseudo-Boolean encodings which generalize SAT. The work is also relevant to backtrack search. In experiments with a SAT backtracker (Chaff [14]) on a hybrid of the reduced and direct encodings which includes some of the at-most-one clauses, we found similar results to those reported here: the hybrid encoding generally used fewer backtracks than the direct encoding to prove non-k-colorability of the same graphs, but for larger k the CPU time was sometimes several times greater.

Acknowledgment This work has received support from Science Foundation Ireland under Grant 00/PI.1/C075.

References

1. C. Bessiere, E. Hebrard, T. Walsh. Local Consistencies in SAT. *Sixth International Conference on Theory and Applications of Satisfiability Testing*, Santa Margherita Ligure, Italy, 2003, pp. 400–407.

2. D. Clark, J. Frank, I. Gent, E. MacIntyre, N. Tomov, T. Walsh. Local Search and the Number of Solutions. *Second International Conference on Principles and Practices of Constraint Programming*, 1996, pp. 119–133.

3. M. Davis, G. Logemann, D. Loveland. A Machine Program for Theorem Proving. *Communications of the ACM* vol. 5, 1962, pp. 394–397.

4. M. Ernst, T. Millstein, D. Weld. Automatic SAT-Compilation of Planning Problems. *Fifteenth International Joint Conference on Artificial Intelligence*, Nagoya, Japan, 1997, pp. 1169–1176.

5. A. Frisch, T. Peugniez. Solving Non-Boolean Satisfiability Problems with Stochastic Local Search. *Seventeenth International Joint Conference on Artificial Intelligence*, Seattle, Washington, USA, 2001.

6. A. van Gelder. Another Look at Graph Coloring via Propositional Satisfiability. *Computational Symposium on Graph Coloring and its Generalizations*, Ithaca, NY, 2002, pp. 48–54.

7. R. Genisson, P. Jegou. Davis and Putnam Were Already Forward Checking. *Twelfth European Conference on Artificial Intelligence*, Budapest, Hungary, 1996, pp. 180–184.

8. I. Gent. Arc Consistency in SAT. *Fifteenth European Conference on Artificial Intelligence*, Lyons, France, 2002.

9. Y. Hanatani, T. Horiyama, K. Iwama. Density Condensation of Boolean Formulas. *Sixth International Conference on the Theory and Applications of Satisfiability Testing*, Santa Margherita Ligure, Italy, 2003, pp. 126–133.

10. H. H. Hoos. SAT-Encodings, Search Space Structure, and Local Search Performance. *Sixteenth International Joint Conference on Artificial Intelligence*, Stockholm, Sweden, Morgan Kaufmann, 1999, pp. 296–302.

11. H. H. Hoos. Stochastic Local Search — Methods, Models, Applications. PhD thesis, CS Department, TU Darmstadt, 1998.

12. K. Iwama, S. Miyazaki. SAT-Variable Complexity of Hard Combinatorial Problems. *IFIP World Computer Congress*, Elsevier Science B. V., North-Holland, 1994, pp. 253–258.

13. S. Kasif. On the Parallel Complexity of Discrete Relaxation in Constraint Satisfaction Networks. *Artificial Intelligence* vol. 45, Elsevier, 1990, pp. 275–286.

14. M. Moskewicz, C. Madigan, Y. Zhao, L. Zhang, S. Malik. Chaff: Engineering an Efficient SAT Solver. *Proceedings of the Thirty Ninth Design Automation Conference,* Las Vegas, June 2001.

15. A. J. Parkes. Lifted Search Engines for Satisfiability. PhD dissertation, University of Oregon, June 1999.

16. S. D. Prestwich. Negative Effects of Modeling Techniques on Search Performance. *Annals of Operations Research* vol. 18, Kluwer Academic Publishers, 2003, pp. 137–150.

17. S. D. Prestwich. Maintaining Arc-Consistency in Stochastic Local Search. *Workshop on Techniques for Implementing Constraint Programming Systems*, Ithaca, NY, 2002.

18. S. D. Prestwich, D. Higgins, O. O'Sullivan. A SAT-Based Approach to Multiple Sequence Alignment. Poster, *Ninth International Conference on Principles and Practice of Constraint Programming*, Kinsale, Ireland, 2003.

19. S. D. Prestwich, S. Bressan. A SAT Approach to Query Optimization in Mediator Systems. *Fifth International Symposium on the Theory and Applications of Satisfiability Testing*, University of Cincinnati, 2002, pp. 252–259. Submitted to *Annals of Mathematics and Artificial Intelligence*.

20. D. Sabin, G. Freuder. Contradicting Conventional Wisdom in Constraint Satisfaction. *Eleventh European Conference on Artificial Intelligence*, 1994, pp. 125–129.
21. B. Selman, H. Kautz, B. Cohen. Noise Strategies for Improving Local Search. *Twelfth National Conference on Artificial Intelligence*, AAAI Press, 1994, pp. 337–343.
22. B. Selman, H. Levesque, D. Mitchell. A New Method for Solving Hard Satisfiability Problems. *Tenth National Conference on Artificial Intelligence*, MIT Press, 1992, pp. 440–446.
23. T. Walsh. SAT v CSP. *Sixth International Conference on Principles and Practice of Constraint Programming, Lecture Notes in Computer Science* vol. 1894, Springer-Verlag, 2000, pp. 441–456.
24. M. Yokoo. Why Adding More Constraints Makes a Problem Easier for Hill-climbing Algorithms: Analyzing Landscapes of CSPs. *Proceedings of the Third International Conference on Principles and Practice of Constraint Programming, Lecture Notes in Computer Science* vol. 1330, Springer-Verlag 1997, pp. 356–370.

A Study of Pure Random Walk on Random Satisfiability Problems with "Physical" Methods

Guilhem Semerjian[1] and Rémi Monasson[1,2]

* CNRS-Laboratoire de Physique Théorique de l'ENS,
24 rue Lhomond, 75005 Paris, France.
* CNRS-Laboratoire de Physique Théorique,
3 rue de l'Université, 67000 Strasbourg, France.

Abstract. The performances of a local search procedure, the Pure Random Walk (PRW), for the satisfiability (SAT) problem is investigated with statistical physics methods. We identify and characterize a dynamical transition for the behavior of PRW algorithm on randomly drawn SAT instances where, as the ratio of clauses to variables is increased, the scaling of the solving time changes from being linear to exponential in the input size. A framework for calculating relevant quantities in the linear phase, in particular the average solving time, is introduced, along with an approximate study of the exponential phase.

1 Introduction

When faced with a computationally hard problem, such as the resolution of a satisfiability formula, one can choose between two main classes of algorithm. On the one hand, complete algorithms of the Davis-Putnam-Loveland-Logeman type performs an exhaustive, yet clever, search of the configuration space, providing either a solution of the formula or a proof of unsatisfiability. On the other hand, a large number of algorithms are based on local search ideas, and explore a part of the configuration space by improving locally the current assignment of variables at each step of their execution. The current theoretical understanding of these incomplete procedures is still far from being satisfactory though some important results were recently obtained [1,2]. The present work, although based on non-rigorous methods issued from theoretical physics, may be of some help in this regard [3].

This paper is organized as follows. In Sec. 2 we define the local search procedure under study and recall the rigorous results previously obtained, then (Sec. 3) a dynamic transition is identified with numerical simulations. The next two sections (4 and 5) are devoted to the two regimes separated by this transition, and we conclude in Sec. 6.

2 Definitions and Known Results

We shall consider K-SAT formulas on N boolean variables drawn from the standard random ensemble [4]. M clauses are generated independently and in the

E. Giunchiglia and A. Tacchella (Eds.): SAT 2003, LNCS 2919, pp. 120–134, 2004.

same way. For each of them, a K-uplet of variables is chosen uniformly, each of which is negated or left unchanged with probability $1/2$. The resulting clause is the disjunction of these K literals. The formula is the conjunction of M clauses. We call $\alpha = M/N$ the ratio of clauses to variables. We concentrate on the asymptotic limit where M and N go to infinity with α kept fixed.

The Pure Random WalkSAT (PRWSAT) algorithm for solving K-SAT instances is defined by the following rules [5]. Choose randomly an assignment of the Boolean variables. If all clauses are satisfied, output "Satisfiable" and the current assignment (which is a solution). If not, choose randomly one of the unsatisfied clauses, one among the K variables of this clause, and reverse its state, from true to false or *vice versa*. Repeat this step until a solution has been found or a limit on the number of flips fixed beforehand has been reached. In the latter case, the output of the procedure is "Don't know": no certainty on the status of the formula is achieved.

Papadimitriou introduced the PRWSAT procedure for $K = 2$, and showed that it solves with high probability any satisfiable 2-SAT instance in a number of steps (flips) of the order of N^2[5]. Recently Schöning was able to prove the following very interesting result for 3-SAT[1]. Call 'trial' a run of PRWSAT consisting of the random choice of an initial configuration followed by $3 \times N$ steps of the procedure. If none of T successive trials on a given instance has been successful (has provided a solution), then the probability that this instance is satisfiable is lower than $\exp(-T \times (3/4)^N)$. This bound turns PRWSAT into a probabilistic algorithm: though it is not complete, the uncertainty on its output can be made as small as desired, with a complexity scaling exponentially with the size of the instance, $T \sim (4/3)^N$, up to non-exponential multiplicative factors. Further works improved on this scaling through a refined choice of the initial configuration [6,7], the best result being, to our knowledge, $T \sim 1.3290^N$ [7].

The above bound is true for any instance. Restriction to special input distributions allows to derive further results. Alekhnovich and Ben-Sasson showed that instances drawn from the random 3-Satisfiability ensemble are solved in polynomial time with high probability when α is smaller than 1.63 [2]. It is remarkable that, despite the fact that the same clauses are seen various times in the course of the search, rigorous results on the operation of PRWSAT can be established.

3   Behavior of the Algorithm

In this section, we briefly sketch the behavior of PRWSAT, as seen from numerical experiments [8] and the analysis of [9,10]. It has been found that a dynamical threshold α_d separates two regimes:

- for $\alpha < \alpha_d \simeq 2.7$ for 3-SAT, the algorithm finds a solution very quickly, namely with a number of flips growing linearly with the number of variables N.

  Figure 1A shows the evolution of the amount of unsatisfied clauses during the execution of a single run of PRWSAT on an instance with ratio $\alpha = 2$

and $N = 500$ variables. It is very important for what follows to stress that the curve is plotted in "reduced" units : the horizontal axis is the number of flips *divided by M*, that we shall call *time* from now on, the vertical one is the *fraction* of unsatisfied clauses, φ_0.

A 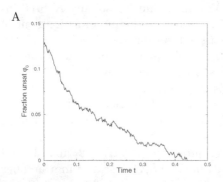   B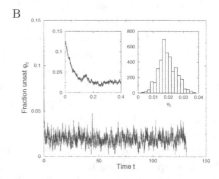

Fig. 1. Fraction φ_0 of unsatisfied clauses as a function of time t (number of flips over M) for two randomly drawn instances of 3-SAT with ratios $\alpha = 2$ (**A**) and $\alpha = 3$ (**B**) with $N = 500$ variables. Note the difference of time scales between the two figures. Insets of figure B: left: blow up of the initial relaxation of φ_0, taking place on the $O(1)$ time scale as in (**A**); right: histogram $p_{500}(\varphi_0)$ of the fluctuations of φ_0 on the plateau $1 \leq t \leq 130$.

The curve shows a fast decrease from the initial value ($\varphi_0(t = 0) = 1/8$ in the large N limit independently of α) down to zero on a time scale $t_{sol} = O(1)$. Fluctuations of the curves plotted with this definition of the axes (in *reduced* units) are smaller and smaller as N grows. As a consequence the distribution of the solving time gets peaked around its mean value t_{sol}, an increasing function of α, when $N \to \infty$. This *relaxation* regime corresponds to the one studied by Alekhnovich and Ben-Sasson, and $\alpha_d > 1.63$ as expected [2].

– for instances in the $\alpha_d < \alpha < \alpha_c$ range, the initial relaxation taking place on $t = O(1)$ time scale is not sufficient to reach a solution (Figure 1B). The fraction φ_0 of unsat clauses then fluctuates around some plateau value for a very long time. On the plateau, the system seems to be trapped at a high level of unsatisfied clauses. The life time of this "trap", that is, the time it takes to escape from the trap and finally reach a solution is so huge that it is possible to define a (quasi) stationary probability distribution $p_N(\varphi_0)$ for the fraction φ_0 of unsat clauses on the plateau (Right inset of Figure 1B). The distribution of fractions is well peaked around some average value (height of the plateau), with left and right tails decreasing exponentially fast with N, $p_N(\varphi_0) \sim \exp(N\bar{\zeta}(\varphi_0))$ with $\bar{\zeta} \leq 0$. Eventually a large negative fluctuation will bring the system to a solution ($\varphi_0 = 0$). Assuming that these fluctuations are independent random events occurring with probability $p_N(0)$ on an interval of time of order 1, the solving time is a stochastic variable with exponential distribution[3]. Its average is, to leading exponential order, the inverse of the probability of finding a solution on the $O(1)$ time scale: $t_{sol} \sim \exp(N\zeta)$ with $\zeta = -\bar{\zeta}(0)$. Escape from the trap therefore takes place on exponentially large–in–N time scales, as confirmed by numerical simulations for different sizes. Schöning's result[1] can be interpreted as a lower bound to the probability $\bar{\zeta}(0) > \ln(3/4)$, true for any instance.

The fraction of unsatisfied clauses on the plateau reached by PRWSAT on the linear time scale is plotted on Figure 2. Notice that the "dynamic" critical value α_d above which the plateau appears (PRWSAT stops finding a solution in linear time) is strictly smaller than the "static" ratio $\alpha_c \simeq 4.3$, where formulas go from satisfiable with high probability to unsatisfiable with high probability. In the intermediate range $\alpha_d < \alpha < \alpha_c$, instances are almost surely satisfiable but PRWSAT needs an exponentially large time to prove so. Interestingly, α_d and α_c coincides for 2-SAT in agreement with Papadimitriou's result[5].

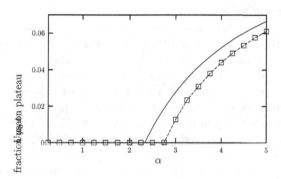

Fig. 2. Fraction φ_0 of unsatisfied clauses on the plateau as a function of the ratio α of clauses per variable. Symbols are the output of numerical experiments, and have been obtained through average of data from simulations at a given size N (nb. of variables) over 1,000 samples of 3-SAT, and extrapolation to infinite sizes (dashed line serves as a guide to the eye). The full line is the prediction of the approximation of Sec. 5.

4   Analysis of the Linear Phase

We present here a method suitable for the study of the low α phase. A detailed presentation of this method has been given in a general case in [12], and applied more specifically to this problem in [9].

[.] This kind of distribution in local search algorithms has been largely discussed, see for instance [11]).

4.1   Averages of Additive Quantities

The formulas studied here are constraints on N boolean variables x_i, from which $2N$ literals $x_i, \bar{x}_i$ are obtained. Given a formula F, two variables will be said *adjacent* if there is a clause of the formula in which both appears, irrespective of the fact that they are negated or not. Two variables will be said *connected* if there is a path of adjacent variables between them. These definitions allow to partition the variables of a formula F into sets of maximal connected components, and to break the formula into the independent sub-formulas F_r, one for each of the components, that we shall call *clusters* below. Note that a variable not appearing in any clause of F forms a cluster on its own.

A function Q from the set of formulas F to the set of real numbers will be said *additive* with respect to cluster decomposition if $Q(F) = \sum_r Q(F_r)$ for any F. We would like to compute the expectation $\mathbb{E}[Q]$ of Q with respect to the random K-SAT ensemble. One can decompose this average in two steps: the K variables belonging to a clause are chosen, and then the corresponding literals are negated or not. These two steps being independent we can perform the average in a convenient way.

Two clusters are said to have the same *shape* if they are isomorphic up to a relabeling of their variables, whether these variables are negated or not (see Fig. 3 for some examples). Distinct shapes of clusters will be labeled by an integer index $s \geq 0$. We denote L_s the number of variables in a cluster. Let us call $[Q]_s$ the average value of $Q(F_r)$ for a cluster of type s, the average being carried out on the choices of the negation or not of the literals in F_r only. $[Q]_s$ is independent of the labeling of the variables forming the cluster for the quantities we consider. Then the "intermediate" expectation of $Q(F)$, for a given choice of the M K-uplet of literals, can be written as $\sum_s n_s[Q]_s$, where n_s is the number of clusters of type s induced by the choice of the variables in the clauses.

To obtain $\mathbb{E}[Q]$, we are left with the calculation of $\mathbb{E}[n_s]$, the average number of type–s clusters in a random K-SAT formula. It is not hard to convince oneself that $\mathbb{E}[n_s] = N P_s/L_s$, where P_s is the probability that a given variable, say x_1, belongs to a cluster of type s. Then,

$$\frac{1}{N}\mathbb{E}[Q] = \sum_s \frac{1}{L_s} P_s[Q]_s \qquad (1)$$

Next, we assume, that the limit (for large N) of Eqn. (1) exists, and that this limit, noted $q(\alpha, K)$, admits a power series expansion around $\alpha = 0$. Calculation of the terms in this expansion is made possible by the simplifications resulting from the large N limit. Indeed the underlying K-hypergraph undergoes a percolation phenomenon at $\alpha_p = 1/(K(K-1))$: for smaller values of α, the clusters have size at most $O(\log N)$, and those of finite size that have a probability of order 1 are trees. In that case the expression for P_s simplifies to $P_s = (\alpha K!)^{m_s} \exp(-L_s \alpha K) V_s$, where m_s is the number of clauses of the cluster, $m_s = (L_s - 1)/(K - 1)$ for tree clusters, and V_s is a symmetry factor independent of α, equal to the number of distinct labellings of the variables of the cluster, divided by $(L_s - 1)!$. Taking into account all these facts, one realizes that the

m-th term in the power series expansion of $q(\alpha, K)$ is a sum of $m+1$ contributions, one for each of the tree clusters with a number of clauses lower or equal than m (one has to expand the exponentials in P_s to reorder the sum as a power series expansion in α). We now apply this method to two quantities of interest for the study of the PRWSAT algorithm.

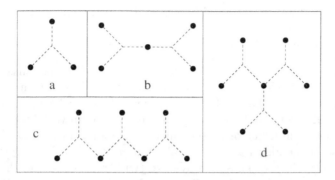

Fig. 3. Tree-like clusters contributing to the expansions, drawn for $K = 3$.

Type	L_s	$V_s(K!)^{m_s}/L_s$	$[T_{sol}]_s$
a	K	1	$\frac{\bullet}{\bullet K}$
b	$2K-1$	$\frac{K^2}{\bullet}$	$\frac{\bullet}{\bullet 2K}\left[2^{K\bullet\ \bullet}+\frac{K\bullet\ \bullet}{K\bullet K-\bullet\bullet}\right]$
c	$3K-2$	$\frac{K^3\bullet K-\bullet\bullet}{\bullet}$	$\frac{\bullet}{\bullet 3K}\left[3\cdot 2^{\bullet K}+2^{K\bullet\ \bullet}\frac{K\bullet\ \bullet}{K\bullet K-\bullet\bullet}+\frac{\bullet K^4\bullet\ \bullet K^3\bullet\ \bullet K^2\bullet\ \bullet K-\bullet}{\bullet K^2\bullet K-\bullet\bullet\bullet\bullet K-\bullet\bullet\bullet K^2-\bullet\bullet}\right]$
d	$3K-2$	$\frac{K^3}{\bullet}$	$\frac{\bullet}{\bullet 3K}\left[3\cdot 2^{\bullet K}+2^K\frac{\bullet\bullet K\bullet\ \bullet\bullet}{K\bullet K-\bullet\bullet}-\frac{\bullet K\bullet\ \bullet}{K^2\bullet K-\bullet\bullet}\right]$

Table 1. Contributions to the cluster expansion for the solving time, see Figure 3.

4.2   Solving Time

The number $T_{sol}(F, C)$ of flips needed by PRWSAT to find a solution is a random variable, with two different sources of randomness: formulas F are drawn randomly, and, for a given formula, the sequence C of choices made by the algorithm is random. Numerical experiments [8] indicate that, for low enough values of α, the distribution of $T_{sol}(F, C)$ gets peaked in the infinite size limit. We make the following

Conjecture: there exists a critical value of the ratio of clauses per variable, $\alpha_d(K)$, such that in the large size $N \to \infty$ limit, a randomly drawn formula with ratio $\alpha < \alpha_d(K)$ is almost surely solved by PRWSAT in linear time. More precisely, the solving time $t_{sol} = T_{sol}(F, C)/M$ is highly concentrated around some value $t_{sol}(\alpha, K)$,

$$\lim_{\epsilon \to 0^+} \lim_{N \to \infty} \text{Proba}\left[t_{sol}(\alpha, K) - \epsilon < t_{sol} < t_{sol}(\alpha, K) + \epsilon\right] = 1 \qquad (2)$$

where the probability is taken with respect to both choices of the formula F and variable flip sequence C. We also conjecture that $t_{sol}(\alpha, K)$ admits a power series expansion in α with a radius of convergence $\alpha_d(K)$.

Provided this conjecture is true, the coefficients of the expansion of t_{sol} in powers of α can be calculated along the lines exposed below. Consider a particular K-SAT formula. The number of flips needed by PRWSAT to solve it is a random variable, that can be written as a sum of random variables, one for each cluster: no matter in which order the different unsatisfied clauses are considered in the course of the algorithm, the total number of flips will be the sum of the flips inside each cluster. Thus, the average (with respect to the choices of the algorithm and the initial configuration) number of flips needed by PRWSAT to solve the formula is additive with respect to the cluster decomposition. One can then apply the generic strategy exposed before, with the quantity Q being the average number of flips to reach a solution. The last task remaining is the computation of the average time to solve the small clusters of finite size. Let us give a few examples. Clusters with no clauses clearly do not contribute to the sum, as their variables are never flipped. Consider now the case of a cluster made of a single clause. All but one initial configurations are already solutions, and the only unsatisfying configuration is corrected in one flip; the average number of steps needed to solve this cluster is then $1/2^K$. The same kind of enumeration can be done on larger sub–formulas. We performed this task for clusters with up to three clauses, with results summarized in Table 1. The expansion of t_{sol} reads,

$$t_{sol}(\alpha, K) = \frac{1}{2^K} + \frac{K(K+1)}{K-1} \frac{1}{2^{2K+1}} \alpha$$
$$+ \frac{4K^6 + K^5 + 6K^3 - 10K^2 + 2K}{3(K-1)(2K-1)(K^2-2)} \frac{1}{2^{3K+1}} \alpha^2 + O(\alpha^3). \qquad (3)$$

This expression compares very well with the results of numerical simulations, see Fig. 4A. As is to be expected for a power-series expansion, the agreement is very good at low values of α and worsens when α increases. Calculation of higher order terms is very cumbersome, but could open the way to an approximate determination of the radius of convergence $\alpha_d(K)$.

4.3 Hamming Distance

The Hamming distance Δ between the initial random configuration chosen by the algorithm and the solution found at the end of its execution is another quantity that can be studied in this expansion framework. Numerical experiments show again a concentration of Δ around its average value $N\,\delta(\alpha, K)$. In addition it is clearly an additive quantity. The expansion of δ can be carried out along the lines of Sec. 4.2, with the result

$$\delta(\alpha, K) = \frac{1}{2^K}\alpha + \frac{1}{2^{2K+1}}\frac{K(K-1)}{K+1}\alpha^2 + \tag{4}$$

$$\frac{1}{2^{3K+1}}\frac{8K^8 - 6K^7 - 33K^6 + 35K^5 + 58K^4 - 24K^3 - 48K^2 - 2K}{3(K+1)^2(4K^2-1)(K^2-2)}\alpha^3 + O(\alpha^4)$$

Comparison with numerical simulations is very satisfactory, see Fig. 4B.

It would be very interesting to understand the behavior of $\delta(\alpha, K)$ at the threshold α_d. As the number of flips (divided by M) needed to find a solution diverges in this limit, one could think that the solution found is completely decorrelated from the initial assignment, and that δ would reach $1/2$ at the threshold. This can however not be strictly the case: for any value of α, a finite fraction of the variables do not belong to any clause, and are thus not flipped during the run of the algorithm. Their values in the initial and final assignments are equal. Similarly, at most one out of the K variables of an isolated clause can take a different value in the two configurations. It would be interesting to understand which part of the formula decorrelates in the vicinity of α_d.

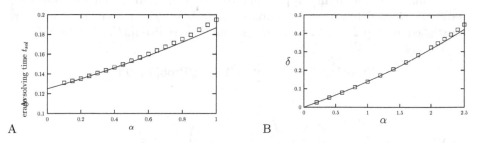

A   B

Fig. 4. A: Average solving time $t_{sol}(\alpha, 3)$ for PRWSAT on 3-SAT. Symbols: numerical simulations, averaged over $1,000$ runs for $N = 10,000$. Solid line: prediction from Eq. (3). B: Average Hamming distance between initial configuration and solution found $\delta(\alpha, 3)$ for PRWSAT on 3-SAT. Symbols: numerical simulations, averaged over 100 runs for $N = 5000$. Solid line: prediction from Eq. (4).

5   Analysis of the Exponential Phase

The method developed above cannot be used for the study of the regime $\alpha_d < \alpha < \alpha_c$, where formulas are satisfiable but the finding of a solution happens through a fluctuation of the number of unsatisfied clauses from its typical value. We now introduce an approximate method adapted to this situation.

5.1   The Simplifying Approximation

Let us denote by $\mathcal{C}$ an assignment of the Boolean variables. PRWSAT defines a Markov process on the set of configurations $\mathcal{C}$, of cardinality 2^N. Tracking the set of probabilities $\text{Prob}[\mathcal{C}, T]$ of each configuration $\mathcal{C}$ as a function of the number of steps T of the algorithm is a formidable task. We will therefore consider a simplified version of the true PRWSAT Markov chain, which retains some essential features of the original process and is much easier to understand analytically. The crucial quantity to consider is the number of clauses unsatisfied by the current assignment of the boolean variable, $M_0(\mathcal{C})$. When $M_0(\mathcal{C})$ vanishes, the algorithm has found a solution and stops. The simplifying approximation that we shall make is the following:

Approximation: after T steps of PRWSAT, the probability that the Boolean configuration is $\mathcal{C}$ is a function of $M_0(\mathcal{C})$ only.

In other words, at each time step, configurations with the same number of unsatisfied clauses are equally likely. This statement cannot be true. Indeed, the Hamming distance between two configurations visited at step T and $T+k$ of the algorithm is at most k, a constraint not enforced in the approximate process. Strikingly, the results obtained within this approximation are much more sensible from a quantitative point of view that one could fear.

Within this simplifying approximation, we can write down the transition matrix A which determines the probability $\text{Prob}[M_0, T+1]$ that M_0 clauses are unsatisfied at step $T+1$ in terms of its counterpart $\text{Prob}[M_0, T]$ at step T,

$$\text{Prob}[M_0', T+1] = \sum_{M_0=0}^{M} A_{M_0' M_0} \text{Prob}[M_0, T] \quad , \tag{5}$$

$$A_{M_0' M_0} = \sum_{Z_u=0}^{M_0} \sum_{Z_s=0}^{M-M_0} \frac{Z_u N}{K M_0} \binom{M_0}{Z_u} \binom{M-M_0}{Z_s} \left(1 - \frac{K}{N}\right)^{M_0 - Z_u} \left(\frac{K}{N}\right)^{Z_u} \times$$

$$\left(1 - \frac{K}{(2^K - 1) N}\right)^{M-M_0-Z_s} \left(\frac{K}{(2^K - 1) N}\right)^{Z_s} \mathbf{1}_{M_0' - M_0 + Z_u - Z_s} \tag{6}$$

where $\mathbf{1}_X = 1$ if $X = 0$, 0 otherwise. Z_u is the number of unsatisfied clauses which contains the variable to be flipped. All these clauses will become satisfied after the flip. The factor $Z_u N/(K M_0)$ represents the probability of flip of the variable. Z_s is the number of clauses satisfied prior to the flip and violated after. The origin of the Binomial laws is the following. Assume that the configuration violates M_0 clauses. In the absence of further information, the variable which is going to flip has probability K/N to be present in a given clause (there are $\binom{N}{K}$ possible K-uplets over N variables, and $\binom{N-1}{K-1}$ which include a given variable). Furthermore, a satisfied clause that contains the flipped variable has a probability $1/(2^K - 1)$ to become unsatisfied later. Z_u (respectively Z_s) is thus drawn from a binomial law with parameter K/N (resp. $K/(N(2^K - 1))$) , over M_0 (resp. $M - M_0$) tries.

As a side remark, let us comment on the similitudes and differences between our approach and the one initiated by Schöning [1]. In Schöning's work, the quantity monitored during the execution of PRWSAT is the Hamming distance d between the current assignment $\mathcal{C}$ of the variables and an arbitrary solution of the formula. Rigorous bounds on the transition probabilities for the evolution of d between steps T and $T + 1$ are then established. For typical random instances, the bound on the running time of the algorithm is far from being optimal : as we have seen before, for $\alpha < \alpha_d$ it is only linear, and for $\alpha > \alpha_d$ it is exponential but with a growth coefficient much lower than $4/3$. This discrepancy is partly due to the fact that there are exponentially numerous solutions, whereas Schöning's bound takes into account only one of them. In a sense the quantity we focus on, that is, the number M_0 of unsatisfied clauses, bypasses this limitation. It would however be much more difficult to obtain rigorous bounds on the transition probabilities between different M_0, which can be true only for typical random instances. It is understood that the two approaches are of different nature, one interested in average-case (over an ensemble of formulas) and the other in worst-case (true for any formula) behaviour of the same probabilistic algorithm.

5.2   Average Fraction of Unsat Clauses

Let us analyze the consequences of the above approximation on the average performances of the algorithm. We define the average fraction of unsat clauses at reduced time $t = T/M$ through

$$\varphi_0(t) = \frac{1}{M} \sum_{M_0=0}^{M} M_0 \, \mathrm{Prob}[M_0, T = tM] \quad . \tag{7}$$

Evolution equation (5) yields the following first-order linear differential equation,

$$\frac{d\varphi_0}{dt} = -1 + \frac{\alpha K}{2^K - 1} - \frac{\alpha K}{1 - 2^{-K}} \varphi_0 \ . \tag{8}$$

This equation, together with the initial condition $\varphi_0(0) = 2^{-K}$ (the initial configuration of the variables is random, thus each of the clauses is unsatisfied with probability 2^{-K}), can be easily solved. There appears a critical value of α,

$$\alpha_d(K) = \frac{2^K - 1}{K} \quad . \tag{9}$$

For $\alpha < \alpha_d(K)$, $\varphi_0(t)$ vanishes for a finite value of its argument, a solution is typically found in linear time. For larger values of α, φ_0 remains positive at all times, which means that typically a large formula will not be solved by PRWSAT, and that the fraction of unsat clauses on the plateau will be the limit when $t \to \infty$ of $\varphi_0(t)$. The predicted value for $K = 3$, $\alpha_d = 7/3$, is in good but not perfect agreement with the estimates from numerical simulations, around 2.7. The fraction of unsatisfied clauses on the plateau reads

$$\varphi_0(t \to \infty) = \frac{1}{2^K}\left(1 - \frac{\alpha_d(K)}{\alpha}\right) \quad , \tag{10}$$

and is shown on Fig. 2.

5.3 Large Deviations and Solving Time

As explained above, when $\alpha > \alpha_d$, the system gets almost surely trapped in a portion of the configuration space with a non zero number of unsatisfied clauses. Numerical experiments indicate the existence of an exponentially small-in-N probability $\sim \exp(N\,\bar\zeta(\alpha))$ with $\bar\zeta < 0$ for the escape of the trap and the discovery of a solution in linear time[4].

We now make use of our approximation to derive an approximate expression for $\bar\zeta$. Contrary to the previous section, we now consider the large deviation of the process with respect to its typical behavior. This can be accessed through the study of the large deviation function $\pi(\varphi_0, t)$ of the fraction φ_0,

$$\pi(\varphi_0, t) = \lim_{N\to\infty} \frac{1}{N} \ln \mathrm{Prob}\big[M_0 = M\,\varphi_0, T = t\,M\big] \tag{11}$$

To compute the large deviation function π, we introduce the generating function of M_0 [13],

$$G[y, T] = \sum_{M_0=0}^{M} \mathrm{Prob}[M_0, T]\, \exp(y\,M_0) \quad , \tag{12}$$

where y is a real-valued number. G is expected to scale exponentially with N with a rate

$$g(y, t) \equiv \lim_{N\to\infty} \frac{1}{N} \ln G[y, T = t\,M] = \max_{\varphi_0}[\pi(\varphi_0, t) + \alpha\,y\,\varphi_0] \,, \tag{13}$$

equal to the Legendre transform of π from insertion of definition (11) into Eqn. (12). Using evolution equations (5,6), we obtain the following equation for g,

$$\frac{1}{\alpha}\frac{\partial g(y, t)}{\partial t} = -y + \frac{\alpha K}{2^K-1}\left(e^y - 1\right) + K\left(e^{-y} - 1 - \frac{1}{2^K-1}\left(e^y - 1\right)\right)\frac{\partial g(y, t)}{\partial y} \,. \tag{14}$$

along with the initial condition

$$g(y, 0) = \alpha \ln\left(1 - \frac{1}{2^K} + \frac{e^y}{2^K}\right) \quad . \tag{15}$$

[.] This statement is directly related to Schöning's bound on the number of "trials" necessary to reach a solution. In practice, PRWSAT is run once, until a solution is found, which happens after an exponentially large time. Intuitively, the two statements "Having an exponentially small probability $\sim \exp(N\,\bar\zeta)$ of finding a solution in $O(N)$ steps" or "Finding a solution in $\sim \exp(-N\,\bar\zeta)$ steps (up to a non-exponential multiplicative factor) with high probability" are equivalent.

The average evolution studied in the previous section can be found again from the location of the maximum of π or, equivalently, from the derivative of g in $y = 0$: $\varphi_0(t) = (1/\alpha) \, \partial g / \partial y(0, t)$. The logarithm of the probability that a solution is reached after a $O(1)$ time ($O(N)$ flips) is given by

$$\bar{\zeta}(\alpha, K) = \lim_{t \to \infty} \pi(\varphi_0 = 0, t) = \int_0^{\tilde{y}(\alpha)} dy \, z(y, \alpha) \,, \tag{16}$$

where $z(y, \alpha) = (y - \alpha K(e^y - 1)/(2^K - 1))/(K(e^{-y} - 1 - (e^y - 1)/(2^K - 1)))$ and $\tilde{y}(\alpha)$ is the negative root of z.

Predictions for $\bar{\zeta}$ in the $K = 3$ case are plotted in Fig. 5. They are compared to experimental measures of ζ, that is, the logarithm (divided by N) of the average solving times. It is expected on intuitive grounds exposed in Sec. 3 that ζ coincides with $-\bar{\zeta}$. Despite the roughness of our approximation, theoretical predictions are in qualitative agreement with numerical experiments.

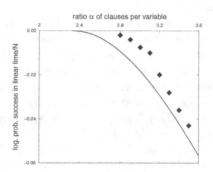

Fig. 5. Large deviations for the 3-SAT problem. The logarithm $\bar{\zeta}$ of the probability of successful resolution (over the linear in N time scale) is plotted as a function of the ratio α of clauses per variables. Prediction for $\bar{\zeta}(\alpha, 3)$ has been obtained within the approximation of Sec. 5. Diamonds corresponds to (minus) the logarithm ζ of the average resolution times (averaged over 2,000 to 10,000 samples depending on the values of α, N, divided by N and extrapolated to $N \to \infty$) obtained from numerical simulations. Error bars are of the order of the size of the diamond symbol. As expected, Schöning's bound $\bar{\zeta} \geq \ln(3/4) \simeq -0.288$, true for any satisfiable instance, lies below the estimate of $\bar{\zeta}$ computed for random 3-SAT instances for all ratios $\alpha < \alpha_c$.

A similar study of the behavior of PRWSAT on XORSAT problems has been also performed in [9,10], with qualitatively similar conclusions: there exists a dynamic threshold α_d for the algorithm, smaller than the satisfiability threshold (known exactly in this case [14,15,16]). For lower values of α, the number of steps needed to find a solution is linear in the size of the formula, between α_d and α_c solutions are found on exponentially large time scales, through fluctuations around a plateau value for the number of unsatisfied clauses. In the XORSAT case, the agreement between numerical experiments and this approximate study (which predicts $\alpha_d = 1/K$) is quantitatively better and seems to improve with growing K.

6   Conclusion and Perspectives

We have studied the dynamics of a simple local search procedure for the satisfaction of Boolean constraints, the Pure Random Walk algorithm. Use of complementary techniques (expansions and approximations) has allowed us to show that, for randomly drawn input instances, PRWSAT may have two qualitatively distinct behaviors. Instances with small ratios α of clauses per variable are almost surely solved in a time growing linearly with their size [2]. On the contrary, for ratios above some "dynamical" threshold α_d, the algorithm gets trapped for an exponentially large time in a region of the assignment space with a positive fraction of unsatisfied clauses. A solution is finally reached through a large "fluctuation" from this trap.

It would be of course of the greatest interest to confirm rigorously the results presented here. At first sight, the low α phase seems easier to tackle [2], due to the apparent concentration of the solving time in the large size limit and to the simplifications occurring in the neighborhood of $\alpha = 0$, as explained in Sec. 4. Another possible direction would be to consider the large K limit, some arguments suggesting that the very crude approximation of Sec. 5 becomes exact in this limit [9]. For "static" properties e.g. the value of the satisfiability threshold, it was recently rigorously shown that the simple "annealed" procedure of statistical physics becomes indeed exact in this limit [17], as was conjectured in [18]. Finally, a challenging problem would be to prove bounds on the transition probabilities for the number of unsatisfied clauses, in the spirit of Schöning's work [1], true w.h.p. on the random distribution of instances.

The PRWSAT local search procedure studied above is a particular case of a general algorithmic architecture known as WalkSAT [19]. In general, one still chooses randomly an unsatisfied clause at each step in an uniform way, but the choice of the variable to be flipped inside this clause is done following a specific heuristic. Many different strategies can, and have, been proposed (see for instance [20,11]). One can classify them into two categories. On the one hand, some of them are "Markovian": the choice of the variable to be flipped requires only the knowledge of their current assignment. For instance, one can choose the variable whose flip will lead to the greatest decrease of the number of unsatisfied clauses (WSAT/G in the classification of [20]), or the one which belongs to the minimal number of clauses only satisfied by the variable to be flipped (WSAT/B). On the other hand, another family of heuristics introduces "memory effects": besides the current status of the boolean variables, one bookkeeps the *age* of each of them, namely the time at which it was flipped for the last time. Heuristics in this family (for instance TABU, NOVELTY and R-NOVELTY) use this age information to choose the variable to be flipped. Most of these heuristics have a tunable parameter that, roughly speaking, controls the amount of randomness in the execution of the algorithm. It has been often empirically observed that this parameter presents an optimum value [20,11]. With the picture of fluctuations around a plateau in mind, the claims of [20] become clearer: it is intuitive that to reach a solution quickly, one would have to favor at the same time a low value of the plateau height, and a large intensity of the fluctuations around it. A compromise must be found between these two quantities to have an optimal tuning of the algorithm. The expansion method for the average solving time in the linear regime can be easily extended to the "Markovian" heuristics. One could thus justify the existence of an optimal tuning, at least in the small α regime.

Further studies of these heuristics is necessary; the first point to investigate is the generality of the scenario presented here, with the appearance of a plateau above some ratio α_d and resolution happening through fluctuations. Next, the dependence of α_d with the the features of the local search heuristic (randomness parameter, ...) should be clarified. This question is all the more of interest that some local search algorithms seem to be able to solve random formulas in a linear number of flips up to, or very close to the sat/unsat threshold α_c [21].

Acknowledgements: We thank S. Cocco, L. Cugliandolo, C. Deroulers, O. Dubois, C. Girit and A. Montanari for numerous and stimulating discussions. Partial financial support was provided by the French Ministry of Research through the ACI Jeunes Chercheurs "Algorithmes d'optimisation et systèmes quantiques désordonnés".

References

1. U. Schöning, A Probabilistic algorithm for k-SAT based on limited local search and restart, *Algorithmica* **32**, 615-623 (2002).
2. M. Alekhnovich, E. Ben-Sasson, Analysis of the Random Walk Algorithm on Random 3-CNFs, preprint (2002).
3. S. Cocco, A. Montanari, R. Monasson, G. Semerjian, Approximate analysis of search algorithms with "physical" methods, *cs.CC/0302003*.
4. D. Mitchell, B. Selman, H. Levesque, Hard and Easy Distributions of SAT Problems, *Proc. of the Tenth Natl. Conf. on Artificial Intelligence (AAAI-92)*, 440-446, The AAAI Press / MIT Press, Cambridge, MA (1992).
5. C.H. Papadimitriou, On Selecting a Satisfying Truth Assignment, *Proceedings of the 32nd Annual IEEE Symposium on Foundations of Computer Science*, 163-169 (1991).
6. T. Hofmeister, U. Schöning, R. Schuler and O. Watanabe, A probabilistic 3-SAT algorithm further improved, *Proceedings of 19th Symposium on Theoretical Aspects of Computer Science*, *LNCS* 2285, 192-203 (2002).
7. S. Baumer and R. Schuler, Improving a probabilistic 3-SAT algorithm by dynamic search and independant clause pairs, in this volume.
8. A.J. Parkes, Scaling Properties of Pure Random Walk on Random 3-SAT, *Lecture Notes in Computer Science* **2470**, 708 (2002).
9. G. Semerjian, R. Monasson, Relaxation and metastability in a local search procedure for the random satisfiability problem, *Phys. Rev. E* **67**, 066103 (2003).

10. W. Barthel, A.K. Hartmann, M. Weigt, Solving satisfiability problems by fluctuations: The dynamics of stochastic local search algorithms, *Phys. Rev. E* **67**, 066104 (2003).
11. H. H. Hoos, T. Stützle, Local Search Algorithms for SAT: An Empirical Evaluation, *J. Automated reasoning* **24**, 421 (2000).
12. G. Semerjian, L.F. Cugliandolo, Cluster expansions in dilute systems: applications to satisfiability problems and spin glasses, *Phys. Rev. E* **64**, 036115 (2001).
13. S. Cocco, R. Monasson, Restarts and exponential acceleration of random 3-SAT instances resolutions: a large deviation analysis of the Davis-Putnam-Loveland-Logemann algorithm, to appear in *Annals of Mathematics and Artificial Intelligence* (2003)
14. O. Dubois, J. Mandler, The 3-XORSAT threshold, *Proc. of the 43rd annual IEEE symposium on Foundations of Computer Science*, Vancouver, 769–778 (2002).
15. S. Cocco, O. Dubois, J. Mandler, R. Monasson, Rigorous decimation-based construction of ground pure states for spin glass models on random lattices, *Phys. Rev. Lett.* **90**, 047205 (2003).
16. M. Mézard, F. Ricci-Tersenghi, R. Zecchina, Alternative solutions to diluted p-spin models and XORSAT problems, *J. Stat. Phys.* **111**, 505 (2003).
17. D. Achlioptas, Y. Peres, The Threshold for Random k-SAT is $2^k(\ln 2 + o(1))$, preprint (2002).
18. R. Monasson, R. Zecchina, Statistical mechanics of the random K-satisfiability model, *Phys. Rev. E* **56**, 1357 (1997).
19. B. Selman, H. Kautz and B. Cohen, Noise Strategies for Improving Local Search, *Proc. AAAI-94*, Seattle, WA, 337-343 (1994).
20. D. McAllester, B. Selman, H. Kautz, Evidence for Invariants in Local Search, *Proc. AAAI-97*, Providence, RI (1997).
21. S. Seitz, P. Orponen, An efficient local search method for random 3-Satisfiability, preprint (2003).

Hidden Threshold Phenomena for Fixed-Density SAT-formulae

Hans van Maaren and Linda van Norden

Technical University Delft
Faculty of Information Technology and Systems
Department of Information Systems and Algorithms
Mekelweg 4, 2628 CD Delft
The Netherlands
H.vanMaaren@ewi.tudelft.nl, L.vanNorden@ewi.tudelft.nl

Abstract. Experimental evidence is presented that hidden threshold phenomena exist for fixed density random 3-SAT and graph-3-coloring formulae. At such fixed densities the average Horn fraction (computed with a specially designed algorithm) appears to be a parameter with respect to which these phenomena can be measured. This paper investigates the effects of size on the observed phenomena.
Keywords: satisfiability, Horn, thresholds, 3-SAT, graph-3-coloring

1   Introduction

It is well-known that the class of random 3-SAT problems is subject to threshold phenomena with respect to the density parameter. Given a random 3-SAT instance, without unit clauses and two-literal clauses, its density d is defined as the ratio of the number of clauses and the number of variables. Low density problem are (with high probability) satisfiable and easily shown to be satisfiable with DPLL techniques. High density problems are (with high probability) unsatisfiable. At a certain value $d_0(n)$, the threshold value for instances with n variables, a phase-transition occurs: instances generated with this particular density are satisfiable with probability 0.5 and solving them with DPLL techniques is very expensive relative to instances with larger or smaller densities. The fact that threshold $d_0(n)$ exists is proven ([10]), but determining the exact position of this threshold for a certain size n is a difficult problem. A large variety of experiments supports the claim that $d_0(n)$ is located in the neighborhood of 4.27.

There is a vaste literature on phase-transitions in the 3-SAT problem and related problems. [13] and [2] study threshold phenomena with respect to the density parameter in 3-SAT and the class of $(2+p)$-SAT problems. In a $(2+p)$-SAT-formula a fraction p of the clauses has three literals while the others have two literals.

In this paper we provide experimental evidence that the class of random 3-SAT problems with fixed density again is subject to similar phenomena with respect to a different parameter: an enhanced version of the Horn fraction. Horn

E. Giunchiglia and A. Tacchella (Eds.): SAT 2003, LNCS 2919, pp. 135–149, 2004.

clauses are clauses in which at most one variable occurs in negated form. The Horn fraction of a CNF-formula is the number of Horn clauses divided by the total number of clauses. Various notions of Horn fraction will be discussed more extensively in Section 3.

Before going into details, we consider Figure 1. At this moment it is enough to know that Figure 1 is based on the use of an algorithm, which we will refer to as average-Horn SAT (aHS for short), described in Section 7. In solving an instance it creates a search tree in which at each node the Horn fraction of the current CNF is computed. On the vertical axis is indicated the number of nodes aHS needs to solve the problem. On the horizontal axis is indicated the average of the Horn fractions over all nodes of the search tree. The unsatisfiable instances are indicated by crosses and the satisfiable ones by dots. The test set is a set of 150 unsatisfiable and 150 satisfiable random 3-SAT instances with 120 variables and density 4.3. All random 3-SAT instances used in this and the next experimental tests are generated by the OK-generator ([11]).

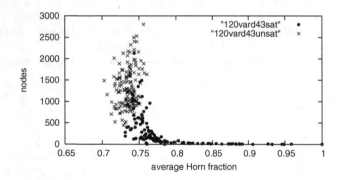

Fig. 1. 120 variables, d=4.3

Examining Figure 1 we can distinguish a few interesting features:

1. A correlation between average Horn fraction and satisfiability
2. A weak form of linear separation between satisfiable and unsatisfiable instances seems possible, in the sense that a line $N = \alpha H + \beta$, in which N is the number of nodes and H the average Horn fraction, can be drawn which constitutes a weak separation.
3. A correlation between average Horn fraction and the number of nodes it took to solve an instance (independent of the satisfiability of the instance).

The obvious questions to ask at this stage are

a. Is it necessary to solve an instance completely to establish a good approximation of its average Horn fraction?
b. Is the correlation between computational effort and average Horn fraction a phenomenon which is typically solver dependent?

c. Are these phenomena size dependent?

d. Do similar pictures arise for models with other densities?

e. Is the phenomenon typical for 3-SAT, or can we expect similar behavior for other classes of CNF-formulae?

We ran several experiments to gain more evidence for the observed features and to get a first impression of what the answers to these questions might be. In this paper we focus on the size independency of the correlations between the Horn fraction, satisfiability and computational effort for random 3-SAT formulae and graph-3-coloring problems at the threshold. We also established density independency and linear separation possibilities for these type of problems, but on a less extensive experimental scale. These latter results were presented at the SAT03 conference. However, because of space limitations we do not present them here.

2 Preliminaries

A clause is a disjunction of literals. A literal is an atomic propositional variable or its negation (indicated by ¬). A CNF-formula is a logical conjunction of clauses. The satisfiability problem is to find an assignment of the truth-values true and false to the propositional variables such that all clauses in the CNF-formula are true or to prove that such truth assignment does not exist. A formula is satisfiable if a truth assignment exists and unsatisfiable otherwise. [5] proved that the satisfiability problem is $\mathcal{NP}$-hard. In this paper, we use 3-SAT to denote the class of problems where the CNF-formula consists only of clauses with exactly three literals (some authors use {3}-SAT for notation). [5] also proved that the class of 3-SAT problems is $\mathcal{NP}$-hard.

3 Horn Fractions

In this section, we will describe different variants of the Horn fraction of an instance. Our aim is to define a concept of 'Horn fraction' of a CNF-formula $\mathcal{F}$ which can be computed efficiently and reveals similar distributions as in Figure 1. A Horn clause is a clause with at most one negated variable.

Definition 1. *The Horn fraction $H(\mathcal{F})$ of a CNF-formula $\mathcal{F}$ is the number of Horn clauses divided by the total number of clauses.*

CNF-formulae with only Horn clauses can be solved in time linear in the number of literals by the algorithm in [7]. Renaming, or flipping, a variable X_i means that every occurrence of X_i is replaced by $\neg X_i$ and every $\neg X_i$ is replaced by X_i. By renaming, the Horn fraction might change.

Definition 2. *The optimal Horn fraction of a CNF-formula $\mathcal{F}$ after renaming, $H_{opt}(\mathcal{F})$ is the largest fraction of Horn clauses that can be obtained by renaming, or flipping, variables in $\mathcal{F}$.*

The problem of computing this optimal Horn fraction after renaming is $\mathcal{NP}$-hard ([6]). Recognizing whether a formula $\mathcal{F}$ is renamable Horn, i.e. $H_{opt}(\mathcal{F}) = 1$, however, can be done in linear time, for example by the algorithm of [1]. [4] proved $\mathcal{NP}$-completeness of the related problem of finding a renamable Horn subproblem with as many variables as possible.

Definition 3. *Given a CNF-formula $\mathcal{F}$ and a deterministic renaming routine $\mathcal{R}$, the sub-optimal Horn fraction after renaming, $H_{subopt}(\mathcal{F}, \mathcal{R})$, is the Horn fraction obtained after applying routine $\mathcal{R}$.*

[3] presents a rounding procedure to approximate $H_{opt}(\mathcal{F})$ that has a guaranteed performance ratio of $\frac{40}{67}$ on 3-SAT problems. This procedure rounds the solution of a linear program or the vector with all 0.5, depending on the objective value of the linear program, to an integer valued solution indicating which variables have to be flipped. The routine $\mathcal{R}$ we use in aHS to sub-optimize the Horn fraction, is a deterministic greedy local search procedure. It is simple and its performance on random 3-SAT problems seems to be quite good: for instances with 40 variables, the sub-optimal Horn fraction after renaming computed by our heuristic is only about 1.9% from the optimal Horn fraction after renaming. Computing the optimal Horn fraction after renaming for larger instances cannot be done in reasonable time.

Definition 4. *Given a CNF-formula $\mathcal{F}$, a deterministic renaming routine $\mathcal{R}$, and a deterministic search tree T which solves $\mathcal{F}$, the sub-optimal average Horn fraction over the search tree (in short average Horn fraction), $AH_{so}(\mathcal{F}, \mathcal{R}, T)$, is the average of the $H_{subopt}(\mathcal{F}_t, \mathcal{R})$ over all nodes t in the search tree T, with $\mathcal{F}_t$ the CNF-formula at node t of T.*

The design of the search tree we use for computing $AH_{so}(\mathcal{F}, \mathcal{R}, T)$ is described in Subsection 7.5. To compute $AH_{so}(\mathcal{F}, \mathcal{R}, T)$, the instance has to be solved completely. As already mentioned in question a in the introduction, we would like to have a method for computing a Horn fraction without having to solve $\mathcal{F}$ completely.

Definition 5. *Given a CNF-formula $\mathcal{F}$, a deterministic renaming routine $\mathcal{R}$, and a deterministic search tree T which solves $\mathcal{F}$, the k-level average Horn fraction, $AH_{so}(\mathcal{F}, \mathcal{R}, T, k)$, is the average Horn fraction over the nodes in T up to level k.*

The above definitions are sound because the routine and search tree are deterministic. As we shall show in Section 7, our choices of $\mathcal{R}$ and T guarantee that $AH_{so}(\mathcal{F}, \mathcal{R}, T, k)$ can be computed in polynomial time for fixed k. An appropriate choice for k depends on the size and nature of the instances. Dealing with problems of the size we investigated in this paper, $k = 2$ or $k = 3$ seem to be appropriate choices.

Figure 2 illustrates the distribution of the same instances as in Figure 1, but now with respect to their 2-level average Horn fractions. On the horizontal axis is indicated the 2-level average Horn fraction of the instance, computed by

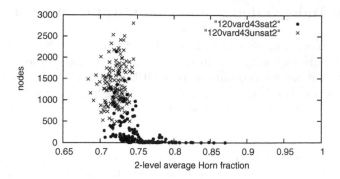

Fig. 2. 120 variables, d=4.3

aHS. On the vertical axis is the number of nodes it took to solve the instance completely by aHS. Figure 2 shows that it is not necessary to solve an instance completely to have an appropriate concept of average Horn fraction which shows a similar distribution.

To test whether this kind of distribution also occurs with respect to the optimal Horn fraction after renaming, we computed this optimal Horn fraction, using the Integer Linear Program proposed by [3] for a set of random 3-SAT instances with 40 variables. Plotting this optimal Horn fraction against the number of nodes it took aHS to solve these instances, revealed no clear correlation. However, if we plot the same instances against the 1-level average Horn fraction the correlations discussed show up. Hence, this type of distribution seems to be specific for the way aHS computes the k-level average Horn fraction (that is, the way aHS decomposes the search space).

3.1   Weighted and Max-Horn Variants

Each of the above notions can be extended to weighted variants. In these variants, the weighted fraction of Horn clauses is optimized or approximated. This weighted fraction is defined as

$$\frac{\sum_{l=1}^{n} w_l H_l}{\sum_{l=1}^{n} w_l S_l}$$

in which n is the number of variables, w_l is a weight of clauses of length l, S_l is the number of clauses of length l and H_l is the number of Horn clauses of length l. Varying weights might influence computation times of our algorithm and the observed distributions significantly, but experiments showed that this is not the case for random 3-SAT problems.

One might consider also the k-level max-Horn fraction which is defined as the maximum of the Horn fractions in the nodes of the search tree up till level k. These variants are used in Section 6.

4 Solver Independency and Computational Effort

In this Section we will present evidence that there is a relation between the
k-level average Horn fraction of an instance and the computational effort necessary to solve it for various current best DPLL techniques. In the remainder,
the k-level average Horn fractions are always established using aHS. We selected
three different SAT-solvers: OKsolver, kcnfs (award winning solvers in the random category of the SAT-competition in 2002 and 2003 respectively) and aHS.
The OKsolver ([12]) is a DPLL-like algorithm, with reduction by failed literals
and autarkies. The branching heuristic chooses a variable creating as many new
two-literal clauses as possible and the first branch is chosen maximizing an approximation of the probability that the subproblem is satisfiable. kcnfs ([8]) is
a complete, deterministic solver optimized for solving random k-SAT formulae.
kcnfs contains a new heuristic searching variables belonging to the backbone of
a CNF-formula.

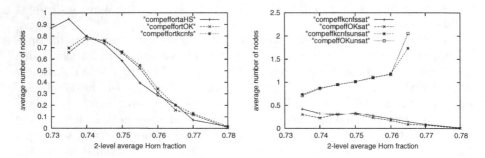

Fig. 3. 400 variables, d=4.25, OKsolver and kcnfs and 150 variables, d=4.25,
aHS

In the first picture in Figure 3 we indicate on the vertical axis the average number of nodes in the search tree created by the OKsolver, kcnfs and aHS,
scaled with the average number of nodes it took the solver concerned to solve the
unsatisfiable instances of the indicated size. On the horizontal axis is indicated
the 2-level average Horn fraction computed with aHS. The instances concerned
are 2000 random 3-SAT instances with 400 variables, for the OKsolver and kcnfs, and 300 instances of 150 variables for aHS. All instances have density 4.25
and are generated by the OKgenerator. Notice that instances of 400 variables
are difficult or even unsolvable in reasonable time for most solvers. The second
figure in Figure 3 represents the average number of nodes for the satisfiable and
unsatisfiable instances separately for the OKsolver and kcnfs.

The figure shows that there is an obvious correlation between the size of the
search tree and the 2-level average Horn fraction established using aHS. From
the first figure it is clear that for all of the three solvers the average size of the
search tree shows an increase/decrease pattern peaking in the vicinity of the

transition point (see Section 5). From the second figure it can be concluded that satisfiable instances get easier when the 2-level average Horn fraction increases while unsatisfiable instances contrarily get more difficult to solve.

5   Size Independency of the Correlation between Horn Fraction and Satisfiability

In this Section we will examine the correlation between Horn fraction and satisfiability for random 3-SAT instances with the number of variables between 150 and 400 and the density equal to 4.25. For each size we use 1000 satisfiable and 1000 unsatisfiable instances. The instances are all close to the threshold. Hence, based on the density alone, an instance has a probability of being satisfiable being equal to 50%. For each of the instances we computed the Horn fraction of the formula in the root node (level 0), the 1-level and 2-level average Horn fraction. From these data we computed the average k-level Horn fraction for both the satisfiable (column μ_s) and unsatisfiable (column μ_u) instances, and their difference (column $\mu_s - \mu_u$). This difference is an indication for the separation between the distributions of the k-level average Horn fraction of satisfiable and unsatisfiable instances. Columns σ_s and σ_u indicate respectively the involved standard deviations. For each size and level, a spline function is fit to the discrete cumulative distribution function, and the intersection point (column 'transition') between both density functions is computed. At this point the probability of an instance being satisfiable is 0.5. The 'satCE' is the probability that a satisfiable instance is classified incorrectly by the test T that classifies an instance as satisfiable if its Horn fraction is larger than or equal to the k-level average Horn fraction in the intersection point and unsatisfiable otherwise. Analogously, the 'unsatCE' is the probability of an unsatisfiable instance being classified incorrectly similarly. The 'lefttail' is the probability that a satisfiable instance has a k-level average Horn fraction smaller than μ_u (the expected value of the k-level average Horn fraction of the unsatisfiable instances).

From Table 1, we can conclude various trends. First, for fixed k, the difference $\mu_s - \mu_u$ decreases, but the probabilities quoted show stable. Further, for fixed size, separation improves with increasing k. To illustrate this somewhat more we added $k = 3$ for the 400 variable instances.

The first picture in Figure 4 shows the probability density functions of satisfiable and unsatisfiable instances with 400 variables and density 4.25 over the 2-level average Horn fraction. The transition point and μ_u are indicated by the two vertical lines. The second picture gives for each value of the 2-level average Horn fraction the probability that an instance with the corresponding fraction is satisfiable. The vertical lines indicate mu_s and mu_u. As mentioned, for the complete data set the probability of being satisfiable is about 50%, but for example for instances with a 2-level average Horn fraction larger than 0.756 this probability of being satisfiable is 80%.

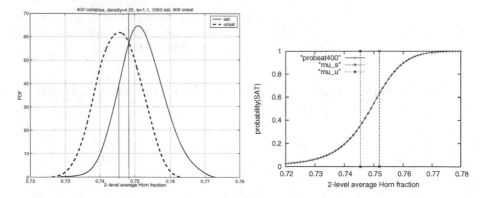

Fig. 4. 400 variables, probability density and probability of being satisfiable

6   Graph 3-Coloring

To investigate whether the phenomena above are typical for random 3-SAT we ran several experiments with graph 3-coloring instances, coded as satisfiability problems. The problem is to check whether the nodes in a graph can be colored with three colors such that no two adjacent nodes have the same color. Let N be the number of nodes in the graph and E the number of edges. To encode the graph-3-coloring problem as a satisfiability problem variables X_{nk} are introduced being true if node n has color k and false otherwise. The encoding of the graph 3-coloring problem has N clauses of length three, expressing that every node has to be assigned one of the three colors, $3N$ clauses of length two formulating for each node and pair of colors that the node does not have both colors, called unicity clauses, and finally there are $3E$ clauses of length two ensuring for every edge that both end points do not have the same color.

For this type of instances the average Horn fraction can predict quite well whether an instance is satisfiable or not, but the optimal Horn fraction in the root node does not give any hint, because the latter is equal for every instance of a certain size. By renaming all variables, all clauses of length two have two positive literals and the clauses of length three have three negative literals. Hence, the number of Horn clauses, after this renaming, is $3N + 3E$, independent of the particular instance. We cannot improve on this, because to get one of the clauses of length three Horn, we need to rename at least two variables. These two variables occur together in one of the $3N$ unicity clauses, so one of the clauses with two literals ceases to be Horn.

Besides the version of the problem described above, we also considered a variant without unicity clauses. These clauses are not strictly necessary to find a feasible coloring of the nodes, because we can always choose one the colors a node gets if it is multiple-colored. This variant turns out to be easier, and the fact that all instances have the same Horn fraction at the root node with weights equal to 1 does not hold, in absence of the unicity clauses.

#var	level	μ_s	σ_s	μ_u	σ_u	$\mu_s - \mu_u$	satCE	unsatCE	lefttail	transition
150	0	0.7102	0.0113	0.7031	0.0112	0.0071	40.0%	32.0%	25.0%	0.708
	1	0.7381	0.0164	0.7260	0.0108	0.0121	39.0%	25.0%	20.3%	0.733
	2	0.7489	0.0155	0.7349	0.0099	0.0140	37.1%	20.5%	15.1%	0.743
200	0	0.7108	0.0094	0.7040	0.0095	0.0068	30.0%	40.0%	22.8%	0.706
	1	0.7378	0.0107	0.7287	0.0094	0.0091	32.0%	32.3%	18.8%	0.7344
	2	0.7489	0.0102	0.7383	0.0085	0.0106	33.5%	23.1%	12.5%	0.7446
250	0	0.7102	0.0086	0.7044	0.0082	0.0058	39.0%	34.0%	25.2%	0.7078
	1	0.7378	0.0092	0.7302	0.0084	0.0076	36.6%	29.7%	20.4%	0.7347
	2	0.7493	0.0085	0.7409	0.0074	0.0084	33.8%	25.5%	16.1%	0.7458
300	0	0.7106	0.0078	0.7050	0.0077	0.0056	36.3%	35.0%	23.3%	0.7079
	1	0.7385	0.0083	0.7315	0.0081	0.0070	32.6%	34.4%	19.5%	0.7557
	2	0.7506	0.0078	0.7428	0.0072	0.0078	32.7%	27.4%	14.4%	0.7471
350	0	0.7105	0.0069	0.7051	0.0067	0.0054	35.9%	33.3%	22.0%	0.708
	1	0.7388	0.0074	0.7321	0.0073	0.0067	31.4%	33.6%	17.6%	0.735
	2	0.7512	0.0072	0.7438	0.0067	0.0074	31.7%	27.4%	15.3%	0.7478
400	0	0.7105	0.0065	0.7059	0.0063	0.0046	37.2%	34.6%	23.9%	0.7084
	1	0.7390	0.0069	0.7333	0.0066	0.0057	35.4%	31.8%	20.0%	0.7364
	2	0.7519	0.0063	0.7454	0.0059	0.0065	31.6%	27.5%	13.8%	0.7489
	3	0.7585	0.0061	0.7517	0.0056	0.0068	27.5%	31.1%	11.9%	0.755

Table 1. Random 3-SAT from 150 to 400 variables

We experimented with the weight of the clauses. It turned out that the weight has a significant influence on both computation times and the shape of the distribution. This influence turned out to be more prominent for the problem without unicity clauses, but it is also present for the problem with unicity clauses. Since the CNF-formulae have roughly seven times as much 2-literal clauses as 3-literal ones, we chose the weighted variant where 'being Horn' of a 3-literal clause is considered seven times more important than 'being Horn' of a 2-literal clause. We confirmed experimentally that this weight indeed showed the best separation properties.

nodes	edges	level	μ_s	σ_s	μ_u	σ_u	$\mu_s - \mu_u$	satCE	unsatCE	lefttail	transition
100	222	1	0.8674	0.0204	0.844	0.0046	0.0234	28.8%	3.3%	11.6%	0.852
		2	0.9029	0.0248	0.8491	0.0053	0.0538	5.6%	1.3%	0.4%	0.86
150	333	1	0.8523	0.0102	0.8445	0.0039	0.0078	43.1%	15.6%	17.5%	0.848
		2	0.8662	0.015	0.8487	0.0041	0.0175	18.9%	7.2%	3.5%	0.854
		3	0.8882	0.0253	0.8532	0.0042	0.035	10.3%	1.8%	0.9%	0.861
150	339	1	0.8503	0.0109	0.8431	0.004	0.0072	59.1%	6.5%	25.5%	0.849
		2	0.8649	0.018	0.8483	0.0041	0.0166	29.9%	7.9%	6.4%	0.854
		3	0.8813	0.0261	0.8505	0.0043	0.0308	11.0%	5.4%	0.6%	0.857

Table 2. Graph-3-coloring with respect to the k-level average Horn fraction

In the next experiment we generated 500 instances with 100 nodes and 222 edges. For this density the probability for an instance being satisfiable is approximately 50%. We generated also 600 instances with 150 nodes with 333 edges, i.e. with the same density as the instances with 100 nodes, but also with 339 edges, seemingly closer to the threshold density for this size. The results are given in the next table, which illustrates that the correlation between k-level average Horn fraction and satisfiability is much more prominent than for random 3-SAT instances. Considering the k-level max-Horn fraction in the search tree at level k, we even get better results, as is shown in Table 3.

nodes	edges	level	μ_s	σ_s	μ_u	σ_u	$\mu_s - \mu_u$	satCE	unsatCE	lefttail	transition
150	333	1	0.8985	0.0536	0.8555	0.0075	0.043	42.2%	10.2%	15.4%	0.864
		2	0.9809	0.0353	0.865	0.0083	0.1159	7.9%	0.6%	1.6%	0.887
		3	0.9989	0.0085	0.869	0.0084	0.1299	0.2%	0.0%	0.2%	0.9
150	339	1	0.8931	0.0542	0.8549	0.0077	0.0382	54.5%	9.0%	26.1%	0.865
		2	0.9688	0.0492	0.8635	0.0079	0.1053	17.1%	1.1%	3.8%	0.881
		3	0.9967	0.0125	0.8682	0.0084	0.1285	0.6%	0.3%	0.0%	0.892

Table 3. Graph-3-coloring with respect to k-level max-Horn fraction

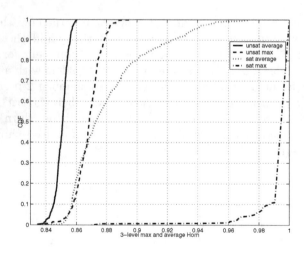

Fig. 5. Graph-coloring, 150 nodes, 339 edges, cumulative distribution functions

In Figure 5 the cumulative distribution functions for the satisfiable and unsatisfiable instances with 150 nodes and 339 edges are shown, both with respect to the 3-level average Horn fraction and the 3-level max-Horn fraction. On the vertical axis are the distribution functions $F(x)$, indicating the probability that an instance has a 3-level average Horn or max-horn fraction smaller than

x. These distributions illustrate that the correlation between satisfiability and 3-level Horn fraction is more prominent (almost complete separation over the sample) for max-Horn than for average Horn. The phenomena show much more convincing than for random 3-SAT.

7   The Algorithm aHS

In this section we describe an algorithm, which is typically designed for the purpose of fast convergence of the average Horn fraction. Before going into the details of our algorithm we first briefly summarize the main steps

1. Read the CNF-formula from the input and identify all unit clauses and propagate the corresponding values through the formula until no unit clauses remain. The resulting formula is $\mathcal{F}$.
2. Apply to $\mathcal{F}$ single look-ahead unit propagation in order to simplify the formula as much as possible. If the formula is proved unsatisfiable or a satisfying assignment is found: STOP. Otherwise the resulting formula is $\mathcal{F}_1$
3. Greedy suboptimize the Horn fraction to obtain $\mathcal{F}_2$
4. Transform $\mathcal{F}_2$ to a linear program and solve this linear program to get a solution vector (splitting point) ω .
5. Define subproblems of $\mathcal{F}_2$ using a decomposition of the search space based on ω
6. Go into breadth-first recursion.

Each of the relevant steps is described in the corresponding subsection.

7.1   Propagating the Unit Clauses

For each of the unit clauses in the input, the variable is fixed to the appropriate truth-assignment. In this step, formulae with only Horn clauses are already solved.

7.2   Single Look-Ahead Unit Propagation

Single-lookahead unit resolution is described in [9]. In full single look-ahead unit resolution, every variable is set temporarily to true. If this does lead to a contradiction the variable can be fixed to false for the current node and all of its children. This value is propagated. If this propagation does lead to a contradiction we can conclude that the formula is unsatisfiable and backtrack. This loop is continued until none of the variables can be concluded to be false in a full loop over the variables. After this loop, an analogous loop for fixing the variables to false is made. We repeat the process until none of the variables could be fixed during the last loop over the positive and negative assignments.

7.3   Greedy Horn Heuristic

Below, we resume a greedy local search heuristic to increase the fraction of
Horn clauses in the formula. After computing the (weighted) Horn fraction, for
every variable X_i, the resulting increase in (weighted) Horn fraction is computed
in case the variable would be flipped. If the largest increase is positive, the
corresponding variable is flipped. If the largest increase is 0, flip, if possible, the
variable with largest positive increase in the number of positive literals.

Notice that the complexity of this heuristic is $\mathcal{O}(nm)$ in which n is the number
of variables and m is the number of clauses.

7.4   Linear Program

Let n be the number of variables and m the number of clauses of a satisfiability
instance. As is common knowledge, an instance of the satisfiability problem can
be described by a binary integer linear program. Let A be a $m \times n$ 0-1 matrix,
with $a_{ij} = 1$ if variable X_j is contained in clause i, $a_{ij} = -1$ if the negation of
variable X_j is contained in clause i and 0 otherwise. Let x be a 0-1 vector with
$x_j = 1$ if variable j is set to true and 0 otherwise. For the vector b of right hand
sides, b_i equals 1 minus the number of negated variables in clause i. With this
notation a satisfying assignment corresponds to a lattice point of the polytope

$$\{Ax \geq b, x \in \{0,1\}^n\}$$

To illustrate this formulation the clause $X_p \vee \neg X_q \vee X_r$ gives the linear constraint
$x_p - x_q + x_r \geq 0$.

As objective function $\sum x_j$ is used. The constraints $x \in \{0,1\}^n$ are replaced
by $0 \leq x \leq 1$.

7.5   Search Tree Design

The search tree design described in this Subsection is based on [14]. The authors
of this paper show that integer programs having only inequalities satisfying the
so-called generalized Horn condition

$$\alpha_1 x_1 + \alpha_2 x_2 + \ldots + \alpha_n x_n \geq \alpha,$$

with at most one α_i negative, can be solved effectively for feasibility using an ho-
motopy procedure and a certain subdivision of Euclidean space. This subdivision
now is precisely what we use in our splitting of the search space.

The region G_i is the cone spanned by the vectors from
$\{-e_0, -e_1, \ldots, -e_n\}\setminus\{-e_i\}$, in which e_j is the unit vector with a 1 in the j-th
row, e_0 is the vector with one in all components and e_i is one of the vectors from
$\{e_0, e_1, \ldots, e_n\}$. The regions that follow from this splitting are described below.

The first region is

$$G_0 = \{x_i \leq \omega_i, \text{ for all } i\}.$$

Since only lattice points are involved, we may replace $x_i \leq \omega_i$ by $x_i \leq \lfloor \omega_i \rfloor$. The clauses that correspond to these linear inequalities are $\neg X_i$ for all i for which $\omega_i < 1$.

Region G_1 is defined by

$$x_1 \geq \omega_1, \tag{1}$$

$$x_1 - x_2 \geq \lceil \omega_1 - \omega_2 \rceil, \tag{2}$$

$$\cdots$$

$$x_1 - x_n \geq \lceil \omega_1 - \omega_n \rceil. \tag{3}$$

$x_1 \geq \omega_1$ may be replaced by $x_1 \geq \lfloor \omega_1 \rfloor + 1$ because $x_1 = \omega_1$ is already included in G_0. Note that this region is empty if $\omega_1 = 1$. Otherwise, X_1 is added to the CNF-formula for this region. Note that or region 1 is empty or $x_1 = 1$, i.e. X_1 is true. In case the region is not empty, we use the fact that $x_1 = 1$ to get $x_k = 0$ if $\omega_1 - \omega_k \in (0, 1]$.

In general, region G_i is defined as

$$x_i \geq \lfloor \omega_i \rfloor + 1, \tag{4}$$

$$x_i - x_k \geq \lfloor \omega_i - \omega_k \rfloor + 1, k < i, \tag{5}$$

$$\cdots$$

$$x_i - x_k \geq \lceil \omega_i - \omega_k \rceil, k > i. \tag{6}$$

Clearly, the region is empty if $\omega_i = 1$. Otherwise, we have $x_i = 1$ or X_i is true. For the case $k > i$ we have $\neg X_k$ if $\lceil \omega_i - \omega_k \rceil = 1$. For the case $k < i$, we have $\neg X_k$ if $\lfloor \omega_i - \omega_k \rfloor = 0$ and we can conclude that the region is empty if it is 1.

Thus, finally, only sets of unit literal clauses are added to the CNF-formula for each of the regions, resulting in a number of at most $n + 1$ child nodes.

Example To illustrate this approach, take for example, $\omega = (1, \frac{3}{4}, \frac{3}{4}, 0)$. In this case the regions are

$$G_0 = \{x_1 \leq 1, x_2 = x_3 = x_4 = 0\} \tag{7}$$

$$G_1 = \emptyset \tag{8}$$

$$G_2 = \{x_2 = 1, x_4 = 0, x_1, x_3 \text{ free}\} \tag{9}$$

$$G_3 = \{x_3 = 1, x_2 = x_4 = 0, x_1 \text{ free}\} \tag{10}$$

$$G_4 = \{x_4 = 1, x_1, x_2, x_3 \text{ free}\} \tag{11}$$

To see that G_1 is empty note that $\omega_1 = 1$. $x_1 \geq \lfloor \omega_1 \rfloor + 1$ implies $x_1 \geq 2$. For example, for G_2, $x_4 = 0$ follows from $\lceil \omega_2 - \omega_4 \rceil = 1$, hence $x_2 - x_4 \geq 1$. Combining with $x_2 = 1$ this gives $x_4 = 0$. Note that G_4 is much larger than the other regions.

In the two dimensional space the splitting of the search space graphically look as follows

Notice that ω is determined by the only two inequalities not satisfying the generalized Horn condition. This is the main intuition behind the splitting: try to eliminate many non-Horn clauses by splitting the search space at ω. Notice

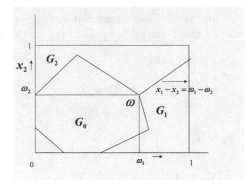

Fig. 6. Splitting the search space into regions

that in higher dimensions ω will be determined by more inequalities being not generalized Horn.

8   Conclusions and Future Research

In this paper, we presented a first experimental evidence that the class of fixed-density random 3-SAT problems is again subject to threshold phenomena with respect to the newly introduced parameter k-level average Horn fraction. We also established similar (even stronger) phenomena for random graph-3-coloring. This feature calls for more extensive experiments to establish more accurate separations, with more accurate statistical methods. Also, theoretical back-up is necessary (sofar, the only back-up comes from [14]). However, the algorithm aHS used to compute the Horn fractions contains elements (greedy optimization, linear programming relaxation) which makes it almost impossible to analyze. Theoretical back-up must come either from a clear mathematical statement showing separability possibilities for an $\mathcal{NP}$-hard problem class or from a much more transparent notion of 'average Horn fraction'. Whether the observed phenomena can be made useful in order to predict satisfiability of a certain 'difficult' and structured CNF-formula of particular interest is under current research.

References

[1] B. Aspvall, M.F. Plass, and R.E. Tarjan. A linear time algorithm for testing the truth of certain quantified boolean formulas. *Information Processing Letters*, 8:121–123, 1979.

[2] G. Biroli, S. Cocco, and R. Monasson. Phase transitions and complexity in computer science: an overview of the statistical physics approach to the random satisfiability problem. *Physica A*, 306:381–394, 2002.

[3] E. Boros. Maximum renable horn sub-CNFs. *Discrete Applied Mathematics*, 96-97:29–40, 1999.

[4] V. Chandru and J.N. Hooker. Detecting embedded horn structure in propositional logic. *Information Processing Letters*, 42:109–111, 1992.

[5] S.A. Cook. The complexity of theorem proving procedures. In *Proceedings of the 3rd annual ACM symposium on the Theory of Computing*, pages 151–158, 1971.

[6] Y. Crama, O. Ekin, and P.L. Hammer. Variable and term removal from boolean formulae. *Discrete Applied Mathematics*, 75:217–230, 1997.

[7] W.F. Dowling and J.H. Gallier. Linear-time algorithms for testing the satisfiability of propositional horn formulae. *Journal of Logic Programming*, 1:267–284, 1984.

[8] O. Dubois and G. Dequen. A backbone-search heuristic for efficient solving of hard 3-sat formulae. IJCAI-01, 2001.

[9] J. Franco. *Relative size of certain polynomial time solvable subclasses of satisfiability*, volume 35 of *DIMACS Series Discrete Math. Theoret. Computer Science*, pages 211–233. Amer.Math.Soc., Piscataway, NJ, 1997.

[10] E. Friedgut. Sharp thresholds of graph proporties and the k-sat problem. *Journal of the American Mathematical Society*, 12(4):1017–1054, 1999.

[11] O. Kullmann. First report on an adaptive density based branching rule for DLL-like SAT-solvers, using a database of mixed random conjunctive normal forms created using the advanced encryption standard (aes). Technical Report CSR 19-2002, University of Wales Swansea, Computer Science Report Series, 2002.

[12] O Kullmann. Investigating the behaviour of a sat solver on random formulas. Submitted to Annals of Mathematics and Artificial Intelligence, 2002.

[13] R. Monasson, R. Zecchina, S. Kirkpatrick, B. Selman, and L. Troyansky. Determining computational complexity from characteristic 'phase transitions'. *Nature*, 400:133–137, July 1999.

[14] H. Van Maaren and C. Dang. Simplicial pivoting algorithms for a tractable class of integer programs. *Journal of Combinatorial Optimization*, 6:133–142, 2002.

Improving a Probabilistic 3-SAT Algorithm by Dynamic Search and Independent Clause Pairs

Sven Baumer[*] and Rainer Schuler

Abt. Theoretische Informatik, Universität Ulm, D-89069 Ulm, Germany

Abstract. The satisfiability problem of Boolean Formulae in 3-CNF (3-SAT) is a well known NP-complete problem and the development of faster (moderately exponential time) algorithms has received much interest in recent years. We show that the 3-SAT problem can be solved by a probabilistic algorithm in expected time $O(1,3290^n)$. Our approach is based on Schöning's random walk algorithm for k-SAT, modified in two ways.

1 Introduction

The satisfiability problem of Boolean formulae is one of the best known NP-complete problems. The problem remains NP-complete if the formulae are restricted to conjunctive normal form, where each clause contains exactly 3 literals (3-CNF). That is, polynomial time algorithms to test whether a formula in 3-CNF is satisfiable do not exist, unless P=NP.

In recent years, many algorithms have been developed which improve upon the trivial 2^n time complexity, where n is the number of variables in the formula. Milestones in the search for better SAT-algorithms are [6,7,9,5,8], with respective runtimes of $O(1,6181^n), O(1,588^n), O(1,579^n), O(1,505^n)$ and $O(1,447^n)$.

In [10] Schöning proposed a probabilistic algorithm to solve the 3-SAT problem. He showed that the success probability of the algorithm is $(\frac{3}{4})^n \cdot \frac{1}{p(n)}$ for some polynomial $p(n)$ in the worst-case, and accordingly has an expected runtime of $O(1,3334^n)$. The algorithm was later derandomized [3], yielding the best currently known deterministic 3-SAT algorithm with a runtime of $O(1,481^n)$.

The randomized algorithm consists of two basic steps:

An initial assignment is chosen randomly, where each variable is set to 0 or 1 independently with probability $\frac{1}{2}$.
In each of following $3n$ steps the truth assignment of one of the variables is flipped (changed). The variable is chosen randomly with probability $\frac{1}{3}$ from a clause which is not satisfied by the current assignment.

The algorithm succeeds if all clauses are satisfied.

---

[*] supported by DFG-project Scho 302/5-2

E. Giunchiglia and A. Tacchella (Eds.): SAT 2003, LNCS 2919, pp. 150–161, 2004.

Consider a (satisfiable) formula F and let a^* denote a fixed satisfying assignment. In the first step of the algorithm, an initial assignment a is chosen randomly. In each iteration of the algorithm, the Hamming distance $d(a, a^*)$ between the current assignment a and a^* decreases by 1 with probability at least $\frac{1}{3}$ and increases by 1 with probability at most $\frac{2}{3}$.

A careful analysis [10] using Markov chains shows that the probability to reach a satisfying assignment from an initial assignment a is at least $(\frac{1}{2})^{d(a,a^*)} \cdot \frac{1}{p(n)}$ for some polynomial $p(n)$. Hence the success probability of the algorithm is bounded by $E[(\frac{1}{2})^{d(a,a^*)} \cdot \frac{1}{p(n)}] = E[(\frac{1}{2})^{d(a,a^*)}] \cdot \frac{1}{p(n)}$, where $E[Y]$ is the expected value of a random variable Y. In the following let $p(n)$ always denote the polynomial factor obtained as in [10]. Further analysis then shows the overall success probability to find a satisfying assignment if one exists to be at least $Pr[\text{Success}] \geq (\frac{3}{4})^n \cdot \frac{1}{p(n)}$ in the case of formulae in 3-CNF, and more generally $(\frac{k}{2k-2})^n \cdot \frac{1}{p(n)}$ for k-CNF.

One way to improve this algorithm is to use a more sophisticated strategy for choosing the initial assignment. For example, it might be a good strategy to increase the probability of choosing initial assignments a with small $d(a, a^*)$, i.e. assignments already close to the satisfying assignment a^*. However, observe that the aim is not just to improve the average Hamming distance $E[d(a, a^*)]$, but instead the weighted average $E[(\frac{1}{2})^{d(a,a^*)}]$. To find such initial assignments, we exploit properties of the structure of the formula F, which is not done in the original algorithm [10].

The first such approach to improve the success probability of Schöning's random walk algorithm was [4] by identifying 'independent clauses' in F, i.e. clauses that are variable disjoint. Hofmeister et al. [4] show that the success probability of the algorithm can be improved if the initial assignment is chosen from assignments that satisfy the independent clauses. Note that the assignment of the variables can be chosen independently for each independent clause. Depending on the number of such clauses, one of two different algorithms is more likely to find a satisfying assignment: in the case of 'many' independent clauses the random walk algorithm will find a satisfying assignment, whereas in the case of 'few' independent clauses, a naive, brute-force search algorithm, which considers all assignments that satisfy the independent clauses, does succeed. Using this strategy, the expected run-time of the random walk algorithm is improved to $O(1,3302^n)$ [4].

We improve on the success probability in two ways. The approach of Hofmeister et al. is extended in this paper by considering pairs of clauses which have (at least) one variable in common. Every clause contains three literals, i.e. restricts the number of assignments to the three variables that satisfy the clause to 7 (out of 8). If two clauses contain a common variable, the fraction of possible assignments decreases further, e.g. only 24 out of 32 assignments satisfy $\{x, y, z\}$ and $\{\neg x, u, v\}$.

A further improvement can be achieved, if a different strategy is used to identify the set of independent clause pairs. Instead of the greedy fashion of [4], we use a dynamic approach: depending on a partial assignment to the clause pairs already selected, new clause pairs are chosen. For example using this strategy only 6 (out of 8) assignments have to be considered for each independent clause. Again we can use a trade off between the success probability of the algorithm selecting the independent clause pairs (referred to as BF in the following, which increases if only few clause pairs are detected) and the success probability of the random walk algorithm (RW, which increases with the number of independent clause pairs). Using the dynamic strategy, the expected run-time to find a satisfying assignment is $O(1,3300^n)$ if only independent clauses are considered and improves to $O(1,3293^n)$ if the approach is extended to independent clause pairs.

We also sketch a further improvement to $O(1,3290^n)$ by a more detailed case analysis.

2   An Improvement of the Brute Force Algorithm for Independent Clause Selection

In the following let F denote a formula in 3-CNF, i.e. F is a conjunction of clauses where every clause contains (at most) 3 literals. Two clauses are called *independent* if the variables of the clauses are disjoint. A set of clauses is called independent if the clauses are pairwise independent. A set of independent clauses is maximal, if every clause contains at least one variable which also occurs in some independent clause. This idea was introduced in [2], where a maximal set of independent clauses is referred to as a type-II prefix.

As shown in [4] the success probability of the random walk algorithm is $(\frac{3}{7})^m \cdot (\frac{3}{4})^{n-3m} \cdot \frac{1}{p(n)}$ and increases with the number m of independent clauses if the initial assignment is chosen randomly from assignments that satisfy the independent clauses. On the other hand, a brute force algorithm can consider every possible assignment to the variables of the independent clauses. If the set of independent clauses is maximal, the simplified formula is in 2-CNF and its satisfiability can be checked in polynomial time [1]. Hence using exhaustive search satisfiability can be checked in time $q(n) \cdot 7^m$ for some polynomial $q(n)$. In the following, a randomized variant of the exhaustive search algorithm is used: For each clause, one of the 7 possible assignments is chosen uniformly at random. This yields a success probability of at least $(\frac{1}{7})^m$, and an expected run-time of $O(q(n) \cdot 7^m)$.

That is, if the number of independent clauses is sufficiently small, the success probability of the brute force algorithm (we assume the assignments are chosen randomly) exceeds the success probability of the random walk. The combined success probability will be minimal if $(\frac{1}{7})^m = (\frac{3}{7})^m \cdot (\frac{3}{4})^{n-3m} \cdot \frac{1}{p(n)}$.

In this section we show that the success probability can be improved further if the independent clauses are selected more carefully. We show that it is sufficient to consider only 6 assignments for each independent clause. This gives a success probability of a brute force algorithm of $(\frac{1}{6})^m$, where m is the number of independent clauses identified on any accepting path.

The algorithm is based on the following simple observation. Each variable x occurs positive as well as negative (otherwise the truth assignment can be fixed to satisfy all occurrences). Assume the truth value of x is chosen randomly. If we choose $x = 0$, only 3 assignments are possible to the remaining two literals of the clause where x occurs positive (otherwise the clause would not be satisfied). Analogously, if we choose $x = 1$, only 3 assignments are possible for the remaining two literals of the clause where x occurs negative. The success probability (i.e. the probability that the correct assignment is chosen) is $\frac{1}{2} \cdot \frac{1}{3} = \frac{1}{6}$ in both cases. Furthermore, in each case a clause can be selected as an independent clause, since all occurrences of the three variables involved are eliminated by fixing the truth values of the according literals.

While this observation is true if all clauses contain exactly three literals, we also need to consider implications due to clauses in 2-CNF (referred to as *chains* in the following). For example, the clauses $\{\mathbf{x}, y, z\}, \{\neg x, u\}, \{\neg\mathbf{u}, v, w\}$ are dependent on x in the same way as $\{\mathbf{x}, y, z\}, \{\neg\mathbf{x}, v, w\}$ are. The procedure is shown in more detail in Figure 1.

Throughout, we assume that the truth values of literals in unit clauses (i.e. clauses that contain exactly one literal) will be fixed accordingly. Furthermore, we assume that the truth value of a variable x is set to $a \in \{0, 1\}$, if the partial assignment $x = \neg a$ leads to a contradiction (i.e. a clause will be empty/falsified).

Repeat
    choose a variable x
    find a clause $C = \{l_1, l_2, l_3\}$ such that $(x = 0)$ implies $(l_1 = 0)$, i.e.
        $l_1 = x$ or there exist clauses $D_i = \{d_i, \neg d_{i+1}\}$, $1 \le i \le k$, and
            $d_1 = x$ and $l_1 = d_{k+1}$
    if C does not exist deterministically set $x = 0$, otherwise:
    find a clause $C' = \{l'_1, l'_2, l'_3\}$ such that $(x = 1)$ implies $(l'_1 = 0)$, i.e.
        $l'_1 = \neg x$ or there exist clauses $D'_i = \{\neg d'_i, d'_{i+1}\}$, $1 \le i \le k$, and
            $d'_1 = \neg x$ and $l'_1 = \neg d_{k+1}$
    if C' does not exist deterministically set $x = 1$, otherwise:
    randomly guess truth assignment x from $\{0, 1\}$
    if $x = 0$ then
        randomly guess truth assignment of l_2, l_3 from $\{01, 10, 11\}$
        select C as independent clause
    if $x = 1$ then
        randomly guess truth assignment of l'_2, l'_3 from $\{01, 10, 11\}$
        select C' as independent clause

Fig. 1. Improved brute force algorithm (BF) for independent clause selection.

Claim 1. Assume that a formula F is satisfiable. Then the algorithm BF finds a satisfying assignment with positive probability.

Observe that the following is true if x is set to 0 deterministically: $\neg x$ does not appear in any clause containing 3 literals. Furthermore, all clauses with 2 literals containing x can be satisfied by assigning 1 to the second literal. Moreover, the literals whose truth assignments are forced by a chain of such 2 literal-clauses do also not appear negatively in any clause with 3 literals. That is, all clauses containing x (positive or negative) or a forced variable are true under this assignment; either because the forced literal occurs positively, or the second literal is also forced to 1 (also note that if this process leads to a contradiction then x is forced to 1 by the simplification step). In particular, if F is satisfiable, then the formula with x set to 0 is also satisfiable.

Claim 2. Assume that a formula F is satisfiable and let a^* denote a satisfying assignment found on some path of the algorithm BF. The probability that a satisfying assignment is found is at least 6^{-m}, where m is the number of independent clauses detected on a path corresponding to the assignment a^*. Furthermore, the probability that $l \leq m$ independent clauses are detected is at least 6^{-l}.

The Claim follows from the observation that in each step of the procedure the truth assignment of C or C' is chosen correctly (according to a satisfying assignment a^*) with probability $\frac{1}{6} = \frac{1}{2} \cdot \frac{1}{3}$.

The brute force algorithm is combined with the random walk RW [4] in an obvious way.

> Repeat
>   let $\mathcal{D} = \emptyset$
>   run BF and
>   let $\mathcal{C}$ denote the set of indep. clauses detected during computation
>   IF $|\mathcal{D}| < |\mathcal{C}|$ then $\mathcal{D} = \mathcal{C}$
>   run RW using $\mathcal{D}$ as the set of independent clauses.

Theorem. The algorithm has expected running time of $O(1,3300^n)$.

The success probability of RW depends on the number of independent clauses on a path to a satisfying assignment. Let c denote some constant. We distinguish two cases.

1. The number of independent clauses on some path of the algorithm BF is less than $c \cdot n$.
   In this case the algorithm BF will find a satisfying assignment in expected time 6^{cn}.
2. The number of independent clauses is larger than $c \cdot n$.
   In this case the algorithm BF finds $\geq c$ independent clauses in expected time 6^{cn}. Afterwards the RW algorithm will find a satisfying assignment in expected time $(7/3)^c \cdot (4/3)^{n-3c} \cdot \frac{1}{p(n)}$.

Choose c such that $6^{cn} = (7/3)^{cn} \cdot (4/3)^{n-3cn} \cdot \frac{1}{p(n)}$, i.e.

$$c = \frac{\log(4/3)}{\log 6 - \log(7/3) + 3\log(4/3)} \approx 0.1591, \quad \text{and} \quad 6^{cn} = 1.3300^n$$

A further improvement on the number of independent clauses might be possible. We note however, that in this way the success probability of the random walk algorithm cannot be improved beyond 0.7539^n which corresponds to an expected run-time of $(7/3)^{n/3} \approx 1,3264^n$.

In the next section we show how the bound in the theorem above can be further improved if the notion of independent clauses is extended to clause pairs.

3   Independent Clause Pairs

We show that it is possible to choose the initial assignment for the RW algorithm from assignments that satisfy pairs of clauses which overlap with one or more variables. We show that the success probability of the random walk algorithm improves further with the number of such clause pairs detected.

Recall two clauses are independent if the variables of the clauses are disjoint. The notion of independent clauses can be extended to clause pairs, or in general to clause tuples. Two clauses that share at least one variable are said to *overlap*. In the following we use *clause pair* to refer to a pair of clauses that overlap. Two clause pairs are called independent, if the sets of variables of the two clause pairs are disjoint. A set of clause pairs is called independent if the clause pairs are mutually independent.

A clause pair can overlap with 1, 2 or 3 literals, where some of the literals may be negated in one clause. Possible overlaps with a clause $\{x, y, z\}$ are shown in the following table, where u, v, w denote variables different from x, y, z. We use the following notation to describe the type of overlap: $-$ indicates a variable with negative overlap, $+$ indicates a positive overlap, and $*$ indicates a new variable not contained in the first clause. Observe that $(***)$ corresponds to two independent clauses. Also note that the case $(+++)$ is not possible since two clauses that agree on all three literals would be identical.

$(***)$	$(+**)$	$(-**)$	$(++*)$	$(--*)$	$(+-*)$	$(---)$	$(++-)$	$(+--)$
$\{u,v,w\}$	$\{x,u,v\}$	$\{\neg x,u,v\}$	$\{x,y,v\}$	$\{\neg x,\neg y,v\}$	$\{x,\neg y,v\}$	$\{\neg x,\neg y,\neg z\}$	$\{x,y,\neg z\}$	$\{x,\neg y,\neg z\}$

Let r_i denote the number of clauses and clause pairs of type i detected (e.g. by some greedy method). In particular we assume that the set of independent clauses and clause pairs is maximal, i.e. all other clauses share at least one variable with a selected clause or clause pair. Let $\hat{n} := n - \sum_i r_i$ denote the number of the remaining variables.

Now we consider the random walk algorithm, where the initial assignment is chosen randomly as follows. For each clause pair and each independent clause the truth assignment to the variables is chosen independently from the set of assignments that satisfy the clause pair and clause, respectively; the exact probabilities are given below. The truth values of the remaining $\hat{n}$ variables are chosen independently from $\{0, 1\}$ with probability $1/2$.

Let a^* denote a fixed satisfying assignment of F and let a denote the randomly chosen initial assignment. Define $W_i^{(j)}$ to be the random variable which is the Hamming distance between a^* and a on the variables of the j-th independent clause or clause pair of type i. Similar, define V to be the random variable which is the Hamming distance of the assignments a and a^* on the variables which are not contained in an independent clause or independent clause pair. The success probability of the algorithm can be calculated using the inequality [10,4]

$$Pr[\text{success}] \geq E[(\frac{1}{2})^{d(a,a^*)}] \cdot \frac{1}{p(n)}$$

The Hamming distance can be rewritten as: $d(a, a^*) = V + \sum_i \sum_{j=1}^{r_i} W_i^{(j)}$, where i ranges over all types of overlaps (see table above). The random variables $W_i^{(j)}$ are independent and equally distributed for each i. Since, as shown in section 3.3, $E[(\frac{1}{2})^{W_i^{(j)}}]$ does not depend on a^*, it follows that $E[(\frac{1}{2})^{W_i^{(j)}}] =: E[(\frac{1}{2})^{W_i}]$ is independent of j. Using $E[(\frac{1}{2})^V] = (\frac{3}{4})^{\hat{n}} \cdot \frac{1}{p(n)}$ [10], we get

$$Pr[\text{succ.}] \geq E[(\frac{1}{2})^V] \cdot \Pi_i \Pi_{j=1}^{r_i} E[(\frac{1}{2})^{W_i^{(j)}}] \cdot \frac{1}{p(n)} = (\frac{3}{4})^{\hat{n}} \cdot \Pi_i (E[(\frac{1}{2})^{W_i}])^{r_i} \cdot \frac{1}{p(n)}$$

$$(1)$$

We note that for every partial truth assignment to the variables of the independent clause pairs and independent clauses, the remaining formula will be in 2-CNF. The running time (or success probability) of a brute force algorithm depends on the number of independent clauses and the number and type of the independent clause pairs. We consider a generalization (see Figure 2) of the brute force algorithm which selects a maximal set of independent clauses and clause pairs.

3.1   Clause Pair Selection

The idea is to extend the independent clause selection method (Figure 1) one step further. Let F be a satisfiable formula in 3-CNF. Each variable x occurs both positively and negatively in some clauses (containing 3 literals). Let a^* denote a satisfying assignment. The truth assignment of x is either 0 or 1. In both cases the truth assignment of the remaining two literals of one of the clauses is 01, 10, or 11. That is, at least one of the two literals is set to 1. Since this literal also occurs positively and negatively, we will find some additional clause (containing 3 literals), where this literal occurs negatively. In this clause, the negated literal is set to 0, and the truth assignment of the remaining literals can be chosen randomly from assignments that satisfy the clause. In this way, the truth assignment of independent clause pairs can be guessed correctly with high probability.

For example, if $\{x, u, v\}$ and $\{\neg u, w, r\}$ are the second and third clause, the assignment to three variables x, u, v is correct with probability $\frac{1}{2} \cdot \frac{1}{3}$ and the assignment to the remaining two variables w, r is correct with probability $\frac{1}{3}$. In this case a correct assignment to five variables is chosen with probability $\frac{1}{2} \cdot \frac{1}{3} \cdot \frac{1}{3} = \frac{1}{18}$, while a pair of overlapping clauses is selected.

On the other hand, we can show that choosing an initial assignment that satisfies clause pairs will guarantee an improved success probability of the random walk algorithm. The exact probabilities for the assignments are calculated in

Section 3.3. We need to distinguish the different types of overlap, depending on the number of common variables, and depending on their *sign* (which is called positive if the variable or the negated variable occurs in both clauses, and called negative if the variable occurs in one clause and the negated variable occurs in the other clause).

The careful reader will notice that the idea could be further extended to clause triplets or clause n-tuples in general, but the calculations involved get rather complex while the benefit seems only minimal.

Repeat
    choose a variable x
    find clauses $C = \{l_1, l_2, l_3\}$ and $C' = \{l'_1, l'_2, l'_3\}$ such that
        ($x = 0$) implies ($l_1 = 0$)   and   ($x = 1$) implies ($l'_1 = 0$)
    if C [respectively C'] does not exist set $x = 0$ [resp. $x = 1$], else
    find $E = \{l_4, l_5, l_6\}$, $F = \{l_7, l_8, l_9\}$, $E' = \{l'_4, l'_5, l'_6\}$, $F' = \{l'_7, l'_8, l'_9\}$ s.t.
        ($l_2 = 1$) implies ($l_4 = 0$),   and   ($l_3 = 1$) implies ($l_7 = 0$),   and
        ($l'_2 = 1$) implies ($l'_4 = 0$),   and   ($l'_3 = 1$) implies ($l'_7 = 0$)
    if E [F, E' or F'] does not exist set $l_2 = 1$ [$l_3 = 1$, $l'_2 = 1$, or $l'_3 = 1$], else
    randomly guess truth assignment of x from $\{0, 1\}$
    if $x = 0$ then
        randomly guess truth assignment of l_2, l_3 from $\{01, 10, 11\}$
        if $l_2 = 1$ then
            randomly guess truth assignment of the remaining literals of E
            select (C, E) as independent clause pair or indep. clauses, resp.
        else if $l_3 = 1$ then
            randomly guess truth assignment of the remaining literals of F
            select (C, F) as independent clause pair or indep. clauses, resp.
    if $x = 1$ then ... (analogously) ...

Fig. 2. Brute force algorithm (BF) for independent clause pair selection.

The success probability of the brute force algorithm (Figure 2) depends on the number and type of the independent clauses and independent clause pairs detected. We have to distinguish between the cases that the clause pair (C, E) is selected via a *direct* overlap (i.e. $\neg l_2 \in E$) and via a *chain* of 2-CNF clauses. Observe that all types of direct overlap require at least one negatively shared variable (denoted $-$). Also observe that all chain cases contain at least one $*$, which denotes the literal which is forced to be 0 by the 2-CNF chain. All other literals denoted by a $*$ are free and can be set to either 0 or 1.

The different types detected are given in the table below. In the second row we give the probability that the correct assignment is chosen by the BF algorithm, both for the direct and chain case. In the third row we give the success probability of the random walk algorithm RW, where the initial assignments are chosen optimally for each type of clause pairs (exact probabilities are given in Section 3.3 below).

In the last row the expected running time of the combined methods is given, if all variables are of the respective type. These running times are obtained by setting the success probabilities of BF and RW to be equal and using $E[\text{time}] = 1/Pr[\text{success}]$:

$$Pr[\text{BF success}]^m = Pr[\text{RW success}]^m \cdot (\frac{3}{4})^{n-m \cdot (\#\ \text{variables})} \cdot \frac{1}{p(n)}$$

where m is the total number of clause pairs of the respective type.

type # variables	(* * *) 3+3	(+ * *) 5	(- * *) 5	(+ + *) 4	(- - *) 4	(+ - *) 4	(- - -) 3	(+ + -) 3	(+ - -) 3
BF direct	—	—	1/18	—	1/12	1/12	1/6	1/6	1/6
BF chain	1/18	1/12	1/12	1/6	1/6	1/6	—	—	—
RW	(3/7)·(3/7)	81/331	27/110	27/83	15/46	27/82	7/16	9/20	45/104
time direct	—	—	$\mathbf{1,3290^n}$	—	$1,3288^n$	$1,3273^n$	$1,3258^n$	$1,3201^n$	$1,3280^n$
time chain	$\mathbf{1,3293^n}$	$1,3287^n$	$1,3283^n$	$1,3276^n$	$1,3271^n$	$1,3251^n$	—	—	—

The run-time of our algorithm is bounded by the maximum of these run-times, $1,3293^n$. The worst case is when all clause pairs identified by the BF algorithm are of the type chain (* * *), i.e. the two selected clauses are two independent clauses for the RW algorithm.

Of course, in practice the BF algorithm will not only identify clause pairs of one single type but a combination of different types. However, such 'mixed' cases have higher success probability and yield a better run-time than the worst case, where all clause pairs are of the same type (also see equation 1).

3.2   Sketch of a Further Improvement

Furthermore, we show that the algorithm BF can be modified to improve in the worst case (* * *) by repeating the previous step once more; this way instead of two independent clauses, either an independent clause and a pair or three independent clauses are selected. This way, we obtain a slightly better run-time of $1,3290^n$ (also highlighted in the above table).

Assume that clauses C, E, F and C', E', F' are detected by the algorithm and that $C = \{l_1, l_2, l_3\}$ and $E = \{l_4, l_5, l_6\}$ are independent, i.e. do not share a variable. W.l.o.g. we assume that $x = 0$ and that $l_2 l_3 \in \{10, 11\}$ and hence (C, E) are chosen as independent clauses. Before choosing the assignment of l_5, l_6 we test each of the three possible assignments $l_5 l_6 \in \{01, 10, 11\}$. If the assignment leads to a contradiction it is not considered any more. Otherwise we test whether the assignment is in fact admissible, that is we test whether it does not change the satisfiability of F. We look for a clause G with 3 literals such that a implies that one of the literals of G is set to 0. If G does not exist, then $l_4 l_5$ is set to a.

We modify the algorithm to distinguish the three cases. (i) If all assignments lead to a contradiction then the algorithm fails. (ii) If one of the assignments is admissible, we choose C and E as independent clauses. The probability that the correct assignment is chosen for the literals of C and E is $\frac{1}{6}$. (iii) Otherwise we extend the algorithm. The assignment of $l_5 l_6$ is chosen randomly (from the assignments that do not lead to a contradiction). The remaining literals of the respective. clause G are chosen randomly from assignments that satisfy G. The clauses C, E and G are selected. C is an independent clause, E and G form an independent clause pair if E and G overlap, otherwise E and G are also independent clauses.

In the case of three independent clause pairs the probability to choose the correct assignment to the 9 variables is $\frac{1}{6} \cdot \frac{1}{3} \cdot \frac{1}{3} = \frac{1}{54}$. Otherwise the probability for the correct assignment will be $\frac{1}{3} \cdot p$, where p denotes the probability that an independent clause pair of the same type is chosen by the (original) algorithm directly.

The success probability is minimal if 3 independent clauses are detected. Using $(1/54)^m = (3/7)^{3m} \cdot (3/4)^{n-9m}$ we obtain a run-time of $1,3289^n$ for this case. Hence the case $(- * *)$ is now the worst case, with a corresponding run-time of $O(1,3290^n)$.

3.3   Calculation of RW Success Probabilities

For sake of brevity, we only investigate the case $(+ * *)$ here, i.e. the two clauses are of type $\{x, y, z\}, \{x, u, v\}$. A more detailed paper containing all cases is available.[1]

The probability for the initial assignment should maximize the expected value $E[(1/2)^X]$ where X is the Hamming distance between the initial assignment a and a satisfying assignment a^*. Since the satisfying assignment a^* is unknown, we choose the probabilities such that the expected value $E[(1/2)^X]$ is maximized for any possible case of a^*. For the given equations it is possible to fix probabilities such that $E[(1/2)^X]$ is identical for all cases of a^*.

To simplify the corresponding equation system, the 32 possible assignments to the literals x, y, z, u, v can be grouped into 12 equivalence classes. All assignments of a class have the same Hamming distances to assignments of the same and any other class. [2]

In the following we denote assignments to these five variables in the format $yzxuv$, i.e. the middle bit represents the truth value of the common variable. E.g. 00100 represents the truth assignment that sets x to 1, and the other variables to 0.

Intuitively, all "mirror images" of an assignment are in the same class, e.g. assignments 00001, 00010, 01000 and 10000 all belong to the same class K_{11}. Note that literal x has a different status, therefore the assignment 00100 belongs to a different group (K_1). The 12 groups are as follows:

[·] S. Baumer, R. Schuler. Improving a probabilistic 3-SAT Algorithm by Dynamic Search and Independent Clause Pairs. *Electronic Colloquium on Computational Complexity*, Report TR03-010.
ftp://ftp.eccc.uni-trier.de/pub/eccc/reports/2003/TR03-010/index.html
[·] Observe that this classification is not a necessary step. We only choose to do so in order to simplify the equation system for presentation. Instead a system with 32 equations could be solved, yielding the same result.

K.	K.	K.	K.	K.	K.	K.	K.	K.	K..	K..	K..
00100	00101	00111	01001	01011	01101	01111	11011	11111	00000	00001	00011
00110	11100	01010	10011	01110	10111				00010	11000	
01100		10001	11001	10101	11101				01000		
10100		10010	11010	10110	11110				10000		

Observe that K_1 to K_9 consist of assignments that satisfy both clauses, whereas assignments from K_{10} to K_{12} don't.

The (fixed) satisfying assignment a^* is from K_1 to K_9. Analyzing each case in the same fashion as [4] yields one equation for every choice of a^*. The fact that the probabilities add up to one yields the 10th equation.

Let p_i denote the probabilities that the 5 literals are set according to group K_i. For every choice of a^* the coefficients α_i of p_i, $1 \leq i \leq 9$ are given in the table below. Assume a^* is fixed. The coefficients are calculated as follows:

$$E[(\frac{1}{2})^{W(+**)}] = \sum_{k=0}^{5}(\frac{1}{2})^k \sum_{i=1}^{9} p_i \cdot |\{a \in K_i \mid d(a,a^*) = k\}|$$

$$= \sum_{i=1}^{9} p_i \cdot \sum_{k=0}^{5}(\frac{1}{2})^k \cdot |\{a \in K_i \mid d(a,a^*) = k\}| = \sum_{i=1}^{9} \alpha_i \cdot p_i$$

a^*	p_1	p_2	p_3	p_4	p_5	p_6	p_7	p_8	p_9	p_{10}	p_{11}	p_{12}	$=$
K_1	$\frac{32}{32}$	$\frac{64}{32}$	$\frac{16}{32}$	$\frac{16}{32}$	$\frac{8}{32}$	$\frac{32}{32}$	$\frac{16}{32}$	$\frac{1}{32}$	$\frac{2}{32}$	$\frac{16}{32}$	$\frac{32}{32}$	$\frac{8}{32}$	$E[(\frac{1}{2})^{W(+**)}]$
K_2	$\frac{16}{32}$	$\frac{56}{32}$	$\frac{20}{32}$	$\frac{20}{32}$	$\frac{13}{32}$	$\frac{40}{32}$	$\frac{26}{32}$	$\frac{2}{32}$	$\frac{4}{32}$	$\frac{8}{32}$	$\frac{28}{32}$	$\frac{10}{32}$	$E[(\frac{1}{2})^{W(|**)}]$
K_3	$\frac{8}{32}$	$\frac{40}{32}$	$\frac{34}{32}$	$\frac{16}{32}$	$\frac{20}{32}$	$\frac{32}{32}$	$\frac{40}{32}$	$\frac{4}{32}$	$\frac{8}{32}$	$\frac{4}{32}$	$\frac{20}{32}$	$\frac{17}{32}$	$E[(\frac{1}{2})^{W(+**)}]$
K_4	$\frac{4}{32}$	$\frac{20}{32}$	$\frac{8}{32}$	$\frac{50}{32}$	$\frac{40}{32}$	$\frac{25}{32}$	$\frac{20}{32}$	$\frac{8}{32}$	$\frac{4}{32}$	$\frac{8}{32}$	$\frac{40}{32}$	$\frac{16}{32}$	$E[(\frac{1}{2})^{W(+**)}]$
K_5	$\frac{2}{32}$	$\frac{13}{32}$	$\frac{10}{32}$	$\frac{40}{32}$	$\frac{56}{32}$	$\frac{20}{32}$	$\frac{28}{32}$	$\frac{16}{32}$	$\frac{8}{32}$	$\frac{4}{32}$	$\frac{26}{32}$	$\frac{20}{32}$	$E[(\frac{1}{2})^{W(+**)}]$
K_6	$\frac{8}{32}$	$\frac{40}{32}$	$\frac{16}{32}$	$\frac{25}{32}$	$\frac{20}{32}$	$\frac{50}{32}$	$\frac{40}{32}$	$\frac{4}{32}$	$\frac{8}{32}$	$\frac{4}{32}$	$\frac{20}{32}$	$\frac{8}{32}$	$E[(\frac{1}{2})^{W(+**)}]$
K_7	$\frac{4}{32}$	$\frac{26}{32}$	$\frac{20}{32}$	$\frac{20}{32}$	$\frac{28}{32}$	$\frac{40}{32}$	$\frac{56}{32}$	$\frac{8}{32}$	$\frac{16}{32}$	$\frac{2}{32}$	$\frac{13}{32}$	$\frac{10}{32}$	$E[(\frac{1}{2})^{W(+**)}]$
K_8	$\frac{1}{32}$	$\frac{8}{32}$	$\frac{8}{32}$	$\frac{32}{32}$	$\frac{64}{32}$	$\frac{16}{32}$	$\frac{32}{32}$	$\frac{32}{32}$	$\frac{16}{32}$	$\frac{2}{32}$	$\frac{16}{32}$	$\frac{16}{32}$	$E[(\frac{1}{2})^{W(+**)}]$
K_9	$\frac{2}{32}$	$\frac{16}{32}$	$\frac{16}{32}$	$\frac{16}{32}$	$\frac{32}{32}$	$\frac{32}{32}$	$\frac{64}{32}$	$\frac{16}{32}$	$\frac{32}{32}$	$\frac{1}{32}$	$\frac{8}{32}$	$\frac{8}{32}$	$E[(\frac{1}{2})^{W(+**)}]$
$\sum$	1	4	2	4	4	4	4	1	1	1	4	2	1

Note that this is an under-determined equation system, with 10 equations (K_1 to K_9 and $\sum$) and 13 variables (p_1 to p_{12} and $E[(\frac{1}{2})^{W(+**)}]$).

Setting $p_{10} = p_{11} = p_{12} = 0$ and solving the corresponding determined system using a computer algebra program (e.g. Maple 7) yields

$$E[(\frac{1}{2})^{W(+**)}] = 81/331,$$

where the probabilities are given by

p_1	p_2	p_3	p_4	p_5	p_6	p_7	p_8	p_9
16/331	16/331	16/331	24/331	12/331	4/331	10/331	6/331	13/331

Solving the full, under-determined system proves the optimal value for p_{10}, p_{11}, p_{12} to be all 0, since positive values for these have a negative contribution to $E[(\frac{1}{2})^{W_{(+**)}}]$.

References

1. B. Aspvall, M.F. Plass, and R.E. Tarjan. A linear-time algorithm for testing the truth of certain quantified boolean formulas. *IPL*, 8(3):121–123, 1979.
2. S. Aida, R. Schuler, T. Tatsukiji, and O. Watanabe. The difference between Polynomial-Time Many-One and Truth-Table Reducibilities on Distributional Problems. *Theory of Computing Systems*, 35:449–463, 2002.
3. E. Dantsin, A. Goerdt, E. Hirsch, R. Kannan, J. Kleinberg, C. Papadimitriou, P. Raghavan, U. Schöning. A deterministic $(2 - 2/(k + 1))^n$ algorithm for k-SAT based on local search. *Theoretical Computer Science*, 289:69–83, 2002.
4. T. Hofmeister, U. Schöning, R. Schuler, and O. Watanabe. A probabilistic 3-SAT algorithm further improved. In *Proceedings 19th Symposium on Theoretical Aspects of Computer Science*, volume 2285 of *LNCS*, pages 192–203. Springer-Verlag, 2002.
5. O. Kullmann. New methods for 3-sat decision and worst-case analysis. *Theoretical Computer Science*, 223:1–72, 1999.
6. B. Monien and E. Speckenmeyer. Solving satisfiability in less than 2^n steps. *Discrete Applied Mathematics*, 10:287–295, 1985.
7. R. Paturi, P. Pudlák and F. Zane. Satisfiability coding lemma. In *38th Ann. IEEE Sympos. on Foundations of Comp. Sci. (FOCS'97)*, pages 566–574, 1997.
8. R. Paturi, P. Pudlák, M.E. Saks, and F. Zane. An improved exponential-time algorithm for k-sat. In *39th Ann. IEEE Sympos. on Foundations of Comp. Sci. (FOCS'98)*, pages 410–414, 1998.
9. I. Schiermeyer. Solving 3-satisfiability in less than $o(1,579^n)$ steps. In *Computer Science Logic (CSL)*, volume 702 of *LNCS*, pages 379–394. Springer, 1992.
10. Uwe Schöning. A probabilistic algorithm for k-SAT and constraint satisfaction problems. In *Proc. 40th FOCS*, pages 410–414. ACM, 1999.

Width-Based Algorithms for SAT and CIRCUIT-SAT

(Extended Abstract)

Elizabeth Broering and Satyanarayana V. Lokam

EECS Department
University of Michigan
Ann Arbor, MI 48105, USA
{ebroerin,satyalv}@eecs.umich.edu

Abstract. We investigate theoretical and practical aspects of algorithms for CIRCUIT-SAT and SAT based on combinatorial parameters. Two such algorithms are given in [1] and [4] based on branch-width of a hypergraph and cut-width of a graph respectively. We give theoretical generalizations and improvements to the cut-width-based algorithm in [4] in terms of many other well-known width-like parameters. In particular, we have polynomial-time backtrack search algorithms for logarithmic cut-width and path-width, $n^{O^{\cdots n}}$-time backtrack search algorithms for logarithmic tree-width and branch-width, and a polynomial-time regular resolution refutation for logarithmic tree-width. We investigate the effectiveness of the algorithm in [1] on practical instances of CIRCUIT-SAT arising in the context of Automatic Test Pattern Generation (ATPG).

1 Introduction

Algorithms for satisfiability of CNF formulas (SAT) and Boolean circuits (CIRCUIT-SAT) are of tremendous practical importance due to their applications in several areas such as testing, verification, and AI. On the other hand, SAT is the quintessential NP-complete problem and CIRCUIT-SAT is polynomially equivalent to it. So, these problems play a central role in the study of computational complexity. Algorithmic approaches and hardness results about SAT have intimate connections to complexity of various proof systems and many deep results in proof complexity contributed to our understanding of such connections. Thus, SAT and CIRCUIT-SAT have been investigated from many different perspectives and this area has emerged as a topic of enormous fundamental significance.

The P $\neq$ NP conjecture leaves us with no hopes of finding a polynomial time algorithm for SAT in general. However, instances of SAT arising in practical applications rarely seem to force worst-case behavior of carefully designed algorithms. In fact, often well-defined symmetries and other structural characteristics of SAT instances enable very efficient solvability of practical instances.

E. Giunchiglia and A. Tacchella (Eds.): SAT 2003, LNCS 2919, pp. 162–171, 2004.

On the other hand, studies on randomly generated instances of SAT yield fascinating insights into the combinatorial structure, algorithmic approaches, and complexity landscape of SAT.

In this paper, we investigate theoretical and practical aspects of algorithms for CIRCUIT-SAT and SAT based on certain graph-theoretic parameters. We are inspired by the recent successes in [1] and [4] that give polynomial time algorithms for SAT and CIRCUIT-SAT on instances with logarithmic *branch-width* (of a hypergraph) and *cut-width* (of a graph), respectively. These results are elegant mathematical statements proving non-trivial upper bounds on the complexity of SAT in terms of nice, well-studied, combinatorial parameters. In the case of [4], their result is used to justify the practical solvability of Automatic Test Pattern Generation (ATPG) for practical instances of VLSI circuits. In the case of [1], they derive consequences to the automatizability of resolution on Tseitin tautologies of small width.

Branch-width of a graph was introduced by Robertson and Seymour [6] in their phenomenal work on Graph Minors. It is closely related to the notion of tree-width. These and other graph parameters like cut-width, path-width, etc., and their generalizations to hypergraphs have been extensively investigated in graph theory and combinatorial optimization; see [2] for a survey. Many NP-complete optimization problems become tractable when restricted to instances of a small width-like parameter. Not surprisingly, computing any of these parameters is also NP-complete. However, many approximation algorithms are available to get good estimates on them. Such algorithms can be used to check whether instances from an application area (with a suspected common structure) have small width and then apply a width-based algorithm or heuristic to efficiently solve an otherwise intractable problem on those instances.

1.1   Our Results

This paper is an attempt to understand the effectiveness of the width-based approach mentioned above to attack SAT and CIRCUIT-SAT problems. An application area we will draw instances to evaluate this approach is that of Automatic Test Pattern Generation (ATPG). We will consider four width parameters: cut-width, path-width, tree-width, and branch-width. Based on known relations among these parameters described in Section 2, for a given graph, branch-width (or tree-width within ±1) has the smallest value whereas cut-width has the largest.

We generalize and improve the algorithm in [4] for CIRCUIT-SAT (with circuits of bounded fan-in) based on cut-width. First, we use the notion of a *weighted* cut-width of a graph to derive a back-track search algorithm (for CNF formulas) of running time exponential in the *weighted* cut-width of a graph derived from the formula. Second, we obtain a slight improvement of the time bound in [4] for CIRCUIT-SAT by removing the dependence on fan-out of the circuit. Finally, we obtain a stronger result by giving a back-track search algorithm with running time exponential in path-width of the graph rather than cut-width. In general, path-width can be smaller than cut-width of a graph. Since branch-width is smallest among the four widths, ideally we would like a back-track search algorithm with complexity exponential in branch-width. However, our results currently yield only a time exponential in (branch-width $\times \log n$). Thus we have polynomial-time back-track search algorithms for logarithmic cut-width and path-width, and $n^{O(\log n)}$-time algorithms for logarithmic tree-width and branch-width.

In [1], branch-width $\mathrm{hbw}(H)$ of the *hypergraph H* defined by the clauses of a CNF formula is used to obtain a SAT algorithm with a running time of $n^{O(1)}2^{\mathrm{hbw}(H)}$. We give a similar SAT algorithm with running time exponential in the tree-width of the *graph G*. Note, that the tree-width referred to here and in the above paragraph, along with the branch-width referred to in the above paragraph is for a *graph* constructed from the formula, in contrast to the branch-width of the *hypergraph* used in [1]. For CIRCUIT-SAT instances arising from applications such as ATPG, the graph construction seems to reflect more closely the physical topology of the circuit. In fact, the graph constructed is very close to the undirected version of the circuit itself. Moreover, certain natural constructions like series-parallel graphs and outer-planar graphs have very small tree-width.

Our on-going experimental investigations will compare the (estimates of) various width parameters for a large number of circuits including the benchmark circuits such as ISCAS89 and MCNC91. These results will shed light on the validity of assuming small (constant, logarithmic etc.) width parameters in these applications and hence the feasibility of width-based algorithms for SAT and CIRCUIT-SAT. Preliminary observations seem to indicate that branch-width algorithm from [1] may be impractical for ATPG instances. A recent result [3] gives an eigenvector heuristic for estimating the branch-width of a graph (not a hypergraph). In their application domain (TSP), the heuristic was much more efficient than known approximation algorithms including the algorithm from [6] used in [1]. We will study the accuracy of estimates by this heuristic and improvement to efficiency of branch-width based SAT algorithms.

2   Definitions and Relations

In this section, we give the definitions of the various width-parameters and state relations among them.

Definition 1. *Let $G = (V, E)$ be a graph and let $f : V \longrightarrow \{1, 2, \ldots, n\}$ be a linear ordering of vertices of G. The cut-width of f is $\max_{1 \leq i \leq n} |\{\{u, v\} \in E : f(u) \leq i < f(v)\}|$, i.e., the maximum number of edges "crossing" an element in the linear order. The cut-width of the graph G, denoted $\mathrm{cw}(G)$, is the minimum cut-width over all linear orderings f.*

We can extend the definition of cut-width to that of a *weighted* cut-width, given a weight function θ on E by simply adding the weights of crossing edges in the above definition.

Definition 2. *Let $G = (V, E)$ be a graph. A tree-decomposition of G is a pair $(T, \mathcal{V})$ where $T = (I, F)$ is a tree (not necessarily a subtree of G) and $\mathcal{V} = \{V_i : i \in I\}$ is a family of subsets of V one per each node i of T such that*

- $\bigcup_{i \in I} V_i = V$;
- *for every edge $\{u, v\} \in E$ of G, there exists an $i \in I$ such that $\{u, v\} \subseteq V_i$;*
- *for all $i, j, k \in I$, if j is on the path between i and k in the tree T, then $V_i \cap V_k \subseteq V_j$ holds.*

The width of a tree-decomposition is defined to be $\max_{i \in I} |V_i| - 1$. The tree-width of G denoted $\mathrm{tw}(G)$ is defined to be the minimum width of any tree-decomposition of G.

Definition 3. *A path-decomposition of a graph G is defined by the special case of the above Definition 2 when the tree is a path. Accordingly the path-width of a graph G, denoted $\mathrm{pw}(G)$ is defined to be the minimum width of any path-decomposition of the graph.*

Clearly, $\mathrm{tw}(G) \leq \mathrm{pw}(G)$ for any graph G. We state below known inequalities. For proofs of these inequalities, we refer to the excellent survey [2].

Lemma 1. *For any graph G on n vertices, $\mathrm{tw}(G) \leq \mathrm{pw}(G) \leq O(\mathrm{tw}(G) \log n)$.*

Lemma 2. *For any graph G on n vertices, $\mathrm{pw}(G) \leq \mathrm{cw}(G) \leq \mathrm{pw}(G) \deg(G)$.*

Definition 4. *A branch-decomposition of a graph $G = (V, E)$ is a pair (T, λ) where T is a binary rooted tree and λ is a bijection between E and the set of leaves of T.*

For a node t of T, let $\mathcal{E}(t) \overset{\text{def}}{=} \{e \in E : \lambda(e)$ is a descendent of $t.\}$. Also, let $\mathcal{E}^{\perp}(t) \overset{\text{def}}{=} E \setminus \mathcal{E}(t)$, $V(t) \overset{\text{def}}{=} \cup\{e : e \in \mathcal{E}(t)\}$, i.e., $V(t)$ is the set of endpoints of edges in $\mathcal{E}(t)$, and $V^{\perp}(t) \overset{\text{def}}{=} \cup\{e : e \in \mathcal{E}^{\perp}(t)\}$. The cut of t is $\mathrm{Cut}(t) \overset{\text{def}}{=} V(t) \cap V^{\perp}(t)$. The order of t is $|\mathrm{Cut}(t)|$.

The branch-width of (T, λ) is maximum order of nodes in T.

Finally, the branch-width of G, denoted $\mathrm{bw}(G)$, is the minimum branch-width of any branch-decomposition of G.

The definition of branch-width is naturally extended to hypergraphs by using the hyperedges instead of edges in the above definition. The branch-width of a hypergraph H is denoted by $\mathrm{hbw}(H)$.

The following relations between branch-width and tree-width of a graph are proved in [6].

Lemma 3. *For any graph G, $\mathrm{bw}(G) - 1 \leq \mathrm{tw}(G) \leq \frac{3}{2}\mathrm{bw}(G)$.*

3   Upper Bounds on SAT and CIRCUIT-SAT Using Various Widths

Definition 5. *Let φ be a CNF formula on variable set X. Let $G_\varphi = (X, E, \theta)$ be a weighted undirected graph where $\{x, y\} \in E$ with weight $\theta(x, y)$ iff variables x and y appear together in $\theta(x, y)$ clauses of φ. Let W_φ be the value of a minimum weighted cut-width of G_φ.*

Theorem 1. *The satisfiability of any CNF formula φ on n variables can be tested in time $O(2^{W_\varphi} n(W_\varphi + \log n))$.*

Proof. W.l.o.g, let $(x_1, x_2, \ldots, x_n)$ be an ordering of the vertices of G_φ (or equivalently, the variables of φ) achieving a minimum weight cut-width. Thus we have, for any i with $1 \leq i \leq n - 1$, that the total weight of edges crossing the cut $(V_i, \overline{V_i})$ is at most W_φ, where $V_i \overset{\text{def}}{=} \{x_1, \ldots, x_i\}$.

As in [4], we will count the number of distinct subformulas of φ over all assignments to variables V_i. Clauses of φ containing only variables from $\overline{V_i}$ will remain intact in all such subformulas. Clauses containing only variables from V_i will contribute either the constant 0 (in which case the φ itself is violated by that assignment) or the constant 1. The remaining clauses are "cut", i.e., they contain variables from both V_i and $\overline{V_i}$. The effect of an assignment to V_i on each such clause is either to turn it into constant 1 or into a (fixed) clause containing only variables of $\overline{V_i}$. It follows that the total number of subformulas of φ induced by all assignments to V_i is at most $2^{\#\text{clauses cut by } (V_i, \overline{V_i})} + 1$.

But each clause cut by $(V_i, \overline{V_i})$ contributes at least one edge of G_φ that crosses the cut $(V_i, \overline{V_i})$. Moreover, each such edge e contributes at most $\theta(e)$ clauses. Thus, the number of clauses cut by $(V_i, \overline{V_i})$ is upper bounded by the total weight of edges crossing that cut, which in turn is upper bounded by W_φ. We conclude that for each i, $1 \leq i \leq n - 1$, the number of subformulas of φ induced by V_i is at most $2^{W_\varphi} + 1$.

Now, as in [4], we can define a backtracking algorithm to test satisfiability of φ by querying the variables according to a minimum weight cut-width ordering of G_φ while keeping track of unsatisfiable subformulas of φ in a cache. The time complexity of such an algorithm is bounded above by the product of the total number of subformulas over all i, $1 \leq i \leq n - 1$, and the time for searching the cache. The total number of subformulas is $O(n2^{W_\varphi})$ by the arguments above. Assuming an efficient data structure to implement the cache, the claimed time bound follows.   ∎

We can now appeal to a well-known construction to reduce a CIRCUIT-SAT instance to a CNF formula and apply Theorem 1 to derive an upper bound on CIRCUIT-SAT in terms of cut-width of the graph derived from the *CNF formula* as in Definition 5. However, there is a more direct upper bound on CIRCUIT-SAT in terms of the cut-width of a graph constructed from the *circuit itself*. The essential ideas of this result are already present in [4]. Here we present a somewhat cleaner proof with a slightly stronger bound. In particular, our bound does not depend on the maximum fan-out of a gate in the circuit as does the bound in [4].

Before we present our proof, we need to recall the construction of a CNF-SAT instance from a CIRCUIT-SAT instance. Given a circuit C, we will construct a CNF formula φ_C such that there is an assignment to the inputs of C producing a 1 output iff the formula φ_C is satisfiable. The formula φ_C will have $n + |C|$ variables, where $|C|$ denotes the number of gates in C; if C acts on inputs $x_1, \ldots, x_n$ and contains gates $g_1, \ldots, g_{|C|}$, then φ_C will have variable set $\{x_1, \ldots, x_n, g_1, \ldots, g_{|C|}\}$. For each gate $g \in C$, we define a set of clauses as follows:

1. if $c = \text{AND}(a, b)$, then add $\{(\overline{c} \vee a), (\overline{c} \vee b), (c \vee \overline{a} \vee \overline{b})\}$
2. if $c = \text{OR}(a, b)$, then add $\{(c \vee \overline{a}), (c \vee \overline{b}), (\overline{c} \vee a \vee b)\}$
3. if $c = \text{NOT}(a)$, then add $\{(c \vee a), (\overline{c} \vee \overline{a})\}$

The formula φ_C is simply the conjunction of all the clauses over all the gates of C. We assume below that C consists of gates from a standard complete basis such as $\{\text{AND}, \text{OR}, \text{NOT}\}$ and that each gate has fan-in at most 2. But our results can easily be generalized to allow other gates and larger fan-in; the final bounds are interesting as long as the number of clauses per gate and the maximum fan-in in the circuit are upper bounded by a constant. Recall that a circuit C is a *directed* acyclic graph (DAG). We define the underlying undirected graph as G_C:

Definition 6. *Given a circuit C with inputs $X = \{x_1, \ldots, x_n\}$ and gates $S = \{g_1, \ldots, g_s\}$, let $G_C = (V, E)$ be the undirected and unweighted graph with $V = X \cup S$ and $E = \{\{x, y\} : x \text{ is an input to gate } y \text{ or vice versa.}\}$. Let W_C be the unweighted cut-width of G_C.*

Note that G_C is somewhat different from G_{φ_C}. The latter is weighted whereas the former is not. But more importantly, G_{φ_C} contains certain edges not in G_C. For instance, if a and b are inputs of, say an AND gate, and a is not an input of gate b, then the edge $\{a, b\}$ appears in G_{φ_C} due to the clause $(c \vee \overline{a} \vee \overline{b})$, but there is no edge between a and b in G_C.

Lemma 4. *For a circuit C with gates from $\{\text{AND}, \text{OR}, \text{NOT}\}$ with fan-in at most 2, CIRCUIT-SAT instance for C can be solved in time $O(n2^{2W_C}(W_C + \log n))$.*

Proof. The proof proceeds along similar lines as that of Theorem 1. However, we bound the number of cut clauses of φ_C directly in terms of the cut-width of G_C instead of G_{φ_C}. W.l.o.g., let $(v_1, \ldots, v_{n+s})$, where $|C| = s$, be an ordering of the vertices of G_C yielding the minimum (unweighted) cut-width. Let $(V_i, \overline{V_i})$ be a cut in this ordering, where $V_i = \{v_1, \ldots, v_i\}$. We use two crucial observations made in [4]: (1) Every cut clause in φ_C can be charged to an edge crossing the cut $(V_i, \overline{V_i})$, and (2) Any edge crossing the cut is charged at most twice. This implies the number of clauses injured by the cut $(V_i, \overline{V_i})$ is at most $2W_C$ and the claimed time bound follows as in Theorem 1.

It remains to justify claims (1) and (2). Referring to sets of clauses for the gates described above, we see that every clause includes an input variable and an output variable of a gate. Thus, every injured clause can be charged to at least one edge of G_C crossing the cut. Also, from 1. 2. and 3. above, observe that each edge (from an input of a gate to its output; *all* edges of G_C are of this kind) appears in at most two clauses. This verifies (1) and (2).   ∎

Remark 1. Note that the above proof does not use the fan-out of any gate in its proof. This is a slight improvement to the result in [4] where the bound depends on the fan-out.

Using path-width, we can improve Theorem 1. The main insight is to observe that in the proof of Theorem 1, a different upper bound on the number of distinct subformulas induced by all assignments to V_i holds: there are no more than 2^{tips} such subformulas where tips is the number of *end points* of edges crossing the cut $(V_i, \overline{V_i})$. Stated differently, this is no more than $2^{|V_i \cap S_i|}$, where S_i is the set of all variables appearing in all the clauses that contain a variable outside V_i.

Theorem 2. *The satisfiability of any CNF formula φ on n variables can be tested in time $O(2^{2\mathrm{pw}(G_\varphi)} n (\mathrm{pw}(G_\varphi) + \log n))$.*

Proof. We first make an observation valid for tree-decompositions, not just path-decompositions, of G_φ. Let $(T, \mathcal{V})$ be any tree-decomposition of $G \overset{\text{def}}{=} G_\varphi$ with width τ as in Definition 2. Removing any edge $\{i_1, i_2\}$ in T will disconnect it into components I_1 and I_2. Let us consider the corresponding subsets of vertices of G: let $U_1 \overset{\text{def}}{=} \cup\{V_j : j \in I_1\}$ and $U_2 \overset{\text{def}}{=} \cup\{V_k : k \in I_2\}$. Let $X \overset{\text{def}}{=} V_{i_1} \cap V_{i_2}$. It is easy to show (see, for example, Lemma 12.3.1 in [5]) that $U_1 \cap U_2 \subseteq X$. Moreover, X *separates* U_1 and U_2 in G, i.e., after removing X there will be no edges in G between a vertex in U_1 and a vertex in U_2. Interpreting what this means for the formula φ, we conclude the following: as we vary over all assignments to variables in U_1, the number of distinct subformluas induced by φ is at most $2^{|X|} \leq 2^{\tau+1}$, since $|V_{i_1}|, |V_{i_2}| \leq \tau + 1$. This can be seen by the intuition outlined before the theorem.

Now, let $(P, \mathcal{V})$ be an optimal *path-decomposition* of $G \overset{\text{def}}{=} G_\varphi$ so that the width of $(P, \mathcal{V})$ is $\mathrm{pw}(G)$. W.l.o.g., let the vertex set of P be $I = \{1, 2, \ldots, p\}$ and edge set $\{\{i, i+1\} : 1 \leq i \leq p - 1\}$. For $i \in I$, let

$$A_i \overset{\text{def}}{=} \cup_{j \leq i} V_j \quad \text{and} \quad B_i \overset{\text{def}}{=} \cup_{k > i} V_k.$$

Applying the argument in the first paragraph to the edge $\{i, i+1\}$ in P, we conclude that the number of subformulas of φ as we vary over all assignments to A_i is bounded above by $2^{\mathrm{pw}(G)+1}$.

We use a backtrack algorithm to test satisfiability of φ with a cache to store subformulas as in the proof of Theorem 1 and in [4]. However, we do this by assigning values to *sets of variables* $V_1, V_2', \ldots, V_p'$ given by the path decomposition $(P, \mathcal{V})$, where $V_i' = V_i \setminus (\cup_{j < i} V_j)$ is the set of variables not already appearing in earlier sets. The degree of any internal node of this tree is clearly bounded above by $2^{\mathrm{pw}(G)}$. An internal node corresponds to a partial assignment to variables in A_i for some i and thus corresponds to a subformula of φ determined by this partial assignment. As soon as a subformula is determined to be unsatisfiable we store it in the cache. When exploring from a new branch, we check if the induced subformula already appears in the cache and if it does, end that branch in a leaf. It follows that each internal node of the search tree is associated with a distinct subformula. Moreover, the number of internal nodes at level i of the search tree is bounded above by the number of subformulas induced by A_i and hence by $2^{\mathrm{pw}(G)+1}$. The number of levels is clearly at most n. Thus the total number of internal nodes in the search tree is no more than $n2^{\mathrm{pw}(G)+1}$. This implies that the total number of nodes in the search tree is $2n2^{2\mathrm{pw}(G)}$ since the degree of the tree is at most $2^{\mathrm{pw}(G)}$.   ∎

Applying Lemma 1, we obtain

Corollary 1. *The satisfiability of any CNF formula φ on n variables can be tested in time $O(2^{2\mathrm{tw}(G_\varphi)\log n}n(\mathrm{tw}(G_\varphi)\log n))$.*

Using Lemma 3, we also obtain

Corollary 2. *The satisfiability of any CNF formula φ on n variables can be tested in time $O(2^{2\mathrm{bw}(G_\varphi)\log n}n^2(\mathrm{bw}(G_\varphi)\log n))$.*

Currently the only way we are able to remove the $\log n$ factor in the exponent of the tree-width bound is with an algorithm which gives a regular resolution refutation on unsatisfiable instances, in contrast to the *tree-like* resolution refutation given by the above back-track search algorithms. This algorithm based on the tree-decomposition of the graph G_φ is similar to the one given in [1] which uses a branch-decomposition of the hypergraph, H_φ.

Theorem 3. *For every unsatisfiable CNF φ and every tree-decomposition $(T, \mathcal{V})$ of G_φ with width τ, a regular resolution refutation of φ can be computed in time $|\varphi|^{O(1)} \cdot 2^{O(\tau)}$.*

Proof. First we will need the following two definitions.

Let $\theta(i)$ be a function which maps each node of the tree T, to a subset of the clauses of φ. A clause C is an element of $\theta(i)$ if for every variable $v \in Vars(C)$, $v \in V_i$. We claim every clause, C of φ, will be in at least one $\theta(i)$. This follows from the fact that every clause will form a clique in G_φ and the following lemma from [2].

Lemma 5. *Let $(T, \mathcal{V})$, where $T = (I, F)$, be a tree decomposition of G, and let $W \subseteq V$ be a clique in G. Then there exists an $i \in I$ with $W \subseteq V_i$.*

For a set of clauses $\mathcal{C}$, let $\ker(\mathcal{C})$ be the subset of minimal clauses of $\mathcal{C}$ with respect to inclusion. Define

$$\mathcal{C}^x \stackrel{\text{def}}{=} \ker\left[\{C \vee D | C \vee x, D \vee \bar{x} \in \mathcal{C}\} \cup \{C \in \mathcal{C} | x \notin Vars(\mathcal{C})\}\right]$$

to be the set of clauses obtained from all possible resolutions of $\mathcal{C}$ on the variable x. For a set of variables $X = \{x_1, \ldots, x_k\}$, we can unambiguously define $\mathcal{C}^X \stackrel{\text{def}}{=} ((\mathcal{C}^{x_1})^{x_2} \ldots)^{x_k}$, since $(\mathcal{C}^x)^y = (\mathcal{C}^y)^x$.

The algorithm can then be defined as follows:

    For all nodes $i \in I$, visited in an upward order in T do:

        $p = $ parent of i

        $V_i' = V_i \setminus V_p$

        $Cl = \theta(i) \cup \{\mathcal{C}(j) : j \text{ is a child of } i\}$

(1) $\mathcal{C}(i) = Cl^{V_i'}$

For every visited node i, the algorithm constructs a set of clauses $\mathcal{C}(i)$ in the variables $V_i \cap V_p$. Since $|V_i \cap V_p| \leq \tau$ for every node i, $|\mathcal{C}(i)| \leq 2^{O(\tau)}$. Therefore the algorithm works in time $|\varphi|^{O(1)} \cdot 2^{O(\tau)}$.

It follows from the third part of the definition of tree decomposition, that if j is not a descendent of i, then $V_i' \cap V_j = \emptyset$. Since j is not a descendent of i, a path from i to j must pass through p, the parent of i. Therefore $V_i \cap V_j \subseteq V_p$. However, V_i' is defined to be $V_i \setminus V_p$, and therefore this intersection must be empty. From this we can see that for any two different nodes i and j, $V_i' \cap V_j' = \emptyset$. Therefore each variable is resolved at most once and the inference is regular.

Finally, we must show that if φ is unsatisfiable then $\emptyset \in \mathcal{C}(root)$. Let $\varphi_{current}$ be a CNF. Initially $\varphi_{current} = \varphi$. Every time the operator (1) is applied in the algorithm, replace the clauses $\theta(i) \cup \{\mathcal{C}(j) : j \text{ is a child of } i\}$ with $\mathcal{C}(i)$ in $\varphi_{current}$.

We claim that $\varphi_{current}$ remains unsatisfiable. The initial value of $\varphi_{current}$ is φ, which is assumed to be unsatisfiable. During each step, $\varphi_{current}$ becomes $\varphi_{current}' \stackrel{\text{def}}{=} \{\varphi_{current} \setminus Cl\} \cup \{Cl^{V_i'}\}$. Since $V_i' \cap V_j = \emptyset$ for any j which is not a descendent of i, $Vars(\varphi_{current} \setminus Cl) \cap V_i' = \emptyset$. Therefore $\varphi_{current}' = (\varphi_{current})^{V_i'}$. The inductive step then follows from the fact that if φ is unsatisfiable and x is a variable, then φ^x is unsatisfiable.

At the end of the algorithm, $\varphi_{current} = \mathcal{C}(root)$. Since $\varphi_{current}$ is unsatisfiable, $\emptyset \in \mathcal{C}(root)$. ∎

4 Future Work

We are currently trying to estimate the various width parameters for practical instances of CIRCUIT-SAT to evaluate the feasibility of width-based approaches on such instances. We tried to assess the performance of (hypergraph) branch-width algorithm of [1] on ATPG benchmarks and compare it with those reported in [4]. Our current implementation of the algorithm from [1] seems to be too slow to test on ATPG instances of size comparable to those tested by [4]. For instance, we can only compute branch-widths of 10-15 on circuits with about 150 inputs and 100 gates in approximately two hours. These observations are consistent with those reported in [3] for this branch-width algorithm. This motivates us to look for more efficient algorithms and/or heuristics to compute branch-width and hence to two experimental tasks: (1) Study efficiency improvements and accuracy estimates of the eigenvector heuristic used in [3] for branch-width of graphs, and (2) Estimate (using approximation algorithms) the four width parameters considered in this paper for a large number of ATPG instances and other classes of circuits (real and generated). It is conceivable that branch-width is too general a graph parameter to apply to CIRCUIT-SAT arising in ATPG applications.

References

1. Alekhnovich, M. and Razborov, A. : Satisfiability, Branch-width, and Tseitin Tautologies, *43rd Symp. Foundations of Computer Science (FOCS)*, 593-603, 2002.
2. Bodlaender, H. L. : A Partial *k*-arboretum of Graphs with Bounded Treewidth, *Theoretical Computer Science*, **209**, 1-45, 1998.
3. Cook, W. and Seymour, P. : Tour Merging via Branch-Decompistion, *manuscript*, December 18, 2002.
4. Prasad, M. R., Chong, P., and Keutzer, K. : Why is Combinatorial ATPG Efficiently Solvable for Practical VLSI Circuits?, *Journal of Elecrtonic Testing: Theory and Applications*, **17**, 509-527, 2001.
5. Diestel, R. : *Graph Theory*, Second Edition, Graduate Texts in Mathematics 173, Springer, 2000.
6. Robertson, N. and Seymour, P. D. : Graph Minors X - Obstructions to Tree-decomposition, *Jl. Combinatorial Theory, Series B*, **52**, 153-190, 1991.

Linear Time Algorithms for Some Not-All-Equal Satisfiability Problems

Stefan Porschen, Bert Randerath, and Ewald Speckenmeyer

Institut für Informatik, Universität zu Köln, D-50969 Köln, Germany.
{porschen,randerath,esp}@informatik.uni-koeln.de

Abstract. In this paper we consider the problem Not-All-Equal satisfiability (NAESAT) for propositional formulas in conjunctive normal form (CNF). It asks whether there is a (satisfying) model for the input formula in no clause assigning truth value 1 to all its literals. NAESAT is NP-complete in general as well as restricted to formulas with either only negated or unnegated variables. After having pointed out that SAT and NAESAT in the unrestricted case are equivalent from the point of view of computational complexity, we propose linear time algorithms deciding NAESAT for restricted classes of CNF formulas. First we show that a NAESAT model (if existing) can be computed in linear time for formulas in which each variable occurs at most twice. Moreover we prove that this computation is in NC and hence can also be solved in parallel efficiently. Secondly, we show that NAESAT can be decided in linear time for monotone formulas in which each clause has length exactly k and each variable occurs exactly k times. Hence, bicolorability of k-uniform, k-regular hypergraphs is decidable in linear time.

Key Words: not-all-equal satisfiability; edge coloring; NP-completeness; hypergraph bicolorability

1 Introduction and Notation

Let CNF denote the set of formulas (free of duplicate clauses) in conjunctive normal form over a set $V = \{x_1, \ldots, x_n\}$ of propositional variables ($n \in \mathbb{N}$). Each formula $C \in$ CNF is considered as a set of its clauses $C = \{c_1, \ldots, c_{|C|}\}$ having in mind that it is a conjunction of these clauses. Each clause $c \in C$ is a disjunction of different literals, and is also represented as a set $c = \{l_1, \ldots, l_{|c|}\}$. Each variable x induces a positive literal (variable x) or a negative literal (negated variable: $\bar{x}$). In general the *complement* of a literal l is denoted as $\bar{l}$. *Setting a literal to (truth value) true (1)* means to set the corresponding variable accordingly. For a given formula C, clause c, by $\mathrm{var}(C), \mathrm{lit}(C)$ resp. $\mathrm{var}(c), \mathrm{lit}(c)$ denote the set of variables, resp. literals occuring in C resp. c. For convenience we allow the empty set to be a formula: $\emptyset \in$ CNF which is satisfiable in any sense. From now on the term formula indicates always an (duplicate free) element of CNF. For $r \in \mathbb{N}$ let CNF($\leq r$) (resp. CNF(r)) denote the subset of formulas C such that each

E. Giunchiglia and A. Tacchella (Eds.): SAT 2003, LNCS 2919, pp. 172–187, 2004.

member of $\text{var}(C)$ occurs at most (resp. exactly) r times in C. By $\text{CNF}(p, q)$ we denote the set of all formulas C such that each clause has length exactly p and each variable occurs exactly q times in C. Let $k - \text{CNF}$ denote the set of formulas C such that $|c| = k$ for each $c \in C$. Moreover CNF_+ denotes the set of *monotone* formulas, i.e., all variables occur negated or all occur unnegated. Obviously from the point of view of satisfiability it suffices only to consider those formulas containing exclusively variables as literals. Finally, a formula C satisfying $(*) : \forall c_i, c_j \in C, i \neq j : |c_i \cap c_j| \leq r$ is called *r-intersecting* and is called *exactly* r-intersecting if we have equality in $(*)$. Let $\text{CNF}^{(r)}$ denote the set of r-intersecting formulas. (Reasonable combinations of the introduced notations have obvious meanings.)

In this paper the problem not-all-equal satisfiability (NAESAT) is studied for formulas in CNF and subsets of it. NAESAT is a special case of the general satisfiability problem (SAT). Given $C \in \text{CNF}$, NAESAT asks whether there is a satisfying truth assignment $t : \text{var}(C) \to \{0, 1\}$ such that there is no $c \in C$ all literals of which are set to 1. In its version as a search problem solving NAESAT for an instance C means to find a NAESAT-assignment for C if existing. Clauses containing a complemented pair of literals can be omitted immediately from the formula, because such a clause is always NAE-satisfiable. Hence, it is assumed in the following that clauses only contain different variables. NAESAT is known to be NP-complete (e.g. [15]). By reducing SET SPLITTING to NAESAT it follows that NAESAT remains NP-complete restricted to formulas in CNF_+. (The same holds for exact satisfiability (XSAT)). In [14] it is shown that SET SPLITTING and also NAESAT in the unrestricted case are decidable in time $O(\|C\| 2^{2/3 |C|})$ where $\|C\|$ is the length of C and $|C|$ is the cardinality of $C \in \text{CNF}$.

On the other hand the NP-completeness of NAESAT for monotone formulas immediately shows that deciding bicolorability of hypergraphs is NP-complete. Simply, identify a formula $C \in \text{CNF}_+$ with the (simple, i.e., having the Sperner property) hypergraph $H_C = (\text{var}(C), C)$, the *formula hypergraph*, where the variables are considered as vertices and the clauses as hyperedges. Then $C \in$ NAESAT holds if and only if H_C exhibits a bicoloring of its vertex set such that there exists no monochromatic hyperedge. Conversely, given a simple hypergraph $H = (V, E)$, assign to each vertex $x \in V$ a propositional variable and interpret the hyperedges in E as clauses yielding a formula $C_H \in \text{CNF}_+$. Again H is bicolorable such that there exists no monochromatic hyperedge if and only if $C_H \in$ NAESAT. Recently it has been shown that bicoloring clique hypergraphs of graphs is in P [13]. For this it was used that NAESAT for planar formulas can be decided in polynomial time by reduction to a MAXCUT problem.

Recall that a formula is by definition a *Horn* formula if each clause contains at most one positive literal. It is well known that testing SAT for Horn formulas can be done efficiently (precisely in linear time [11]). However NAESAT remains NP-complete even if restricted to Horn formulas, since the special class of negative Horn formulas (each clause contains only negative variables) already corresponds to NAESAT for monotone formulas.

part of Fig. 1 we randomly generated NAE3-SAT instances and used the Davis-Putnam algorithm to check whether the instances are satisfiable or not. By usage of these random NAE3-SAT instances we constructed the corresponding 3-SAT instances with up to twice as many clauses. In both parts of the figure every point represents the average of 100 trials. In the right hand part of Fig. 1 we compared randomly generated 3-SAT instances with 3-SAT instances constructed by randomly generated NAE3-SAT instances.

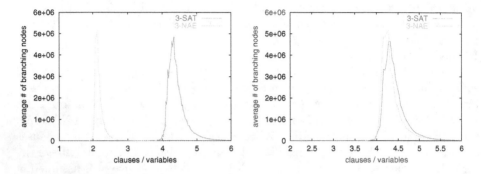

Fig. 1. 3-SAT vs. NAE 3-SAT (350 variables).

3   Graphs Assigned to Monotone Formulas

Let $C \in \mathrm{CNF}_+$ and let H_C be the formula hypergraph as introduced in the introduction, then the *intersection graph* G_C of H_C has a vertex for each clause (hyperedge) $c \in C$ and two vertices are joined by an edge if the corresponding clauses (hyperedges) have a non-empty intersection. We call G_C the *clause graph* corresponding to C. In hypergraph terminology a 1-intersecting, monotone formula $C \in \mathrm{CNF}_+^{(1)}$ corresponds to a *linear* hypergraph H_C. Further, assign to each $C \in \mathrm{CNF}_+$ the *variable graph* $G_{V(C)}$: Introduce a vertex for each $x \in \mathrm{var}(C)$ and join two vertices by an edge if there is a clause in C containing the corresponding variables (cf. also [9]). Clearly for each $c \in C$ the subgraph $G_{V(C)}|c$ of $G_{V(C)}$ is isomorphic to the complete graph $K_{|c|}$ we call it the *clause clique* for c or simply the *c-clique*. If $C \in \mathrm{CNF}_+^{(1)}(2)$ then G_C has an edge for exactly each variable in C and $G_{V(C)}$ is a line graph, namely $G_{V(C)} \cong L(G_C)$ i.e., it is the line graph of G_C. The *incidence graph* of a formula $C \in \mathrm{CNF}$ is the bipartite graph whose vertex set consists of one partition corresponding to its variables and another consisting of its clauses. An edge is introduced for each pair $x \in \mathrm{var}(C), c \in C$ satisfying $x \in \mathrm{var}(c)$. For $C \in \mathrm{CNF}$ and $x \in \mathrm{var}(C)$ let $C_x := \{c \in C : x \in \mathrm{var}(c)\}$ be the set of clauses containing variable x regardless whether positive or negative. For the number of occurences of x in C we write $\omega_C(x) := |C_x|$. Recalling $\mathrm{CNF}(p, q) = \{C \in \mathrm{CNF} : \forall c \in C : |c| = p, \forall x \in \mathrm{var}(C) : \omega_C(x) = q\}$ we observe that each $C \in \mathrm{CNF}(p, q)$ satisfies

$q|\text{var}(C)| = \sum_{x \in \text{var}(C)} \omega_C(x) = \sum_{c \in C} |c| = p|C|$. Hence $|\text{var}(C)| = |C|$ if and only if $p = q$. The class $\text{CNF}(k,k), k \in \mathbb{N}, k \geq 2$, is called the *diagonal class* ($k = 1$ is trivial).

Lemma 3. *Let* $C \in \text{CNF}_+^{(1)}(k,k)$. *Then* G_C *and* $G_{V(C)}$ *are* d_k-*regular with* $d_k := k(k-1)$. *Moreover,* C *is exactly 1-intersecting if and only if* $G_C \cong K_{n(k)} \cong G_{V(C)}$ *and* $n(k) = k(k-1) + 1$.

PROOF. Each clause of a formula $C \in \text{CNF}_+^{(1)}(k,k)$ contains k different variables. Each variable x selects exactly k different clauses from C, namely those in C_x. The corresponding vertices are pairwise joined by an edge in G_C delivering a K_k. Let these edges be labeled by x. Then clearly none of these edges can be labeled by another variable y too, otherwise the formula would not be 1-intersecting. It follows that each variable of a clause contributes $k - 1$ different neighbours for that clause yielding the total degree d_k. Now let $c \in C_x$, then x is connected in $G_{V(C)}$ to each variable in $\text{var}(C_x) \setminus \{x\}$. Since C_x is 1-intersecting these neighbours must be mutually distinct for all $c \in C_x$, hence $G_{V(C)}$ is d_k-regular. To prove the second claim, let $C \in \text{CNF}_+(k,k)$ be exactly 1-intersecting. Then clearly $G_C \cong K_{|C|}$. Thus $d_k = k(k-1) = |C| - 1 \Leftrightarrow |C| = d_k + 1 =: n(k)$. On the other hand $G_{V(C)}$ is d_k-regular, and is therefore isomorphic to $K_{n(k)}$. Conversely, consider a formula $C \in \text{CNF}_+(k,k)$ of $n(k)$ variables resp. clauses such that $G_C \cong K_{n(k)} \cong G_{V(C)}$. Then $G_{V(C)}$ is $n(k) - 1 = d_k$-regular. That is each x has pairwise different neighbours in each clause containing it, hence no pair of variables can be contained in two different clauses meaning that C is 1-intersecting. As G_C is a clique we have $|c_i \cap c_j| \geq 1$ for all clauses $c_i, c_j \in C$ from which the lemma follows.   □

Note that the formula hypergraph H_C for an exactly 1-intersecting formula $C \in \text{CNF}_+^{(1)}(k,k)$ constitutes a (symmetric) Steiner system $S(2,k,n(k))$ where $|C| = n(k) = k(k-1) + 1$.

4   A Linear Time Algorithm Deciding NAESAT for the Class CNF(≤ 2)

Suppose that $C \in \text{CNF}_+(\leq 2)$ contains two clauses $c \neq c'$ such that $|c \cap c'| > 1$, then clearly $C \in \text{NAESAT} \Leftrightarrow C \setminus \{c,c'\} \in \text{NAESAT}$. This helps us to obtain a 1-intersecting formula recursively. For eliminating negative literals, thus for obtaining a monotone formula we use the following resolution principle:

Lemma 4. *Let* $C = \{c_1, \ldots, c_m\} \in \text{CNF}(\leq 2)$ *such that there is a literal* $l \in \text{lit}(C)$ *and two clauses* $c_i, c_j \in C$ *with* $l \in c_i, \bar{l} \in c_j$ *(then* $c \cap \{l, \bar{l}\} = \emptyset$ *for* $i \neq k \neq j$). *Defining* $c_{ij} := c_i \setminus \{l\} \cup c_j \setminus \{\bar{l}\}$ *and* $C_{ij} := C \setminus \{c_i, c_j\} \cup \{c_{ij}\}$ *then* $C_{ij} \in \text{CNF}(\leq 2)$ *and* $C \in \text{NAESAT}$ *if and only if* $C_{ij} \in \text{NAESAT}$.

Note, that the resolution principle does not hold for NAESAT, if variables occur more than twice.

2   NAESAT and SAT for CNF Have Essentially the Same Complexity

At first glance one might hope that there are algorithms solving NAESAT for arbitrary $C \in \text{CNF}$ essentially faster in the worst case than in time $O(2^n)$. This is because the further restriction on the truth assignments should make available certain branching rules. But this impression is wrong as pointed out now. Let $\gamma : c \mapsto c^\gamma$ denote the map that flips all literals in a clause c, thus $l \in \text{lit}(c)$ $\Leftrightarrow \bar{l} \in \text{lit}(c^\gamma)$. This induces also a flipping map for formulas $C \mapsto C^\gamma$, where $C^\gamma = \{c_1^\gamma, \ldots, c_{|C|}^\gamma\}$ for $C = \{c_1, \ldots, c_{|C|}\}$. Let us introduce some notions:

Definition 1. For $C \in \text{CNF}$ a subformula of C is any collection of subsets of clauses in C. A subformula is called strict if it contains only whole clauses of C. The strict subformula $\text{Core}(C) := \{c \in C : c^\gamma \notin C\} \in \text{CNF}$ is called the core of C (which is always unique and can be the empty formula). The complement $\text{Sym}(C) := C \setminus \text{Core}(C) \in \text{CNF}$ is called the symmetric subformula of C. Let $\text{Core}(\text{CNF}) := \{C \in \text{CNF} : \text{Core}(C) = C\}$, $\text{Sym}(\text{CNF}) := \{C \in \text{CNF} : C = C^\gamma\}$.

Defining the map $\tau : \text{CNF} \ni C \mapsto C^\tau := C \cup C^\gamma \in \text{CNF}$ we obviously have:

Lemma 1. Let $C \in \text{CNF}$, then $C \in \text{NAESAT}$ if and only if $C^\tau \in \text{SAT}$, i.e. NAESAT is SAT restricted to $\text{Sym}(\text{CNF})$.

The lemma shows that NAESAT reduces to SAT. The other direction is as follows (cf. [12]): For $C \in \text{CNF}$ let $\text{var}(C) = \{x_1, \ldots, x_n\}$. Defining $C_{x_0} := \{c \cup \{x_0\} : c \in C\} \in \text{Core}(\text{CNF})$ it is not hard to see that:

Lemma 2 (Knuth, '92). $C \in \text{SAT} \Leftrightarrow C_{x_0}^\tau \in \text{NAESAT}$.

By Lemmas 1 and 2 one obtains that SAT and NAESAT have essentially the same computational complexity for the unrestricted class CNF. More precisely:

Theorem 1. For a formula $C \in \text{CNF}$ let $m(C) := m = |C|$ be the number of its clauses and let $n(C) := n = |\text{var}(C)|$ be the number of its variables. Let $0 < \varepsilon < 1$ and p be an arbitrary but fixed polynomial of (maximal) degree $d(p)$. There is an algorithm $A_{\text{SAT}} : \text{CNF} \to \{0, 1\}$ such that $A_{\text{SAT}}(C) = 1 \Leftrightarrow C \in \text{SAT}$ having time complexity $O(m^{d(p)} 2^{\varepsilon n})$ if and only if there is an algorithm $A_{\text{NAESAT}} : \text{CNF} \to \{0, 1\}$ such that $A_{\text{NAESAT}}(C) = 1 \Leftrightarrow C \in \text{NAESAT}$ with time complexity $O(m^{d(p)} 2^{\varepsilon n})$.

We close this section with some remarks on random instances of NAE3-SAT. Achlioptas et al. presented in [1] a sharp threshold c_3^{NAE} for NAE3-SAT. Moreover they locate the threshold c_3^{NAE} between 1.514 and 2.215. As already mentioned in Lemma 1 a NAE3-SAT formula is equivalent to a 3-SAT formula with up to twice as many clauses. The experiments of Achlioptas et al. with formulas containing up to 200 variables support the possibility that $2c_3^{NAE} \sim c_3^{SAT} (\sim 4.2)$ might hold (which indeed is true in a k-asymptotic manner for the case of k-clauses [2]). Here c_3^{SAT} is the sharp threshold of 3-SAT. Our experiments on instances with up to 350 variables also support this possibility. In the left hand

Lemma 5. *For $C \in \mathrm{CNF}_+(2)$ holds $C \in \mathrm{NAESAT}$ if and only if G_C admits a bicoloring of its edges such that each vertex of G_C is incident to an edge of either color.*

Given a simple graph an edge coloring as defined in the lemma can be found in linear time exploiting Euler tour techniques [6]. We adapt this approach for our purposes resulting in procedure EdgeColor. The following procedure recursively decides NAESAT for $\mathrm{CNF}(\leq 2)$ and returns an explicit NAESAT model for the input instance when existing otherwise returns NIL.

Procedure $\mathrm{NAESAT}_{\leq 2}(C, t)$: Boolean array t of length $|\mathrm{var}(C)|$
input: $C \in \mathrm{CNF}(\leq 2)$
output: NAESAT model t for C, if $C \in \mathrm{NAESAT}$, else **NIL**
if $c \in C : |c| = 1$ **then return NIL** ($*C \notin \mathrm{NAESAT}*$)
if $C = \emptyset$ **then return** $t \rightarrow \mathbf{0}$ ($*C \in \mathrm{NAESAT}*$)
if $\emptyset \in C$ **then return NIL** ($*C \notin \mathrm{NAESAT}*$)
(1) Simplify(C) (now $*C \in \mathrm{CNF}_+^{(1)}(\leq 2)*$)
(2) EdgeColor(C, t)

In the sequel we discuss the procedures in steps (1),(2):

Procedure Simplify(C)
remove all duplicate clauses from C
if C contains a variable x occurring only negated **then**
    substitute $\bar{x} \leftarrow x$
if there is $c \in C$ containing $r \geq 2$ unique literals **then**
    remove c; $\mathrm{NAESAT}_{\leq 2}(C, t)$
if there is $c = \{l_1, \bar{l}_2\} \in C$ **then**
    evaluate C according to $l_2 \leftarrow l_1$; $\mathrm{NAESAT}_{\leq 2}(C, t)$
if there is $c = \{l_1, l_2\} \in C$ **then**
    substitute $l_2 \leftarrow \bar{l}_1$ in C; remove c; $\mathrm{NAESAT}_{\leq 2}(C, t)$
if there are $c_i, c_j \in C, x \in \mathrm{var}(C) : x \in c_i, \bar{x} \in c_j$ **then**
    $C \rightarrow C \setminus \{c_i, c_j\} \cup c_{ij}$; $\mathrm{NAESAT}_{\leq 2}(C, t)$
if there are $c_i, c_j \in C : |c_i \cap c_j| > 1$ **then**
    $C \rightarrow C \setminus \{c_i, c_j\}$; $\mathrm{NAESAT}_{\leq 2}(C, t)$

Lemma 6. *Let $C \in \mathrm{CNF}(\leq 2)$, then $C \in \mathrm{NAESAT}$ if and only if* Simplify$(C) \in$ NAESAT. *Each step of* Simplify *(ignoring statement $\mathrm{NAESAT}_{\leq 2}(C, t)$) needs linear time.*

There is a basic existence result also used in [6] to prove the Vizing theorem on edge colorings.

Lemma 7. *Let G be a connected graph with minimal degree $\delta(G) \geq 2$ that is not an odd cycle. Then G has a 2-edge-coloring in which both colors are represented at each vertex.*

PROOF. Assume that G is Eulerian. If G is an even cycle, then the proper 2-edge-coloring has the required property. Otherwise, G has a vertex v_0 of degree > 2. Now let $v_0 e_1 v_1 e_2 ... e_r v_0$ be an Euler tour of G, and set $E_{odd} = \{e_i | i \text{ odd }\}$ and $E_{even} = \{e_i | i \text{ even }\}$. Then the 2-edge coloring (E_{odd}, E_{even}) of G has the desired property, since each vertex of G is an internal vertex of the given Euler tour of G. Now suppose that $G = (E, V)$ is not Eulerian. Then due to Euler [7] and Hierholzer [10] it is well-known that G contains vertices of odd degree. The handshake-argument implies that in any graph there is an even number of vertices of odd degree. Now we construct a new graph G^* by adding a new vertex v_0 and joining it to each vertex of odd degree in G. Clearly, G^* is Eulerian. Now let $v_0 e_1 v_1 e_2 ... e_r v_0$ be an Euler tour of G^*, and again set $E_{odd} = \{e_i | i \text{ odd}\}$ and $E_{even} = \{e_i | i \text{ even}\}$. Then the 2-edge coloring $(E_{odd} \cap E, E_{even} \cap E)$ of G has the required property. $\square$

The only gap in order to obtain a linear time procedure is a linear time subprocedure for finding an Euler tour in an Eulerian graph. Here we use the classical Hierholzer algorithm (HA) [10] executing this task in linear time. In conclusion if the clause graph G_C of an instance $C \in \mathrm{CNF}_+^{(1)}(2)$ has no component isomorphic to an odd cycle, then G_C has a 2-edge-coloring in which both colors are represented at each vertex. But now there is a 1:1-correspondence between this special 2-edge-coloring of G_C and a truth assignment NAE-satisfying C. Likewise we need linear time if we want to preprocess unique variables of a formula $C \in \mathrm{CNF}_+^{(1)}(\leq 2)$. We recursively push unique variables on a stack and remove the associated clauses from the current formula C until the current formula C contains no unique variable anymore, i.e., $C \in \mathrm{CNF}_+^{(1)}(2)$. Having found a NAESAT-model t_0 for the current formula we can pop the unique variables from the stack properly assigning truth values to them such that the input formula is likewise NAE-satisfiable.

Procedure EdgeColor(C, t) : Boolean array t
input: $C \in \mathrm{CNF}_+^{(1)}(\leq 2) \cap \mathrm{NAESAT}$
output: NAESAT model t for C
repeat
        mark all unique variables (if exist) and push them on a stack
        remove the associated clauses from the formula
until the current formula C contains no unique variable
$(* \text{ now } C \in \mathrm{CNF}_+^{(1)}(2)*)$
compute a NAESAT assignment t_0 for the current formula (as described below)
pop the unique variables from the stack and extend t_0 to these unique variables
such that it results to a NAESAT assignment t for the input formula

When entering EdgeColor the current formula C cannot contain a clause of length 2, hence no vertex in G_C has degree less than 3 and C is always in NAESAT:

Computation of t_0:
if $C = \emptyset$ return $t_0 \leftarrow \mathbf{0}$
By DFS compute the connected components of G_C
For each component $G_C^k = (V_C^k, E_C^k)$ $(k \in \mathbb{N})$ **do:**
   if G_C^k has only vertices of even degree **then**
      compute by HA an Euler tour starting at a vertex v_0
      $(*$ let this be $v_0 e_1 v_1 e_2 ... e_r v_0 *)$
      set t_0 according to the 2-edge coloring $(\{e_i : i \text{ odd}\}, \{e_i : i \text{ even}\})$
   if G_C^k has a vertex of odd degree **then**
      $(*$ let V_{odd}^k be the set of odd degree vertices in G_C^k $*)$
      introduce a vertex $v_0^* \notin V_C^k$ and join it to each $v \in V_{\text{odd}}^k$
      $(*$ let G_C^{k*} be the resulting graph which is Eulerian $*)$
      compute by HA an Euler tour of G_C^{k*} starting at v_0^*
      $(*$ let this be $v_0^* e_1^* v_1^* e_2^* ... e_r^* v_0^* *)$
      set t_0 according to $(\{e_i^* : i \text{ odd}\} \cap E_C^k, \{e_i^* : i \text{ even}\} \cap E_C^k)$
end do

Lemma 8. *For* $C \in \mathrm{CNF}_+^{(1)}(\leq 2) \cap \mathrm{NAESAT}$, $\mathrm{EdgeColor}(C, t)$ *returns a model* t *for* C. *Furthermore,* $\mathrm{EdgeColor}(C, t)$ *needs linear time.*

Theorem 2. *For* $\mathrm{CNF}(\leq 2)$ *the search version of* NAESAT *is decidable in linear time.*

It is not hard to see that procedure Simplify can be parallelized efficiently on a PRAM. Moreover, computing connected components is in NC [16]. Precisely: for a finite simple graph G with n vertices and m edges holds $\mathrm{Components}(G) \in \mathrm{CRCW}(\frac{n+m}{\log n}, \log n)$. Also computing Euler tours is in NC [4]: For G connected holds $\mathrm{EulerTour}(G) \in \mathrm{CRCW}(m\alpha(m, m), \log n)$, where α denotes the inverse Ackermann function. Thus we obtain:

Corollary 1. *The search problem* NAESAT *can be solved efficiently for* $\mathrm{CNF}(\leq 2)$ *on a CRCW-PRAM. More precisely:*

$$\mathrm{NAESAT\_CNF}(\leq 2) \in \mathrm{CRCW}((n + m)\alpha(m, m), \log n)$$

with $n = |\mathrm{var}(C)|, m = |C|$ *for* $C \in \mathrm{CNF}(\leq 2)$.

5   NAESAT for the Class $\mathrm{CNF}_+ (k, k), k \geq 2$

Recall that $\mathrm{CNF}(p, q)$ is the set of all formulas such that each clause has length exactly p and each variable occurs exactly q times. It has been shown [17] (by using the König-Hall theorem on the incidence graph of a formula) that SAT restricted to the diagonal class $\mathrm{CNF}(k, k), k \in \mathbb{N}$, is trivial in the sense that such formulas are always satisfiable.

Proposition 1 (Tovey, '84). $\mathrm{CNF}(k,k) \subset \mathrm{SAT}$ *for each* $k \in \mathbb{N}$.

In fact by using the same argument, Tovey obtained the assertion also for the case in which the number of occurences is at most as high as the fixed length of the clauses. There are some immediate consequences of Proposition 1 and Lemma 1 concerning NAESAT:

Corollary 2. *Let* $k \in \mathbb{N}$ *be even, then* $\mathrm{CNF}_+(k, k/2) \subset \mathrm{NAESAT}$.

Corollary 3. *For* $k \in \mathbb{N}$ *even holds* $\mathrm{Core}(\mathrm{CNF}(k, k/2)) \subset \mathrm{NAESAT}$.

As mentioned in the introduction a k-uniform hypergraph H_C can be assigned to a *monotone* formula $C \in k - \mathrm{CNF}_+$; and NAE-satisfiability of C corresponds to bicolorability of H_C. An old (probabilistic) result of Beck [5] states that every k-uniform hypergraph with less than $m(k) = \Omega(2^k k^{1/3})$ hyperedges is bicolorable (has property B) (cf. also [3]). The hypergraphs corresponding to formulas in the monotone diagonal class $\mathrm{CNF}_+(k, k)$ in addition have the k-regularity property, i.e., each vertex occurs in exactly k different hyperedges.

To clarify the question what can be said concerning NAESAT for the diagonal class we state an equivalent characterization for formulas to be in NAESAT:

Proposition 2. *For* $C \in \mathrm{CNF}$ *holds* $C \in \mathrm{NAESAT}$ *if and only if there is a subformula* $P \in 2 - \mathrm{CNF}$ *such that for each* $c \in C$ *there is a* $p \in P$ *with* $p \subseteq c$ *and* $P \in \mathrm{X2SAT}$.

For $C \in \mathrm{CNF}_+$ define a *transversal graph* of its variable graph $G_{V(C)}$ as a subgraph T whose edge set contains an edge of each c-clique for $c \in C$. Note that restricted to CNF_+ the last Proposition is equivalent to:

Proposition 3. *If* $C \in \mathrm{CNF}_+$ *then* $C \in \mathrm{NAESAT}$ *if and only if the variable graph* $G_{V(C)}$ *contains a bipartite transversal graph.*

Definition 2. *For* $k \geq 3$ *a formula* $C \in \mathrm{CNF}_+(k, k)$ *is called* autonomous *if it contains no proper subformula* $C' \subset C$ *with* $C' \in \mathrm{CNF}_+(k, k)$. *Let* $\mathcal{A}(k)$ *denote the set of all autonomous formulas* $C \in \mathrm{CNF}_+(k, k)$.

As it will turn out it is sufficient to restrict the consideration to the class $\mathcal{A}(k)$.

Lemma 9. *Let* $k \geq 3$ *be fixed. For* $C \in \mathrm{CNF}_+(k, k)$ *holds* $C \in \mathcal{A}(k)$ *if and only if* $G_{V(C)}$ *is connected.*

PROOF. For $k \geq 3$ let $C \in \mathcal{A}(k)$ be fixed and let $G := G_{V(C)}$ be its variable graph such that the edges are labeled by the corresponding clauses. Assume that G has at least two (maximal) components G_1, G_2. We assign a fomula C_1 to G_1 as follows $\mathrm{var}(C_1) := V(G_1)$ is the set of variables and the clauses are given by the labels of the edges in G_1 accordingly. First note that C_1 is a strict proper subformula of C. Indeed, if there would be a clause c' in C_1 being a proper subset of its pendant c in C then there must be edges in G joining the variables of c' to all variables in c. Hence G_1 would be no maximal component. Next we prove

$(*)$: $\forall x \in \mathrm{var}(C_1)$ holds $\omega_{C_1}(x) = k$ meaning $C_1 \in \mathrm{CNF}_+(k,k)$ and thus C is not autonomous contradicting the assumption. To verify property $(*)$ assume that there is $x \in \mathrm{var}(C_1)$: $\omega_{C_1}(x) < k$. But $\omega_C(x) = k$ hence x has k edges labeled by the k different clauses containing it, thus G_1 would be no maximal component proving $(*)$.

Conversely, let $C \in \mathrm{CNF}_+(k,k), k \geq 3$, be a fixed formula with a connected variable graph $G_{V(C)}$ and assume that C contains a proper strict subformula $C' \subset C$. W.l.o.g. we can assume that C' is a smallest such formula and hence is a member of $\mathcal{A}(k)$. Indeed, we must have that $x \in \mathrm{var}(C')$: $\omega_{C'}(x) = k$ because otherwise C' cannot be a strict subformula. It follows that the variable graph $G_{V(C')}$ of C' is a maximal connected component different from $G_{V(C)}$ contradicting the assumption by the assertion already proven. □

Definition 3. *Let* $k \in \mathbb{N}, k \geq 3$ *be fixed. A formula* $B \in \mathrm{CNF}_+(k,k)$ *is called a* k*-block, if it is exactly 1-intersecting. Let* $\mathcal{B}(k)$ *denote the set of all* k*-blocks in* $\mathrm{CNF}_+(k,k)$.

Lemma 10. *For each* $k \geq 3$ *holds* $\mathcal{B}(k) \subset \mathcal{A}(k)$.

PROOF. By Lemma 3 we have that $G_{V(B)}$ is a clique for each $B \in \mathcal{B}(k)$ and hence B is a member of $\mathcal{A}(k)$ by Lemma 9. □

For fixed $k \geq 3$ there is up to variable renaming and permutation only one k-block. We call this the *canonical block* R_k whose clauses are considered as k-tuples. It has a *leading clause* $\xi = (x_1, \ldots, x_k)$. For each x_i we have clauses $c_i(j) \in C_{x_i} \setminus \{\xi\}, 1 \leq j \leq k-1$, each containing x_i in the first position. The other positions are occupied by the elements of a $k-1 \times k-1$ variable matrix $A_k = (a_{pq})$ as follows. Define $c_1(j) = (x_1, z_j(A_k))$, where $z_j(A_k)$ is the jth row vector of $A_k, 1 \leq j \leq k-1$. Clearly, none of the remaining clauses can contain two elements of the same row of A_k. Thus we can use the transpose A_k^T to build the clauses $c_2(j) = (x_2, s_j(A_k))$, where $s_j(A_k) = z_j(A_k^T)$ is the jth column of $A_k, 1 \leq j \leq k$. To build the remaining clauses $c_i(j), 3 \leq i \leq k$, the columns of A_k have to be permuted accordingly to ensure that each pair of different clauses is exactly 1-intersecting. We need $k-2$ matrices B_i consisting of the accordingly permuted columns of A_k^T and the rows of B_i form the clauses $c_i(j) = (x_i, z_j(B_i)), 3 \leq i \leq k, 1 \leq j \leq k-1$. Since the first positions in each $c_i(j)$ are occupied by x_i we can leave unchanged the first column of A_k^T, hence $s_1(B_i) = s_1(A_k^T)$. But now all other columns in $B_i, B_{i'}$ must be permuted mutually different. Clearly, such permutations exist for each fixed k and it is assumed that such a set is fixed for building R_k. Obviously R_k contains $k(k-1) + 1 = n(k)$ variables and equally many clauses. For example in the case $k = 3$ let

$$A_3 = \begin{pmatrix} a & b \\ c & d \end{pmatrix}$$

and let R_3 consist of the clauses:

$$
\begin{aligned}
\xi &= \{x_1, x_2, x_3\} \\
c_1(1) &= \{x_1, a, b\} \\
c_1(2) &= \{x_1, c, d\} \\
c_2(1) &= \{x_2, a, c\} \\
c_2(2) &= \{x_2, b, d\} \\
c_3(1) &= \{x_3, a, d\} \\
c_3(2) &= \{x_3, b, c\}
\end{aligned}
$$

Clearly, by construction holds $R_k \in \mathcal{B}(k)$ for each fixed k and:

Lemma 11. *For $k \geq 3$ fixed each $B \in \mathcal{B}(k)$ is up to variable renaming and permutation equal to R_k.*

Recall that a transversal graph of $G_{V(C)}$ (for $C \in \mathrm{CNF}_+$) is a subgraph T such that $E(T)$ contains an edge of each clause clique.

Lemma 12. *For $k \geq 3$ let $B \in \mathcal{B}(k)$, then each transversal graph T of $G_{V(B)}$ contains a cycle.*

PROOF. Since B is a k-block we have $|E(T)| \geq n(k) \geq |V(T)|$. Suppose T contains no cycle, hence it is a forest of at least one tree. Thus we have $|E(T)| \leq |V(T)| - 1 \leq n(k) - 1 < n(k)$ yielding a contradiction. $\qquad\square$

As it turns out only blocks $B \in \mathcal{B}(3)$ have the property that any transversal graph of $G_{V(B)}$ necessarily contains an odd cycle.

Lemma 13. $\mathcal{B}(3) \cap \mathrm{NAESAT} = \emptyset$, $B \setminus \{b\} \in \mathrm{NAESAT}$ *for any $b \in B$, $B \in \mathcal{B}(3)$.*

PROOF. Let $B \in \mathcal{B}(3)$ and $b \in B$ be arbitrary. Transform B into R_3 as in the Example above such that b gets the leading clause $\xi = \{x_1, x_2, x_3\}$. Let $G_3 = (V, E)$ denote the variable graph of R_3 and $G_3' = (V, E \setminus E(\xi))$ that one of $R_3 - \{\xi\}$. Then one verifies immediately that $T = (V, E(T))$ with $E(T) = \{x_2a, x_2b, x_3b, x_3c, x_1a, x_1c\}$ is a bipartite transversal graph (a path) of G_3' hence it yields an NAESAT solution for R_3' thus for $B \setminus \{b\}$.

Let $T' = (V', E')$ be any bipartite transversal graph of G_3' with $|E'| = 6$, hence having in common exactly one edge with each c-clique for $c \in R_3'$. Let $V' = V_1' \cup V_2'$ be the vertex set partition. No edge of the ξ-clique in G_3 is contained in E'. We claim that there is no way to extend T' to a bipartite transversal graph T of G_3. We distinguish two main cases. Case I: $\xi \subset V'$. Then T' is extendable if and only if two elements of ξ are in one partition, say V_1', and one element is contained in the other V_2'. Then the two elements $u, v \in V_1' \cap \xi$ must have a common neighbour in V_2' otherwise they could be coulored differently and hence would be contained in distinct partitions. Since each $x_i \in \xi$ is contained in only two clauses different from ξ we have for the set of neighbours $1 \leq |N(x_i)| \leq 2$ otherwise we would have two edges of the same clause clique in T', thus $1 \leq |N(u) \cap N(v)| \leq 2$.

If $N(u) \cap N(v) = \{w\}$ then $w \in V' \setminus \xi$ can have no other neighbour in T'. Indeed the only possibility would be the remaining element $z \in \xi$ but this is in the same partition as w. Hence the clause containing z and w can only be covered by the edge zr where r is the last element of that clause. There remaining variables s, t of the matrix A_3 must occur together in the second clause containing z. On the other hand, since u, v have no other neighbours the two left clauses containing u, v must be covered by sr and tr. That is z, s, t form a clause but appear in the same partition V_2'. Hence no edge of that clause clique is in T' contradicting the assumption. In the remaining case we have $N(u) \cap N(v) = \{w, r\}$. The element r is determined as the last element in the matrix A_3 appearing not in the two clauses containing u, r resp. v, r. Let the two elements in these clauses be s, t. Then they must occur together with $z \in \xi$ in one clause. The second clause containing z is $\{w, r, z\}$, from which no edge can be contained in T' since all these lie in the same partition V_2' contradicting that T' contains an edge of each clause in R_3'.

Case II: $\xi \not\subset V'$. Suppose the intersection is empty, then a clause containing x_1 is covered by ab one containing x_2 must be covered by ac and finally one containing x_3 must be covered by bc yielding a triangle. Suppose $\xi \cap V' = \{u\}$. Then all four variables in the clauses containing the other two elements of ξ must form a 4-cycle. Clearly two cover the remaining clauses either two of them have to be connected yielding a triangle or u has to be connected to two of them neighboured in the 4-cycle (because each other pair would lead two covering only one clause by two edges) yielding also a triangle. Finally, suppose $\xi \cap V' = \{u, v\}$ and let z be the last element in ξ. Then we get two independent edges covering the clause cliques containing z. W.l.o.g. we suppose that $z = x_1$ (otherwise rename and permute accordingly) and thus we have the edges ab, cd. Suppose x_2, x_3 both have degree exactly one in T'. Then we must either have the edge ac or bd with respect to the clause cliques containing x_2 and either ad or bc for those containing x_3. In any case a triangle is produced, hence at least one of x_2, x_3 must have degree two in T'. If both have degree two, then we either have x_3a, x_3c or x_3b, x_3d for the clause cliques containing x_3 (all other choices immediately yield a triangle). In the same way we either have x_2a, x_2d or x_2b, x_2c for the clause cliques containing x_2 (otherwise we cover only one x_1-clique). Each combination leads to a cycle C_5. The last possibility is that one of x_2, x_3 has degree one and the other has degree two in T'. W.l.o.g. let x_2 have degree one. Then we must either have the edge ac or the edge bd in T'. In both cases we obtain a triangle when trying to connect x_3 to two of the vertices in the clause cliques containing x_3. This completes the proof that R_3 and thus any $B \in \mathcal{B}(3)$ is not NAE-satisfiable.   □

Remark 1. *The variable graph of a 2-block simply is a triangle and therefore not NAE-satisfiable. And for formulas $C \in \mathcal{A}(2) \setminus \mathcal{B}(2)$ obviously holds $C \in$ NAESAT if and only if $G_{V(C)}$ is bipartite, i.e., if and only if $G_{V(C)}$ is not an odd cycle. Indeed, since each vertex has degree 2 and $G_{V(C)}$ is connected it is a cycle in any case.*

Proposition 4. *If $k \in \mathbb{N}, k \geq 3$, then $\mathcal{B}(k) \subset$ NAESAT if and only if $k \geq 4$.*

PROOF. Because of Lemma 13 we only have to show that $\mathcal{B}(k) \subset$ NAESAT for $k \geq 4$. Let $k \geq 4$ be fixed and let $B \in \mathcal{B}(k)$ be represented in canonical form R_k with leading clause $\xi = \{x_1 \ldots, x_k\}$ and variable matrix $A = (a_{pq})_{1 \leq p,q \leq k-1}$. Let G'_k be obtained from the variable graph $G_k = (V, E)$ of R_k by removing the vertex x_1 and all edges $E(\xi)$ from the ξ-clique, i.e., $G'_k = (V \setminus \{x_1\}, E \setminus [E(\xi) \cup N(x_1)])$, where $N(x_1)$ denotes the set of neighbours of x_1 in G_k. Define $C'_{x_i} := C_{x_i} \setminus \{\xi\}, 1 \leq i \leq k$. For $2 \leq i \leq k$ we cover each set of clauses C'_{x_i} by a star S_i whose root is x_i and the set of its neighbours $N_{S_i}(x_i)$ consists of $k-1$ elements of the matrix B_i chosen such that no two elements appear in the same row or column of B_i and such that the vertices in the sets $N_{S_i}(x_i)$ are mutually variable disjoint. Note that this is possible only if $k > 3$. Let C''_{x_1} be obtained from the set of clauses C'_{x_1} by removing x_1 from each clause. By construction we can cover the ith clause of C''_{x_1} by an edge $e_i = u_i v_{i+1}$ where $u_i \in N_{S_i}(x_i)$ and $v_{i+1} \in N_{S_{i+1}}(x_{i+1})$, i.e., by an edge joining a neighbour of x_i to a neighbour of x_{i+1} for $2 \leq i \leq k-1$. The first clause of C''_{x_1} can be covered by another edge $e_1 = u_1 v_2$ from an element $u_1 \in N_{S_2}(x_2)$ to an element $v_2 \in N_{S_3}(x_3)$. Let the resulting transversal graph of G'_k be T'. Observe that the only cycle contained in T' is a C_6 defined by $x_2 u_1 v_2 x_3 v_3 u_2 x_2$ or a C_4 if $v_2 = v_3 \in N_{S_3}(x_3)$. Thus T' is bipartite. Obviously T' can be extended to a bipartite transversal T of G_k by introducing the edge $x_1 x_2$ yielding an NAESAT solution for R_k and hence for B.   □

Proposition 5. *For $k \geq 3$ let $C \in \mathcal{A}(k)$ be 1-intersecting, then $C \in$ NAESAT if and only if $C \notin \mathcal{B}(3)$.*

PROOF. For $k \geq 3$ let $C \in \mathcal{A}(k)$ ($n := |C|$) be fixed and 1-intersecting with variable graph $G_{V(C)} =: G = (V, E)$. For each $x \in \text{var}(C) = V$ we have $|C_x| = k$. By Lemma 3 G is $d_k = k(k-1)$-regular. We distinguish two cases.
I. C is exactly 1-intersecting, hence $C \in \mathcal{B}(k)$ is the clique K_{d_k+1} and we are done by Proposition 4.
II. $C \notin \mathcal{B}(k)$. Hence G is no clique, but is d_k-regular. Note, that C also can contain no block as $C \in \mathcal{A}(k)$. On the other hand the *maximal exactly* 1-intersecting subformulas of C are subformulas of pairwise variable disjoint k-blocks. By Lemma 13 we know that even in the case $k = 3$ such a *subblock* (which always exists because of d_k-regularity) has an NAESAT-solution. Suppose that for these subblocks we have bipartite transversal graphs as constructed in the proof of Proposition 4. It is not hard to see that each clause not contained in such a subblock in C must have two variables of different subblocks (otherwise it is contained in the variable set of a subblock and hence would be a member of it contradicting the maximal exactly 1-intersectedness). That is such clauses can be covered by introducing an edge joining these two variables in the bipartite transversal graphs of the corresponding subblocks. Observe that by this construction as shown in the proof of Proposition 4 we only produce (if at all) even cycles. Thus the claim is proven.   □

Theorem 3. *For $k \geq 3$ a formula $C \in \mathcal{A}(k)$ is in* NAESAT *if and only if $C \notin \mathcal{B}(3)$.*

PROOF. For an arbitrary $k \geq 3$ let $G_V(C) =: G = (V, E)$ be the variable graph of $C \in \mathcal{A}(k)$. We define $n := |V|, m := |E|$ and for each $e \in E$ let C_e be the set of clauses in C containing both variables of e. With $\pi(e) := |C_e|$ the *weight* of e we observe that $\forall e \in E : 1 \leq \pi(e) \leq k$ and $\sum_{e \in E} \pi(e) = nk(k-1)/2 =: q$. Hence $m \leq q$. Leading us to distinguish two basic cases.

I. $m = q \Leftrightarrow \forall e \in E : \pi(e) = 1$, which is equivalent to the case that C is 1-intersecting. Indeed, assume that C is 1-intersecting and that there is $e \in E : \pi(e) > 1$, then there are clauses $c \neq c'$ such that $e \subset c \cap c'$ yielding a contradiction. The other direction is obvious. The assertion of the theorem then follows immediately from Proposition 5.

II. $m < q \Leftrightarrow \forall e \in E : \pi(e) \geq 1$ and there is an $e \in E : \pi(e) > 1$. Let $V = \{x_1, \ldots, x_n\}$. First suppose that each edge in G has weight at least two. Then it is not hard to see that there is a collection of mutually vertex disjoint stars T_i (which may be degenerated to single edges) with roots $x_i, i \in I$, resp. for an appropriate $I \subseteq \{1, \ldots, n\}$ such that C_{x_j} is covered for $1 \leq j \leq n$, and thus C is covered. In the remaining case there is at least one edge of weight one (and at least one edge of weight two). Let $C' \subseteq C$ be the subformula consisting of all clauses containing only edges of weight one. If C' is empty, then each clause contains an edge of weight at least two. We can proceed as in the previous case as for the star transversal used there we need only one edge of weight at least two in each clause. If C' is not empty, then it is 1-intersecting and solved by a bipartite transversal graph as in the proof of Proposition 5. This is possible since there are $x_i \in V$ such that subsets of C_{x_i} constitute C' and there can be no block at all. Clearly the clauses in C' can have in common no edge of the part of G corresponding to $C \setminus C'$ each clause clique of which has an edge of weight at least two. Thus solving this as in the previous case yields a bipartite transversal graph for G where if at all only even cycles occur.   □

Let $C \in \mathrm{CNF}_+(k, k)$ be an arbitrary formula. As the subformulas of C corresponding to the components of $G_{V(C)}$ are autonomous thus, mutually variable disjoint they can be treated independently. Thus we are able to conclude:

Corollary 4. *For $k \in \mathbb{N}, k \geq 3$ fixed let $C \in \mathrm{CNF}_+(k, k)$, then $C \in$ NAESAT if and only if $G_{V(C)}$ has no component isomorphic to K_7 in the case $k = 3$.*

Remark 2. *Clearly, a component isomorphic to K_7 in the variable graph of a formula $C \in \mathrm{CNF}_+(k, k)$ may occur also for $k \geq 4$, but is not critical.*

Theorem 4. *Given $C \in \bigcup_{k \geq 2} \mathrm{CNF}_+(k, k)$ it is decidable in linear time $O(\|C\|)$ whether $C \in$ NAESAT.*

PROOF. If $k = 2$ then $\mathrm{CNF}_+(2, 2) \subset 2-\mathrm{CNF}$ and hence NAESAT is decidable in linear time by iterating the procedure Simplify in Section 4. The cases $k \geq 4$ are trivial in the sense that a corresponding formula always has an NAESAT-solution by Theorem 3. It remains the case $k = 3$, hence let $C \in \text{CNF}_+(3,3)$ and G be its variable graph. The only crucial case is by Corollary 4 that G has a component K_7. Since $k = 3$ is a constant we can compute the connected components of G thereby checking whether one of it is a K_7 in linear time $O(\|C\|)$, where $\|C\|$ is the length of C. □

Corollary 5. *Bicolorability of k-uniform, k-regular hypergraphs is decidable in linear time.*

Consider the class of formulas $\text{CNF}_+(k, \leq k), k \in \mathbb{N}$, i.e., each clause has length k and each variable occurs *at most* k times, then $|\text{var}(C)| \geq |C|$. Observe that it is possible by mimicking the above argumentations to show that also a formula $C \in \text{CNF}_+(k, \leq k), k \geq 2$, is in NAESAT if and only if its variable graph in the case $k = 3$ has no component K_7 corresponding to a subformula being a 3-block. Thus also for this class NAESAT is decidable in linear time.

Acknowledgement: We would like to thank J. Rühmkorf for running the experiments and I. Schiermeyer for bringing to our knowledge the reference [6]. We also would like to thank the anonymous referees for helpful comments.

References

1. D. Achlioptas, A. Chtcherba, G. Istrate and C. Moore, The Phase Transition in 1-in-k SAT and NAE 3-SAT, in: Proceedings of 12th Annual ACM-SIAM Symp. on Discrete Algorithms (SODA'01), 2001, 721-722.
2. D. Achlioptas and C. Moore, The asymptotic order of the random $k-$SAT threshold, in: 43th Annual Symposium on Foundations of Computer Science (Vancouver, BC, 2002), IEEE Comput. Soc. Press, Los Alamitos, CA, 2002, to appear.
3. N. Alon, J. H. Spencer and P. Erdős, The Probabilistic Method, Wiley-Interscience, Series in Discrete Mathematics and Optimization, New York, 1992.
4. M. Atallah and U. Vishkin, Finding Euler tours in parallel, J. of Computer and System Sci. 29(30) (1984) 330-337.
5. J. Beck, On 3-chromatic hypergraphs, Discrete Math. 24 (1978) 127-137.
6. J. A. Bondy and U. S. R. Murty, Graph Theory and Applications, Macmillan, London and Elsevier, New York, 1976.
7. L. Euler, Solutio problematis ad geometriam situs pertinentis, Commentarii Academiae Petropolitanae 8 (1736), 128-140.
8. M. R. Garey and D. S. Johnson, Computers and Intractability: A Guide to the Theory of NP-Completeness, W. H. Freeman and Company, San Francisco, 1979.
9. M. Golumbic, A. Mintz and U. Rotics, Factoring and Recognition of Read-Once Functions using Cographs and Normality, in: Proceedings of the 38th ACM Design Automation Conference, DAC 2001, Las Vegas, USA, 18-22 June 2001, 109-114.
10. C. Hierholzer, Über die Möglichkeit, einen Linienzug ohne Wiederholung und ohne Unterbrechung zu umfahren, Math. Ann. 6 (1873) 30-32.
11. H. Kleine Büning and T. Lettman, Aussagenlogik: Deduktion und Algorithmen, B. G. Teubner, Stuttgart, 1994.
12. D. E. Knuth, Axioms and Hulls, LNCS vol. 606, Springer, New York, 1992.

13. J. Kratochvil and Zs. Tuza, On the complexity of bicoloring clique hypergraphs of graphs, J. Algorithms 45 (2002) 40-54.
14. B. Monien, E. Speckenmeyer and O. Vornberger, Upper Bounds for Covering Problems, Methods of Operations Research 43 (1981) 419-431.
15. T. J. Schaefer, The complexity of satisfiability problems, in: Conference Record of the Tenth Annual ACM Symposium on Theory of Computing, San Diego, California, 1-3 May 1978, 216-226.
16. Y. Shiloach and U. Vishkin, An $O(\log n)$ parallel connectivity algorithm, J. Algorithms 2 (1981) 57-63.
17. C. A. Tovey, A Simplified NP-Complete Satisfiability Problem, Discrete Appl. Math. 8 (1984) 85-89.

On Fixed-Parameter Tractable
Parameterizations of SAT

Stefan Szeider[*]

Department of Computer Science, University of Toronto,
M5S 3G4 Toronto, Ontario, Canada
szeider@cs.toronto.edu

Abstract. We survey and compare parameterizations of the propositional satisfiability problem (SAT) in the framework of Parameterized Complexity (Downey and Fellows, 1999). In particular, we consider (a) parameters based on structural graph decompositions (tree-width, branch-width, and clique-width), (b) a parameter emerging from matching theory (maximum deficiency), and (c) a parameter defined by translating clause-sets into certain implicational formulas (falsum number).

1   Introduction

The framework of Parameterized Complexity, introduced by Downey and Fellows [12], provides a means for coping with computational hard problems: It turned out that many intractable (and even undecidable) problems can be solved efficiently "by the slice", that is, in time $\mathcal{O}(f(k) \cdot n^\alpha)$ where f is any function of some parameter k, n is the size of the instance, and α is a constant independent from k. In this case the problem is called *fixed-parameter tractable (FPT)*. If a problem is FPT, then instances of large size can be solved efficiently.

The objective of this paper is to survey and compare known results for fixed-parameter tractable SAT decision. Although the SAT problem has been considered in more general works on parameterized complexity (e.g., [9]) and FPT results have been obtained focusing on a single parameterization of SAT (e.g., [2, 18]), it appears that no broader approach has been devoted to this subject.

We suggest the following concept of fixed-parameter tractability for SAT. Consider a parameter π for clause-sets; i.e., π is a function which assigns some non-negative integer $\pi(F)$ to any given clause-set F. We say that *"satisfiability of clause-sets with bounded π is fixed-parameter tractable"* if there is an algorithm which answers correctly for given clause-sets F and $k \geq 0$

$$\text{``}F\text{ is satisfiable'' or ``}F\text{ is unsatisfiable'' or ``}\pi(F) > k\text{''}$$

in time $\mathcal{O}(f(k) \cdot l^\alpha)$; here l denotes the length (i.e., sum of clause widths) of F, f is any function, and α is a constant independent from k. (Being aware of the phenomenon of so-called "robust algorithms" [27, 13], we do not require (i) that the

---

[*] Supported by the Austrian Science Fund (FWF) projects J2111 and J2295.

E. Giunchiglia and A. Tacchella (Eds.): SAT 2003, LNCS 2919, pp. 188–202, 2004.

algorithm actually computes $\pi(F)$, nor (ii) that the algorithm actually decides whether $\pi(F) \leq k$.)

A trivial example for such parameter can be obtained by defining $\pi(F)$ as the length of the clause-set F' which results in applying some of the usual polynomial-time simplifications to a given clause-set F, say elimination of unit and binary clauses, and of clauses which contain pure literals.

1.1 New Contributions of This Paper

Besides a review of known results (FPT algorithms for clause-sets with bounded primal tree-width and branch-width), we obtain the following new results.

We introduce the notion of *incidence tree-width* of clause-sets, and we show the following.

- *Satisfiability of clause-sets with bounded incidence tree-width is FPT.*
- *Incidence tree-width is more general than primal tree-width*; i.e., bounded primal tree-width implies bounded incidence tree-width, but there are clause-sets of bounded incidence tree-width and arbitrarily high primal tree-width.

Recently it could be shown that clause-sets of bounded *maximum deficiency*, a parameter defined via matchings in incidence graphs, allow fixed-parameter tractable SAT decision [29]. We compare tree-width with maximum deficiency, and we obtain the following result.

- *Incidence/primal tree-width and maximum deficiency are incomparable*; i.e., there are clause sets of bounded primal tree-width (and so of bounded incidence tree-width) with arbitrarily high maximum deficiency; on the other hand, there are clause-sets of arbitrarily high incidence tree-with (and so of arbitrarily high primal tree-width) with bounded maximum deficiency. (Actually we show incomparability of maximum deficiency and *clique-width*; the latter is a more general parameter than tree-width; see, e.g., [10].)

Finally, we consider a known FPT result on satisfiability for a certain class of non-CNF formulas [15], and we formulate a transformation scheme which makes this result applicable to clause-sets. This transformation enables us to define the parameter *falsum number* for clause-sets. Our results for this parameter are as follows.

- *Satisfiability of clause-sets with bounded falsum number is FPT.*
- *Maximum deficiency is more general than falsum number*; i.e., the falsum number of a clause-set without pure literals is at least as large as its maximum deficiency.

1.2 Notation

A *literal* is a variable x or a negated variable $\neg x$; we write $\overline{x} = \neg x$ and $\overline{\neg x} = x$. A finite set of literals without a complementary pair $x, \neg x$ is a *clause*. A *clause-set* is a finite set of clauses. A variable x *occurs* in a clause C if either $x \in C$ (x occurs positively) or $\neg x \in C$ (x occurs negatively); $\mathsf{var}(C)$ denotes the set of variables occurring in a clause C; for a clause-set F we put $\mathsf{var}(F) = \bigcup_{C \in F} \mathsf{var}(C)$. A literal x is a *pure literal* of F if $\{x, \overline{x}\} \cap \bigcup_{C \in F} C = \{x\}$. The *width* of a clause is its cardinality; the *width* $w(F)$ of a clause-set F is the width of a largest clause of F (or 0 if $F = \emptyset$). The *length* of F is $\sum_{C \in F} |C|$. Semantically, a clause-set F is considered as a propositional formula in conjunctive normal form (CNF): an *assignment* $\tau : \mathsf{var}(F) \to \{0, 1\}$ satisfies F if it evaluates to 1 in the usual sense for CNFs. A clause-set F is *satisfiable* if it has a satisfying assignment; otherwise it is *unsatisfiable*. F is *minimal unsatisfiable* if it is unsatisfiable and every proper subset $F' \subsetneq F$ is satisfiable.

2   From Clause-Sets to Graphs and Hypergraphs

Several parameters originally defined for graphs and hypergraphs can be applied to clause-sets via transformations of clause-sets to (hyper)graphs.

Some of the following definitions are illustrated in Figure 1.

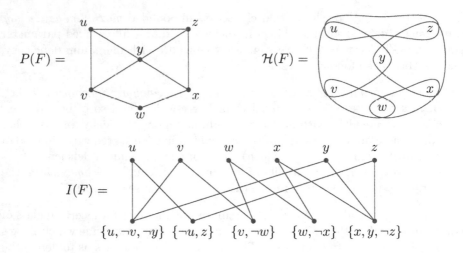

Fig. 1. (Hyper)graphs associated to the clause-set $F = \{\{u, \neg v, \neg y\}, \{\neg u, z\},$ $\{v, \neg w\}, \{w, \neg x\}, \{x, y, \neg z\}\}$; the primal graph $P(F)$, the hypergraph $\mathcal{H}(F)$, and the incidence graph $I(F)$.

The *primal graph* $P(F)$ of a clause-set F is the graph whose vertices are the variables of F, and where two variables are joined by an edge if both variables occur together in a clause. The *incidence graph* $I(F)$ is a bipartite graph: one vertex class consists of the variables of F, the other vertex class consists of the clauses of F; a variable x and a clause C are joined by an edge if x occurs in C. The *directed incidence graph* $I_d(F)$ arises from $I(F)$ by orienting edges from C to x if $x \in C$, and from x to C if $\neg x \in C$.

A clause-set F gives rise to a hypergraph $\mathcal{H}(F)$, *the hypergraph of F*, in a natural way: the vertices of $\mathcal{H}(F)$ are the variables of F, and to every clause $C \in F$ there is a hyperedge which is incident to exactly the variables in $\mathsf{var}(C)$. Note that $\mathcal{H}(F)$ may contain "parallel" hyperedges, i.e., different clauses C, C' always yield different hyperedges E, E', even if $\mathsf{var}(C) = \mathsf{var}(C')$.

3   Tree-Width of Primal Graphs and Branch-Width

Tree-width, a popular parameter for graphs, was introduced by Robertson and Seymour in their series of papers on graph minors; see, e.g., [6] for references. Let G be a graph, $T = (V, E)$ a tree, and χ a labeling of the vertices of T by sets vertices of G. Then (T, χ) is a *tree decomposition* of G if the following conditions hold:

(T1) Every vertex of G belongs to $\chi(t)$ for some vertex t of T;
(T2) for every edge (v, w) of G there is some vertex t of T such that $v, w \in \chi(t)$;
(T3) for any vertices t_1, t_2, t_3 of T, if t_2 lies on a path from t_1 to t_3, then
  $\chi(t_1) \cap \chi(t_3) \subseteq \chi(t_2)$.

The *width* of a tree decomposition (T, χ) is the maximum $|\chi(t)| - 1$ over all vertices t of T. The *tree-width* $tw(G)$ of G is the minimum width over all its tree-decompositions.

Note that trees have tree-width 1 (the only purpose of "-1" in the definition of tree-width is to make this statement true).

For fixed $k \geq 1$, deciding whether a given graph has tree-width at most k (and computing a tree-decomposition of width $\leq k$, if it exists) can be done efficiently (in quadratic time by Robertson and Seymour [24], and even in linear time by Bodlaender [5]; the latter algorithm, however, imposes large hidden constants and is not well-suited for practical applications). Computing the tree-width of a given graph, however, is an NP-hard problem [3].

In order to consider clause-sets of bounded tree-width, one can either bound the tree-width of the corresponding primal graphs or the tree-width of the corresponding incidence graphs: for a clause-set F we call $tw(P(F))$ the *primal tree-width* of F, and $tw(I(F))$ the *incidence tree-width* of F.

Theorem 1 (Gottlob, et al. [18]) *Satisfiability of clause-sets with bounded primal tree-width is fixed-parameter tractable.*

The proof of this result relies on the fact that clause-sets of bounded primal tree-width can be transformed into equivalent *acyclic* constraint satisfaction problems (CSPs) which in turn can be solved efficiently by a classical algorithm due to Yannakakis [31].

The following lemma is well-known; see, e.g., [6].

Lemma 1 *Let (T, χ) be a tree-decomposition of a graph G and let $K \subseteq V(G)$ be a set of vertices which induces a complete subgraph in G. Then $K \subseteq \chi(t)$ holds for some vertex t of T.*

The next lemma follows directly from Lemma 1 (recall from Section 1.2 that $w(F)$ denotes the width of F).

Lemma 2 $w(F) \leq tw(P(F)) + 1 \leq |\mathsf{var}(F)|$ *holds for every clause-set F.*

Hence Theorem 1 is impractical for clause-sets of large width. For example, the simple minimal unsatisfiable clause-set $\{\{x_1, \ldots, x_n\}, \{\neg x_1\}, \ldots, \{\neg x_n\}\}$ has primal tree-width $n - 1$; however, its incidence tree-width is 1. Thus, it would be desirable to extend Theorem 1 to incidence graphs. We will accomplish this in the next section applying general results on clique-width.

The notion of "branch-width" for (hyper)graphs has been introduced by Robertson and Seymour; it is based on the following decomposition scheme: Let $\mathcal{H}$ be a hypergraph, $T = (V, E)$ a ternary tree (i.e., all vertices of T have either degree 0 or 3), and τ a bijection from the set of leaves of T to the set of hyperedges of $\mathcal{H}$; (T, τ) is called a *branch decomposition* of $\mathcal{H}$. The *order* of an edge e of T is the number of vertices of $\mathcal{H}$ which are incident to hyperedges $\tau(t_1), \tau(t_2)$ such that t_1 and t_2 belong to different components of $T - e$. The *width* of a branch decomposition (T, τ) is the maximum order of all edges of T; the *branch-width* $bw(\mathcal{H})$ of a hypergraph $\mathcal{H}$ is the smallest width over all its branch decompositions.

The branch-width of a clause-set F is the branch-width of its hypergraph, $bw(F) := bw(\mathcal{H}(F))$. In [2] Alekhnovich and Razborov show the following.

Theorem 2 (Alekhnovich and Razborov [2]) *Satisfiability of clause-sets with bounded branch-width is fixed-parameter tractable.*

This result is obtained via a modification of Robertson and Seymour's algorithm for computing branch-decompositions [26]; from a branch-decomposition of $\mathcal{H}(F)$ one can extract efficiently either a satisfying assignment (if F is satisfiable) or a regular resolution refutation (if F is unsatisfiable). Further results and algorithms for SAT and #SAT with bounded branch-width can be found in [4].

Note that if every vertex of a hypergraph $\mathcal{H}$ is incident with at least two hyperedges of $\mathcal{H}$, and if some hyperedge of $\mathcal{H}$ contains k vertices, then $k \leq bw(\mathcal{H})$. However, if a vertex of the hypergraph $\mathcal{H}(F)$ of a clause-set F is incident with exactly one hyperedge, then v is necessarily a pure literal of F. Hence $w(F) \leq bw(F)$ holds for clause-sets without pure literals. In particular, the simple clause-set $\{\{x_1, \ldots, x_n\}, \{\neg x_1\}, \ldots, \{\neg x_n\}\}$ as considered above has branch-width n. We can state state [25, Lemma 5.1] as follows.

Lemma 3 *For clause-sets F without pure literals we have*

$$bw(F) \leq tw(P(F)) + 1 \leq \frac{3}{2} bw(F).$$

Hence a class of clause-sets without pure literals has bounded primal tree-width if and only if it has bounded branch-width.

4   Tree-Width and Clique-Width of Incidence Graphs

The next result (which seems to be known, [17]) indicates that incidence tree-width is the more general parameter than primal tree-width.

Lemma 4 *For every clause-set F we have*

$$tw(I(F)) \leq \max(tw(P(F)), w(F)) \leq tw(P(F)) + 1.$$

Proof. Let (T, χ) be a width k tree-decomposition of $P(F)$. By Lemma 1 we can choose for every clause $C \in F$ some vertex t_C of T such that $\mathsf{var}(C) \subseteq \chi(t_C)$. We obtain a tree T' from T by adding for every clause $C \in F$ a new vertex t'_C and the edge (t_C, t'_C). Finally, we extend the labeling χ to T' defining $\chi(t'_C) = \mathsf{var}(C) \cup \{C\}$. We can verify that (T', χ) is a tree-decomposition of $I(F)$ by checking the conditions (T1)–(T3). Since $|\chi(t'_C)| = |C| + 1$, the width of (T', χ) is at most the maximum of k and $w(F)$. However, Lemma 1 also implies that $tw(P(F)) \geq w(F) - 1$, hence the lemma is shown true.   □

On the other hand, as observed above, there are clause-sets whose primal graphs have arbitrarily high tree-width and whose incidence graphs are trees.

The question rises whether Theorem 1 can be generalized to incidence tree-width. Below we answer this question positively, deploying a strong model-theoretic result of [9] which generalizes "Courcelle's Theorem" (see, e.g., [12, Chapter 6]) to graphs of bounded clique-width.

First we give some definitions taken from [10]. Let k be a positive integer. A *k-graph* is a graph whose vertices are labeled by integers from $\{1, \ldots, k\}$. We consider an arbitrary graph as k-graph with all vertices labeled by 1. We call the k-graph consisting of exactly one vertex v (say, labeled by $i \in \{1, \ldots, k\}$) an *initial k-graph* and denote it by $i(v)$. Let $\mathcal{C}(k)$ denote the class of k-graphs which can be constructed from initial k-graphs by means of the following three operations.

(C1) If $G, H \in \mathcal{C}(k)$ and $V(G) \cap V(H) = \emptyset$, then the union of G and H, denoted by $G \oplus H$, belongs to $\mathcal{C}(k)$.

(C2) If $G \in \mathcal{C}(k)$ and $i, j \in \{1, \ldots, k\}$, then the k-graph $\rho_{i \to j}(G)$ obtained from G by changing the labels of all vertices which are labeled by i to j belongs to $\mathcal{C}(k)$.

(C3) If $G \in \mathcal{C}(k)$, $i, j \in \{1, \ldots, k\}$, and $i \neq j$, then the k-graph $\eta_{i,j}(G)$ obtained from G by connecting all vertices labeled by i with all vertices labeled by j belongs to $\mathcal{C}(k)$.

The *clique-width* $cw(G)$ of a graph G is the smallest integer k such that $G \in \mathcal{C}(k)$. Constructions of a k-graph using the above steps (C1)–(C3) can be represented by *k-expressions*, terms composed of $i(v)$, $G \oplus H$, $\eta_{i,j}(G)$ and $\rho_{i \to j}(G)$. Thus, a k-expression certifies that a graph has clique-width $\leq k$. For example, the 4-expression

$$\rho_{2 \to 1}(\eta_{1,2}(2(y) \oplus \rho_{2 \to 1}(\eta_{1,2}(2(x) \oplus \rho_{2 \to 1}(\eta_{1,2}(1(v) \oplus 2(w)))))))$$

represents a construction of the complete graph K_4 on $\{v, w, x, y\}$, hence $cw(K_4) \leq 2$. In view of this example it is easy to see that any complete graph has clique-width ≤ 2, hence a result similar to Lemma 1 does not hold for clique-width.

The above definitions apply also to *directed graphs* except that in construction (C3) the added edges are directed from label i to label j. Thus, we can consider k-expressions for a directed graph D and we can define the *directed clique-width $dcw(D)$* of D as the smallest k such that D has a k-expression. Let D be a directed graph and G_D its *underlying undirected graph* (i.e., G is obtained from D by "forgetting" the direction of edges and by identifying possible parallel edges); since every k-expression for D is also a k-expression for G_D, $cw(G_D) \leq dcw(D)$ follows.

The next result is due to Courcelle and Olariu [10] (see also [9]).

Theorem 3 (Courcelle and Olariu [10]) *Let D be a directed graph and (T, χ) a width k' tree-decomposition of G_D. Then we can obtain in polynomial time a k-expression for D with $k \leq 2^{2k'+1} + 1$. Thus $dcw(D) \leq 2^{2tw(G_D)+1} + 1$.*

Courcelle, Makowsky and Rotics [9] show the following (recall from Section 2 that $I_d(F)$ denotes the *directed incidence graph* of F).

Theorem 4 (Courcelle, et al. [9]) *Given a clause-set F of length l and a k-expression for $I_d(F)$ (thus $dcw(I_d(G)) \leq k$). Then the number of satisfying total truth assignments of F can be counted in time $\mathcal{O}(f(k) \cdot l)$ where f is some function which does not depend on F.*

In [9] it is shown that if a k-expression for a directed graph D is given (k is some constant), then statements formulated in a certain fragment of monadic second-order logic (MS_1) can be evaluated on D in linear time. Satisfiability of F can be formulated as an MS_1 statement on $I_d(F)$: F is satisfiable if and only if there exists a set of variables V_0 such that for every clause $C \in F$, $I_d(F)$ contains either an edge directed from C to some variable in V_0, or it contains an edge directed from some variable in $\text{var}(F) \setminus V_0$ to C.

Before we can apply Theorem 4 to a given clause-set we have to find a k-expression for its directed incidence graph; though, it is not known whether k-expressions can be found in polynomial time for constants $k \geq 4$ (see, e.g., [9]). Anyway, in view of Theorem 3, we can use the previous result to improve on Theorem 1 by considering incidence graphs instead of primal graphs.

Corollary 1 *Satisfiability of clause-sets with bounded incidence tree-width is fixed-parameter tractable.*

Note, however, that a practical use of Theorem 4 is very limited because of large hidden constants and high space requirements; cf. the discussion in [9]. Nevertheless, it seems to be feasible to develop algorithms which decide satisfiability directly by examining a given tree-decomposition of the incidence graph, without calling on the general model-theoretic results of [9].

Even for the case that it turns out that recognition of graphs with bounded clique-width is NP-complete, it remains possible that satisfiability of clause-sets with bounded clique-width is fixed-parameter tractable (by means of a "robust algorithm", see the discussion in Section 1).

5   Maximum Deficiency

The *deficiency* of a clause-set F on n variables and m clauses is $\delta(F) := m - n$; its *maximum deficiency* is

$$\delta^*(F) = \max_{F' \subseteq F} \delta(F'),$$

i.e., the maximum deficiency over the subsets of F. Since $\delta(\emptyset) = 0$, the maximum deficiency of a clause-set is always positive. This parameter is strongly connected with matchings in bipartite graphs, see, e.g., [14].

Lemma 5 *A maximum matching of the incidence graph of a clause-set F exposes exactly $\delta^*(F)$ clauses.*

Since maximum matchings can be found efficiently, $\delta^*(F)$ can be calculated efficiently as well. Note also that $\delta^*(F) = \delta(F)$ holds for minimal unsatisfiable clause-sets [22, 14].

In [22, 14], algorithms are presented which decide satisfiability of clause-sets F in time $n^{\mathcal{O}(\delta^*(F))}$; this time complexity does not constitute fixed-parameter tractability. However, in [29] the author of the present paper develops a DLL-type[1] algorithm which decides satisfiability of clause-sets with n variables in time $\mathcal{O}(2^{\delta^*(F)} n^3)$; hence we have:

Theorem 5 (Szeider [29]) *Satisfiability of clause-sets with bounded maximum deficiency is fixed-parameter tractable.*

The key to the new result of [29] is an efficient procedure for reducing any clause-set F into an equisatisfiable clause-set F' with the property that setting any variable of F' to true or false decreases its maximum deficiency ("F' is δ^*-critical"). Applying this reduction at every node of the binary search tree traversed by the DLL-type algorithm ensures that the height of the search tree does not exceed the maximum deficiency of the input clause-set.

Next we construct clause-sets with small maximum deficiency and large primal tree-width.

Theorem 6 *For every $k \geq 1$ there are minimal unsatisfiable clause-sets F such that $\delta^*(F) = 1$ and $tw(P(F)) = k$.*

Proof. We consider clause-sets used by Cook ([8], see also [30]) for deriving exponential lower bounds for the size of tableaux refutations. Let k be any

---

* Davis, Logemann, and Loveland [11].

positive integer and consider the complete binary tree T of height $k+1$, directed from the root to the leaves. Let $v_1, \ldots, v_m$, $m = 2^{k+1}$, denote the leaves of T. For each non-leaf v of T we take a new variable x_v, and we label the outgoing edges of v by x_v and $\overline{x_v}$, respectively. For each leaf v_i of T we obtain the clause C_i consisting of all labels occurring on the path from the root to v_i. Consider $F = \{C_1, \ldots, C_m\}$. It is not difficult to see that F is minimal unsatisfiable (in fact, it is "strongly minimal unsatisfiable" in the sense of [1]). Moreover, since $|\mathsf{var}(F)| = 2^{k+1} - 1$, we have $\delta^*(F) = \delta(F) = 1$. Since $|C_i| = k+1$, $tw(P(F)) \geq k$ follows from Lemma 2. On the other hand, $tw(P(F)) \leq k$, since we can define a tree-decomposition (T, χ) of width k for F as follows (T is the binary tree used above to define F). For each leaf v_i of T we put $\chi(v) = \mathsf{var}(C_i)$; for each non-leaf w we define $\chi(w)$ as the set of variables x_v such that v lies on the path from the root of T to w (in particular, $x_w \in \chi(w)$). $\qquad \square$

Conversely, there are clause-sets with small primary tree-width and large maximum deficiency:

Theorem 7 *For every $k \geq 1$ there are minimal unsatisfiable clause-sets H such that $\delta^*(H) = k$ and $tw(P(H)) \leq 2$.*

Proof. We consider the clause-set $H := \bigcup_{i=0}^k H_i$ where $H_0 = \{\{z_0\}\}$, $H_k = \{\{\overline{z_{k-1}}\}\}$, and for $i = 1, \ldots, k-1$,

$$H_i := \{\{\overline{z_{i-1}}, x_i, y_i\}, \{\overline{x_i}, y_i\}, \{x_i, \overline{y_i}\}, \{\overline{x_i}, \overline{y_i}, z_i\}\}.$$

It follows by induction on k that $\delta(H) = k$ and that H is minimal unsatisfiable. Hence $\delta^*(H) = k$. We define a tree-decomposition (T, χ) of H taking the path $v_0, \ldots, v_k$ for T and setting $\chi(v_i) = \mathsf{var}(H_i)$. The width of this tree-decomposition is at most 2, hence $tw(H) \leq 2$ follows. $\qquad \square$

Next we show a result similar to Theorem 6.

Theorem 8 *For every $k \geq 1$ there are clause-sets F such that $\delta^*(F) = 1$ and $dcw(I_d(F)) \geq cw(I(F)) \geq k$.*

Proof. Let k be a positive integer and let q be the smallest odd integer with $q \geq \max(3, k-1)$. We consider the $q \times q$ grid G_q (see Figure 2 for an example). We denote by $v_{i,j}$ the vertex of row i and column j. Evidently, G_q is bipartite; let V_1, V_2 be the bipartition with $v_{1,1} \in V_2$ (in Figure 2, vertices in V_1 are drawn black, vertices in V_2 are drawn white). Since q is odd, we have $|V_1| = (q^2+1)/2-1$ and $|V_2| = (q^2 + 1)/2$. Next we obtain a clause-set F_q with $I(F_q) = G_q$: We consider vertices in V_1 as variables, and we associate to every vertex $v_{i,j} \in V_2$ the clause $\{v_{i,j-1}, \overline{v_{i,j+1}}, v_{i-1,j}, \overline{v_{i+1,j}}\} \cap (V_1 \cup \overline{V_1})$. As shown in [16], any $q \times q$ grid, $q \geq 3$, has exactly clique-width $q + 1$; hence $dcw(I_d(F_q)) \geq cw(I(F_q)) = cw(G_q) \geq k$.

Consider the matching M_q of G_q consisting of all the edges $(v_{i,2j}, v_{i,2j+1})$ for $i = 1, \ldots, q$ and $j = 1, \ldots, (q-1)/2$, and the edges $(v_{2i,1}, v_{2i+1,1})$ for $i =$

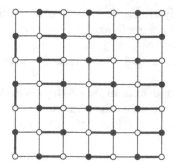

Fig. 2. The grid G_7; bold edges indicate the maximum matching M_7.

$1, \ldots, (q-1)/2$ (in Figure 2, edges of M_q are indicated by bold lines). Since $|M_q| = |V_1|$, M_q is a maximum matching and F_q is 0-expanding. By Lemma 5 $\delta^*(F_q) = \delta(F_q) = 1$ follows.   □

It can be shown that every clause-set whose incidence graph is a square grid is satisfiable (i.e., such clause-sets are "var-satisfiable" [28]); hence the clause-sets F_q constructed in the preceding proof are satisfiable. Since for a directed graph D the directed clique-width of any induced subgraph of D does not exceed the directed clique-width of D, it is not difficult to obtain from F_q *unsatisfiable* clause-sets of high directed clique-width and constant maximum deficiency. However, it would be interesting to find *minimal unsatisfiable* clause-sets with such a property.

6   Falsum Number

A propositional formula α is called **f**-*implicational* if $\rightarrow$ (implication) is the only connective of α; however, α may contain the constant **f** (falsum).

Theorem 9 (Franco, et al. [15]) *Satisfiability of* **f**-*implicational formulas of length l with at most two occurrences of each variable and k occurrences of* **f** *can be decided in time $\mathcal{O}(k^k l^2)$. Hence satisfiability of such formulas is fixed-parameter tractable.*

This result has been recently improved to $\mathcal{O}(3^k l^2)$, $k \geq 4$, using dynamic programming techniques [20].

Our objective is to apply Theorem 9 to clause-sets by means of a procedure that translates any given clause-set F into an equisatisfiable **f**-implicational formula $F^{\rightarrow}$. In Fig. 3 we state a slight generalization of the procedure used by Heusch [19] (Heusch considers only clause-sets where every variable occurs at most three times). We call the resulting **f**-implicational formula $F^{\rightarrow}$ a *standard translation* of the given clause-set F.

Step 1. We recursively eliminate clauses containing pure literals.

Step 2. If a variable x occurs in exactly one clause negatively and in more than one clause positively, we perform a renaming; i.e., we replace each occurrence of x by $\neg x$ and vice versa. We repeat this step as often as possible.

Step 3. If a variable x occurs in more than one clause positively, say in clauses $C_1, \ldots, C_r$, we take a new variable x' and replace the clause C_i by $(C_i \setminus \{x\}) \cup \{\neg x'\}$, $i = 1, \ldots, r$, and we add the clause $\{x, x'\}$. We repeat this step as often as possible.

Now each variable occurs exactly once positively.

Step 4. If a variable x occurs in more than one clause negatively, say in clauses $C_1, \ldots, C_r$, we take new variables $x_1, \ldots, x_r$, and replace the clause C_i by $(C_i \setminus \{\neg x\}) \cup \{\neg x_i\}$, $i = 1, \ldots, r$. Moreover, we introduce the formula $x \to (x_1 \wedge \cdots \wedge x_r)$. We repeat this step as often as necessary.

We end up with a set F' of clauses and a set S of formulas of the shape $x \to (x_1 \wedge \cdots \wedge x_r)$.

Step 5. For each clause $C \in F'$, choose an ordering $L_1, \ldots, L_s$ of its literals and replace C by the formula $L_1 \vee \cdots \vee L_s$.

Step 5 yields a set S' of disjunctions (originating from the clauses of F').

Step 6. We apply to formulas of S and S' the equivalences

(E1) $\neg x = x \to \mathbf{f}$ (E3) $\varphi \wedge \psi = (\varphi \to (\psi \to \mathbf{f})) \to \mathbf{f}$

(E2) $\varphi \vee \psi = (\varphi \to \mathbf{f}) \to \psi$ (E4) $(\varphi \to \mathbf{f}) \to \mathbf{f} = \varphi$

and obtain a set of $\mathbf{f}$-implicational formulas T and T', respectively.

Step 7. We choose an ordering $\alpha_1, \ldots, \alpha_p$ of the formulas in $T' \cup T$ and obtain the $\mathbf{f}$-implicational formula $F^\to := (\alpha_1 \to \ldots \to \alpha_p \to \mathbf{f}) \to \mathbf{f}$. Note that Step 5 can be performed by applying (E3) to $\alpha_1 \wedge \cdots \wedge \alpha_p$.

Fig. 3. Transformation of a clause-set F into an $\mathbf{f}$-implicational formula $F^\to$.

We state some properties of this construction which can be easily verified.

1. F and $F^\to$ are equisatisfiable.
2. Every variable of $F^\to$ occurs at most twice.
3. The length of $F^\to$ is polynomially bounded by the length of F.

Since the translation procedure contains some nondeterministic steps, a clause-set may have several standard translations. We define the *falsum number* $\#_{\mathbf{f}}(F)$ of a clause-set F as the smallest number of $\mathbf{f}$-occurrences over all its standard translations.

Lemma 6 *Let $C = \{L_1, \ldots, L_n\}$ be a clause with r negative literals, π a permutation of $\{1, \ldots, n\}$, and let $C^\to$ be an $\mathbf{f}$-implicational formula obtained from $L_{\pi(1)} \vee \cdots \vee L_{\pi(r)}$ by the equivalences (E2) and (E4). Then $C^\to$ contains at least $|n-r-1|$ occurrences of $\mathbf{f}$. Such $C^\to$ which contains exactly $|n-r-1|$ occurrences of $\mathbf{f}$ can be found in polynomial time.*

Proof. We proceed by induction on n. For $n \leq 1$ the statement holds by trivial reasons. Assume $n \geq 2$ and consider $L_{\pi(1)} \vee \cdots \vee L_{\pi(r)}$. We put $C_0 = C \setminus \{L_{\pi(1)}\}$.

First assume that $L_{\pi(1)}$ is a negative literal. By induction hypothesis, $C_0^{\rightarrow}$ contains at least $|(n-1)-(r-1)-1| = |n-r-1|$ occurrences of **f**. We cannot do better than setting $C^{\rightarrow} = L_{\pi(1)} \rightarrow C_0^{\rightarrow}$. Hence the first part of the lemma holds if $L_{\pi(1)}$ is a negative literal. Now assume that $L_{\pi(1)}$ is positive literal. By induction hypothesis, $C_0^{\rightarrow}$ contains at least $|n - r - 2|$ occurrences of **f**. Since $r \leq n - 1$, $n-r-2$ is negative if and only if $r = n-1$. We obtain $C^{\rightarrow} = (L_{\pi(1)} \rightarrow \mathbf{f}) \rightarrow C_0^{\rightarrow}$ by equivalence (E2) (equivalence (E4) cannot be applied, since neither $L_{\pi(1)}$ nor $C_0^{\rightarrow}$ has the form $\beta \rightarrow \mathbf{f}$). Thus $C^{\rightarrow}$ contains at least $|n-r-2|+1 = |n-r-1|$ occurrences of **f**. Hence the first part of the lemma holds in any case.

Next we show by induction on n that we can actually find some $C^{\rightarrow}$ which contains exactly $|n-r-1|$ occurrences of **f**. If C contains a negative literal L, then we put $C_0 = C \setminus \{L\}$. By induction hypothesis we find $C_0^{\rightarrow}$ with exactly $|(n-1)-(r-1)-1| = |n-r-1|$ occurrences of **f**. We put $C^{\rightarrow} = L_{\pi(1)} \rightarrow C_0^{\rightarrow}$ as above. However, if all literals of C are positive (i.e., r=0), then only equivalence (E2) applies, and we obtain a translation $C^{\rightarrow}$ with $n-1 = |n-r-1|$ occurrences of **f**. $\qquad\square$

Note that the previous lemma holds even if we allow arbitrary groupings, e.g., $(L_{\pi(1)} \vee (L_{\pi(2)} \vee L_{\pi(3)})) \vee L_{\pi(4)}$. We also note that $C^{\rightarrow}$ contains no **f**-occurrences if and only if C is a *definite Horn* clause (i.e., C contains exactly one positive literal).

Lemma 7 *For a clause-set F we can find a standard translation $F^{\rightarrow}$ with minimal number of **f**-occurrences in polynomial time. Hence the falsum number of a clause-set can be computed in polynomial time.*

Proof. Consider the sets of **f**-implicational formulas T, T' as obtained within the procedure of Fig. 3 (see Step 6). In view of Lemma 6, we can assume that the total number of **f**-occurrences in T is minimal. We choose an ordering $\alpha_1, \ldots, \alpha_p$ of the formulas in $T \cup T'$ and put $F^{\rightarrow} := (\alpha_1 \rightarrow \ldots \rightarrow \alpha_p \rightarrow \mathbf{f}) \rightarrow \mathbf{f}$. If some formula α in $T \cup T'$ has the form $\alpha' \rightarrow \mathbf{f}$, then we assure that α comes last, and we can save two **f**-occurrences by equivalence (E4), and $F^{\rightarrow}$ reduces to $(\alpha_1 \rightarrow \ldots \rightarrow \alpha_{p-1} \rightarrow \alpha') \rightarrow \mathbf{f}$. Thus, $\#_{\mathbf{f}}(F)$ equals the total number of **f**-occurrences in $T \cup T'$ plus $j \in \{0, 2\}$, where $j = 0$ if some formula of $T \cup T'$ has the form $\alpha \rightarrow \mathbf{f}$, and $j = 2$ otherwise. $\qquad\square$

By means of this lemma, Theorem 9 immediately yields the following result.

Theorem 10 *Satisfiability of clause-sets with bounded falsum number is fixed-parameter tractable.*

Our next result indicates that falsum number for clause-sets is outperformed by maximum deficiency.

Theorem 11 $\#_{\mathbf{f}}(F) \geq \delta(F)$ *holds for clause-sets F without pure literals. Consequently, $\#_{\mathbf{f}}(F) \geq \delta^*(F)$ for minimal unsatisfiable clause-sets.*

Proof. Let $F = \{C_1, \ldots, C_m\}$ and $\mathsf{var}(F) = \{x_1, \ldots, x_n\}$. We apply the first four steps of the translation to F, and we are left with a set of clauses $F' = \{C'_1, \ldots, C'_m, \{x_1, x'_1\}, \ldots, \{x_r, x'_r\}\}$, $r \leq n$, and a set S of implications. No variable except $x_{r+1}, \ldots, x_n$ occurs positively in $C'_1, \ldots, C'_m$, hence at most $n-r$ clauses of $C'_1, \ldots, C'_m$ are definite Horn (note that each variable occurs exactly once positively in F'). It follows now by Lemma 6 that by applying Steps 5 and 6, we introduce at least $m - r \geq m - n \geq \delta(F)$ occurrences of $\mathbf{f}$. $\square$

It remains open whether other translations yield a significantly smaller falsum number than the standard translation.

7   Discussion and Open Questions

Parameterized complexity is a fast growing research area, and we expect that several new FPT results for SAT will be obtained in the years to come. We hope that this paper provides a starting point for further developments and comparative results.

The parameters considered above depend on the chosen transformation of clause-sets to other combinatorial objects (graphs, hypergraphs, directed graphs, $\mathbf{f}$-implicational formulas); therefore it is natural to ask (a) for new transformations which yield smaller values for the considered parameters, and (b) for transformations to other known FPT problems (see, e.g., [7]) which possibly give rise to natural parameterizations for SAT.

Furthermore, it might be interesting to study recursively defined SAT hierarchies (see [23, 21]) in the framework of parameterized complexity. Known algorithms decide satisfiability of clause-sets belonging to the k'th level of these hierarchies in time $n^{\mathcal{O}(k)}$; this does not constitute fixed-parameter tractability. However, fixed-parameter *intractability* results (i.e., $W[1]$-hardness, [12]) are apparently not known.

Acknowledgment

The author wishes to thank Prof. J. A. Makowsky for explaining the theory behind Theorem 4 and for many stimulating discussions during the author's visit in Haifa in April 2001.

References

[1] R. Aharoni and N. Linial. Minimal non-two-colorable hypergraphs and minimal unsatisfiable formulas. *J. Combin. Theory Ser. A*, 43:196–204, 1986.

[2] M. Alekhnovich and A. A. Razborov. Satisfiability, branch-width and Tseitin tautologies. In *43rd Annual IEEE Symposium on Foundations of Computer Science (FOCS'02)*, pages 593–603, 2002.

[3] S. Arnborg, D. G. Corneil, and A. Proskurowski. Complexity of finding embeddings in a k-tree. *SIAM J. Algebraic Discrete Methods*, 8(2):277–284, 1987.

[4] F. Bacchus, S. Dalmao, and T. Pitassi. Algorithms and complexity results for #SAT and Bayesian Inference. In *44th Annual IEEE Symposium on Foundations of Computer Science (FOCS'03)*. To appear.

[5] H. L. Bodlaender. A linear-time algorithm for finding tree-decompositions of small treewidth. *SIAM J. Comput.*, 25(6):1305–1317, 1996.

[6] H. L. Bodlaender. A partial k-arboretum of graphs with bounded treewidth. *Theoret. Comput. Sci.*, 209(1-2):1–45, 1998.

[7] M. Cesati. Compendium of parameterized problems. http://bravo.ce.uniroma2.it/home/cesati/research/, 2001.

[8] S. A. Cook. An exponential example for analytic tableaux. Manuscript, 1972.

[9] B. Courcelle, J. A. Makowsky, and U. Rotics. On the fixed parameter complexity of graph enumeration problems definable in monadic second-order logic. *Discr. Appl. Math.*, 108(1-2):23–52, 2001.

[10] B. Courcelle and S. Olariu. Upper bounds to the clique width of graphs. *Discr. Appl. Math.*, 101(1-3):77–114, 2000.

[11] M. Davis, G. Logemann, and D. Loveland. A machine program for theorem-proving. *Comm. ACM*, 5:394–397, 1962.

[12] R. G. Downey and M. R. Fellows. *Parameterized Complexity*. Springer Verlag, 1999.

[13] H. Fleischner, S. Földes, and S. Szeider. Remarks on the concept of robust algorithm. Technical Report RRR 26-2001, *Rutgers Center for Operations Research* (RUTCOR), Apr. 2001.

[14] H. Fleischner, O. Kullmann, and S. Szeider. Polynomial-time recognition of minimal unsatisfiable formulas with fixed clause-variable difference. *Theoret. Comput. Sci.*, 289(1):503–516, 2002.

[15] J. Franco, J. Goldsmith, J. Schlipf, E. Speckenmeyer, and R. P. Swaminathan. An algorithm for the class of pure implicational formulas. *Discr. Appl. Math.*, 96/97:89–106, 1999.

[16] M. C. Golumbic and U. Rotics. On the clique-width of some perfect graph classes. *Internat. J. Found. Comput. Sci.*, 11(3):423–443, 2000. Selected papers from the Workshop on Graph-Theoretical Aspects of Computer Science (WG'99), Part 1 (Ascona).

[17] G. Gottlob and R. Pichler. Hypergraphs in model checking: Acyclicity and hypertree-width versus clique-width. In F. Orejas, P. G. Spirakis, and J. van Leeuwen, editors, *28th International Colloquium on Automata, Languages and Programming (ICALP'01)*, volume 2076 of *Lecture Notes in Computer Science*, pages 708–719, 2001.

[18] G. Gottlob, F. Scarcello, and M. Sideri. Fixed-parameter complexity in AI and nonmonotonic reasoning. *Artificial Intelligence*, 138(1-2):55–86, 2002.

[19] P. Heusch. The complexity of the falsifiability problem for pure implicational formulas. *Discr. Appl. Math.*, 96/97:127–138, 1999.

[20] P. Heusch, S. Porschen, and E. Speckenmeyer. Improving a fixed parameter tractability time bound for the shadow problem. Technical Report 2001-425, Universität zu Köln, 2001.

[21] O. Kullmann. Investigating a general hierarchy of polynomially decidable classes of CNF's based on short tree-like resolution proofs. Technical Report TR99–041, *Electronic Colloquium on Computational Complexity* (ECCC), 1999.

[22] O. Kullmann. An application of matroid theory to the SAT problem. In *Fifteenth Annual IEEE Conference on Computational Complexity*, pages 116–124, 2000.

[23] D. Pretolani. Hierarchies of polynomially solvable satisfiability problems. *Ann. Math. Artif. Intell.*, 17(3-4):339–357, 1996.

[24] N. Robertson and P. D. Seymour. Graph minors. II. Algorithmic aspects of tree-width. *J. Algorithms*, 7(3):309–322, 1986.

[25] N. Robertson and P. D. Seymour. Graph minors. X. Obstructions to tree-decomposition. *J. Combin. Theory Ser. B*, 52(2):153–190, 1991.

[26] N. Robertson and P. D. Seymour. Graph minors. XIII. The disjoint paths problem. *J. Combin. Theory Ser. B*, 63(1):65–110, 1995.

[27] J. P. Spinrad. Representations of graphs. Book manuscript, Vanderbilt University, 1997.

[28] S. Szeider. Generalizations of matched CNF formulas. *Ann. Math. Artif. Intell.* Special issue with selected papers from the 5th Int. Symp. on the Theory and Applications of Satisfiability Testing (SAT'02), to appear.

[29] S. Szeider. Minimal unsatisfiable formulas with bounded clause-variable difference are fixed-parameter tractable. In T. Warnow and B. Zhu, editors, *The 9th International Computing and Combinatorics Conference (COCOON'03)*, volume 2697 of *Lecture Notes in Computer Science*, pages 548–558. Springer Verlag, 2003.

[30] A. Urquhart. The complexity of propositional proofs. *Bull. of Symbolic Logic*, 1(4):425–467, Dec. 1995.

[31] M. Yannakakis. Algorithms for acyclic database schemes. In C. Zaniolo and C. Delobel, editors, *Very Large Data Bases, 7th International Conference, Sep. 9–11, 1981, Cannes, France*, pages 81–94. IEEE Computer Society, 1981.

On the Probabilistic Approach to the Random Satisfiability Problem

Giorgio Parisi

Dipartimento di Fisica, Sezione INFN, SMC and UdRm1 of INFM,
Università di Roma "La Sapienza",
Piazzale Aldo Moro 2, I-00185 Rome (Italy)
giorgio.parisi@roma1.infn.it,
http:chimera.roma1.infn.it

Abstract. In this note I will review some of the recent results that have been obtained in the probabilistic approach to the random satisfiability problem. At the present moment the results are only heuristic. In the case of the random 3-satisfiability problem a phase transition from the satisfiable to the unsatisfiable phase is found at $\alpha = 4.267$. There are other values of α that separates different regimes and they will be described in details. In this context the properties of the survey decimation algorithm will also be discussed.

1 Introduction

Recently many progresses [1,2] have been done on the analytic and numerical study of the random K-satisfiability problem [3,4,5,6], using the approach of survey-propagation that generalizes the more old approach based on the belief-propagation algorithm [1] [9,7,10,11] . Similar results have also been obtained for the coloring a of random graph [12].

In the random K-sat problem there are N variables $\sigma(i)$ that may be true of false (the index i will sometime called a node). An instance of the problem is given by a set of $M \equiv \alpha N$ clauses. In this note we will consider only the case $K = 3$. In this case each clause c is characterized by set of three nodes (i_1^c, i_2^c, i_3^c), that belong to the interval $1 - N$ and by three Boolean variables (b_1^c, b_2^c, b_3^c, i.e. the signatures in the clause). In the random case the i and b variables are random with flat probability distribution. Each clause c is true if the expression

$$E^c \equiv (\sigma(i_1^c)XOR\,b_1^c)\ OR\ (\sigma(i_2^c)XOR\,b_2^c)\ OR\ (\sigma(i_3^c)XOR\,b_3^c) \qquad (1)$$

is true [2].

- The belief propagation algorithm (sometimes called "Min-Sum") is the the zero temperature limit of the "Sum-Product" algorithm. In the statistical mechanics language [7] the belief propagation equations are the extension of the TAP equations for spin glasses [8,13] and the survey-propagation equations are the TAP equations generalized to the broken replica case.
- When all the b^c are false $E^c = \overline{\sigma(i_.^c)}\ OR\ \sigma(i_.^c)\ OR\ \sigma(i_.^c)$ while when all the b^c are true $E^c = \overline{\sigma(i_.^c)}\ OR\ \overline{\sigma(i_.^c)}\ OR\ \overline{\sigma(i_.^c)}$.

E. Giunchiglia and A. Tacchella (Eds.): SAT 2003, LNCS 2919, pp. 203–213, 2004.

The problem is satisfiable iff we can find a set of the variables σ such that all the clauses are true (i.e. a legal configuration); in other words we must find a truth value assignment. The entropy [11] of a satisfiable problem is the logarithm of the number of the different sets of the σ variables that make all the clauses true, i.e the number of legal configurations.

The goal of the analytic approach consists in finding for given α and for large values of N the probability that a random problem (i.e. a problem with random chosen clauses) is satisfiable. The $0 - 1$ law [4,6,16] is supposed to be valid: for $\alpha < \alpha_c$ all random systems (with probability one when N goes to infinity) are satisfiable and their entropy is proportional to N with a constant of proportionality that does not depend on the problem. On the other hand, for $\alpha > \alpha_c$ no random system (with probability one) is satisfiable. An heuristic argument[1,2] suggests that $\alpha_c = \alpha^* \approx 4.267$ where α^* can be computed using the survey-propagation equations defined later. There is already a proof [17] that the value of α^* computed with the techniques of survey-propagation is a rigorous upper bound to α_c (the proof has been obtained only for even K, the extension to odd K is technically difficult).

2   Results

Generally speaking we are interested to know not only the number of legal configurations, but also the properties of the set of all legal configurations. At this end it is convenient to say that two configurations are adjacent if their Hamming distance is less than ϵN, where ϵ is a small number.

We can argue that in the limit of large N:

1. In the interval $\alpha < \alpha_d \approx 3.86$ the set of all legal configurations is connected, i.e. there is a path of mutually adjacent configurations that joins two configurations of the set. In this region the belief-propagation equations (to be define later) have an unique solution.
2. In the interval $\alpha_d < \alpha < \alpha_c \approx 4.267$ the set of all the legal configurations breaks in an large number of different disconnected regions that are called with many different names in the physical literature [14,15] (states, valleys, clusters, lumps...). Roughly speaking the set of all the legal configurations can be naturally decomposed into clusters of proximate configurations, while configurations belonging to different clusters (or regions) are not close. This phenomenon is called in spontaneous replica symmetry breaking in the physical literature. The core of the approach of this note is the analysis of this phenomenon and of the methods used to tame its consequences [3]. The precise definition of these regions is rather complex [18]; roughly speaking we could say that two legal configurations belongs to the same region if they are

---

[*] Other models, where this phenomenon is not present, like random bipartite matching can be analyzed in a much simple way, although 15 years have been needed from the statements of the main result (i.e. the length of the shortest matching in the infinite N limit is $\zeta(2)$) to the rigorous proof of this fact.

in some sense adjacent, i.e. they belongs to a different region if their Hamming distance is greater than ϵN. In this way the precise definition of these regions depends on ϵ, however it can be argued that there is an interval in ϵ where the definition is non-trivial and is independent from the value of ϵ: for a rigorous definition of these regions see [20,21,19]. The number of these regions is given by $\exp(\Sigma^N(\alpha))$, where $\Sigma^N(\alpha)$ is the total complexity; for large N the total complexity is asymptotically given by $\Sigma^N(\alpha) = N\Sigma(\alpha)$ where $\Sigma(\alpha)$ is the complexity density. In this interval the belief-propagation equations have many solutions and each of these solution is associated to a different cluster. The statistical properties of the set of the solutions of the belief-propagation equations can be studied using the belief-propagation equations (to be defined later).

3. Only in the interval $\alpha_b \approx 3.92 < \alpha < \alpha_c$ there are literals σ that are frozen, i.e. they take the same value in all the legal configurations of a region [4]. We could say that the frozen variables form the backbone of a region. It is important to realize that a given clause may simultaneously belong to the backbone of one region and not belong to the backbone of an other region.

The arguments run as follow. Let us start with a given instance of the problem. We first write the belief propagation equations. For each clause that contains the node i (we will use the notation $c \in i$ although it may be not the most appropriate) $p_T(i, c)$ is defined to be the probability that the variable $\sigma(i)$ would be true in absence of the clause c when we average over the set of all the legal configuration ($p_F(i, c) = 1 - p_T(i, c)$ is the probability to be false). If the node i_1^c were contained in only one clause, we would have that

$$p_T(i_1^c) = u_T(p_T(i_2^c, c), p_T(i_3^c, c), b_1^c, b_2^c, b_3^c) \equiv u_T(i_1, c) \ ,$$
$$p_F(i_1^c) = 1 - u_T(i_1, c) \ , \tag{2}$$

where u_T is an appropriate function that is defined by the previous relation. An easy computation shows that when all the b are false, the variable $\sigma(i_1^c)$ must be true if both variable $\sigma(i_2^c)$ and $\sigma(i_3^c)$ are false, otherwise it can have any value. Therefore we have in this case that

$$u_T(i, c) = \frac{1}{2 - p_F(i_2^c, c)p_F(i_3^c, c)} \ . \tag{3}$$

In a similar way, if some of the b variable are true, we should exchange the indices T and F for the corresponding variables, i.e., if b_1^c is true, then $u_T(u_1)$ becomes $u_F(u_1)$. Finally we have that

---

The distinction between α_d [5] a and α_b [1] is not usually done in the literature and sometimes it is wrongly assumed that $\alpha_b = \alpha_d$.

$$p_T(i,c) = \frac{\prod_{d \in i, d \neq c} u_T(i,d)}{Z_0(i,c)}$$

$$p_F(i,c) = \frac{\prod_{d \in i, d \neq c} u_F(i,d)}{Z_0(i,c)}$$

$$Z_0(i,c) = \prod_{d \in i, d \neq c} u_T(i,d) + \prod_{d \in i, d \neq c} u_F(i,d). \tag{4}$$

We note the previous formulae can be written in a more compact way if we introduce a two dimensional vector $\boldsymbol{p}$, with components p_T and p_F. We define the product of these vector

$$c_T = a_T \, b_T \qquad c_F = a_F \, b_F, \tag{5}$$

if $\boldsymbol{c} = \boldsymbol{a} \cdot \boldsymbol{b}$.

If the norm of a vector is defined by

$$|\boldsymbol{a}| = a_T + a_F \,, \tag{6}$$

the belief propagation equations are defined to be

$$\boldsymbol{p}(i,c) = \frac{\prod_{d \in i, d \neq c} \boldsymbol{u}(i,d)}{|\prod_{d \in i, d \neq c} \boldsymbol{u}(i,d)|} \,. \tag{7}$$

In total there are $3M$ variables $p_T(i,c)$ and $3M$ equations. These equations in the limit of large N should have an unique solution in the interval $\alpha < \alpha_d$ and the solution should give the correct values for the probabilities $p_T(i,c)$. In this interval the entropy (apart corrections that are subleading when N goes to infinity) is given by

$$S = -\sum_{i=1,N} \log(Z_1(i)) + 2 \sum_{c=1,M} \log(Z_2(c)) \,. \tag{8}$$

Here the first sum runs over the nodes and the second one runs over the clauses; $Z_2(c)$ is the probability that the clause c would be satisfied in a system where the validity of the clause c is not imposed. One finds that in the case where all the b variables are false

$$Z_2(c) = 1 - p_F(i_1^c) p_F(i_2^c) p_F(i_2^c) \,. \tag{9}$$

In a similar way $Z_1(i)$ is the probability that all we can find a legal configuration containing the site i starting from a configuration where the site i is not present and it is given by:

$$Z_1(i) = |\prod_{d \in i} \boldsymbol{u}(i,d)| \tag{10}$$

The belief propagation equations can also be written as :

$$\frac{\partial S}{\partial \boldsymbol{p}(i,c)} = 0 \,. \tag{11}$$

The belief-propagation equations can be formally derived by a local analysis by assuming that, in the set of the legal configurations of the system where all the clauses $c \in i$ are removed, the variables $\sigma(k)$ that would enter in these clauses are not correlated. This cannot is not true for finite N, but this statement may be correct in the limit $N \to \infty$ with the appropriate qualifications.

Generally speaking for a given sample these equations may not have an exact solution, but they do have quasi-solutions [24] (i.e. approximate solutions, where the approximation becomes better and better when N goes to infinity [5]): these equations have been derived using a local analysis that is correct only in the limit $N \to \infty$.

In the interval $\alpha_d < \alpha < \alpha_b$ the variables p_F and p_T are different from 0 and 1; however in the region $\alpha_b < \alpha < \alpha_c$ there solutions (or quasi-solutions) of the belief equations have a fraction of the variables p_F and p_T that are equal to 0 or 1.

When the number of solutions of the belief propagation equations is large, the properties of the sets of solutions of the belief propagation equations can be obtained by computing the solution of the survey propagation equations defined as follows. In the general approach in each node we introduce the probability $(\mathcal{P}_{i,c}(p))$ to find a solution of the belief-propagation equations with $p_T(i, c) = p$. With some effort we can write down local equations for this probability. These are the full survey equations that allow the computation of the total entropy.

This approach is computationally heavy. As far as the computation of the complexity is concerned, we can use a simpler approach, where we keep only a small part of the information contained in the function $\mathcal{P}_{i,c}(p)$, i.e. the weight of the two delta function at $p = 0$ and $p = 1$. More precisely we introduce the quantity $s_T(i, c)$ that is defined as the probability of finding $p_T(i, c) = 1$, in the same way $s_F(i, c)$ is the probability of finding $p_T(i, c) = 0$ and $s_I(i, c)$ is the probability of finding $0 < p_T(i, c) < 1$. It is remarkable that it is possible to write closed equations also for these probabilities (these equations are usually called the survey propagation equations [1]).

We can use a more compact notation by introducing a three dimensional vector s given by

$$s = \{s_T, s_I, s_F\} . \tag{12}$$

Everything works as before with the only difference that we have a three component vector instead of a two component vector. Generalizing the previous arguments one can introduce the quantity $u(i, c)$ that is the value that the survey at i would take if only the clause c would be present in i (in other words $u(i, c)$ is the message that arrives to the site i coming from the clause c). In the case where all the b are false, a simple computation gives

---

* More precisely if we have N equations $E_i[\sigma] = 0$ for $i = 1, N$ a solution σ of this system of equation satisfies the condition $N^* sum_{i \cdots, N})(E_i[\sigma])^* = 0$; a quasi-solution satisfies the weaker condition $N^* sum_{i \cdots, N})(E_i[\sigma])^* < h(N)$, where $h(N)$ is a function that goes to zero when N goes to infinity. The definition of a quasi-solution depends on the properties of the function $h(N)$ and this point must be further investigated: it may turn out at the end that quasi-solutions are not needed.

$$\boldsymbol{u}(i,c) = \{s_F(i_2^c, c)s_F(i_3^c, c),\ 1 - s_F(i_2^c, c)s_F(i_3^c, c),\ 0\}\ . \tag{13}$$

The formula can generalized as before [6] to the case of different values of b. One finally finds the survey propagation equations:

$$\boldsymbol{s}(i,c) = \frac{\prod_{d\in i, d\neq c} \boldsymbol{u}(i,d)}{|\prod_{d\in i, d\neq c} \boldsymbol{u}(i,d)|}\ , \tag{14}$$

where we have defined product in such a way that

$$\boldsymbol{ab} = \{a_T b_T + a_I b_T + a_T b_I, a_I b_I,\ a_F\ b_F + a_I\ b_F + a_F\ b_I\}. \tag{15}$$

It is convenient to introduce the reduced complexity $(\Sigma_R(\alpha))$, that counts the number of solutions of the belief equations where two different solutions are considered equals if they coincide in the points where the beliefs are 0 or 1 [7]. In other words two solutions of the beliefs equations with an identical backbone enters only once in the counting that leads to the reduced complexity.

If there is an unique solution to the survey propagation equations, it is possible to argue that the reduced total complexity should be given by

$$\Sigma_R = -\sum_{i=1,N} \ln(Z_1(i)) + 2\sum_{c=1,M} \ln(Z_2(c)) \tag{16}$$

where now the definition of the Z's is changed and it is done using the surveys, not the beliefs:

$$Z_1(i) = \ln(|\prod_{d\in i} \boldsymbol{u}(i,d)|), \quad Z_2(c) = \ln(|\boldsymbol{s}(i,c)\boldsymbol{u}(i,c)|) \tag{17}$$

The reduced complexity $\Sigma_R(\alpha)$ it is particularly interesting because it hat been conjecture that it should vanishes at the critical point α_c. This allow the computation of the point α_c.

It is interesting that also in this case the survey propagation equations can be written in a simple form:

$$\frac{\partial \Sigma_R}{\partial \boldsymbol{s}(i,c)} = 0\ . \tag{18}$$

One finally finds that the survey-propagation equations do not have an unique solution when $\alpha > \alpha_U \approx 4.36$. The fact is not of direct importance because $\alpha_U > \alpha_c$. Indeed in the region $\alpha > \alpha_c$ the complexity is negative so that the there are no solutions of the belief-propagation equations associated to the solution of the survey-propagation equation.

---

* It always happens that the vector $\boldsymbol{u}$ has only one zero component ($u_T u_F = 0$). This fact may be used to further simplify the analysis.

* It is not clear at the present moment if there are different solutions of the belief equations that coincide in all the points where the probability is 0 or 1: in other words we would like to know if $\Sigma_R(\alpha) = \Sigma(\alpha)$, where $\Sigma(\alpha)$ counts the total number of solutions of the belief equations.

It is evident that for $\alpha > \alpha_c$ there are no more legal configurations and $\Sigma_R(\alpha)$ is not well defined. A negative value $\Sigma_R(\alpha)$ can be interpreted by saying that the probability of finding a legal configuration goes to zero exponentially with N. We stress that the entropy density remains finite at α_c, the conjecture that $\Sigma_R(\alpha)$ vanishes at α_c implies that the reduced complexity is captures the number of essentially different regions of legal configuration. A priori a finite value $\Sigma_R(\alpha_c)$ cannot be excluded.

3   Methods

We now show how to obtain the above results on the solutions of the belief propagation equations (and of the survey propagation equations) for a large random system in the limit of large N. These equations are interesting especially in the infinite N limit where the factor graph does not contain short loop. For finite N in the random case, (or in the infinite N limit for a non-random case) the belief equations may have solutions, but the properties of these solutions do not represent exactly the properties of the systems. If the number of short loops is small, perturbative techniques may be used to compute the corrections to the belief equations. If short loops are very common (e.g. if the literals are on the sites of an f.c.c. lattice and the clauses are on faces of the same lattice), it is rather likely that the beliefs equations are useless and they could only used as starting point of much more sophisticated approaches.

We attack this problem by studying the solution of the belief propagation equations (and of the survey propagation equations) on an random infinite tree. Sometimes the solution is unique, i.e. it does not depends on the boundary conditions at infinity, and sometimes it is not not unique. The statistical properties of the solution of these equations can be studied with probabilistic method. One arrives to integral equations that can be solved numerically using the method of population dynamics [1,2,14,15]. The numerical solutions of these integral equations can be used to compute α_d, α_b, α_c, and α_U.

The generalization of the Aldous construction [22,23] of an infinite rooted tree associated to a graph can play the role of a bridge between a finite instance of the problems and the infinite random tree where analytic computations [1,2,14,15] are done. For example it could be used to prove the existence and the uniqueness of the beliefs and survey propagation equation in the appropriate intervals.

In this way one can argue that the properties on an infinite random tree are relevant for the behaviour of a given random system in the limit of large N.

We can check that the results we have obtained in these way for the solution of the belief and survey propagation equations are correct by computing in an explicit way the solution (when it is unique) of the equations for a given sample for large N (e.g $N = 10^4 - 10^6$). For example we may compare the distribution of the beliefs or of the surveys in a large system with the one obtained by solving the integral equations for the probabilities: the agreement is usually very good.

In the same spirit the validity of the result for α_d may be checked by studying the convergence of the iterative procedure for finding the solution of the belief-propagation equations on a given large problem One finds that just at α_d the iterative procedure for finding a solution does not converge anymore and this is likely a sign of the existence of many solutions to the belief-propagation equations. In a similar way we can check the correctness of α_U.

4   Survey Decimation Algorithm

The survey decimation algorithm has been proposed [1,2,25,26,27] for finding the solution of the random K-satisfiability problem [4,5,6].

We start by solving the survey propagation equation. If a survey $(s(i))$ is very near to $(1,0,0)$ (or to $(0,0,1)$) in most of the legal solutions of the beliefs equations (and consequently in the legal configurations) the corresponding local variables will be true (or false).

The main step in the decimation procedure consists is starting from a problem with N variables and to consider a problem with $N-1$ variables where $s(i)$ is fixed to be true (or false). We denote

$$\Delta(i) = \Sigma^N - \Sigma^{N-1} . \tag{19}$$

If $\Delta(i)$ is small, the second problem it is easier to solve: it has nearly the same number of solutions of the belief equations and one variable less. (We assume that the complexity can be computed by solving the survey propagation equations).

The decimation algorithm proceeds as follows. We reduces by one the number of variables choosing the node i in the appropriate way, e.g. by choosing the node with minimal $\Delta(i)$. We recompute the solutions of the survey equations and we reduce again the number of variables. At the end of the day two things may happen:

1. We arrive to a negative complexity (in this case the reduced problem should have no solutions and we are lost),
2. The denominator in equation (14) becomes zero, signaling the presence of a contradiction (and also in this case we are lost),
3. The non-trivial solution of the survey equation disappears. If this happens the reduced problem is now easy to be solved.

The quantity $\Delta(i)$ may be estimated analytically so that it is possible to choose the variable with minimal $\Delta(i)$. A careful analysis of the results for large, but finite N [27] shows that the algorithm works in the limit of infinite N up to $\alpha_A \approx 4.252$, that is definite less, but very near to α_c.

Unfortunately at the present moment this result for α_A can be obtained only analyzing how the argument works on a finite sample and we are unable to write down integral equations for the probability distributions of the solution of the survey propagation equations. This drawback leads to the impossibility of computing analytically α_A: it is a rather difficult task to understand in details why for α_A it so near to α_c. It is interesting to note that for $\alpha < \alpha_A$ the survey decimation algorithm takes a time that is polynomial in N and using a smart implementation the time is nearly linear in N. It is important to stress that survey algorithm is an incomplete search procedure which may not be able to find any solution to a satisfiable instance. This actually happens with a non-negligible probability e.g. for sizes of the order of a few thousands of variables also when $\alpha < \alpha_A$, however in the limit $N \to \infty$, it should work as soon $\alpha < \alpha_A$.

In the interval $\alpha_A < \alpha < \alpha_c$ the decimation procedure leads to a regime of negative complexity so that the algorithm does not work. Unfortunately there is no analytic computation of α_A. It is likely that the fact that $\alpha_U = 4.36$ is not far from α_c is related to the fact the $\alpha_c - \alpha_A$ iss small.

It would be very interesting to understand better if such a relation is true and to put in a quantitative form. In this regard it would be be important to study the K dependence of $\alpha_c - \alpha_A$ and $\alpha_U - \alpha_c$. This analysis may give some hints why $\alpha_c - \alpha_A$ is so small in 3-SAT.

5   Conclusions

Similar problems have been studied by physicists in the case of infinite range spin glasses [7]: here the problem consists in finding the minimum E_J of the quantity:

$$H_J[\tau] \equiv \sum_{i,k=1,N} J_{i,k} \tau_i \tau_k \qquad (20)$$

where the minimum is done respect to the variables $\tau_i = \pm 1$ and the J are independent Gaussian variables with zero average and variance $N^{-1/2}$. Physical intuition tells us that in the limit N goes to infinity the intensive quantity

$$e_J = \frac{E_J}{N} \qquad (21)$$

should be (with probability 1) independent from N and it will be denoted by e_∞. In 1979 it was argued using the so called replica method (that will be not discussed in this note) that e_∞ was equal to the maximum of a certain functional $F[q]$, where $q(x)$ is a function defined in the interval $[0-1]$. Later on, 1985 the same results were rederived using heuristical probabilistic consideration, similar to those presented here (but much more complex). In this note we have introduced a hierarchical construction, where three levels (configurations, beliefs, surveys) are presents: in the case of spin glasses an infinite number of levels is needed (in the spin glass case the survey equations do not have an unique solution and we have to consider higher and higher levels of abstraction). Only very recently Talagrand [28], heavily using Guerra's ideas and results, was able to prove that the value for e_∞, computed 24 year before, was correct.

The possibility of using these techniques for deriving eventually exact results on the K-SAT problem is a very recent one: only a few year ago [14,15] the previous techniques has been extended to more complex spin glass models where the matrix J is sparse (it has an average number of elements per raw that does not increase with N).

The field is very young and rigorous proofs of many steps of the construction (also those that would likely be relatively simple) are lacking. We only know, as a consequence of general theorems, that this methods give an upper bound to the value of α_c, and this upper bound should be computed by maximizing an appropriate functional of the probability distribution of the surveys. This upper bound is rigorous one [17]: it essentially use positivity arguments (the average of a non-negative function is non-negative) in a very smart way and it does not depend on the existence or uniqueness of the solutions of the equations for the probability distribution of the survey. On the contrary the way we have followed to compute this upper bound (i.e. α^*) require some extra work before becoming fully rigorous. I stress that this upper bounds and the Talagrand's exact result do not need in any way considerations on the solutions of the survey propagation equations (or of their generalization) on a finite sample. The survey propagation equations are crucial for giving an intuitive image of the situation (i.e. at a metaphoric level) and for constructing the survey decimation algorithm. The heuristic derivation could have been done using the replica method, where survey propagation equations are never mentioned, but the argument is much more difficult to follow and to transform in a rigorous one.

The other results come from empirical (sometimes very strong) numerical evidence and from heuristic arguments. For example at my knowledge there is no proof that the integral equations for the probability of the surveys (or of the beliefs) have an unique solution and that the population dynamics algorithm converges (to that unique solution). Proofs in this direction would be very useful and are a necessary step to arrive to a rigorous quantitative upper bounds and eventually exact results. On the other hand the proof of existence of an unique solutions (or quasi-solutions) of the surveys (or beliefs) propagation equations in the large N limit is lacking for any value of α, although the analysis with the population dynamics (whose precise mathematical properties have to be clarified) tell us which should the maximum values (α_U and α_b respectively, below which these uniqueness properties hold. Many steps have to be done, but a rigorous determination of α_c seems to be a feasible task in a not too far future.

References

1. M. Mézard, G. Parisi and R. Zecchina, Science **297**, 812 (2002).
2. M. Mézard and R. Zecchina *The random K-satisfiability problem: from an analytic solution to an efficient algorithm* cond-mat 0207194.
3. S.A. Cook, D.G. Mitchell, *Finding Hard Instances of the Satisfiability Problem: A Survey*, In: *Satisfiability Problem: Theory and Applications.* Du, Gu and Pardalos (Eds). DIMACS Series in *Discrete Mathematics and Theoretical Computer Science*, Volume 35, (1997)
4. S. Kirkpatrick, B. Selman, *Critical Behaviour in the satisfiability of random Boolean expressions*, Science 264, 1297 (1994)
5. Biroli, G., Monasson, R. and Weigt, M. *A Variational description of the ground state structure in random satisfiability problems, Euro. Phys. J.* **B 14** 551 (2000),

6. Dubois O. Monasson R., Selman B. and Zecchina R. (Eds.), *Phase Transitions in Combinatorial Problems*, Theoret. Comp. Sci. 265, (2001), G. Biroli, S. Cocco, R. Monasson, Physica A 306, 381 (2002).

7. Mézard, M., Parisi, G. and Virasoro, M.A. *Spin Glass Theory and Beyond*, World Scientific, Singapore, (1987).

8. D.J. Thouless, P.A. Anderson and R. G. Palmer, *Solution of a 'solvable' model*, Phil. Mag. 35, 593 (1977)

9. J.S. Yedidia, W.T. Freeman and Y. Weiss, *Generalized Belief Propagation*, in *Advances in Neural Information Processing Systems 13* eds. T.K. Leen, T.G. Dietterich, and V. Tresp, MIT Press 2001, pp. 689-695.

10. F.R. Kschischang, B.J. Frey, H.-A. Loeliger, *Factor Graphs and the Sum-Product Algorithm, IEEE Trans. Infor. Theory* 47, 498 (2002).

11. Monasson, R. and Zecchina, R. *Entropy of the K-satisfiability problem, Phys. Rev. Lett.* 76 3881–3885 (1996).

12. R. Mulet, A. Pagnani, M. Weigt, R. Zecchina *Phys. Rev. Lett.* 89, 268701 (2002).

13. C. De Dominicis and Y. Y. Goldschmidt: *Replica symmetry breaking in finite connectivity systems: a large connectivity expansion at finite and zero temperature*, J. Phys. A (Math. Gen.) 22, L775 (1989).

14. M. Mézard and G. Parisi: Eur.Phys. J. B 20 (2001) 217;

15. M. Mézard and G. Parisi: *'The cavity method at zero temperature'*, cond-mat/0207121 (2002) to appear in J. Stat. Phys.

16. O. Dubois, Y. Boufkhad, J. Mandler, *Typical random 3-SAT formulae and the satisfiability threshold*, in *Proc. 11th ACM-SIAM Symp. on Discrete Algorithms*, 124 (San Francisco, CA, 2000).

17. S. Franz and M. Leone, *Replica bounds for optimization problems and diluted spin systems*, cond-mat/0208280.

18. G. Parisi, *Glasses, replicas and all that* cond-mat/0301157 (2003).

19. S. Cocco, O. Dubois, J. Mandler, R. Monasson. Phys. Rev. Lett. 90, 047205 (2003).

20. M. Talagrand, *Rigorous low temperature results for the p-spin mean field spin glass model, Prob. Theory and Related Fields* 117, 303–360 (2000).

21. Dubois and Mandler, FOCS 2002, 769.

22. D. Aldous, *The zeta(2) Limit in the Random Assignment Problem*, Random Structures and Algorithms 18 (2001) 381-418.

23. G. Parisi: cs.CC/0212047 *On local equilibrium equations for clustering states* (2002).

24. G. Parisi: cs.CC/0212009 *On the survey-propagation equations for the random K-satisfiability problem* (2002).

25. A. Braustein, M. Mezard, M. Weigt, R. Zecchina: cond-mat/0212451 *Constraint Satisfaction by Survey Propagation* (2002).

26. A. Braunstein, M. Mezard, R. Zecchina; cs.CC/0212002 *Survey propagation: an algorithm for satisfiability* (2002).

27. G. Parisi: cs.CC/0301015 *Some remarks on the survey decimation algorithm for K-satisfiability* (2003).

28. M. Talagrand, private communication (2003).

Comparing Different Prenexing Strategies for Quantified Boolean Formulas*

Uwe Egly, Martina Seidl, Hans Tompits, Stefan Woltran, and Michael Zolda

Institut für Informationssysteme 184/3, Technische Universität Wien,
Favoritenstraße 9–11, A–1040 Vienna, Austria
[uwe,seidl,tompits,stefan,zolda]@kr.tuwien.ac.at

Abstract. The majority of the currently available solvers for quantified Boolean formulas (QBFs) process input formulas only in prenex conjunctive normal form. However, the natural representation of practicably relevant problems in terms of QBFs usually results in formulas which are not in a specific normal form. Hence, in order to evaluate such QBFs with available solvers, suitable normal-form translations are required. In this paper, we report experimental results comparing different prenexing strategies on a class of structured benchmark problems. The problems under consideration encode the evaluation of nested counterfactuals over a propositional knowledge base, and span the entire polynomial hierarchy. The results show that different prenexing strategies influence the evaluation time in different ways across different solvers. In particular, some solvers are robust to the chosen strategies while others are not.

1 Introduction

The use of quantified Boolean formulas (QBFs) as a viable host language to capture different reasoning tasks from the area of Artificial Intelligence has gained increasing interest in recent years [5,14]. The overall principle in such approaches is to encode a given problem efficiently in terms of QBFs, and then to evaluate the resultant formulas by existing QBF solvers. The availability of sophisticated QBF solvers, like, e.g., the systems ssolve [8], QuBE [9], Quaffle [16], and semprop [11], makes such an approach practically appealing. Besides the BDD-based prover boole and the system QUBOS [1], most of the currently available QBF-solvers require that input formulas are in prenex conjunctive normal form.

Since the "natural" representation of many reasoning tasks in terms of QBFs usually results in formulas which do not adhere to a specific normal form, suitable normal-form translation schemas are required in order to apply existing QBF solvers. In general, such a translation can be divided into two steps: (i) the generation of a prenex form with an arbitrary quantifier-free matrix, and (ii) the translation of the matrix into normal form (e.g., CNF). We focus here on the first step, i.e., on *prenexing* a given QBF.

Similar to the situation in classical first-order logic, there are different ways how quantifiers can be shifted for obtaining formulas in prenex normal form. In this paper,

---

* This work was partially supported by the Austrian Science Fund (FWF) under grant P15068.

E. Giunchiglia and A. Tacchella (Eds.): SAT 2003, LNCS 2919, pp. 214–228, 2004.

we take an abstract view on the quantifier structure, allowing for an intuitive formalisation of general prenexing strategies. In particular, we focus here on four specific prenexing strategies and report experimental results comparing these strategies on structured benchmark problems spanning the entire polynomial hierarchy. Our results show that the chosen strategy crucially influences the resultant evaluation time. In fact, this phenomenon mirrors a similar observation from classical theorem proving, where, e.g., there exist classes of formulas for which different quantifier shifting strategies yield a non-elementary difference of proof size and search-space size [2,4].

The QBFs used for our experiments are encodings of the problem of checking derivability of a nested counterfactual [7] from a given propositional knowledge base. The use of QBFs resulting from this problem is appealing because they are not in prenex form and admit a large number of different prenex forms.

Informally, a counterfactual $p > q$ is a conditional query stating that if p would be true in the knowledge base, then q would also be true. Accordingly, a nested counterfactual $p_1 > (p_2 \cdots (p_n > q) \cdots)$ corresponds to conditional queries involving a sequence of revisions. As shown in [7], by varying the nesting depth of such statements and by allowing negations, we obtain problems which span the entire polynomial hierarchy. Furthermore, as an aside, our QBF encodings realise also, to the best of our knowledge, the first implementation for computing inference problems for nested counterfactuals.

2 Background

We deal with a quantified propositional language $\mathcal{L}$ over an alphabet $\mathcal{P}$ using $\bot$, $\neg$, $\wedge$, and $\forall p$ (for any $p \in \mathcal{P}$) as primitive logical operators. The operators $\top$, $\vee$, $\rightarrow$, and $\exists p$ are defined in the usual way in terms of other connectives. Such a language of *quantified Boolean formulas* (QBFs) generalises classical propositional logic by additional quantifications over propositional atoms. The *propositional skeleton*, sk(Φ), of a QBF Φ is given by deleting all quantifiers in Φ.

An occurrence of a variable p in a QBF Φ is *free* iff it does not appear in the scope of a quantifier Qp (Q $\in \{\forall, \exists\}$), otherwise the occurrence of p is *bound*. The set of variables occurring free in a QBF Φ is denoted by *free*(Φ). If *free*(Φ) $= \emptyset$, then Φ is *closed*, otherwise Φ is *open*. For Q $\in \{\forall, \exists\}$, we define the *complementary quantifier*, $\bar{Q}$, by $\bar{Q} = \exists$ if Q $= \forall$, and $\bar{Q} = \forall$ otherwise.

We use the notions of a *formula tree*, *subformula*, and a *proper subformula* in the usual way. Additionally, we introduce a *direct quantified subformula* (DQS) of a QBF Φ, denoted $dqs(\Phi)$, as follows: Ψ is a DQS of Φ iff (i) Ψ is a proper subformula of Φ of the form Q$p\,\Theta$, and (ii) there is no proper subformula Q$'q\,\Theta'$ of Φ such that Ψ is a subformula of Θ'. Finally, a subformula occurrence Ψ of a given formula Φ is said to have *positive* (*negative*) *polarity* (*in* Φ) iff Ψ occurs in the scope of an even (odd) number of explicit or implicit negations in Φ.

The semantics of QBFs is defined as follows, where $\Phi[p_1/\psi_1, \ldots, p_n/\psi_n]$ denotes the result of uniformly substituting the free variable occurrences of each p_i (with $i \in \{1, \ldots, n\}$) in Φ by a formula ψ_i. By an *interpretation*, M, we understand a set of variables. In general, the truth value, $\nu_M(\Phi)$, of a QBF Φ under an interpretation M is recursively defined as follows:

1. if $\Phi = \bot$, then $\nu_M(\Phi) = 0$;
2. if $\Phi = p$ is a variable, then $\nu_M(\Phi) = 1$ if $p \in M$, and $\nu_M(\Phi) = 0$ otherwise;
3. if $\Phi = \neg\Psi$, then $\nu_M(\Phi) = 1 - \nu_M(\Psi)$;
4. if $\Phi = (\Phi_1 \wedge \Phi_2)$, then $\nu_M(\Phi) = min(\{\nu_M(\Phi_1), \nu_M(\Phi_2)\})$;
5. if $\Phi = \forall p\,\Psi$, then $\nu_M(\Phi) = \nu_M(\Psi[p/\top] \wedge \Psi[p/\bot])$.

The truth conditions for $\top$, $\vee$, $\rightarrow$, and $\exists$ follow from the given ones in the usual way. We say that Φ is *true under* M iff $\nu_M(\Phi) = 1$, otherwise Φ is *false under* M. Observe that a *closed* QBF is either true under each interpretation or false under each interpretation. Hence, for closed QBFs, there is no need to refer to particular interpretations. Two QBFs, Φ_1 and Φ_2, are (logically) equivalent iff $\nu_M(\Phi_1) = \nu_M(\Phi_2)$, for every interpretation M.

In the sequel, we use the following abbreviations in the context of QBFs. First, for a set $P = \{p_1, \ldots, p_n\}$ of propositional variables and a quantifier Q $\in \{\forall, \exists\}$, we let $QP\Phi$ stand for the formula $Qp_1 Qp_2 \ldots Qp_n \Phi$. Similarly, for sets P_i, we also write $QP_1 P_2 \ldots P_n$ for $QP_1 QP_2 \ldots QP_n$. Moreover, let $S = \{\phi_1, \ldots, \phi_n\}$ and $T = \{\psi_1, \ldots, \psi_n\}$ be indexed sets of formulas. Then, $S \leq T$ abbreviates $\{\phi_i \rightarrow \psi_i \mid 1 \leq i \leq n\}$ and is identified with $\bigwedge_{i=1}^n (\phi_i \rightarrow \psi_i)$. Moreover, $S < T$ stands for $(\bigwedge_{i=1}^n (\phi_i \rightarrow \psi_i)) \wedge \neg(\bigwedge_{i=1}^n (\psi_i \rightarrow \phi_i))$.

A QBF Φ is *cleansed* iff (i) for any two distinct quantifier occurrences Qp and Qq in Φ, it holds that $p \neq q$, and (ii) if $Qp\,\Psi$ is a quantified subformula occurrence in Φ, then $p \notin free(\Phi)$ and $p \in free(\Psi)$. We assume in the following that all QBFs are cleansed. Observe that this is no restriction of the general case, as any QBF can be cleansed by simple manipulations, viz. by renaming of bound variables and deleting quantifiers Qp which do not have an occurrence of p in their scope.

A closed QBF Ψ of form $Q_1 P_1 \ldots Q_m P_m \psi$ is said to be in *prenex normal form* iff ψ is purely propositional. If, additionally, ψ is in conjunctive normal form, then Φ is said to be in *prenex conjunctive normal form* (PCNF). QBFs in prenex form lead to the prototypical decision problems for the polynomial hierarchy (cf. [13,15]). In particular, the evaluation problem of QBFs of form $\exists P_1 \forall P_2 \exists P_3 \ldots QP_i\phi$ is Σ_i^P-complete, where Q $= \exists$ if i is odd and Q $= \forall$ if i is even, and, dually, the evaluation problem of QBFs of form $\forall P_1 \exists P_2 \forall P_3 \ldots \bar{Q}P_i\phi$ is Π_i^P-complete, with Q as before.

The basis for the prenex strategies discussed in the next section is the following result, which gives a number of equivalence-preserving manipulations for QBFs.

Proposition 1. *Let Ψ_{Φ_1} be a cleansed QBF containing a subformula occurrence Φ_1, and let Ψ_{Φ_2} be the result of replacing that occurrence of Φ_1 in Ψ_{Φ_1} by Φ_2. Then, Ψ_{Φ_1} and Ψ_{Φ_2} are equivalent whenever Φ_1 and Φ_2 are one of the following forms:*

1. $\Phi_1 = (\neg Qp\,\Phi)$ *and* $\Phi_2 = \bar{Q}p\,(\neg\Phi)$;
2. $\Phi_1 = (Qp\,\Phi) \circ \Psi$ *and* $\Phi_2 = Qp\,(\Phi \circ \Psi)$ *(for $\circ \in \{\wedge, \vee\}$)*;
3. $\Phi_1 = (Qp\,\Phi) \rightarrow \Psi$ *and* $\Phi_2 = \bar{Q}p\,(\Phi \rightarrow \Psi)$;
4. $\Phi_1 = \Phi \circ (Qp\,\Psi)$ *and* $\Phi_2 = Qp\,(\Phi \circ \Psi)$ *(for $\circ \in \{\wedge, \vee, \rightarrow\}$)*;
5. $\Phi_1 = (\forall p\,\Phi) \wedge (\forall q\,\Psi)$ *and* $\Phi_2 = \forall p\,(\Phi \wedge \Psi[q/p])$;
6. $\Phi_1 = (\exists p\,\Phi) \vee (\exists q\,\Psi)$ *and* $\Phi_2 = \exists p\,(\Phi \vee \Psi[q/p])$.

Observe that the usual condition "*p is not free in* ..." is not necessary in Proposition 1 because we assume cleansed formulas.

The operation corresponding to 1.–4. is called *quantifier shifting*, whilst the operation corresponding to 5. and 6. is called *quantifier fusion*.

3   Strategies for Translations into Normal Form

In this section, we present the prenexing strategies used for our experiments. To this end, we first discuss prenexing strategies from a rather abstract point of view, and afterwards we introduce the particular strategies we are interested in.

We start with the following concepts.

Let Φ be a cleansed QBF. According to the definition of a cleansed QBF, if an atom p occurs bound in Φ, then Φ has exactly one occurrence of a subformula of the form $Qp\,\Psi$, where $Q \in \{\forall, \exists\}$. We denote this subformula occurrence by $\sigma(p, \Phi)$. Furthermore, we define

$$quant(\Phi) = \{Qp \mid p \text{ bound in } \Phi \text{ and } \sigma(p, \Phi) \text{ is positive in } \Phi\} \cup$$
$$\{\bar{Q}p \mid p \text{ bound in } \Phi \text{ and } \sigma(p, \Phi) \text{ is negative in } \Phi\}.$$

Definition 1. *Let $\mathcal{L}_c$ be the set of all cleansed QBFs over language $\mathcal{L}$ and let $\Phi \in \mathcal{L}_c$. The* set of q-paths *of Φ, $\mathcal{Q}(\Phi)$, is defined inductively as follows:*

1. *If Φ is a propositional variable, $\top$, or $\bot$, then $\mathcal{Q}(\Phi) = \{\epsilon\}$.*
2. *If $\Phi = \neg\Phi_1$, then $\mathcal{Q}(\Phi) = \{\bar{Q}_1 p_1 \ldots \bar{Q}_k p_k \mid Q_1 p_1 \ldots Q_k p_k \in \mathcal{Q}(\Phi_1)\}$.*
3. *If $\Phi = \Phi_1 \circ \Phi_2$ ($\circ \in \{\wedge, \vee\}$), then $\mathcal{Q}(\Phi) = \mathcal{Q}(\Phi_1) \cup \mathcal{Q}(\Phi_2)$.*
4. *If $\Phi = \Phi_1 \rightarrow \Phi_2$, then $\mathcal{Q}(\Phi) = \mathcal{Q}(\neg\Phi_1 \vee \Phi_2)$.*
5. *If $\Phi = Qp\,\Phi_1$ ($Q \in \{\forall, \exists\}$), then $\mathcal{Q}(\Phi) = \{Qps \mid s \in \mathcal{Q}(\Phi_1)\}$.*

Let $alt(s)$ be the number of quantifier alternations on a q-path s. Moreover, let $m_\Phi = \max\{alt(s) \mid s \in \mathcal{Q}(\Phi)\}$. The *set of critical paths* of Φ is given by $\mathcal{C}(\Phi) = \{s \mid alt(s) = m_\Phi, s \in \mathcal{Q}(\Phi)\}$.

Definition 2. *Let $\mathcal{L}_c$ be the set of all cleansed QBFs over language $\mathcal{L}$ and let $Q_\mathcal{L}$ be the set of all finite strings $Q_1 p_1 \ldots Q_n p_n$ of quantifiers over $\mathcal{L}$. A q-permutation (over $\mathcal{L}$) is a function $\pi: \mathcal{L}_c \rightarrow Q_\mathcal{L}$ such that, for each $\Phi \in \mathcal{L}_c$, $\pi(\Phi)$ is a string containing all the elements from $quant(\Phi)$ without repetitions. A pseudo q-permutation (over $\mathcal{L}$) is a function $\pi_1: \mathcal{L}_c \rightarrow Q_\mathcal{L}$, along with a function $\pi_1': \mathcal{P} \rightarrow \mathcal{P}$, such that, for each $\Phi \in \mathcal{L}_c$, $\pi_1(\Phi)$ is a string containing a proper subset S of all the elements from $quant(\Phi)$ without repetitions, and $\pi_1'(p) \in \{q \mid Qq \in S\}$, for each p with $Qp \in quant(\Phi) \setminus S$.*

We write $\pi[\![\Phi]\!]$ to denote the prenex QBF $\pi(\Phi)\mathrm{sk}(\Phi)$, resulting from $\mathrm{sk}(\Phi)$ by concatenating the quantifier string $\pi(\Phi)$. Likewise, we write $\pi_1[\![\Phi]\!]$ to denote the QBF $\pi_1(\Phi)(\mathrm{sk}(\Phi)[p_1/\pi_1'(p_1), \ldots, p_n/\pi_1'(p_n)])$, where $\{p_1, \ldots, p_n\}$ is the set of variables free in $\pi_1(\Phi)\mathrm{sk}(\Phi)$ but not in Φ. In what follows, we use Υ as a meta-variable standing for a q-permutation or a pseudo q-permutation.

Informally, pseudo q-permutations incorporate a renaming of variables in order to express quantifier fusion. More precisely, we define the following concepts:

Definition 3. *Let Υ be a q-permutation or a pseudo q-permutation. Then, Υ is a* prenexing strategy *iff, for each $\Phi \in \mathcal{L}_c$, $\Upsilon[\![\Phi]\!]$ and Φ are logically equivalent. We also say that a prenexing strategy Υ is based on quantifier shifting if Υ is a q-permutation, and, correspondingly, a prenexing strategy Υ is based on quantifier shifting and quantifier fusion if Υ is a pseudo q-permutation.*

Definition 4. *Let Φ be a cleansed QBF. Then, the binary relation $\rhd_\Phi \subseteq quant(\Phi) \times quant(\Phi)$, given by the condition that $Q_1 p_1 \rhd_\Phi Q_2 p_2$ iff $\sigma(p_2, \Phi)$ is a direct quantified subformula of $\sigma(p_1, \Phi)$, is called the* quantifier ordering *for Φ. We say that a quantifier ordering $\rhd_\Psi$ is* compatible *with a quantifier ordering $\rhd_\Phi$ iff $\mathsf{sk}(\Psi) = \mathsf{sk}(\Phi)$ and $\rhd_\Psi \subseteq (\rhd_\Phi)^+$, where $(\rhd_\Phi)^+$ denotes the transitive closure of $\rhd_\Phi$.*

Since the computational effort to evaluate QBFs is related to the number of quantifier alternations in a given QBF, one aims, in general, to use strategies which yield prenex QBFs containing a minimal number of quantifier alternations. One can show that, for a given cleansed QBF Φ containing n quantifier alternations, there exists a prenex form having either n or $n + 1$ quantifier alternations. In fact, the latter case happens providing Φ has at least two critical paths with different starting quantifiers. QBFs of this kind are called *D-formulas* in [6]. We note that the evaluation problem of D-formulas with n quantifier alternations is complete for the complexity class D_{n+1}^P rather than for Σ_{n+1}^P or Π_{n+1}^P.

In the light of the above discussion, we call a prenexing strategy Υ *minimal* iff, for any cleansed QBF Φ having n quantifier alternations, the number of quantifier alternations in $\Upsilon[\![\Phi]\!]$ is either n or $n + 1$, depending on the structure of the formula.

In what follows, we restrict our attention to (closed) QBFs which are *not* D-formulas. If we have a closed D-formula, then it is more appropriate to use a decomposition of the D-formula into Σ-formulas and Π-formulas and to solve each of these formulas independently. The truth value for the D-formula can then be computed from the solutions using truth tables.

We introduce four elementary prenexing strategies based on quantifier shifting which were used for our experiments; in fact, each of these strategies is a minimal q-permutation. The strategies we are interested in are denoted by $\exists^\uparrow \forall^\uparrow$, $\exists^\uparrow \forall^\downarrow$, $\exists^\downarrow \forall^\uparrow$, and $\exists^\downarrow \forall^\downarrow$, and rely on a different treatment of existential and universal quantifiers. Intuitively, the four strategies work as follows. Consider the formula tree for a given QBF. Roughly speaking, $\exists^\uparrow$ (resp., $\exists^\downarrow$) means that existential quantifiers are shifted in such a way that they are placed in the resultant quantifier prefix as high (resp., low) as possible (under the proviso that we just apply the replacements for quantifier shifting from Proposition 1) yielding a quantifier ordering which is compatible with the original one; a similar meaning applies to $\forall^\uparrow$ and $\forall^\downarrow$.

For illustration, consider the cleansed QBF

$$\Phi = \exists p \Big(\forall q \, \exists r \, \forall s \, \exists t \, \varphi_0 \wedge \forall q' \, \exists r' \, \varphi_1 \wedge \neg \forall q'' \, \exists r'' \, \varphi_2 \Big),$$

where $\varphi_0, \varphi_1, \varphi_2$ are propositional formulas. Hence, the propositional skeleton of Φ is given by $\mathsf{sk}(\Phi) = \varphi_0 \wedge \varphi_1 \wedge \neg \varphi_2$. Furthermore, the quantifier ordering $\rhd_\Phi$ implies the following chains:

$$\exists p \rhd_{\Phi} \forall q \rhd_{\Phi} \exists r \rhd_{\Phi} \forall s \rhd_{\Phi} \exists t, \quad \exists p \rhd_{\Phi} \forall q' \rhd_{\Phi} \exists r', \text{ and } \quad \exists p \rhd_{\Phi} \exists q'' \rhd_{\Phi} \forall r''.$$

Observe that the quantifiers $\forall q''$ and $\exists r''$ in Φ are inverted in $\rhd_{\Phi}$ due to their negative polarity in Φ. The set of critical paths of Φ is $\mathcal{C}(\Phi) = \{\exists p \forall q \exists r \forall s \exists t\}$. Finally, as the resultant prenex forms of Φ we obtain:

$$\exists^{\uparrow} \forall^{\uparrow} [\![\Phi]\!] = \exists p q'' \; \forall q q' r'' \; \exists r r' \; \forall s \; \exists t \; (\varphi_0 \wedge \varphi_1 \wedge \neg \varphi_2);$$

$$\exists^{\uparrow} \forall^{\downarrow} [\![\Phi]\!] = \exists p q'' \; \forall q q' \; \exists r r' \; \forall s r'' \; \exists t \; (\varphi_0 \wedge \varphi_1 \wedge \neg \varphi_2);$$

$$\exists^{\downarrow} \forall^{\uparrow} [\![\Phi]\!] = \exists p \; \forall q q' \; \exists r q'' \; \forall s r'' \; \exists t r' \; (\varphi_0 \wedge \varphi_1 \wedge \neg \varphi_2);$$

$$\exists^{\downarrow} \forall^{\downarrow} [\![\Phi]\!] = \exists p \; \forall q \; \exists r q'' \; \forall s q' r'' \; \exists t r' \; (\varphi_0 \wedge \varphi_1 \wedge \neg \varphi_2).$$

Note that the overall structure of the above strategies leaves room for different variants. Let us consider the strategy $\exists^{\uparrow} \forall^{\downarrow}$ as an example. The formula $\exists^{\uparrow} \forall^{\downarrow} [\![\Phi]\!]$ is constructed with the preference to place $\exists$ quantifiers as high as possible, even if some $\forall$ quantifiers are not placed in an optimal way. Another preference is to place $\forall$ quantifiers as low as possible, even if some $\exists$ quantifiers are not placed in an optimal way. Under the latter method, we get

$$\exists^{\uparrow} \forall^{\downarrow} [\![\Phi]\!] = \exists p q'' \; \forall q \; \exists r \; \forall s q' r'' \; \exists t r' \; (\varphi_0 \wedge \varphi_1 \wedge \neg \varphi_2).$$

For our specific class of benchmarks employed in our experiments, however, these differences do not matter.

4   Benchmarks

We now describe a class of benchmark problems used for evaluating the prenexing strategies introduced above. These problems are given as QBF encodings for reasoning with nested counterfactuals, obtained from a model of counterfactual reasoning as investigated in [7]. In contrast to QBF encodings of other nonmonotonic inference tasks (cf., e.g., [5]), the current problems span the entire polynomial hierarchy.

Intuitively, a counterfactual, $p > q$, where p and q are atoms, is true over a propositional theory T iff the minimal change of T to incorporate p entails q. This is naturally generalised for nested counterfactuals in a straightforward way.

Formally, following [7], the language of (right-)nested counterfactuals is recursively defined as follows: (i) if ϕ and ψ are propositional formulas, then $\phi > \psi$ is a nested counterfactual, and (ii) if c is a nested counterfactual, then $\neg c$ and $(\phi > c)$ are nested counterfactuals, where ϕ is a propositional formula. For brevity, we usually write $\phi \not> d$ instead of $\neg(\phi > d)$, where d is a propositional formula or a counterfactual. The *nesting depth*, $nd(c)$, of a counterfactual c is given by the number of occurrences of the symbol ">" in c minus one.

For defining the semantics of nested counterfactuals, let T be a propositional theory and ϕ a propositional formula. First, let

$$\mathcal{F}(\phi, T) = \{S \mid S \subseteq T, S \not\models \neg \phi, S \subset U \subseteq T \Rightarrow U \models \neg \phi\},$$

which gives the maximal subsets of T consistent with ϕ. Then for a counterfactual $c = \phi > d$ (with d being either a formula or a nested counterfactual), we define $T \models c$ iff, for every $S \in \mathcal{F}(\phi, T)$, $S \cup \{\phi\} \models d$. Finally, $T \models \neg c$ iff $T \not\models c$.

As shown in [12], the complexity of deciding whether $T \models \phi > \psi$ (ϕ and ψ propositional formulas) holds is Π_2^P-complete. Eiter and Gottlob [7] extended this result by showing that Π_2^P-completeness also holds for nested counterfactuals without negation, whereas, in the general case, deciding $T \models c$ is Π_{k+2}^P-complete if $nd(c) = k$ and PSPACE-complete if the nesting depth of c is unbounded.

This model of counterfactual reasoning can be described in terms of QBFs in the following way.

Definition 5. *Let* $c = \phi_0 \succ_0 (\phi_1 \succ_1 (\ldots \succ_{k-1} (\phi_k \succ_k \psi) \ldots))$ *be a nested counterfactual with* $nd(c) = k$ *and* $\succ_i \in \{>, \not\succ\}$, *and let* $T = \{\varphi_1, \ldots, \varphi_n\}$ *be a theory. Furthermore, let* $S_i = \{s_{i,1}, \ldots, s_{i,n+i}\}$ *and* $U_i = \{u_{i,1}, \ldots, u_{i,n+i}\}$ *be sets of new atoms, for* $0 \leq i \leq k$, *and define* $\mathcal{M}_0[T, c] = S_0 \leq T$ *and, for* $0 < j \leq k$,

$$\mathcal{M}_j[T, c] = S_j \leq \left(\mathcal{M}_{j-1}[T, c] \cup \{\phi_{j-1}\} \right).$$

For $0 \leq i \leq k$, *we furthermore define*

$$\Phi_i = \exists V_i \left(\mathcal{M}_i[T, c] \wedge \phi_i \right) \wedge \forall U_i \left((S_i < U_i) \rightarrow \forall V_i((\mathcal{M}_i[T, c])[S_i/U_i] \rightarrow \neg \phi_i) \right),$$

where each V_i *denotes the set of atoms occurring in* $T \cup \{\phi_0, \ldots, \phi_i\}$, *as well as*

$$\Psi_i = \begin{cases} \forall S_i(\Phi_i \rightarrow \Psi_{i+1}) & \text{if } \succ_i = >; \\ \exists S_i(\Phi_i \wedge \neg \Psi_{i+1}) & \text{if } \succ_i = \not\succ, \end{cases}$$

with $\Psi_{k+1} = \forall W((\mathcal{M}_k[T, c] \wedge \phi_k) \rightarrow \psi)$, *where* W *contains all atoms from* V_k *and those occurring in* ψ. *Then the desired encoding,* $\mathcal{T}[T, c]$, *is given by* Ψ_0.

Theorem 1. $T \models c$ *iff* $\mathcal{T}[T, c]$ *is true.*

Observe that $\mathcal{T}[T, c]$ is not a cleansed QBF. However, by renaming the atoms in V_k and W, it is straightforward to get a cleansed version. From now on, we tacitly assume that $\mathcal{T}[T, c]$ refers to such a cleansed formula.

For our experiments, we used, on the one hand, the encodings for the inference tasks

$$T \models p_0 \not\succ (p_1 \not\succ (\ldots \not\succ (p_k \not\succ q) \ldots)), \tag{1}$$

with k odd, for obtaining QBFs with complexity in Σ_{k+2}^P, and, on the other hand, the encodings for the tasks

$$T \models p_0 > (p_1 \not\succ (\ldots \not\succ (p_k \not\succ q) \ldots)), \tag{2}$$

with k even, for QBFs with complexity in Π_{k+2}^P, where $\{p_0, \ldots, p_k, q\}$ are atoms occurring in T.

For illustration, let us consider the encoding $\mathcal{T}[T, c]$ of Task (1) with $k = 3$, which can be written as

$$\exists S_0(\Phi_0 \wedge \forall S_1(\Phi_1 \rightarrow \exists S_2(\Phi_2 \wedge \forall S_3(\Phi_3 \rightarrow \forall W((\mathcal{M}_3[T, c] \wedge p_3) \rightarrow q))))).$$

The quantifier ordering for the QBF $\mathcal{T}[T, c]$ is graphically represented as follows, where $\exists S_0 \forall S_1 \exists S_2 \forall S_3 \exists U_3 V_3'$ is the (only) critical path:

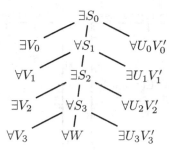

$$\exists S_0$$
$$\exists V_0 \qquad \forall S_1 \qquad \forall U_0 V_0'$$
$$\forall V_1 \qquad \exists S_2 \qquad \exists U_1 V_1'$$
$$\exists V_2 \qquad \forall S_3 \qquad \forall U_2 V_2'$$
$$\forall V_3 \qquad \forall W \qquad \exists U_3 V_3'$$

Applying our four basic shifting strategies to the above QBF yields prenex QBFs of form $\exists P_0 \forall P_1 \exists P_2 \forall P_3 \exists P_4 \psi$, where the sets P_i are given as follows:

Strategy	P_0	P_1	P_2	P_3	P_4
$\exists^{\downarrow}\forall^{\downarrow}$	S_0	S_1	S_2	$S_3 U_0 V_0' V_1 U_2 V_2' V_3 W$	$V_0 U_1 V_1' V_2 U_3 V_3'$
$\exists^{\uparrow}\forall^{\uparrow}$	$S_0 V_0$	$S_1 U_0 V_0' V_1$	$S_2 U_1 V_1' V_2$	$S_3 U_2 V_2' V_3 W$	$U_3 V_3'$
$\exists^{\uparrow}\forall^{\downarrow}$	$S_0 V_0$	S_1	$S_2 U_1 V_1' V_2$	$S_3 U_0 V_0' V_1 U_2 V_2' V_3 W$	$U_3 V_3'$
$\exists^{\downarrow}\forall^{\uparrow}$	S_0	$S_1 U_0 V_0' V_1$	S_2	$S_3 U_2 V_2' V_3 W$	$V_0 U_1 V_1' V_2 U_3 V_3'$

5 Experimental Evaluation

In this section, we report experimental results based on the benchmark problems discussed above. We apply the four strategies $\exists^{\downarrow}\forall^{\downarrow}$, $\exists^{\uparrow}\forall^{\uparrow}$, $\exists^{\uparrow}\forall^{\downarrow}$, and $\exists^{\downarrow}\forall^{\uparrow}$ to the encodings for Tasks (1) and (2). In particular, the encoded tasks are obtained from theories T consisting of randomly generated formulas in CNF, containing 4 to 10 clauses of length 3. The atoms for the theories are selected from a set of 4 to 8 variables and are negated with probability 0.5. Moreover, all variables p_i in the counterfactual are selected from this set as well. The sample size for each parameter setting is fixed to 100 instances.

In total, we used 10 500 inference problems for nested counterfactuals. Each instance of such a problem gives rise to four different prenex QBFs, corresponding to the four chosen prenexing strategies. In order to obtain QBFs in PCNF, we translate the propositional skeleton of the encodings into a CNF using a structure-preserving normal-form translation. Variables which are introduced by the latter translation are existentially quantified, which does not change the overall number of quantifier alternations. The resultant QBFs contain 369 to 952 variables and their matrices consist of 662 to 1832 clauses. For the tests we focus on the number of variables rather than on the number of clauses, since the former number turned out to be in a closer relation to the resulting running times of the solvers. We group the results into two main categories, viz. the results for false instances and the results for true instances. Within each main category, we distinguish between the number of quantifier alternations (ranging between 7 and 9) stemming from the nesting depth of the counterfactual used in (1) and (2). Finally, we mention that around 75% of the 10 500 instances evaluate to true.

In the diagrams (Fig. 1–4), the x-axis gives the number of variables in the QBF and the y-axis gives the running time (in seconds). Observe that the y-axis is scaled logarithmically. For each number of quantifier alternations considered, we give the running times for true and false instances separately. QBFs which encode the tasks with smaller

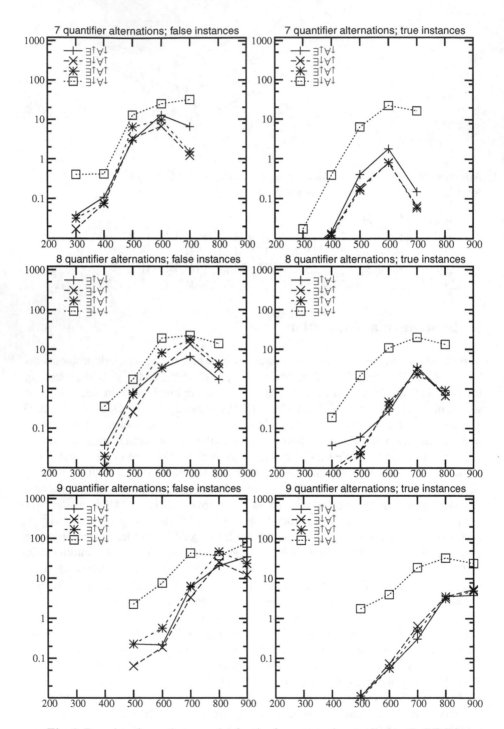

Fig. 1. Running times (in seconds) for the four strategies applied to QuBE-BJ.

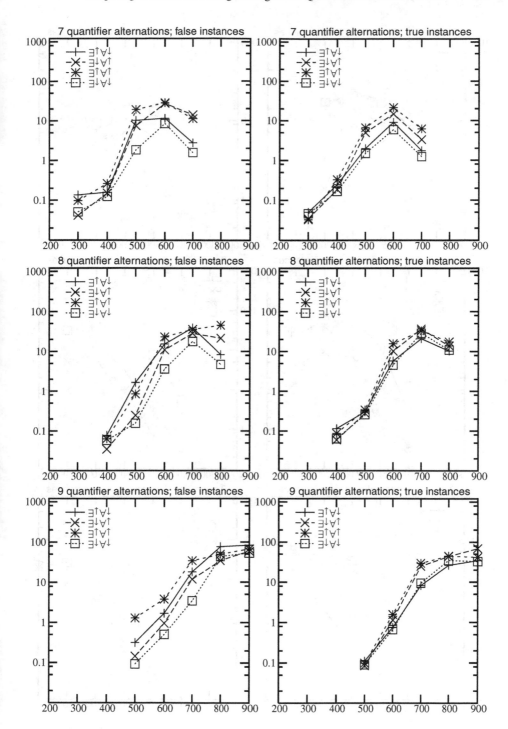

Fig. 2. Running times (in seconds) for the four strategies applied to `semprop`.

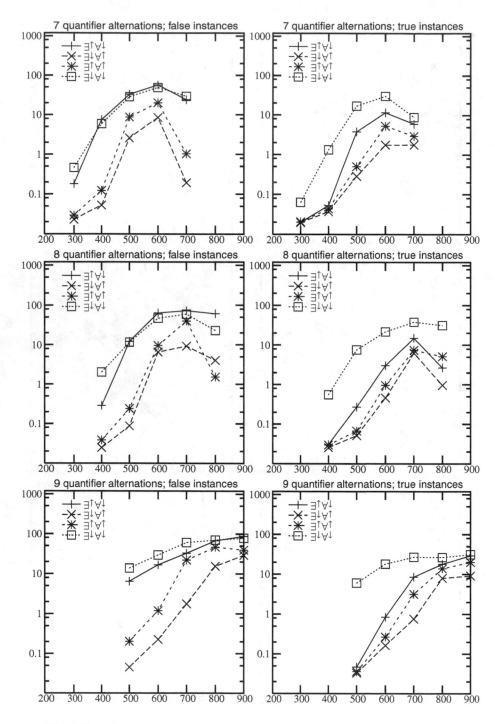

Fig. 3. Running times (in seconds) for the four strategies applied to `ssolve`.

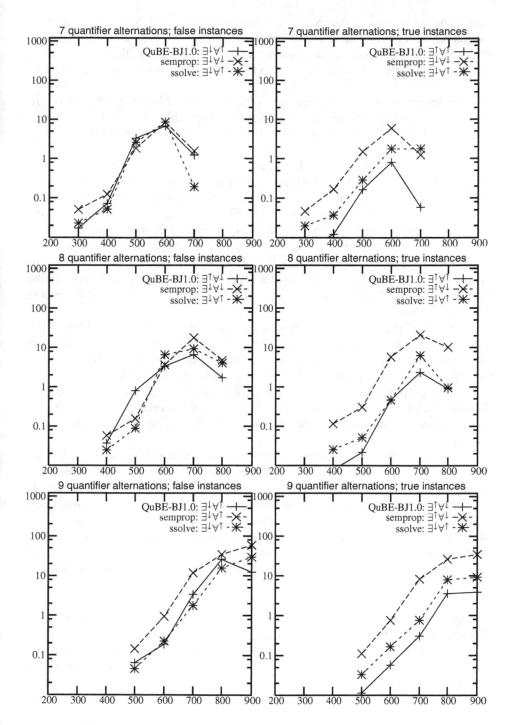

Fig. 4. Running times (in seconds) for each solver's best strategy.

nesting depths possess in general less variables than QBFs encoding tasks of counter-factual reasoning with a greater depth. In the diagrams, we clustered the benchmarks with respect to the number of variables occurring in the *resulting* QBFs. For instance, the data for the point "400 variables" results from the average running time for all QBFs having 350 to 449 variables.

Our experiments were carried out with the QBF-solvers[1] QuBE-BJ1.0 [9], semprop v24.02.02 [11], and ssolve version 5 [8], using a Linux PC with a Pentium III 733 processor and 256MB of main memory. The parameters for the solvers were set to default except for the timeout, which was set to 1200 seconds and was reached by 1906 instances; this is less than 1,5% of the number of test runs which is given by $126\,000 = 10\,500 \times 3$ (solvers) $\times 4$ (strategies). In Fig. 1–3, we compare the impact of the different prenexing strategies to the running time of the solvers QuBE-BJ1.0, semprop, and ssolve, respectively. Finally, in Fig. 4, we give a comparison between the three solvers by choosing, for each solver, the best strategy within each of the six groups of tests. From these figures, we get the following basic observations.

Starting with Fig. 1, we first remark that, for QuBE-BJ, the strategy $\exists^\downarrow\forall^\downarrow$ performs significantly worse than the other strategies. Among them, $\exists^\downarrow\forall^\uparrow$ is on the average the best choice. We also see that, except for $\exists^\downarrow\forall^\downarrow$, the true instances are solved faster than the false instances.

When invoking semprop, we get a quite different picture in Fig. 2. Here, $\exists^\downarrow\forall^\downarrow$ turns out to be the most efficient way to solve the instances. However, here we do not have a significant difference between all the four strategies. Moreover, the previously observed shorter running times on the true instances is not achieved by semprop.

Finally, the performance of ssolve is depicted in Fig. 3. Like in the case of QuBE-BJ, the strategy $\exists^\downarrow\forall^\downarrow$ performs bad but also $\exists^\uparrow\forall^\downarrow$ is not a good choice, at least for the false instances. In general, $\exists^\downarrow\forall^\uparrow$ outperforms the other strategies, similar as it is the case for QuBE-BJ.

For a final comparison between the QBF-solvers considered, we compare the running times of the best strategy of each of the three solvers in Fig. 4. While for the false instances, we see the solvers head-to-head, semprop loses for the true instances. In fact, in each curve we depicted the running times for that strategy which was the best for the respective solver on this set of instances. Observe that for ssolve, the selected strategy was always $\exists^\downarrow\forall^\uparrow$, while this picture is not so clear for the other solvers. More specifically, for semprop, $\exists^\downarrow\forall^\downarrow$ was best on three sets, $\exists^\uparrow\forall^\downarrow$ was best on two sets, and $\exists^\downarrow\forall^\uparrow$ was on the average the best choice on the remaining set. For QuBE-BJ, each of the strategies $\exists^\downarrow\forall^\uparrow$, $\exists^\uparrow\forall^\downarrow$, and $\exists^\uparrow\forall^\uparrow$ was on the average the best strategy on two test sets.

As one can easily recognise from the results, different prenexing strategies strongly influence the time for deciding the resultant QBFs; one order of magnitude between a "good" and a "bad" prenexing strategy is not an exception. Generally, our benchmark results do not indicate *the* best prenexing strategy; the results for a strategy depend also on the employed solver. In particular, we have that strategy $\exists^\downarrow\forall^\uparrow$ seems to be best for QuBE-BJ (at least on the true instances, for the false instances the picture is not that

---

[1] These solvers were the most successful in the QBF evaluation at SAT-03. See the URL http://www.lri.fr/~simon/qbf03/live/ for the results.

clear), while $\exists^{\downarrow}\forall^{\downarrow}$ is best suited if semprop is applied, for both the true and false instances. Finally, ssolve performs best with $\exists^{\downarrow}\forall^{\uparrow}$ for true and false instances.

Broadly spoken, the results are intriguing to some extent as the prenexing strategy influences the solving time *in different ways across different solvers*. While the performances of ssolve and QuBE-BJ seem to be influenced by different strategies, semprop is rather unsensitive to different prenexing methods. A partial explanation for this behaviour may be given by the fact that semprop involves, besides back-jumping, smart learning techniques, whereas QuBE-BJ implements back-jumping and trivial truth, but no learning, and ssolve uses yet another form of trivial truth and trivial falsity, but no advanced look-back. Clearly, further distinguishing classes of benchmark problems are needed in order to obtain a clearer picture on the interplay between different prenexing strategies and different solvers.

6   Conclusion and Future Work

In this paper, we addressed the influence of different prenexing strategies for quantified Boolean formulas on the evaluation time for a class of structured benchmark problems. To this end, we defined a general framework for expressing prenexing strategies in which various techniques like quantifier shifting, quantifier fusion, or mini-scoping can be easily expressed. For our experiments, we chose four very elementary prenexing strategies, based on quantifier shifting, which can be seen as core techniques for more advanced methods. Furthermore, we used three well-known QBF solvers as underlying evaluation engines, viz. the systems QuBE-BJ, semprop, and ssolve.

The considered benchmark formulas encode reasoning tasks for nested counterfactuals, spanning the entire polynomial hierarchy. As an aside, these (polynomial-time constructible) encodings realise also, to best of our knowledge, the first implementation for computing inference problems for nested counterfactuals.

The rationale of our experiments is the observation that the "natural" representation of many reasoning tasks in terms of QBFs usually results in formulas which do not adhere to a specific normal form. Hence, in order to apply existing QBF solvers, suitable normal-form translation schemas (and especially prenexing strategies) are required. Our results show that, in general, there is not a single best strategy. Rather, the prenexing strategy influences the solving time in different ways across different solvers. Interestingly, one of the considered solvers, semprop, is rather robust under the chosen benchmark set and prenexing strategies.

Future work will focus on four topics. First, we will have to investigate more problem classes (like, e.g., nested circumscription [3]) which span also the polynomial hierarchy and which yield QBFs of highly non-prenex nature. We expect that a broader range of benchmarks gives us a clearer picture of the impact of using different strategies and allows for a better judgment of the observed effects.

Second, we will compare the approach using prenex normal forms with the approach using provers which do not require the input formula being in a specific normal form. Currently, such a comparison is hampered by the fact that the available successful system QUBOS [1] showed some inconsistencies with our benchmark formulas. An improved release will therefore be required.

Third, we will extend the experiments to prenexing strategies based on quantifier shifting *and* quantifier fusion. While quantifier shifting linearises the quantifiers distributed in the formula *without* changing the number of quantifiers, applications of quantifier fusion *decrease* this number.

Fourth, we will also investigate how quantifier shifting and fusion can be integrated into the translation to conjunctive normal form. Additionally, it is interesting to see how more involved techniques for reducing the quantifier dependencies can be incorporated. One example of such a technique (for classical first-order logic) is Goubault's advanced skolemisation technique [10], which yields smaller Skolem function symbols (i.e., Skolem functions depending on less variables).

Acknowledgements. The authors would like to thank the anonymous referees for their valuable comments which helped improving the paper.

References

1. A. Ayari and D. Basin. QUBOS: Deciding Quantified Boolean Logic Using Propositional Satisfiability Solvers. In *Proc.* FMCAD-02, vol. 2517 of *LNCS*, pp. 187–201. Springer, 2002.
2. M. Baaz and A. Leitsch. On Skolemization and Proof Complexity. *Fundamenta Informaticae*, 20:353–379, 1994.
3. M. Cadoli, T. Eiter, and G. Gottlob. Complexity of Nested Circumscription and Abnormality Theories. In *Proc.* IJCAI-01, pp. 169–174. Morgan Kaufmann, 2001.
4. U. Egly. Quantifiers and the System KE: Some Surprising Results. In *Proc.* CSL-98, vol. 1584 of *LNCS*, pp. 90–104. Springer, 1999.
5. U. Egly, T. Eiter, H. Tompits, and S. Woltran. Solving Advanced Reasoning Tasks Using Quantified Boolean Formulas. In *Proc.* AAAI-00, pp. 417–422. AAAI/MIT Press, 2000.
6. U. Egly, H. Tompits, and S. Woltran. On Quantifier Shifting for Quantified Boolean Formulas. In *Proc. SAT-02 Workshop on Theory and Applications on QBFs*, pp. 48–61, 2002.
7. T. Eiter and G. Gottlob. The Complexity of Nested Counterfactuals and Iterated Knowledge Base Revisions. *Journal of Computer and System Sciences*, 53(3):497–512, 1996.
8. R. Feldmann, B. Monien, and S. Schamberger. A Distributed Algorithm to Evaluate Quantified Boolean Formulas. In *Proc.* AAAI-00, pp. 285–290. AAAI/MIT Press, 2000.
9. E. Giunchiglia, M. Narizzano, and A. Tacchella. QuBE: A System for Deciding Quantified Boolean Formulas Satisfiability. In *Proc.* IJCAR-01, vol. 2083 of *LNCS*, pp. 364–369. Springer, 2001.
10. J. Goubault. A BDD-Based Simplification and Skolemization Procedure. *Logic Journal of the IGPL*, 3(6):827–855, 1995.
11. R. Letz. Lemma and Model Caching in Decision Procedures for Quantified Boolean Formulas. In *Proc.* TABLEAUX-02, vol. 2381 of *LNCS*, pp. 160–175. Springer, 2002.
12. B. Nebel. Belief Revision and Default Reasoning: Syntax-Based Approaches. In *Proc.* KR-91, pp. 417–428. Morgan Kaufmann, 1991.
13. C. H. Papadimitriou. *Computational Complexity*. Addison-Wesley, Reading, Mass., 1994.
14. J. Rintanen. Constructing Conditional Plans by a Theorem Prover. *Journal of Artificial Intelligence Research*, 10:323–352, 1999.
15. L. J. Stockmeyer. The Polynomial-Time Hierarchy. *Theoretical Computer Science*, 3(1):1–22, 1977.
16. L. Zhang and S. Malik. Towards a Symmetric Treatment of Satisfaction and Conflicts in Quantified Boolean Formula Evaluation. In *Proc.* CP-02, vol. 2470 of *LNCS*, pp. 200–215. Springer, 2000.

Solving Error Correction for Large Data Sets by Means of a SAT Solver

Renato Bruni

Università di Roma "La Sapienza", Dip. di Informatica e Sistemistica,
Via Michelangelo Buonarroti 12 - 00185 Roma, Italy,
bruni@dis.uniroma1.it

Abstract. The paper is concerned with the problem of automatic detection and correction of erroneous data into large datasets. The adopted process should therefore be computationally efficient. As usual, errors are here defined by using a rule-based approach: all and only the data records respecting a set of rules are declared correct. Erroneous records should afterwards be corrected, by using as much as possible the correct information contained in them. By encoding such problem into propositional logic, for each erroneous record we have a propositional logic formula, for which we want a model having particular properties. Correction problems can therefore be quickly solved by means of a customized SAT solver. Techniques for an efficient encoding of difficult cases are presented.

1 Introduction

Despite (or, perhaps, due to) the easiness with which data are nowadays available, their correctness often is a very relevant problem. Erroneous data should in fact be *detected* and possibly *corrected* by means of automated techniques. When dealing with large amount of information, moreover, not only the correctness of the correction procedure is needed, but also its computational efficiency.

As customary for structured information, data are organized into *records*. The problem of error detection is generally approached by formulating a set of *rules* that every record must respect in order to be declared *correct*. Records not respecting all the rules are declared *erroneous*. Such rule-based approach has several advantages (e.g. flexibility, intelligibility) on other methods. In the field of database theory, rules are also called *integrity constraints* [18], whereas in the field of statistics, rules are also called *edits* [8], and express the error condition. The study of such rules is a central problem for the areas of data mining and data analysis [4,13,14]. The problem of *error correction* is usually tackled by changing some of the values of an erroneous record, in order to obtain a *corrected record* which satisfies the above rules, according to the two following criteria: a) the corrected record should be as *similar* as possible to the erroneous record; b) the correction operation should respect the *frequency distributions* of the data. This is deemed to produce a record which should be as close as possible to the (unknown) *original record* (the record that would be present in absence of errors). In the field of statistics, the correction process is also called *data imputation.* Several different rules encoding and solution algorithm have been proposed (e.g. [3,17]). A very well-known approach to the problem, which implies the generation of all rules logically implied by the initial set of rules, is due to Fellegi and Holt [8]. In practical case, however, such methods suffer from severe computational limitations [17]. In the field of computer science, the correction process is also called *data cleaning.* Errors may be detected as inconsistencies in knowledge representation, and corrected with consistency restoring techniques [1,15,19]. Another approach to error correction, in database theory, consists in performing a cross validation among multiple sources of the same information [18]. A previous work [5] describes the case when rules are encoded into *linear inequalities.* The correction is therefore achieved by solving MILP problems by means of a *branch-and-cut* procedure (ILOG Cplex). In another work [6] error localization and error correction problems are modeled as *set covering* problems and solved by means of the *volume algorithm* compared to *branch-and-bound.*

An automatic procedure for performing generic data correction by using only *propositional logic* is here presented. The above described problems are therefore solved by solving a variant of *propositional satisfiability* (SAT) problems. (see e.g. [7,10,12,22] for extensive references). SAT solvers available today are the result of decades of research work and are deemed to be among the fastest NP-complete problem-specific solvers (see for instance computational results in [20,21]). For this reason, the computational issue appears to be tackled in a satisfactory manner.

A formalization of data and rules is given in Sect. 2. In the same section is also sketched the basic procedure for propositional logic encoding of rules, already outlined in [6]. There are, however, cases when the basic procedure suffers from an unpleasant growth of the number of clauses and variables. We therefore analyze, in Sect. 3, such cases, and present new techniques to overcome the above problems. After this, the correction problem is formalized, and a complexity analysis is given, in Sect. 4. An efficient new correction procedure, implementable by introducing minor modifications into a generic DPLL SAT solver, is then reported. Such modifications are actually introduced in the recent SAT solver named SIMO (Satisfiability Internal Module Object oriented) [11]. Computational results on datasets of two different origins (*population* data and *television transmitters* data) are reported in Sect. 5.

2 General Propositional Logic Encoding

In Database theory, a *record schema* is a finite set of fields f_i, with $i = 1 \ldots l$. A *record instance* is a set of values v_i, one for each of the above fields [18]. In order to help exposition, we will focus on records representing *persons,* which are somehow more intuitively understandable. Note, however, that the proposed procedure is completely general, being not influenced by the meaning of treated data. The record scheme will be denoted by P, whereas a corresponding record instance will be denoted by p.

$$P = \{f_1, \ldots, f_l\} \qquad p = \{v_1, \ldots, v_l\}$$

Example 2.1. For persons, fields can be for instance `age` or `marital status`, and corresponding examples of values can be 18 or `single`.

Each field f_i has a *domain* D_i, with $i = 1 \ldots l$, which is the set of possible values for that field. Since we are dealing with errors, the domains include all values that can be found in data, even the erroneous ones. Fields are usually distinguished in *quantitative* and *qualitative* ones. A quantitative field is a field on whose values are applied (at least some) mathematical operators (e.g. $>$, $+$), hence such operators should be defined on its domain. Examples of quantitative fields are numbers (real or integer), or even the elements of an ordered set. A qualitative field simply requires its domain to be a discrete set with finite number of elements. The case of fields having a not-finite number of not-ordered values is generally ignored. The proposed approach is able to deal with both qualitative and quantitative values.

Example 2.2. For the qualitative field `marital status`, answer can vary on a discrete set of values, or, due to errors, be missing or not meaningful (`blank`).

$$D_{\texttt{marital status}} = \{\texttt{single}, \texttt{married}, \texttt{separate}, \texttt{divorced}, \texttt{widow}, \texttt{blank}\}$$

For the quantitative field `age`, due to errors, the domain is $D_{\texttt{age}} = \mathbb{Z} \cup \{\texttt{blank}\}$

A record instance p is declared correct if and only if it respects a *set of rules* denoted by $R = \{r_1, \ldots, r_t\}$. Each rule can be seen as a mathematical function r_h from the Cartesian product of all the domains to the Boolean set $\{0,1\}$.

$$r_h : D_1 \times \ldots \times D_l \to \{0, 1\}$$

Rules are such that p is a correct record if and only if $r_h(p) = 1$ for all $h = 1 \ldots t$.

Example 2.3. An error within a person record can be the following one:

$$\texttt{marital status} = \texttt{married} \quad \text{and} \quad \texttt{age} = 10 \text{ years old}$$

The rule to detect this kind of errors could be: if `marital status` is `married`, `age` must be not less than, say, 14.

Rules can be obtained from various sources (e.g. human expertise, machine learning, see also [4,13,14]). We consider the set of rules already given, and intuitively expressed as logical connections (representable with $\wedge$, $\vee$, $\Rightarrow$, etc.) among *statements*. Each statement involves a single field f_i, and consist in a set of values from its domain D_i.

Example 2.4. A statement can be (`marital status` = `married`). Another statement can be (`age` $\geq$ 14).

We now describe the basic procedure for encoding rules into clauses. A *clause* C_u is a disjunction of possibly negated propositional variables x_v. We assume the reader familiar with propositional logic. By denoting with π_u the set of indices of the positive variables of C_u, and with ν_u that one of negative variables of C_u, the clause is:

$$\bigvee_{v \in \pi_u} x_v \vee \bigvee_{v \in \nu_u} \neg x_v$$

Values appearing in the rules are called *breakpoints*, or *cut points*, for the domains. All breakpoints concerning domain D_i represent logical *watershed* between the values of domain D_i. Their set will be denoted with B_i, as follows:

$$B_i = \{b_i^1, \ldots, b_i^{k'_i}\}$$

The set B_i determines a partition of domain D_i into subsets S_i^s. By furthermore merging possibly equivalent subsets, we obtain the following collection of subsets:

$$\mathcal{S}_i = \{S_i^1, \ldots, S_i^{k_i}\}$$

All and only values belonging to the same subset are *equivalent* from the rules' point of view. We congruently have $D_i = (S_i^1 \cup \ldots \cup S_i^{k_i})$. A subset for the *out-of-range* values is always present, while other subsets form the *feasible* domain.

Example 2.5. Suppose that, by scanning a given set of rules R, the following set of breakpoints B_{age} is obtained for the field `age` of a person.

$$\{b_{\text{age}}^1 = 0,\ b_{\text{age}}^2 = 14,\ b_{\text{age}}^3 = 18,\ b_{\text{age}}^4 = 26,\ b_{\text{age}}^5 = 110,\ b_{\text{age}}^6 = \texttt{blank}\}$$

By using B_{age} and R, the subsets are obtained. S_{age}^5 is the out-of-range one.

$$S_{\text{age}}^1 = \{0, \ldots, 13\},\ S_{\text{age}}^2 = \{14, \ldots, 17\},\ S_{\text{age}}^3 = \{18, \ldots, 25\}$$
$$S_{\text{age}}^4 = \{26, \ldots, 110\},\ S_{\text{age}}^5 = \{\ldots, -1\} \cup \{111, \ldots\} \cup \{\texttt{blank}\}$$

So far, subsets can be encoded with propositional variables in several manners (for instance, k_i subsets can be encoded by $\lceil \log_2 k_i \rceil$ variables). We choose to encode the k_i subsets of domain D_i with $n_i = k_i - 1$ variables, with the aim of obtaining a more comprehensible encoding, as follows. When value v_i of field f_i belongs to subset S_i^s, it means $x_{is} = True$ and $x_{ih} = False$, for $h = 1, \ldots, n_i$, $h \neq s$. The same holds for all other subsets of f_i, except for the out-of-range subset, which is encoded by putting all variables x_{ih} at $False$, for $h = 1, \ldots, n_i$.

Example 2.6. The field `marital status` has 6 subsets, hence $6-1 = 5$ variables

$$x_{\text{mar.st.}=\text{single}},\ x_{\text{mar.st.}=\text{married}},\ x_{\text{mar.st.}=\text{separate}},\ x_{\text{mar.st.}=\text{divorced}},\ x_{\text{mar.st.}=\text{widow}}$$

We define $n' = n_1 + \ldots + n_l$, obtaining so n' propositional variables $\{x_{is}\}$. Now, every statement, (i.e. every set of values of a domain appearing in a rule) can be substituted by the corresponding propositional variable (or by the corresponding logic connection of propositional variables). Since each rule is itself a logic connection of statements, each rule can be expressed as a logic connection of propositional variables, which is put in the form of clauses.

Example 2.7. Consider the rule (`marital status` = `married`) $\Rightarrow$ (`age` ≥ 14).
By substituting the introduced propositional variables, we have the logic formula
$x_{\text{mar.st.}=\text{married}} \Rightarrow \neg x_{\text{age}\in\{0,13\}}$, easily coverted into the following clause.

$$\neg x_{\text{mar.st.}=\text{married}} \vee \neg x_{\text{age}\in\{0,13\}}$$

In addition to information given by rules, there is other information that a human
operator would consider obvious, but which must be provided. With our choice
for variables, we need to express that each field f_i must have only one value, and
therefore $\binom{n_i}{2}$ (number of unordered pairs from n_i possibilities) clauses, named
congruency clauses, are added.

3   Special Cases for Encoding and Complete Procedure

The above described basic procedure is suitable for a large variety of situations.
There are, however, cases where it suffers from an undesirable growth in the
number of clauses and/or variables. We now analyze the most relevant ones.

3.1   Denominative Fields

A first problem arise when a qualitative field f_i has a feasible domain containing a
large number of values. This happens for instance in the case of `name` of a person.
One can theoretically state a rule for every correct name, and so detect errors
like Jhon instead of John. This would, however, require the use of a large number
of rules and, consequently, of clauses and variables. It is instead convenient, in
such cases, to consider a list L_i of values which are *feasible* for f_i

$$L_i = \{v_i^1, \ldots, v_i^{a_i}\}$$

Such list could be for instance obtained from other correct data sets of similar
nature. So far, the (possibly erroneous) value v_i is *identified* as a value $v_i^b \in L_i$
and substituted with it. If v_i is so erroneous that cannot be identified as any
feasible value, it is then identified as the `blank` value and substituted with it.
The identification process involves *linkage* problems that cannot be discussed
here. After such substitution, there is no need of rules for every correct name
anymore. Note that other rules for the field f_i may still be written (for instance
absence of `blank`). Such rules would just determine, as usual, a partition in the
domain D_i.

3.2   Corresponding Fields

A second problem arise when two fields f_i, f_j having a *large* number of feasible
values must be *congruent*. This happens for instance in the case of a `location`
and its geographical `coordinates` (latitude and longitude). Writing the rules
defining all the correct couples (`location, coordinates`) would produce a large
number of clauses and variables. A tentative for recognizing such couple within a

list would ignore a global correction of the record. It is instead convenient (after identifying denominative fields such as `location` by using a list) to produce the correct couple of values

$$\{v_i^i, v_j^i\}$$

which is obtainable on the basis of f_i, and the correct couple of values

$$\{v_i^j, v_j^j\}$$

which is obtainable on the basis of f_j. One of such couples must be chosen, and there is no need of rules defining which are the correct couples anymore. In order to allow such *choice* during the correction process, instead, further propositional variables need to be introduced. By pursuing the aim of a more comprehensible encoding (as in Sect. 2), the 2 choices are encoded here by 2 variables y_i and y_j. Clauses expressing their mutual exclusion are also added: $(y_i \vee y_j), (\neg y_i \vee \neg y_j)$.

Such y variables must be connected to the x variables, by expressing that $y_i \Rightarrow (v_i = v_i^i, v_j = v_j^i)$, and analogously for y_j, which means:

$$y_i \Rightarrow x_{is} \text{ for } s \text{ such that } v_i^i \in S_i^s \qquad y_j \Rightarrow x_{is} \text{ for } s \text{ such that } v_i^j \in S_i^s$$
$$y_i \Rightarrow x_{js} \text{ for } s \text{ such that } v_j^i \in S_j^s \qquad y_j \Rightarrow x_{js} \text{ for } s \text{ such that } v_j^j \in S_j^s$$

Such logical expressions are trivially converted into clauses. Note that also for the corresponding fields f_i and f_j other rules may be written. The above can be easily generalized to the case of any number q of corresponding fields, by introducing the q variables $\{y_1, \ldots, y_q\}$ and the suitable clauses.

Example 3.1. Consider a record having `location = Rome` and `coordinates = 40.5N74W`. By using a Geographic Information System (GIS) we have the couples

location = Rome

$$\downarrow$$

$$\begin{cases} \texttt{location}^{\text{loc}} = \text{Rome} \\ \texttt{coordinates}^{\text{loc}} = 41.5\text{N}12.5\text{E} \end{cases}$$

coordinates = 40.5N74W

$$\downarrow$$

$$\begin{cases} \texttt{location}^{\text{coor}} = \text{NewYork} \\ \texttt{coordinates}^{\text{coor}} = 40.5\text{N}74\text{W} \end{cases}$$

We introduce two new variables y_{loc} and y_{coor}, respectively indicating whether the `location` or the `coordinates` are chosen as correct. Suppose that the field `location` has a subset for `Rome` and a subset for `NewYork`, and that the field `coordinates` has a subset for `40-50N10-20E` and a subset for `40-50N70-80W`. We therefore generate the following clauses:

$$\begin{cases} (y_{\text{loc}} \vee y_{\text{coor}}), (\neg y_{\text{loc}} \vee \neg y_{\text{coor}}), \\ (\neg y_{\text{loc}} \vee x_{\text{loc=Rome}}), (\neg y_{\text{loc}} \vee x_{\text{coor=40\_50N10\_20E}}), \\ (\neg y_{\text{coor}} \vee x_{\text{loc=NewYork}}), (\neg y_{\text{coor}} \vee x_{\text{coor=40\_50N70\_80W}}) \end{cases}$$

3.3   Contained Fields

A third problem arise in the case of a field f_i representing an object, and another field f_j representing a set containing many such objects. This happens for instance in the case of `city` and `state`. Writing the rules for all correct couples (`city`, `state`) would again produce a large number of clauses and variables. It is therefore convenient to proceed similarly to the former case, with the only difference that the values obtained on the basis of the containing field f_j will have a `blank` for the contained field f_i.

Example 3.2. For a record having `city` = `Rome` and `state` = `France`, we have:

$$\text{city} = \text{Rome} \qquad\qquad \text{state} = \text{France}$$
$$\downarrow \qquad\qquad\qquad\qquad \downarrow$$
$$\begin{cases} \text{city}^{\text{city}} = \text{Rome} \\ \text{state}^{\text{city}} = \text{Italy} \end{cases} \qquad \begin{cases} \text{city}^{\text{state}} = \text{blank} \\ \text{state}^{\text{state}} = \text{France} \end{cases}$$

3.4 Mathematical Rules

There are, however, cases for which the logic encoding is not convenient. A typical example are rules containing mathematical operations between fields. In such case, a propositional logic encoding is possible, e.g. [23], but, due to the very large growth in the size of the problem, a mathematical encoding of the problem seems preferable. This clearly prevent using a SAT solver, and requires the use of a MILP procedure, as in [5].

3.5 Complete Encoding Procedure

We can now state the complete procedure for propositional encoding of rules.

1. Identification of domain D_i for each field f_i, considering that we are dealing with errors
2. Generation of list L_i and identification of v_i for each denominative field f_i
3. Identification of k_i subsets $\{S_i^s\}$ in each domain D_i, by using breakpoints B_i and rules R, and by merging (possible) equivalent subsets within D_i
4. Definition of n_i variables $\{x_{is}\}$ to encode subsets $\{S_i^s\}$ of each domain D_i
5. Expression of each rule r_h by means of clauses defined over the introduced variables $\{x_{is}\}$
6. Definition of variables $\{y_j\}$ and clauses for each special field f_j
7. Generation of congruency clauses to supply information not present in rules

By writing rules according to a precise syntax, the above encoding was performed by an automatic procedure. Altogether, the set of rules produces a set of m clauses over $n = n' + n''$ propositional variables $\{x_{is}, y_j\}$, hence a CNF formula $\mathcal{F}$. Each record p produces a truth assignment $\{x_{is} = t_{is}, y_j = t_j\}$ for such variables (t_{is} and t_j denoting a truth value in $\{True, False\}$). We say, briefly, that p must *satisfy* $\mathcal{F}$, i.e. make it *True*, to be correct. Erroneous record detection simply consists in checking the truth assignment given by each record on $\mathcal{F}$.

4   The Correction Procedure

Given an *erroneous record* $p^e = \{v_1^e, \ldots, v_l^e\}$, the *correction* process consists in changing some of its values, obtaining a *corrected record* $p^c = \{v_1^c, \ldots, v_l^c\}$ which satisfies the formula $\mathcal{F}$ and is as close as possible to the (unknown) *original record* p^o, that is the one we would have in absence of errors. Note that, generally, not all values involved in failed rules need to be changed, and that there are several alternative sets of values that, if changed, can make p^c such as to satisfy $\mathcal{F}$.

Example 4.1. Suppose that the following record p^e is declared erroneous using the two following rules: *i*) it is impossible that anybody not having a car lives in city A and works in city B; *ii*) the minimum age for driving is 18.

{... age = 17, car = no, city of residence = A, city of work = B ...}

Values involved in failed rules are here those of fields {car, city of residence, city of work}. Nevertheless, the record could be corrected either by changing values of the set {city of residence}, or by changing those of the set {city of work}, or by changing those of the set {age, car}.

Two principles should be followed during the correction process: to apply the minimum changes to erroneous data, and to modify as little as possible the original frequency distributions of values for each field (see e.g. [8]). The first aspect is generally settled by assigning a cost for changing each value v_i^e. Such cost is based on the reliability of the field f_i. It is assumed that, when error is something unintentional, the erroneous fields are the minimum-cost set of fields that, if changed, can satisfy $\mathcal{F}$. The second statistical aspect is currently settled by using, for each erroneous record p^e, a donor [3]. A *donor record* $p^d = \{v_1^d, \ldots, v_l^d\}$ is a correct record which, according to some (easily computable) distance function $\sigma(p^e, p^d) \in \mathbb{R}_+$, is the nearest one to the erroneous record p^e, hence it represents a record having similar characteristics. Such donor is used for providing values for the fields to be changed, according to the so-called *data driven* approach (see e.g. [3]).

We now turn back to propositional logic. In order to simplify notation, we will hereafter denote by p_e the truth assignment $\{x_{is} = t_{is}^e, y_j = t_j^e\}$ (where some t_j^e may be undefined) corresponding to the erroneous record p^e and similarly for p^d, p^c, etc. The record correction problem can be logically defined as follows.

Definition 4.2. *Given a logic formula $\mathcal{F}$ (encoding of the set of rules R), a truth assignment p_e not satisfying $\mathcal{F}$ (encoding of the erroneous record p^e), a truth assignment p_d satisfying $\mathcal{F}$ (encoding of the donor record p^d), the record correction problem consists is finding a truth assignment for the $\{x_{is}, y_j\}$ variables such that*

$$\min\{\delta(p_e, \{x_{is}, y_j\}) : \{x_{is}, y_j\} \text{ satisfies } \mathcal{F}\}$$

where $\delta(p_e, \{x_{is}, y_j\}) \in \mathbb{R}_+$ is a linear (or other easily computable) distance function measuring the cost of the changes introduced in p^e for obtaining p^c.

Theorem 4.3. *The above defined record correction problem is NP-hard.*

Proof: Given a propositional logic formula $\mathcal{F}$ and a model m_1, the problem of deciding whether it exists another model m_2 such that the Hamming distance between m_1 and m_2 is less than or equal to a given number k has been proved NP-complete in [2]. Therefore, given a satisfiable $\mathcal{F}$ and a generic truth assignment t_1, the problem of deciding whether it exists a model within a given Hamming distance from t_1 has the same complexity. If the adopted distance function is instead a generic easily (i.e. polynomially) computable function, the problem clearly remains NP-hard [9]. Since this latter problem constitutes the decision version of the above defined record correction problem, the thesis follows.

Moreover, the minimization version of the error correction problem has strong similarities with the combinatorial optimization problem[1].

The solution of the defined record correction problem is searched by means of a DPLL style procedure. Very basically, a DPLL SAT solver explores the solution space by generating a search tree, repeating the following steps until a satisfying solution is reached or the solution space is completely explored.

1. *Choose a variable x_s to be fixed*
2. *Fix x_s to a truth value t_s and propagate such fixing*
3. *If needed, backtrack and cut that branch*

The modifications introduced for solving our problem are basically the following:

a) Variables $\{x_{is}, y_j\}$ are initially progressively fixed to values $\{t_{is}^e, t_j^e\}$. After the first backtrack, truth values corresponding to subsets S_i^s in progressively increasing distance from those of p^e will be chosen, according to the minimum change principle.

b) The distance $\delta(p_e, \bar{p})$ of current (partial) solution $\bar{p}$ from p_e is computed at every node of the branching tree, on the basis of the truth assignments $\{t_{is}^e, t_j^e\}$ and $\{\bar{t}_{is}, \bar{t}_j\}$. The donor's truth assignment p_d is used as current optimal solution, hence backtrack is performed not only when an empty clause is obtained, but also when the distance $\delta(p_e, \bar{p})$ of current (partial) solution is such that

$$\delta(p_e, \bar{p}) \geq \sigma(p_e, p_d)$$

· Let $A = \{a., \ldots, a_f\}$ be a finite ground set, $\mathcal{D} = \{D., \ldots, D_g\}$ be a collection of subsets of A, given a linear cost function $w(D) \in \mathbb{R}$, the combinatorial optimization problem is $\{\min w(D) : D \in \mathcal{D}\}$ [16]. It is quite immediate to see the correspondence between the set p^e and the ground set A. By denoting with $\{D., \ldots, D_t\}$ the alternative sets of values of p^e that, if changed, can satisfy $\mathcal{F}$ (Such collection of sets is nonempty because by changing at most all fields we are able to satisfy $\mathcal{F}$, and finite because so it is the number of possible subsets of fields in p^e. Compare also to Example 4.1.) and by considering as $w(D)$ the function $\delta(p_e, p_c)$ measuring the cost of the values changed in p^e for satisfying $\mathcal{F}$, an instance of record correction problem becomes an instance of combinatorial optimization problem.

c) After reaching a (complete) solution $p^\star$ such that $\delta(p_e, p^\star) < \delta(p_e, p_d)$ (which means that $p^\star$ is more similar than p_d to p_e), the search does not stop, but continues using now $p^\star$ as current optimal solution, and updating $p^\star$ when possible.

d) The search stops either when a solution $p^\star$ having a distance $\delta(p_e, p^\star)$ being small *enough* is reached, or when the search tree is completely explored without finding a solution having acceptable distance.

Once the truth assignment corresponding to the optimal solution $p^\star$

$$\{x_{is} = t_{is}^\star, y_j = t_j^\star\}$$

is obtained, that corresponds to knowing, for each field f_i, which subset S_i^s the corrected value v_i^c must belong to. The actual value v_i^c will then be the initial one v_i^e if $v_i^e \in S_i^s$, otherwise the donor's one v_i^d if $v_i^d \in S_i^s$, or otherwise generated by using a data driven or a probabilistic approach. Depending on the field, in the latter cases, one may report that the information cannot be reconstructed, although known belonging to subset S_i^s.

5   Implementation and Results

The procedure is implemented by introducing minor modifications in in the recent SAT solver named SIMO 2.0 (Satisfiability Internal Module Object oriented) [11]. Such solver was chosen also because of its original purpose of providing an open-source library for SAT solvers. The proposed procedure has been applied to the correction of datasets arising from two different origins: *population* data and *television transmitters* data. The correction of such kind of data actually represents a very relevant problem for people working in the respective fields. As observable, each single record correction problem is quite small-sized for a modern SAT solver. However, very large datasets should often be corrected, and the whole process is therefore computationally demanding. Reported results are obtained on a Pentium IV 1.7GHz PC with 1Gb RAM. In both cases, results are very encouraging both from the computational and from the data quality point of view.

5.1   Population Data

The process of error correction in the case of simulated population data sets using 258 real rules, kindly provided by experts of the Italian National Statistic Institute (Istat), is performed. About 400 congruency clauses were also needed. Four different data sets (p_050, p_100, p_150, p_200) are obtained by randomly changing the values of 5 demographic variables (`relation to head of the house, sex, marital status, age, years married`) into 182,864 correct simulated person records. Such values are changed by introducing either blank values or other valid values with a base probability η respectively multiplied by 0.5, 1, 1.5,

2, in order to obtain the above data sets. The resulting erroneous record content of such four data sets after the described perturbation respectively became 22.5%, 40.5%, 54.0%, 64.4%. Since the original (correct) records were in these cases known, the accuracy of the correction process could be evaluated.

Results of the correction of each whole data set are reported in Table 1. Number of variables and number of clauses of the formula $\mathcal{F}$ considered when solving each single record correction problem differs from record to record. This because, depending on the errors, one could know in advance that some (few) fields are not interested by the correction process. We therefore generated variables only for those fields which could be interested by the correction process, and consequently only the clauses containing them. Error content of the datasets is generally enough for allowing the reconstruction, and the record correction process was successful in more than the 99.9% of the cases. There are, however, records for which no solution was found, because the best solution obtained has a distance higher than the maximum acceptable distance.

Number of records	182,864
Number of fields	20
Number of rules	258
Error content before the correction process in p_050	22.5%
Error content before the correction process in p_100	40.5%
Error content before the correction process in p_150	54.5%
Error content before the correction process in p_200	64.4%
Error content after the correction process (all)	< 0.01%
Total correction time for p_050	50.4 min.
Total correction time for p_100	79.3 min.
Total correction time for p_150	112.6 min.
Total correction time for p_200	113.8 min.
Average time per record for p_050	0.01 sec.
Average time per record for p_100	0.02 sec.
Average time per record for p_150	0.04 sec.
Average time per record for p_200	0.04 sec.
Average number of variables for each record correction	315
Average number of clauses for each record correction	633 (258+0+375)

Table 1: Results on the four datasets p_050, p_100, p_150, p_200 of simulated persons data using real rules.

5.2 Television Transmitters Data

The correction process for data representing Television Transmitters data is performed. The processed dataset has 22,504 records, each having 209 fields describing the geographical, orographical and radio-electrical characteristics of each transmitter. Differently from the former cases, this dataset already contained errors, and they have not been artificially introduced. A set of 75 explicit rules has been written by experts of Frequency Planning for Broadcasting. The resulting number of clauses is 224, 75 of which derive from the initial explicit rules, 99 from the logical connections between the x variables and the y variables, and 50 are congruency clauses. Results of the correction of the whole dataset are reported in Table 2. Proposed techniques for dealing with corresponding fields have been of crucial importance for such dataset. Error content of the data is higher than in the former datasets, and in some cases the information is not enough for allowing a reconstruction. In such cases, in fact, the best solution obtained has a distance higher than the maximum acceptable distance, and therefore a certain record percentage could not be corrected.

Number of records	22,504
Number of fields	209
Number of rules	75
Error content before the correction process	62.5%
Error content after the correction process	2.1%
Total correction time	4.0 min.
Average time per record	0.01 sec.
Average number of variables for each record correction	76
Average number of clauses for each record correction	224 (75+99+50)

Table 2: Results on the television transmitters dataset using real rules.

6   Conclusions

Data correction problems are of great relevance in almost every field were an automatic data processing is used. When cleaning large scale databases, moreover, computational efficiency is essential. A propositional logic encoding is, in many cases, the most direct and effective representation both for records and for the set of rules. Erroneous records detection is carried out with an inexpensive procedure, while erroneous records correction is solved by customizing a generic DPLL style SAT solver (SIMO 2.0 for the reported results). The record correction problem is in fact NP-hard. Additional techniques for handling difficult cases for encoding are presented. Computational results on datasets having different origins (*population* data and *television transmitters* data) are very encouraging.

Acknowledgments. The author whish to thank Daniele Praticò for his contribution to present paper, and SIMO developers for providing their solver.

References

1. M. Ayel and J.P. Laurent (eds.). *Validation, Verification and Testing of Knowledge-Based Systems.* J. Wiley & Sons, Chichester, 1991.
2. O. Bailleux and P. Marquis. DISTANCE-SAT: Complexity and Algorithms. In *Proceedings* AAAI/IAAI 1999, 642-647.

3. M. Bankier. Canadian Census Minimum change Donor imputation methodology. In *Proceedings of the Workshop on Data Editing*, UN/ECE, Cardiff, 2000.
4. E. Boros, P.L. Hammer, T. Ibaraki, A. Kogan. Logical analysis of numerical data. *Mathematical Programming*, 79, 163-190, 1997.
5. R. Bruni, A. Reale, R. Torelli. Optimization Techniques for Edit Validation and Data Imputation. In *Proceedings of Statistics Canada International Symposium: Achieving Data Quality in a Statistical Agency*, Ottawa, Canada, 2001.
6. R. Bruni and A. Sassano. Error Detection and Correction in Large Scale Data Collecting. In *Advances in Intelligent Data Analysis*, LNCS 2189, Springer, 2001.
7. V. Chandru and J.N. Hooker. *Optimization Methods for Logical Inference*. Wiley, New York, 1999.
8. P. Fellegi and D. Holt. A Systematic Approach to Automatic Edit and Imputation. *Journal of the American Statistical Association* 71(353), 17-35, 1976.
9. M.R. Garey and D.S. Johnson. *Computers and Intractability*. Freeman, New York, 1979.
10. I.P. Gent, H. van Maarcn, T. Walsh (eds.) *SAT 2000*. IOS Press, Amsterdam, 2000.
11. E. Giunchiglia, M. Maratea, A. Tacchella. Look-Ahead vs. Look-Back techniques in a modern SAT solver. In *Proceedings of the Sixth International Conference on Theory and Applications of Satisfiability Testing* (SAT2003), Portofino, Italy, 2003.
12. J. Gu, P.W. Purdom, J. Franco, and B.W. Wah. Algorithms for the Satisfiability (SAT) Problem: A Survey. *DIMACS Series in Discrete Mathematics* American Mathematical Society, 1999.
13. D.J. Hand, H. Mannila, P. Smyth. *Principles of Data Mining*. MIT Press, London, 2001.
14. T. Hastie, R. Tibshirani, J. Friedman. *The Elements of Statistical Learning*. Springer, New York, Berlin, Heidelberg, 2002.
15. T. Menzies. Knowledge Maintenance: The State of the Art. *Knowledge Engineering Review*, 14(1), 1-46, 1999.
16. G.L. Nemhauser, L.A. Wolsey. *Integer and Combinatorial Optimization*. Wiley, New York, 1988.
17. C. Poirier. A Functional Evaluation of Edit and Imputation Tools. *UN/ECE Work Session on Statistical Data Editing*, Working Paper n.12, Rome, Italy, 2-4 June 1999.
18. R. Ramakrishnan, J. Gehrke. *Database Management System*. McGraw Hill, 2000.
19. N. Rescher, R. Brandom. *The Logic of Inconsistency*. B. Blackwell, Oxford, 1980.
20. L. Simon and P. Chatalic. SAT-Ex: the experimentation web site around the satisfiability problem. http://www.lri.fr/~simon/satex.
21. L. Simon, D. Le Berre (and E.A. Hirsch for SAT2002). Sat competition web site http://www.satlive.org/SATCompetition
22. K. Truemper. *Effective Logic Computation*. Wiley, New York, 1998.
23. J. Warners. A Linear-time Transformation of Linear Inequalities into Conjunctive Normal Form. *Information Processing Letters*, 68, 63-69, 1998.

Using Problem Structure for Efficient Clause Learning*

Ashish Sabharwal, Paul Beame, and Henry Kautz

Computer Science and Engineering
University of Washington, Seattle WA 98195-2350
{ashish,beame,kautz}@cs.washington.edu

Abstract. DPLL based clause learning algorithms for satisfiability testing are known to work very well in practice. However, like most branch-and-bound techniques, their performance depends heavily on the variable order used in making branching decisions. We propose a novel way of exploiting the underlying problem structure to guide clause learning algorithms toward faster solutions. The key idea is to use a higher level problem description, such as a graph or a PDDL specification, to generate a good branching sequence as an aid to SAT solvers. The sequence captures hierarchical structure that is lost in the CNF translation. We show that this leads to exponential speedups on grid and randomized pebbling problems. The ideas we use originate from the analysis of problem structure recently used in [1] to study clause learning from a theoretical perspective.

1 Introduction

The NP-complete problem of determining whether or not a given CNF formula is satisfiable has been gaining wide interest in recent years as solutions to larger and larger instances from a variety of problem classes become feasible. With the continuing improvement in the performance of satisfiability (SAT) solvers, the field has moved from playing with toy problems to handling practical applications in a number of different areas. These include hardware verification [2, 3], group theory [4], circuit diagnosis, and experiment design [5, 6].

The Davis-Putnam-Logemann-Loveland procedure (DPLL) [7, 8] forms the backbone of most common complete SAT solvers that perform a recursive backtrack search, looking for a satisfying assignment to the variables of the given formula. The key idea behind it is the pruning of the search space based on falsified clauses. Various extensions to the basic DPLL procedure have been proposed, including smart branching selection heuristics [9], randomized restarts [10], and clause learning [11, 12, 13, 14, 15]. The last of these, which this paper attempts to exploit more effectively, originated from earlier work on explanation based learning (EBL) [16, 17, 18, 19] and has resulted in tremendous improvement in performance on many useful problem classes. It works by adding a new clause to the set of given clauses ("learning" this clause) whenever the DPLL procedure fails on a partial assignment and needs to backtrack. This typically saves work later in the search process when another assignment fails for a similar reason.

Both random CNF formulas and those encoding various real-world problems are hard for current SAT solvers. However, while DPLL based algorithms with lookahead

---

* Research supported by NSF Grant ITR-0219468

E. Giunchiglia and A. Tacchella (Eds.): SAT 2003, LNCS 2919, pp. 242–256, 2004.

but no learning (such as `satz` [20]) and those that try only one carefully chosen assignment without any backtracks (such as `SurveyProp` [21]) are our best tools for solving random formula instances, formulas arising from various real applications seem to require clause learning as a critical ingredient. The key thing that makes this second class of formulas different is the inherent structure, such as dependence graphs in scheduling problems, causes and effects in planning, and algebraic structure in group theory.

Trying to understanding clause learning from the theoretical point of view of proof complexity has lead to many useful insights. In [1], we showed that on certain classes of formulas, clause learning is provably exponential stronger than a proof system called regular resolution. This in turn implies an even larger exponential gap between the power of DPLL and that of clause learning, thus explaining the performance gains observed empirically. It also shows that for such structured formulas, our favorite non-learning SAT solvers for random formulas such as `satz` and `SurveyProp` are doomed to fail, whereas clause learning provides potential for small proofs. However, in order to leverage their strength, clause learning algorithms must use the "right" variable order for their branching decisions for the underlying DPLL procedure. While a good variable order may result in a polynomial time solution, a bad one can make the process as slow as basic DPLL without learning. This leads to a natural question: can such insights from theoretical analysis of problem structure help us further? For example, for the domains where we can deduce analytically that small solutions exist, can we guide clause learning algorithms to *find* these solutions efficiently?

As we mentioned previously, most theoretical and practical problem instances of satisfiability problems originate, not surprisingly, from a higher level description, such as a graph or Planning Domain Definition Language (PDDL) specification [22]. Typically, this description contains more structure of the original problem than is visible in the flat CNF representation in DIMACS format [23] to which it is converted before being fed into a SAT solver. This structure can potentially be used to gain efficiency in the solution process. While there has been work on extracting structure after conversion into a CNF formula by exploiting variable dependency [24], constraint redundancy [25], symmetry [26], binary clauses [27] and partitioning [28], using the original higher level description itself to generate structural information is likely to be more effective. The latter approach, despite its intuitive appeal, remains largely unexplored, except for suggested use in bounded model checking [29] and the separate consideration of cause variables and effect variables in planning [30].

In this paper, we further open this line of research by proposing an effective method for exploiting problem structure to guide the branching decision process of clause learning algorithms. Our approach uses the original high level problem description to generate not only a CNF encoding but also a *branching sequence* [1] that guides the SAT solver toward an efficient solution. This branching sequence serves as auxiliary structural information that was possibly lost in the process of encoding the problem as a CNF formula. It makes clause learning algorithms learn useful clauses instead of wasting time learning those that may not be reused in future at all. We give an exact sequence generation algorithm for pebbling formulas. The high level description used is a pebbling graph. Our sequence generator works for the 1UIP learning scheme [15], which is one of the best known. Our empirical results are based on our extension of the popular SAT solver zChaff [14].

We show that the use of branching sequences produced by our generator leads to exponential speedups for the class of grid and randomized pebbling formulas. These formulas, more commonly occurring in theoretical proof complexity literature [31, 32, 33, 34], can be thought of as representing precedence graphs in dependent task systems and scheduling scenarios. They can also be viewed as restricted planning problems. Although admitting a polynomial size solution, both grid and randomized pebbling problems are not so easy to solve deterministically, as is indicated by our experimental results for unmodified zChaff. From a broader perspective, our result for pebbling formulas serves as a proof of concept that analysis of problem structure can be used to achieve dramatic improvements even in the current best clause learning based SAT solvers.

2   Preliminaries

A Conjunctive Normal Form (CNF) formula F is an AND ($\wedge$) of *clauses*, where each clause is an OR ($\vee$) of *literals*, and a literal is a variable or its negation ($\neg$). The Davis-Putnam-Logemann-Loveland (DPLL) procedure [7, 8] for testing satisfiability of such formulas works by *branching* on variables, setting them to TRUE or FALSE, until either an initial clause is *violated* (*i.e.* has all literals set to FALSE) or all variables have been set. In the former case, it backtracks to the last branching decision whose other branch has not been tried yet, reverses the decision, and proceeds recursively. In the latter, it terminates with a satisfying assignment. If all possible branches have been unsuccessfully tried, the formula is declared unsatisfiable. To increase efficiency, *pure literals* (those whose negation does not appear) and *unit clauses* (those with only one unset literal) are immediately set to true. In this paper, by DPLL we will mean this basic procedure along with additions such as randomized restarts [10] and local or global branching heuristics [9], but no learning.

Clause learning (see *e.g.* [11]) can be thought of as an extension of the DPLL procedure that caches causes of assignment failures in the form of learned clauses. It proceeds by following the normal branching process of DPLL until there is a "conflict" after unit propagation. If this conflict occurs without any branches, the it declares the formula unsatisfiable. Otherwise, it analyzes the "conflict graph" and learns the cause of the conflict in the form of a "conflict clause" (see Fig. 1). It now backtracks and continues as in ordinary DPLL, treating the learned clause just like initial ones. One expects that such cached causes of conflicts will save computation later in the process when an unsatisfiable branch due to fail for a similar reason is explored.

Different implementations of clause learning algorithms vary in the strategy they use to choose the clause to learn from a given conflict [15]. For instance, grasp [12] uses the *Decision scheme* among others, zChaff [14] uses the *1UIP* scheme, and we proposed in [1] a new learning scheme called *FirstNewCut*. The results in this paper are for the 1UIP scheme, but can be obtained for certain other schemes as well, including FirstNewCut.

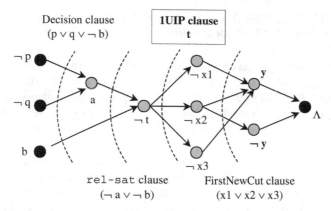

Fig. 1. A conflict graph with various conflict clauses that can potentially be learned

2.1 Branching Sequence

The notion of branching sequence was used in [1] to prove exponential separation between DPLL and clause learning. It generalizes the idea of a static *variable order* by letting it differ from branch to branch in the underlying DPLL procedure. In addition, it also specifies which branch (TRUE or FALSE) to explore first. This can clearly be useful for satisfiable formulas, and can also help on unsatisfiable ones by making the algorithm learn useful clauses earlier in the process.

Definition 1 ([1]). *A* branching sequence *for a CNF formula F is a sequence $\sigma = (l_1, l_2, \ldots, l_k)$ of literals of F, possibly with repetitions. A DPLL based algorithm $\mathcal{A}$ on F branches according to σ if it always picks the next variable v to branch on in the literal order given by σ, skips it if v is currently assigned a value, and otherwise branches further by setting the chosen literal to FALSE and deleting it from σ. When σ becomes empty, $\mathcal{A}$ reverts back to its default branching scheme.*

Definition 2. *A branching sequence σ is* complete *for F under an algorithm $\mathcal{A}$ if $\mathcal{A}$ branching according to σ terminates before or as soon as σ becomes empty.*

Clearly, how well a branching sequence works for a formula depends on the specifics of the clause learning algorithm used, such as its learning scheme and backtracking process. One needs to keep these in mind when generating the sequence. It is also important to note that while the size of a variable order is always the same as the number of variables in the formula, that of an effective branching sequence is typically much more. In fact, the size of a branching sequence complete for an unsatisfiable formula F is equal to the size of an unsatisfiability proof of F, and when F is satisfiable, it is proportional to the time needed to find a satisfying assignment.

2.2 Pebbling Formulas

Pebbling formulas are unsatisfiable CNF formulas whose variations have been used repeatedly in proof complexity to obtain theoretical separation results between different proof systems [31, 32, 33, 34]. The version we will use in this paper is known to be easy for regular resolution but hard for tree-like resolution (and hence for DPLL without learning) [33].

A *Pebbling formula* pbl_G is an unsatisfiable CNF formula associated with a directed, acyclic *pebbling graph* G (see Fig. 2). Nodes of G are labeled with disjunctions of variables. A node labeled $(x_1 \lor x_2)$ with, say, three predecessors labeled $(p_1 \lor p_2 \lor p_3)$, q_1 and $(r_1 \lor r_2)$ generates six clauses $(\neg p_i \lor \neg q_j \lor \neg r_k \lor x_1 \lor x_2)$, where $i \in \{1, 2, 3\}, j \in \{1\}$ and $k \in \{1, 2\}$. Intuitively, a node labeled $(x_1 \lor x_2)$ is thought of as *pebbled* under a (partial) variable assignment σ if $(x_1 \lor x_2) =$ TRUE under σ. The clauses mentioned above state that if all predecessors of a node are pebbled, then the node itself must also be pebbled. For every indegree zero *source node* of G labeled $(s_1 \lor s_2)$, pbl_G has a clause $(s_1 \lor s_2)$, stating that all source nodes are pebbled. For every outdegree zero *target node* of G labeled $(t_1 \lor t_2)$, pbl_G has clauses $\neg t_1$ and $\neg t_2$, saying that target nodes are not pebbled, and thus providing a contradiction.

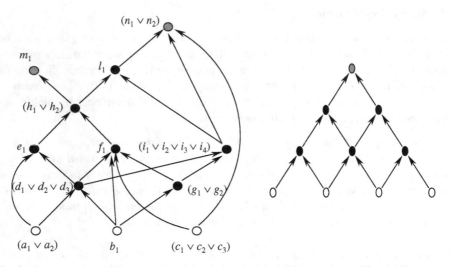

Fig. 2. A general pebbling graph with distinct node labels, and a grid pebbling graph with 4 layers

Grid pebbling formulas are based on simple pyramid shaped layered pebbling graphs with distinct variable labels, 2 predecessors per node, and disjunctions of size 2 (see Fig. 2). *Randomized pebbling formulas* are more complicated and correspond to random pebbling graphs. In general, they allow multiple target nodes. However, the more the targets, the easier it is to produce a contradiction because we can focus only on the (relatively smaller) subgraph under the lowest target. Hence, for our experiments, we add a simple grid structure at the top of randomly generated pebbling formulas to make them have exactly one target.

All pebbling formulas with a single target are minimally unsatisfiable, *i.e.* any strict subset of their clauses admits a satisfying assignment. For each formula Pbl_G we use for our experiments, we also use a satisfiable version of it, called Pbl_G^{SAT}, obtained by randomly choosing a clause of Pbl_G and deleting it. When G is viewed as a task graph, Pbl_G^{SAT} corresponds to a single fault task system, and finding a satisfying assignment for it corresponds to locating the fault.

3   Branching Sequence for Pebbling Formulas

In this section, we will give an efficient algorithm to generate an effective branching sequence for pebbling formulas. This algorithm will take as input the underlying pebbling graph (which is the high level description of the pebbling problem), and not the pebbling formula itself. As we will see in Section 4, the generated branching sequence gives exponential empirical speedup over zChaff for both grid and randomized pebbling formulas.

zChaff, despite being one of the current best clause learners, by default does not perform very well on seemingly simple pebbling formulas, even on the uniform grid version. Although clause learning should ideally need only polynomial (in fact, linear in the size of the formula) time to solve these problem instances, choosing a good branching order is critical for this to happen. Since nodes are intuitively pebbled in a bottom up fashion, we must also learn the right clauses (*i.e.* clauses labeling the nodes) in a bottom up order. However, branching on variables labeling lower nodes before those labeling higher ones prevents any DPLL based learning algorithm to backtrack the right distance and proceed further. To make this clear, consider the general pebbling graph of Fig. 2. Suppose we branch on and set d_1, d_2, d_3 and a_1 to FALSE. This will lead to a contradiction through unit propagation by implying a_2 is TRUE and b_1 is FALSE. We will learn $(d_1 \lor d_2 \lor d_3 \lor \neg a_2)$ as the associated 1UIP conflict clause and backtrack. There will still be a contradiction without any further branches, making us learn $(d_1 \lor d_2 \lor d_3)$ and backtrack. At this stage, we have learned the correct clause but are *stuck* with the two branches on d_1 and d_2. Unless we already branched on e_1, there is no way we can now learn it as a clause corresponding to the next higher node.

3.1   Sequence Generation: GenSeq1UIP

Algorithm 1, GenSeq1UIP, describes a way of generating a good branching sequence for pebbling formulas. It works on any pebbling graph G with distinct label variables as input and produces a branching sequence linear in the size of the associated pebbling formula. In particular, the sequence size is linear in the number of variables as well when the indegree as well as label size are bounded by a constant.

GenSeq1UIP starts off by handling the set U of all nodes labeled with unit clauses. Their outgoing edges are deleted and they are treated as pseudo sources. The procedure first generates branching sequence for non-target nodes in U in increasing order of height. The key here is that when zChaff learns a unit clause, it fast-backtracks to decision level zero, effectively restarting at that point. We make use of this fact to learn these unit clauses in a bottom up fashion, unlike the rest of the process which proceeds top down in a depth-first way.

Input   : Pebbling graph G with no repeated labels
Output : Branching sequence for G for the 1UIP learning scheme
begin

> **foreach** v **in** *BottomUpTraversal* (G) **do**
> > $v.height \leftarrow 1 + max_{u \in v.preds}\{u.height\}$
> > Sort $(v.preds,\ increasing\ order\ by\ height)$
>
> // First handle unit clause labeled nodes and generate their sequence
> $U \leftarrow \{v \in G.nodes : |v.labels| = 1\}$
> $G.edges \leftarrow G.edges \setminus \{(u,v) \in G.edges : u \in U\}$
> Add to $G.sources$ any new nodes with now 0 preds
> Sort $(U,\ increasing\ order\ by\ height)$
> **foreach** $u \in U \setminus G.targets$ **do**
> > Output $u.label$
> > GenSubseq1UIPWrapper (u)
>
> // Now add branching sequence for targets by increasing height
> Sort $(G.targets,\ increasing\ order\ by\ height)$
> **foreach** $t \in G.targets$ **do**
> > GenSubseq1UIPWrapper (t)

end

GenSubseq1UIPWrapper $(node\ v)$ **begin**

> **if** $|v.preds| > 0$ **then**
> > GenSubseq1UIP $(v, |v.preds|)$

end

GenSubseq1UIP $(node\ v,\ int\ i)$ **begin**

> $u \leftarrow v.preds[i]$
>
> // If this is the lowest predecessor ...
> **if** $i = 1$ **then**
> > **if** $!u.visited$ **and** $u \notin G.sources$ **then**
> > > $u.visited \leftarrow$ TRUE
> > > GenSubseq1UIPWrapper (u)
> >
> > **return**
>
> // If this is not the lowest one ...
> Output $u.labels \setminus \{u.lastLabel\}$
> **if** $!u.visitedAsHigh$ **and** $u \notin G.sources$ **then**
> > $u.visitedAsHigh \leftarrow$ TRUE
> > Output $u.lastLabel$
> > **if** $!u.visited$ **then**
> > > $u.visited \leftarrow$ TRUE
> > > GenSubseq1UIPWrapper (u)
>
> GenSubseq1UIP $(v, i - 1)$
>
> **for** $j \leftarrow (|u.labels| - 2)$ **downto** 1 **do**
> > Output $u.labels[1], \ldots, u.labels[j]$
> > GenSubseq1UIP $(v, i - 1)$
>
> GenSubseq1UIP $(v, i - 1)$

end

Algorithm 1: GenSeq1UIP

GenSeq1UIP now adds branching sequences for the targets. Note that for an unsatisfiability proof, we only need the sequence corresponding to the first (lowest) target. However, we process all targets so that this same sequence can also be used when the formula is made satisfiable by deleting enough clauses. The subroutine GenSubseq1UIP runs on a node v, looking at its ith predecessor u in increasing order by height. No labels are output if u is the lowest predecessor; the negations of these variables will be indirectly implied during clause learning. However, it is recursed upon if not already visited. This recursive sequence results in learning something close to the clause labeling this lowest node, but not quite that exact clause. If u is a higher predecessor (it will be marked as $visitedAsHigh$), GenSubseq1UIP outputs all but one variables labeling u. If u is not a source and has not already been visited as high, the last label is output as well, and u recursed upon if necessary. This recursive sequence results in learning the clause labeling u. Finally, GenSubseq1UIP generates a recursive pattern, calling the subroutine with the next lower predecessor of v. The precise structure of this pattern is dictated by the 1UIP learning scheme and fast backtracking used in zChaff. Its size is exponential in the degree of v with label size as the base.

Example To clarify the algorithm, we describe its execution on a small example. Let G be the pebbling graph in Fig. 3. Denote by t the node labeled $(t_1 \vee t_2)$, and likewise for other nodes. Nodes c, d, f and g are at height 1, nodes a and e at height 2, node b at height 3, and node t at height 4. $U = \{a, b\}$. The edges (a, t) and (b, t) originating from these unit clause labeled nodes are removed, and t, with no predecessors anymore, is added to the list of sources. We output the label of the non-target unit nodes in U in increasing order of height, and recurse on each of them in order, *i.e.* we output a_1, setting $B = (a_1)$, call GenSubseq1UIPWrapper on a, and then repeat this process for b. This is followed by a recursive call to GenSubseq1UIPWrapper on the target node t.

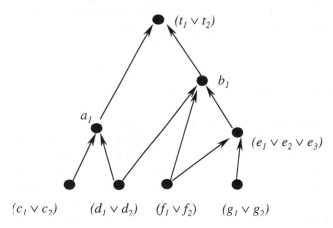

Fig. 3. A simple pebbling graph to illustrate branch sequence generation

The call GenSubseq1UIPWrapper on a in turn invokes GenSubseq1UIP with parameters $(a, 2)$. This sorts the predecessors of a in increasing order of height to, say, d, c, with d being the lowest predecessor. v is set to a and u is set to the second predecessor c. We output all but the last label of u, *i.e.* of c, making the current branching sequence $B = (a_1, c_1)$. Since u is a source, nothing more needs to be done for it and we make a recursive call to GenSubseq1UIP with parameters $(a, 1)$. This sets u to d, which is the lowest predecessor and requires nothing to be done because it is also a source. This finishes the sequence generation for a, ending at $B = (a_1, c_1)$. After processing this part of the sequence, zChaff will have a as a learned clause.

We now output b_1, the label of the unit clause b. The call, GenSubseq1UIPWrapper on b, proceeds similarly, setting predecessor order as (d, f, e), with d as the lowest predecessor. Procedure GenSubseq1UIP is called first with parameters $(b, 3)$, setting u to e. This adds all but the last label of e to the branching sequence, making it $B = (a_1, c_1, b_1, e_1, e_2)$. Since this is the first time e is being visited as high, its last label is also added, making $B = (a_1, c_1, b_1, e_1, e_2, e_3)$, and it is recursed upon with GenSubseq1UIPWrapper(e). This recursion extends the sequence to $B = (a_1, c_1, b_1, e_1, e_2, e_3, f_1)$. After processing this part of B, zChaff will have both a and $(e_1 \vee e_2 \vee e_3)$ as learned clauses. Getting to the second highest predecessor f of b, which happens to be a source, we simply add another f_1 to B. Finally, we get to the third highest predecessor d of b, which happens to be the lowest as well as a source, thus requiring nothing to be done. Coming out of the recursion, back to $u = f$, we generate the pattern given by the last for loop, which is empty because the label size of f is only 2. Coming out once more of the recursion to $u = e$, the for loop pattern generates e_1, f_1 and is followed by a call to GenSubseq1UIP with the next lower predecessor f as the second parameter, which generates f_1. This makes the current sequence $B = (a_1, c_1, b_1, e_1, e_2, e_3, f_1, f_1, e_1, f_1, f_1)$. After processing this, zChaff will also have b as a learned clause.

The final call to GenSubseq1UIPWrapper with parameter t doesn't do anything because both predecessors of t were removed in the beginning. Since both a and b have been learned, zChaff will have an immediate contradiction at decision level zero. This gives us the complete branching sequence $B = (a_1, c_1, b_1, e_1, e_2, e_3, f_1, f_1, e_1, f_1, f_1)$ for the pebbling formula Pbl_G.

3.2 Complexity Analysis

Let graph G have n nodes, indegree of non-source nodes between d_{min} and d_{max}, and label size between l_{min} and l_{max}. For simplicity of analysis, we will assume that $l_{min} = l_{max} = l$ and $d_{min} = d_{max} = d$ ($l = d = 2$ for a grid graph).

Let us first compute the size of the pebbling formula associated with G. The running time of GenSeq1UIP and the size of the branching sequence generated will be given in terms of this size. The number of clauses in the pebbling formula Pbl_G is nl^d, ignoring a slight counting error for the source and target nodes. Taking clause sizes into account, the size of the formula, $|Pbl_G|$, is $n(l + d)l^d$. Note that size of the CNF formula itself grows exponentially with the indegree and gets worse as label size increases. The best case is when G is the grid graph, where $|Pbl_G| = \Theta(n)$. This explains the degradation in performance of zChaff, both original and modified, as we move from grid graphs to random graphs (see section 4). Since we construct Pbl_G^{SAT} by deleting exactly one randomly chosen clause from Pbl_G (assuming G has only one target node), the size $|Pbl_G^{SAT}|$ of the satisfiable version is also essentially the same.

Let us now compute the running time of GenSeq1UIP. Initial computation of heights and predecessor sorting takes time $\Theta(nd \log d)$. Assuming n_u unit clause labeled nodes and n_t target nodes, the remaining node sorting time is $\Theta(n_u \log n_u + n_t \log n_t)$. Since GenSubseq1UIPWrapper is called at most once for each node, the total running time of GenSeq1UIP is $\Theta(nd \log d + n_u \log n_u + n_t \log n_t + n T_{wrapper})$, where $T_{wrapper}$ denotes the running time of GenSubseq1UIPWrapper without including recursive calls to itself. When n_u and n_t are much smaller than n, which we will assume as the typical case, this simplifies to $\Theta(nd \log d + n T_{wrapper})$. Let $T(v, i)$ denote the running time of GenSubseq1UIP(v, i), again without including recursive calls to the wrapper method. Then $T_{wrapper} = T(v, d)$. However, $T(v, d) = lT(v, d-1) + \Theta(l)$, which gives $T_{wrapper} = T(v, d) = \Theta(l^{d+1})$. Substituting this back, we get that the running time of GenSeq1UIP is $\Theta(nl^{d+1})$, which is about the same as $|Pbl_G|$.

Finally, we consider the size of the branching sequence generated. Note that for each node, most of its contribution to the sequence is from the recursive pattern generated near the end of GenSubseq1UIP. Let us denote that by $Q(v, i)$. Then $Q(v, i) = (l-2)(Q(v, i-1) + \Theta(l))$, which gives $Q(v, i) = \Theta(l^{d+2})$. Hence, the size of the sequence generated is $\Theta(nl^{d+2})$, which again is about the same as $|Pbl_G|$.

Theorem 1. *Given a pebbling graph G with label size at most l and indegree of non-source nodes at most d, algorithm* GenSeq1UIP *produces a branching sequence σ of size at most S in time $\Theta(dS)$, where $S = |Pbl_G| \approx |Pbl_G^{SAT}|$. Moreover, the sequence σ is complete for Pbl_G as well as for Pbl_G^{SAT} under any clause learning algorithm using fast backtracking and 1UIP learning scheme (such as* zChaff*).*

Proof. The size and running time bounds follow from the previous discussion in this section. That this sequence is complete can be verified by a simple hand calculation simulating clause learning with fast backtracking and 1UIP learning scheme.

4   Experimental Results

We conducted experiments on a Linux machine with a 1600 MHz AMD Athelon processor, 256 KB cache and 1024MB RAM. Time limit was set to 1 day and memory limit to 512MB; the program was set to abort as soon as either of these was exceeded. We took the base code of zChaff [14] and modified it to incorporate branching sequence given as part of the input, along with a CNF formula. When a branching sequence is specified but gets exhausted before a satisfying assignment is found or the formula is proved to be unsatisfiable, the code reverts to the default variable selection scheme of zChaff. We analyzed the performance with random restarts turned off. For all other parameters, we used the default values of zChaff.

Table 1 shows the performance on grid pebbling formulas. Results are reported for zChaff with no learning or specified branching sequence (DPLL), with specified branching sequence only, with clause learning only (original zChaff), and both. Table 2 shows similar results for randomized pebbling formulas. In both cases, the

Table 1. zChaff on grid pebbling formulas. Note that problem size substantially increases as we move down the table. ‡ denotes out of memory

	Grid formula		Runtime in seconds	
Solver	Layers	Variables	Unsatisfiable	Satisfiable
DPLL	5	30	0.24	0.12
	6	42	110	0.02
	7	56	> 24 hrs	0.07
	8	72	> 24 hrs	> 24 hrs
Branching	5	30	0.20	0.00
sequence	6	42	105	0.00
only	7	56	> 24 hrs	0.00
	9	90	> 24 hrs	> 24 hrs
Clause	20	420	0.12	0.05
learning	40	1,640	59	36
only (original	65	4,290	‡	47
zChaff)	70	4,970	‡	‡
Clause	100	10,100	0.59	0.62
learning +	500	250,500	254	288
branching	1,000	1,001,000	4,251	5,335
sequence	1,500	2,551,500	21,097	‡

branching sequence used was generated according to Algorithm 1, GenSeq1UIP. Note that randomized pebbling graphs typically have a more complex structure. In addition, higher indegree and larger disjunction labels make both the CNF formula size as well as the required branching sequence larger. This explains the difference between the performance of zChaff, both original and modified, on grid and randomized pebbling instances. For all instances considered, the time taken to generate the branching sequence from an input graph was substantially less than that for generating the pebbling formula itself.

5   Discussion and Future Work

This paper has developed the idea of using a high level description of a satisfiability problem for generating auxiliary information that can guide a SAT algorithm trying to solve it. Our preliminary experimental results show a clear exponential improvement in performance when such information is used to solve both grid and randomized pebbling problems. Although somewhat artificial, these problems are interesting in their own right and provide hard instances for some of the best existing SAT solvers like zChaff. This bolsters our belief that high level structure can be recovered and exploited to make clause learning more efficient.

Extending our results to more practical problems such as generalized STRIPS planning [35] is an obvious direction for future work. These problems induce a natural

Table 2. zChaff on random pebbling formulas with distinct labels, indegree ≤ 5, and disjunction label size ≤ 6. Note that problem size increases as we move down the table. ‡ denotes out of memory

Solver	Random pebbling formula			Runtime in seconds	
	Nodes	*Variables*	*Clauses*	*Unsatisfiable*	*Satisfiable*
	9	33	300	0.00	0.00
DPLL	10	29	228	0.58	0.00
	10	48	604	> 24 hrs	> 24 hrs
Branching	10	29	228	0.09	0.00
sequence	10	48	604	115	0.02
only	12	42	2,835	> 24 hrs	8.88
	12	43	1,899	> 24 hrs	> 24 hrs
Clause	50	154	3,266	0.91	0.03
learning	87	296	9,850	‡	65
only (original	109	354	11,106	584	0.78
zChaff)	110	354	18,467	‡	‡
Clause	110	354	18,467	0.28	0.29
learning +	4,427	14,374	530,224	48	49
branching	7,792	25,105	944,846	181	‡
sequence	13,324	43,254	1,730,952	669	‡

layered graph structure similar to but more complicated than pebbling graphs. A similar layered structure is seen in bounded model checking problems [36]. We hope that some of the ideas mentioned in this paper will help relate pebbling with planning and bounded model checking, and allow one to use our solution for the former to create an effective strategy for the latter. On another front, there has been a lot of work on generating variable orders for BDD (Binary Decision Diagram) based algorithms (see *e.g.* [37, 38]), where using a good order is perhaps even more critical. Some of the ideas there extend to the BED (Boolean Expression Diagram) model [39] which combines BDDs with propositional satisfiability for model checking. There has also been work on using BDD variable orders for DPLL algorithms without learning [40]. It would be interesting to see if any of these variable ordering strategies provide new ways of capturing structural information in our context.

The form in which we extract and use problem structure is a branching sequence. Although capable of capturing more information than a static variable order, branching sequences suffer from a natural drawback. The exactness they seem to require for pebbling formulas might pose problems when we try to generate branching sequences for harder problems where we know that a polynomial size sequence is unlikely to exist. The usefulness of an incomplete or approximately perfect branching sequence is still unclear. It is not unlikely that we get substantial (but not exponential) improvement as long as the approximate sequence makes correct decisions most of the time, especially near the top of the underlying DPLL tree. However, this needs to be tested experimentally.

Finally, the entire approach of generating auxiliary information by analyzing the problem domain has the inherent disadvantage of requiring the knowledge of higher level problem description. This makes it different from blackbox approaches that try to extract structure after the problem has been translated into a CNF formula. This precluded us, for example, from testing the approach on most of the standard CNF benchmarks for which such a description is not available. However, given that the source of non-random formulas encoding real-world problems is always a high level description, this, we believe, is not a real drawback.

References

[1] Beame, P., Kautz, H., Sabharwal, A.: Understanding the power of clause learning. In: Proceedings of the 18th International Joint Conference on Artificial Intelligence, Acapulco, Mexico (2003) To appear.

[2] Velev, M., Bryant, R.: Effective use of boolean satisfiability procedures in the formal verification of superscalar and vliw microprocessors. In: Proceedings of the 38th Design Automation Conference. (2001) 226–231

[3] Biere, A., Cimatti, A., Clarke, E.M., Fujita, M., Zhu, Y.: Symbolic model checking using SAT procedures instead of BDDs. In: Proceedings of the 36th Design Automation Conference, New Orleans, LA (1999) 317–320

[4] Zhang, H., Hsiang, J.: Solving open quasigroup problems by propositional reasoning. In: Proceedings of the International Computer Symp., Hsinchu, Taiwan (1994)

[5] Konuk, H., Larrabee, T.: Explorations of sequential ATPG using boolean satisfiability. In: 11th VLSI Test Symposium. (1993) 85–90

[6] Gomes, C.P., Selman, B., McAloon, K., Tretkoff, C.: Randomization in backtrack search: Exploiting heavy-tailed profiles for solving hard scheduling problems. In: Proceedings of the 4th International Conference on Artificial Intelligence Planning Systems, Pittsburgh, PA (1998)

[7] Davis, M., Putnam, H.: A computing procedure for quantification theory. Communications of the ACM 7 (1960) 201–215

[8] Davis, M., Logemann, G., Loveland, D.: A machine program for theorem proving. Communications of the ACM 5 (1962) 394–397

[9] Li, C.M., Anbulagan: Heuristics based on unit propagation for satisfiability problems. In: IJCAI (1). (1997) 366–371

[10] Gomes, C.P., Selman, B., Kautz, H.: Boosting combinatorial search through randomization. In: Proceedings, AAAI-98: 15th National Conference on Artificial Intelligence, Madison, WI (1998) 431–437

[11] Bayardo Jr., R.J., Schrag, R.C.: Using CST look-back techniques to solve real-world SAT instances. In: Proceedings, AAAI-97: 14th National Conference on Artificial Intelligence. (1997) 203–208

[12] Marques-Silva, J.P., Sakallah, K.A.: GRASP – a new search algorithm for satisfiability. In: Proceedings of the International Conference on Computer Aided Design, San Jose, CA, ACM/IEEE (1996) 220–227

[13] Zhang, H.: SATO: An efficient propositional prover. In: Proceedings of the International Conference on Automated Deduction, LNAI. Volume 1249. (1997) 272–275

[14] Moskewicz, M.W., Madigan, C.F., Zhao, Y., Zhang, L., Malik, S.: Chaff: Engineering an efficient SAT solver. In: Proceedings of the 38th Design Automation Conference, Las Vegas, NV, ACM/IEEE (2001) 530–535

[15] Zhang, L., Madigan, C.F., Moskewicz, M.H., Malik, S.: Efficient conflict driven learning in a boolean satisfiability solver. In: Proceedings of the International Conference on Computer Aided Design, San Jose, CA, ACM/IEEE (2001) 279–285

[16] de Kleer, J., Williams, B.C.: Diagnosing multiple faults. Artificial Intelligence **32** (1987) 97–130

[17] Stallman, R., Sussman, G.J.: Forward reasoning and dependency-directed backtracking in a system for computer-aided circuit analysis. Artificial Intelligence **9** (1977) 135–196

[18] Genesereth, R.: The use of design descriptions in automated diagnosis. Artificial Intelligence **24** (1984) 411–436

[19] Davis, R.: Diagnostic reasoning based on structure and behavior. Artificial Intelligence **24** (1984) 347–410

[20] Li, C.M., Anbulagan: Heuristics based on unit propagation for satisfiability problems. In: Proceedings of the 15th International Joint Conference on Artificial Intelligence, Nagoya, Japan (1997) 366–371

[21] Mézard, M., Zecchina, R.: Random k-satisfiability problem: From an analytic solution to an efficient algorithm. Physical Review E **66** (2002) 056126

[22] Ghallab, M., Howe, A., Knoblock, C., McDermott, D., Ram, A., Veloso, M., Weld, D., Wilkins, D.: PDDL – the planning domain definition language. Technical report, Yale University, New Haven, CT (1998)

[23] Johnson, D.S., Trick, M.A., eds.: Cliques, Coloring and Satisfiability: Second DIMACS Implementation Challenge. Volume 26 of DIMACS Series in Discrete Mathematics and Theoretical Computer Science. American Mathematical Society (1996)

[24] Giunchiglia, E., Maratea, M., Tacchella, A.: Dependent and independent variables in propositional satisfiability. In: Proceedings of the 8th European Conference on Logics in Artificial Intelligence (JELIA). Volume 2424 of Lecture Notes in Computer Science., Cosenza, Italy, Springer-Verlag (2002) 296–307

[25] Ostrowski, R., Grégoire, E., Mazure, B., Sais, L.: Recovering and exploiting structural knowledge from cnf formulas. In: 8th Principles and Practice of Constraint Programming. Volume 2470 of Lecture Notes in Computer Science., Ithaca, NY, Springer-Verlag (2002) 185–199

[26] Aloul, F.A., Ramani, A., Markov, I.L., Sakallah, K.A.: Solving difficult SAT instances in presence of symmetry. In: Proceedings of the 39th Design Automation Conference, New Orleans, LA (2002) 731–736

[27] Brafman, R.I.: A simplifier for propositional formulas with many binary clauses. In: Proceedings of the 17th International Joint Conference on Artificial Intelligence, Seattle, WA (2001) 515–522

[28] Amir, E., McIlraith, S.A.: Partition-based logical reasoning. In: Proceedings of the 7th International Conference on Principles of Knowledge Representation and Reasoning, Breckenridge, CO (2000) 389–400

[29] Shtrichman, O.: Tuning SAT checkers for bounded model checking. In: Proceedings of the 12th International Conference on Computer Aided Verification, Chicago, IL (2000) 480–494

[30] Kautz, H.A., Selman, B.: Pushing the envelope: Planning, propositional logic, and stochastic search. In: Proceedings, AAAI-96: 13th National Conference on Artificial Intelligence, Portland, OR (1996) 1194–1201

[31] Bonet, M.L., Esteban, J.L., Galesi, N., Johansen, J.: On the relative complexity of resolution refinements and cutting planes proof systems. SIAM Journal on Computing **30** (2000) 1462–1484

[32] Bonet, M.L., Galesi, N.: A study of proof search algorithms for resolution and polynomial calculus. In: Proceedings 40th Annual Symposium on Foundations of Computer Science, New York, NY, IEEE (1999) 422–432

[33] Ben-Sasson, E., Impagliazzo, R., Wigderson, A.: Near-optimal separation of treelike and general resolution. Technical Report TR00-005, Electronic Colloquium in Computation Complexity, http://www.eccc.uni-trier.de/eccc/ (2000)

[34] Beame, P., Impagliazzo, R., Pitassi, T., Segerlind, N.: Memoization and DPLL: Formula caching proof systems. In: Proceedings 18th Annual IEEE Conference on Computational Complexity, Aarhus, Denmark (2003) To appear.

[35] Kautz, H.A., Selman, B.: Planning as satisfiability. In: Proceedings of the 10th European Conference on Artificial Intelligence, Vienna, Austria (1992) 359–363

[36] Biere, A., Cimatti, A., Clarke, E.M., Zhu, Y.: Symbolic model checking without BDDs. In: Proceedings of the 5th International Conference on Tools and Algorithms for Construction and Analysis of Systems, Amsterdam, The Netherlands (1999) 193–207

[37] Aziz, A., Tasiran, S., Brayton, R.K.: BDD variable orderings for interacting finite state machines. In: Proceedings of the 31th Design Automation Conference, San Diego, CA (1994) 283–288

[38] Harlow, J.E., Brglez, F.: Design of experiments in BDD variable ordering: Lessons learned. In: Proceedings of the International Conference on Computer Aided Design, San Jose, CA (1998) 646–652

[39] Hulgaard, H., Williams, P.F., Andersen, H.R.: Equivalence checking of combinational circuits using boolean expression diagrams. IEEE Transactions on Computer-Aided Design of Integrated Circuits **18** (1999) 903–917

[40] Reda, S., Drechsler, R., Orailoglu, A.: On the relation between SAT and BDDs for equivalence checking. In: Proceedings of the International Symposium on Quality Electronic Design, San Jose, CA (2002) 394–399

Abstraction-Driven SAT-based Analysis of Security Protocols*

Alessandro Armando and Luca Compagna

DIST – Università degli Studi di Genova, Viale Causa 13 – 16145 Genova, Italy,
{armando,compa}@dist.unige.it

Abstract. In previous work we proposed an approach to the automatic translation of protocol insecurity problems into propositional logic
with the ultimate goal of building an automatic model-checker for security protocols based on state-of-the-art SAT solvers. In this paper we
present an improved procedure based on an abstraction/refinement strategy which, by interleaving the encoding and solving phases, leads to a
significant improvement of the overall performance of our model-checker.

1 Introduction

In spite of their apparent simplicity, security protocols are notoriously error-
prone. Many published protocols have been implemented and deployed in real
applications only to be found flawed years later. For instance, the Needham-
Schroeder authentication protocol [28] was found vulnerable to a serious attack
17 years after its publication [22]. The problem is that security protocols are
intended to be used in open, hostile environments (such as, e.g., the Internet)
and therefore steps carried out by honest participants interplay with the ability
of malicious agents to eavesdrop, divert, and even forge new messages. This results in a combinatorially high number of actual protocol runs whose analysis
stretches traditional validation techniques (e.g. testing). This problem has stim-
ulated the development of formal method techniques for this novel application
domain either by adapting existing techniques (such as, e.g., in [16, 29, 30]) or
by devising new ones (such as, e.g., in [5, 7, 24, 12, 14, 20, 25]).

In [3] we proposed an approach to the automatic translation of protocol
insecurity problems into propositional logic with the ultimate goal of building
a model-checker for security protocols based on state-of-the-art SAT solvers.
The approach combines a reduction of protocol insecurity problems to planning
problems with well-known SAT-reduction techniques developed for planning.
We have built SATMC, a model-checker for security protocols based on these

---

* We wish to thank Enrico Giunchiglia, Marco Maratea, Massimo Narizzano, and Armando Tacchella for many useful and stimulating discussions and their support in
the use of the SIM solver. This work was partially funded by the FET Open EC
Project AVISPA (IST-2001-39252), and by MIUR. Moreover, the second author was
partially supported by the IHP-RTN EC project CALCULEMUS (HPRN-CT-2000-
00102).

E. Giunchiglia and A. Tacchella (Eds.): SAT 2003, LNCS 2919, pp. 257–271, 2004.

ideas, which readily finds attacks on a set of well-known authentication protocols. Motivated by the observation that the time spent in generating the propositional formulae largely dominates the time spent by the SAT solver, in this paper we introduce an abstraction/refinement strategy that significantly improves the encoding time and hence the overall performance of the model-checker. We show that the performance of SATMC depends not only on the speed of the SAT solver but also on the ability of the SAT solver to return partial models.

Structure of the paper. In Section 2 we introduce the notion of protocol insecurity problem and show its relationship with the notion of planning problem. In Section 3 we introduce techniques for encoding bounded planning problems into SAT. In Section 4 we present our improved model-checking procedure based on abstraction and refinement. In Section 5 we present experimental results obtained with our model-checker which confirm the better performance of the improved procedure. In Section 6 we discuss the related work and finally, in Section 7, we draw some conclusions and outline the future work.

2   Protocol Insecurity Problems as Planning Problems

To illustrate, consider the following one-way authentication protocol:

$$(1) \quad A \to B : \{N\}_K$$
$$(2) \quad B \to A : \{f(N)\}_K$$

where N is a nonce[1] generated by Alice, K is a symmetric key, f is a function known to Alice and Bob, and $\{X\}_K$ denotes the result of encrypting text X with key K. Successful execution of the protocol should convince Alice that she has been talking with Bob, since only Bob could have formed the appropriate response to the message issued in (1). In fact, Ivory can deceit Alice into believing that she is talking with Bob whereas she is talking with her. This is achieved by executing concurrently two sessions of the protocol and using messages from one session to form messages in the other as illustrated by the following protocol trace:

$$(1.1) \quad A \to I(B) : \{N\}_K$$
$$(2.1) \quad I(B) \to A : \{N\}_K$$
$$(2.2) \quad A \to I(B) : \{f(N)\}_K$$
$$(1.2) \quad I(B) \to A : \{f(N)\}_K$$

Alice starts the protocol with message (1.1). Ivory intercepts the message and (pretending to be Bob) starts a second session with Alice by replaying the received message—cf. step (2.1). Alice replies to this message with message (2.2). But this is exactly the message Alice is waiting to receive in the first protocol session. This allows Ivory to finish the first session by using it—cf. (1.2). At the end of the above steps Alice believes she has been talking with Bob, but this is obviously not the case.

[.] *Nonces* are numbers generated by principals that are intended to be used *only once*.

A problem with the rule-based notation used above to specify security protocols is that it leaves implicit many important details such as the shared information and how the principals should react to messages of an unexpected form. This kind of description is usually supplemented with explanations in natural language which in our case explain that N is a nonce generated by Alice, that f is a function known to the honest participants only, and that K is a shared key. To cope with the above difficulties and pave the way to the formal analysis of protocols a set of models and specification formalisms have been put forward [8, 23]. In this paper we use a planning language *a la* STRIPS to this end.

A *planning problem* is a tuple $\Pi = \langle \mathcal{F}, \mathcal{A}, Ops, I, \mathcal{G} \rangle$, where $\mathcal{F}$ and $\mathcal{A}$ are disjoint sets of variable-free atomic formulae of a sorted first-order language called *fluents* and *actions* respectively; Ops is a set of expressions of the form $(Pre \xrightarrow{Act} Add ; Del)$ where $Act \in \mathcal{A}$ and Pre, Add, and Del are finite sets of fluents such that $Add \cap Del = \emptyset$; I is a set of fluents representing the initial state and $\mathcal{G}$ is a boolean combinations of fluents representing the final states. A state is represented by a set S of fluents, meaning that all the fluents in S hold in the state, while all the fluents in $\mathcal{F} \setminus S$ do not hold in the state (close-world-assumption). An action is applicable in a state S iff the action preconditions (fluents in Pre) occur in S and the application of the action leads to a new state obtained from S by removing the fluents in Del and adding those in Add. A *solution to a planning problem Π*, called *plan*, is a sequence of actions whose execution leads from the initial state to a final state and the preconditions of each action appears in the state to which it applies. Plans can be represented in a flexible way by means of partial-order plans. A *partial-order plan* is a pair $\langle \Lambda, \prec \rangle$ where $\Lambda \subseteq \mathcal{A} \times N$ and $\prec$ is a partial order on the set of natural numbers N. A plan $\mathcal{P}$ is in the set of the plans denoted by a partial-order plan $\langle \Lambda, \prec \rangle$ iff there exists a bijection $\pi : \Lambda \to \{1, \ldots, |\Lambda|\}$ such that if $i_1 \prec i_2$ then $\pi(\alpha_1, i_1) < \pi(\alpha_2, i_2)$; if $\pi(\alpha, i) = k$ then α is the k-th action in $\mathcal{P}$. The *length* of the partial-order plan $\langle \Lambda, \prec \rangle$ is the cardinality of the set $\{i \mid \exists \alpha \in \mathcal{A} \text{ s.t. } \langle \alpha, i \rangle \in \Lambda\}$. For instance, the partial-order plan $\langle \{\langle a, 0 \rangle, \langle b, 0 \rangle, \langle c, 3 \rangle, \langle d, 5 \rangle, \langle a, 5 \rangle\}, \{0 \prec 3, 0 \prec 5, 3 \prec 5\} \rangle$ has length 3 and represents the set of plans $\{\langle a, b, c, d, a \rangle, \langle b, a, c, d, a \rangle, \langle a, b, c, a, d \rangle, \langle b, a, c, a, d \rangle\}$.

A protocol insecurity problem is given by a transition relation specifying the run allowed by the protocol plus an initial state and a set of bad states (i.e. states whose reachability implies the violation of the desired security properties). Protocol insecurity problems can be easily recast as planning problems. The states of the transition system model the state of the honest principals, the knowledge of the intruder, as well as the messages sent over the channel but not yet processed by the intended recipient (or diverted by the intruder). Actions model the legal transitions that can be performed by honest participants as well as the abilities of the intruder.

For the simple protocol above, fluents are of the form:

- $i(t)$, meaning that the intruder knows the term t;
- *fresh*(n), meaning that the term (usually is a constant representing a nonce) n has not been used yet;

- $m(j, s, r, t)$, meaning that a message t has been sent (supposedly) from principal s to principal r at protocol step j, and
- $w(j, s, r, [t_1, \ldots, t_k])$, representing the state of execution of principal r at step j; it means that r knows the terms $t_1, \ldots, t_k$ at step j, and (if $j = 1, 2$) that r is waiting for a message from s for step j to be executed.

The initial state of the system is:[2]

$$w(0, a, a, [\,]) \centerdot w(1, a, b, [\,]) \centerdot w(0, b, b, [\,]) \centerdot w(1, b, a, [\,])$$
$$\centerdot \mathit{fresh}(n_1) \centerdot \mathit{fresh}(n_2) \centerdot i(a) \centerdot i(b)$$

Fluents $w(0, a, a, [\,])$, $w(1, a, b, [\,])$, $w(0, b, b, [\,])$, and $w(1, b, a, [\,])$ state that principals a and b are ready to play both the role of the initiator and of the responder. Fluents $\mathit{fresh}(n_1)$ and $\mathit{fresh}(n_2)$ state that n_1 and n_2 are fresh numbers (i.e. they have never been used before) and then that they can be used as nonces. Finally $i(a)$ and $i(b)$ state that the identities of a and b are known to the intruder.

The behavior of the honest principals and of the intruder is specified by means of operators. The activity of sending the first message is modeled by:[3]

$$w(0, A, A, [\,]) \centerdot \mathit{fresh}(N) \xrightarrow{\mathit{step}_1(A,B,N)} m(1, A, B, \{N\}_k) \centerdot w(2, B, A, [f(N)]) \ ;$$
$$w(0, A, A, [\,]) \centerdot \mathit{fresh}(N)$$

Intuitively, this operator says that, if in the current global state there is an initiator A ready to start the protocol with somebody, and a fresh term N is available, then A can send the first message of the protocol to a responder B. By doing this, A will update its state and, of course, N will not be usable as fresh term anymore. Notice that nonce $f(N)$ is added to the acquired knowledge of A for subsequent use. The receipt of the message and the reply of the responder is modeled by:

$$m(1, A, B, \{N\}_k) \centerdot w(1, A, B, [\,]) \xrightarrow{\mathit{step}_2(A,B,N)} m(2, B, A, \{f(N)\}_k) \centerdot w(3, B, B, [\,]) \ ;$$
$$m(1, A, B, \{N\}_k) \centerdot w(1, A, B, [\,])$$

The final step of the protocol is modeled by:

$$m(2, B, A, \{f(N)\}_k) \centerdot w(2, B, A, [f(N)]) \xrightarrow{\mathit{step}_3(A,B,N)} w(4, A, A, [\,]) \ ;$$
$$m(2, B, A, \{f(N)\}_k) \centerdot w(2, B, A, [f(N)])$$

where steps 3 and 4 occurring as first parameter in w-fluents are used to denote the final state of the responder and of the initiator, respectively.

The following rule models the ability of the intruder of diverting the information exchanged by the honest participants:

$$m(J, S, R, T) \xrightarrow{\mathit{divert}(J,R,S,T)} i(R) \centerdot i(S) \centerdot i(T) \ ; \ m(J, S, R, T)$$

[.] To improve readability we use the "$\centerdot$" operator as set constructor. For instance, we write "$x \centerdot y \centerdot z$" to denote the set $\{x, y, z\}$.

[.] Here and in sequel we use capital letters to denote variables.

The ability of encrypting and decrypting messages is modeled by:

$$i(T) \centerdot i(K) \xrightarrow{encrypt(K,T)} i(\{T\}_K) \; ; \tag{1}$$

$$i(\{T\}_K) \centerdot i(K) \xrightarrow{decrypt(K,T)} i(T) \; ; \tag{2}$$

Finally, the intruder can send arbitrary messages possibly faking somebody-else's identity in doing so:

$$i(T) \centerdot i(S) \centerdot i(R) \xrightarrow{fake(J,R,S,T)} m(J, S, R, T) \; ;$$

In particular, this operator states that, if the intruder knows a term T and two agent identities S and R, then it can send on the network channel the message T to R impersonating the identity of S. Notice that with the above rules we represent the most general intruder based on the Dolev-Yao model [15]. In this model the intruder has the ability of eavesdropping and diverting messages as well as that of composing, decomposing, encrypting, and decrypting messages (provided the encryption keys are known).[4] Finally, he can send those messages to other participants with a false identity.

A security protocol is intended to enjoy a specific security property. In our example this property is the ability of authenticating Bob to Alice. A security property can be specified by providing a set of "bad" states, i.e. states whose reachability implies a violation of the property. For instance, any state containing a subset of fluents of the form $w(4, A, A, [\,]) \centerdot w(1, A, B, [\,])$ (i.e. A has finished a run of the protocol as initiator and B is still at the beginning of the protocol run as responder) witnesses a violation of the expected authentication property and therefore it should be considered as a bad state. It is easy to build a propositional formula $\mathcal{G}$ such that each model of $\mathcal{G}$ represents a bad state. For the above example $\mathcal{G} \equiv (w(4, a, a, [\,]) \wedge w(1, a, b, [\,])) \vee (w(4, b, b, [\,]) \wedge w(1, b, a, [\,]))$.

3   Automatic SAT-compilation of Planning Problems

Let $\Pi = \langle \mathcal{F}, \mathcal{A}, Ops, I, \mathcal{G} \rangle$ be a planning problem with finite $\mathcal{F}$ and $\mathcal{A}$ and let n be a positive integer, then it is possible to build a propositional formula Φ_Π^n such that any model of Φ_Π^n corresponds to a partial-order plan of length n representing solutions of Π. The encoding of a planning problem into a set of SAT formulae can be done in a variety of ways (see [21, 17] for a survey). The basic idea is to add an additional time-index to actions and fluents to indicate the state at which the action begins or the fluent holds. Fluents are thus indexed by 0 through n and actions by 0 through $n-1$. If p is a fluent or an action and i is an index in the appropriate range, then $i\!:\!p$ is the corresponding time-indexed propositional variable.

---

[4] In other words we do the *perfect encryption assumption*. It turns out that many protocols are flawed even under this strong assumption.

It is worth pointing out that the reduction of planning problems to SAT paves the way to an automatic SAT-compilation of protocol insecurity problems. However a direct application of the approach is not viable, because the resulting encodings are of unmanageable size even for simple protocols. To overcome this difficulty in [3] we put forward a number of optimizing transformations on protocols insecurity problems that make the approach both feasible and effective on many protocols of interest. For instance, certain rules can be merged together without loosing any possible attack on the protocol thereby decreasing the number of actions and hence the size of the resulting SAT encodings. A detailed description of the optimizing transformations goes beyond the scope of this paper and the interested reader should consult [3].

In the rest of this section we will formally describe the standard linear encoding and a combination of the previous one with the bitwise representation of actions.

3.1   Standard Linear Encoding

By using the standard linear encoding, Φ_Π^n is the smallest set (intended conjunctively) such that:

- **Initial State Axioms:** for all fluents f, if $p \in I$ then $0\!:\!p \in \Phi_\Pi^n$, else $\neg 0\!:\!p \in \Phi_\Pi^n$;
- **Goal State Axioms:** $\mathcal{G}_n \in \Phi_\Pi^n$, where $\mathcal{G}_n$ is the formula $\mathcal{G}$ in which each fluent occurring in it is indexed by n;
- **Universal Axioms:** for each $(Pre \xrightarrow{\alpha} Add \,;\, Del) \in Ops$ and $i = 0, \ldots, n-1$:

$$(i\!:\!\alpha \supset \bigwedge\{i\!:\!f \mid f \in Pre\}) \in \Phi_\Pi^n$$
$$(i\!:\!\alpha \supset \bigwedge\{(i+1)\!:\!f \mid f \in Add\}) \in \Phi_\Pi^n$$
$$(i\!:\!\alpha \supset \bigwedge\{\neg(i+1)\!:\!f \mid f \in Del\}) \in \Phi_\Pi^n$$

- **Explanatory Frame Axioms:** for all fluents f and $i = 0, \ldots, n-1$:

$$(i\!:\!f \wedge \neg(i+1)\!:\!f) \supset \bigvee \left\{ i\!:\!\alpha \mid (Pre \xrightarrow{\alpha} Add \,;\, Del) \in Ops, f \in Del \right\} \in \Phi_\Pi^n$$
$$(\neg i\!:\!f \wedge (i+1)\!:\!f) \supset \bigvee \left\{ i\!:\!\alpha \mid (Pre \xrightarrow{\alpha} Add \,;\, Del) \in Ops, f \in Add \right\} \in \Phi_\Pi^n$$

- **Conflict Exclusion Axioms (CEA):** for $i = 0, \ldots, n-1$:

$$\neg(i\!:\!\alpha_1 \wedge i\!:\!\alpha_2) \in \Phi_\Pi^n$$

for all $\alpha_1 \neq \alpha_2$ such that $(Pre_1 \xrightarrow{\alpha_1} Add_1 \,;\, Del_1) \in Ops$, $(Pre_2 \xrightarrow{\alpha_2} Add_2 \,;\, Del_2) \in Ops$, and $Pre_1 \cap Del_2 \neq \emptyset$ or $Pre_2 \cap Del_1 \neq \emptyset$.

It is immediate to see that the number of literals in Φ_Π^n is in $O(n|\mathcal{F}|+n|\mathcal{A}|)$. Moreover the number of clauses generated by the Universal Axioms is in $O(nP_0|\mathcal{A}|)$ where P_0 is the maximal number of fluents mentioned in a list of an operator (usually a small number); the number of clauses generated by the Explanatory Frame Axioms is in $O(n|\mathcal{F}|)$; finally, the number of clauses generated by the CEA is in $O(n|\mathcal{A}|^2)$. For instance, the application of the above encoding to the *EKE* protocol (with $n = 5$) generates a propositional formula with 61,508 atoms and 13,948,534 clauses of whom 13,165,050 (about 94 %) are due to the CEA.

3.2   Linear Encoding Combined with Bitwise Action Representation

Since Conflict Exclusion Axioms (CEA) grow quadratically in the number of actions, the application of the standard linear encoding to asynchronous concurrent systems can generate huge propositional formulae. The CEA are useful to restrict which actions may occur simultaneously and therefore to guarantee that any model of the propositional formula would correspond to a partial-order plan representing solutions (plans) of the original planning problem.

The bitwise action representation does not require CEA. In order to represent univocally each action in $\mathcal{A} = \{\alpha^1, \ldots, \alpha^{|\mathcal{A}|}\}$ $k = \lceil log_2|\mathcal{A}|\rceil$ propositional variables $(b^1, \ldots, b^k)$ are used. Let $\|\alpha^j\|$ the propositional formula whose model identifies the bitwise representation of the action α^j, then the combination of the standard linear encoding with the Bitwise Action Representation (BAR) results in the propositional formula Φ_Π^n defined as in 3.1 where the conflict exclusion axioms are neglected and each occurrence of $i{:}\alpha$ is replaced by $i{:}\|\alpha\|$.[5] The number of literals in Φ_Π^n decreases, wrt the standard linear encoding, to $O(n|\mathcal{F}| + nk)$. However, while the number of clauses generated by the Universal Axioms does not change and the conflict exclusion axioms are not required, the number of clauses generated by the Explanatory Frame Axioms combined with the bitwise action representation becomes, in the worst case, exponential in the number of actions. Precisely, it is in $O(nk^{|\mathcal{A}|}|\mathcal{F}|)$.

To avoid this exponential growth it is sufficient to restore in the Explanatory Frame Axioms the propositional variables associated with the actions in place of their bitwise representation and to extend the formula as follow:

– **Bitwise Axioms:** for each $(Pre \xrightarrow{\alpha} Add \,;Del) \in Ops$ and $i = 0, \ldots, n-1$:

$$i{:}\|\alpha\| \equiv i{:}\alpha \in \Phi_\Pi^n$$

With respect to the standard linear encoding, the number of literals in Φ_Π^n increases to $O(n|\mathcal{F}| + n|\mathcal{A}| + nk)$, and the number of clauses in Φ_Π^n is neither quadratic nor exponential in the number of actions. In fact the number of clauses generated by the Bitwise Axioms is in $O(n|\mathcal{A}| + nk|\mathcal{A}|)$. For instance, the application of this encoding to the *EKE* protocol with $n = 5$ generates a propositional formula with 61,578 atoms (among these 70 atoms are used to represent the actions) and 1,630,044 clauses (among these 846,560, i.e. about 52 %, are results from the bitwise axioms). From here on we call *bitwise encoding* this "smart" variant of the standard linear encoding combined with the bitwise action representation.

An important difference between the standard linear encoding and the bitwise encoding is that the former allows for some actions to be executed in parallel (at the same step), while the latter imposes that one (and only one) action can be executed at each step. Hence solutions found at step $\overline{n}$ by applying the standard linear encoding will be found at step $n \geq \overline{n}$ if the bitwise encoding is applied.

---

[5] Notice that $i{:}\|\alpha\|$ is the formula $\|\alpha\|$ in which each propositional variable occurring in it, is indexed by i.

4   The Abstraction/Refinement Strategy

Abstraction [11] is a powerful techniques for the analysis of large-state systems. The idea is to abstract the system under consideration into simpler one that that has all the behaviors of the original one, but may also exhibit some spurious behaviors. Thus—by construction—if a safety property holds for the abstract system, it will also hold for the concrete one. On the other hand, a counterexample for the abstract system does not always correspond to a counterexample for the concrete one. If the behavior that violates the safety property in the abstract system cannot be reproduced in the concrete system, the counterexample is said to be *spurious*. When such a counterexample is found, the abstraction must be refined in order to eliminate the spurious behavior. The procedure is then iterated until either a non spurious counterexample is found, or the abstract system satisfies the safety property.

More precisely, let P be a set of propositional letters, we define a *labeled transition system* (LTS) to be a tuple $\langle \Sigma, \mathcal{I}, \mathcal{L}, \mathcal{R}, v \rangle$, where Σ is a set of states, $\mathcal{I} \subseteq \Sigma$ is the set of initial states, $\mathcal{L}$ is a set of transition labels, $\mathcal{R} \subseteq \{s_0 \overset{l}{\rightsquigarrow} s_1 \mid s_0, s_1 \in \Sigma; \ l \in \mathcal{L}\}$ is the labeled transition relation[6], and v is a total function mapping states into 2^P. A *computation path of length k* of a LTS is a sequence $s_0 \overset{l_1}{\rightsquigarrow} s_1 \cdots s_k \overset{l_{k+1}}{\rightsquigarrow} s_{k+1}$ such that $k \geq 0$, $s_0 \subseteq \mathcal{I}$, and $s_i \overset{l_{i+1}}{\rightsquigarrow} s_{i+1} \in \mathcal{R}$, for each $i = 0, \ldots, k$. Let $M = \langle \Sigma, \mathcal{I}, \mathcal{L}, \mathcal{R}, v \rangle$ and $M^a = \langle \Sigma^a, \mathcal{I}^a, \mathcal{L}^a, \mathcal{R}^a, v^a \rangle$ be two LTSs and let $h : \Sigma \rightarrow \Sigma^a$ be a total (abstraction) function, then M^a *is an abstraction of M* w.r.t. h, iff the following conditions hold:

– for every state s in $\mathcal{I}$ there exists a state s^a in $\mathcal{I}^a$ such that $h(s) = s^a$;
– for all states $s_0, s_1 \in \Sigma$ and $s_0^a \in \Sigma^a$ with $h(s_0) = s_0^a$, if $s_0 \overset{l}{\rightsquigarrow} s_1 \in \mathcal{R}$ then there exists a state $s_1^a \in \Sigma^a$ such that $s_0^a \overset{l^a}{\rightsquigarrow} s_1^a \in \mathcal{R}^a$ and $h(s_1) = s_1^a$.

If M^a is an abstraction of M, then it is simple to prove that for every computation path $s_0 \overset{l_1}{\rightsquigarrow} \cdots \overset{l_{k+1}}{\rightsquigarrow} s_{k+1}$ of M there exists a computation path $s_0^a \overset{l_1^a}{\rightsquigarrow} \cdots \overset{l_{k+1}^a}{\rightsquigarrow} s_{k+1}^a$ of M^a such that $h(s_i) = s_i^a$, for each $i = 0, \ldots, k+1$. As a consequence, the following fact (called *preservation theorem*) holds: given a propositional formula φ respecting the abstraction function h,[7] if $M^a \vDash AG\varphi$ then $M \vDash AG\varphi$.[8]

Since in the standard linear encoding, the number of CEA is the main source of difficulty is tempting to avoid their generation altogether. It turns out that the resulting problem is an abstraction of the original problem. To show this, we must define the model-checking problem associated with a planning problem. Let $\Pi = \langle \mathcal{F}, \mathcal{A}, Ops, I, \mathcal{G} \rangle$ be a planning problem, then the model-checking problem associated with Π is $M \vDash AG\neg\mathcal{G}$ where $AG\neg\mathcal{G}$ is the safety property to be checked and $M = \langle \Sigma, \mathcal{I}, \mathcal{L}, \mathcal{R}, v \rangle$ is the concrete LTS such that $v : \Sigma \rightarrow 2^{\mathcal{F}},$[9]

. We restrict our attention to LTS with a total transition relation.
. Following [11] we say that *a propositional formula φ respects an abstraction function h* if for each $s \in \Sigma$, $v(h(s)) \vDash \varphi$ implies $v(s) \vDash \varphi$.
. Note that A and G represent the "for all computation" path quantifier and the "globally" temporal operator, respectively.
. Notice that, given a set S with 2^S we indicate the powerset of S.

$\mathcal{I} = \{s \mid s \in \Sigma, v(s) = I\}$, $\mathcal{L} \subseteq 2^{\mathcal{A}}$ and $\mathcal{R}$ is the set of labeled transitions $s_0 \overset{\{\alpha_1,\dots,\alpha_m\}}{\rightsquigarrow} s_1$ with $m \geq 0$ and such that the following conditions hold:[10]

(1) for each $i = 1, \dots, m$, $(Pre_i \overset{\alpha_i}{\longrightarrow} Add_i \ ; Del_i) \in Ops$;
(2) $\bigcup_{i=1}^{m} Pre_i \subseteq v(s_0)$;
(3) $v(s_1) = (v(s_0) \setminus \bigcup_{i=1}^{m} Del_i) \cup (\bigcup_{i=1}^{m} Add_i)$;
(4) for each $\alpha_i, \alpha_j \in \{\alpha_1, \dots, \alpha_m\}$ such that $i \neq j$, then $Pre_i \cap Del_j = \emptyset$.

Intuitively, from a state s_0 it is possible to reach a state s_1 iff there exists a set of actions such that their preconditions hold in s_0 and for each pair of actions in that set, the conflict exclusion constraint is satisfied (i.e. the precondition list of the first action does not intersect with the delete list of the second and viceversa). By abstracting away the CEA we obtain the abstract LTS $M^a = \langle \Sigma, \mathcal{I}, \mathcal{L}^a, \mathcal{R}^a, v^a \rangle$ whose main difference from the concrete system M lies in the labeled transition relation. In particular, since in the abstract system conflict exclusion is not enforced, $\mathcal{R}^a$ must satisfy only conditions (1), (2), and (3). Hence, all the computation paths allowed for the concrete system are also allowed in the abstract system (that is $\mathcal{R} \subseteq \mathcal{R}^a$) and the preservation theorem holds. On the other hand, if a counterexample $s_0 \overset{\Lambda_1}{\rightsquigarrow} \dots \overset{\Lambda_{k+1}}{\rightsquigarrow} s_{k+1}$ ($s_0 \in \mathcal{I}$, Λ_i is a set of actions for $i = 1, \dots, k+1$, and s_{k+1} is a bad state i.e. it violates $\neg \mathcal{G}$) is found in the abstract system, then, in order to not be spurious, it has to be validated in the concrete system. The validation procedure checks that each step of the counterexample does not involve conflicting actions i.e. it checks condition (4) over the sets Λ_i ($i = 1, \dots, k+1$). When the validation procedure fails, the abstract LTS must be refined by adding to its transition relation the conflict exclusion constraints thereby avoiding the conflicting actions discovered so far in spurious counterexamples. The whole procedure is repeated until either a counterexample without conflicting actions is found, or the abstract system satisfies the safety property.

5   A SAT-based Model-Checker for Security Protocols

We have implemented the above ideas in SATMC, a SAT-based model-checker for security protocol analysis. Given a protocol insecurity problem Ξ, SATMC starts by applying the aforementioned optimizing transformations to Ξ thereby obtaining a new protocol insecurity problem Ξ' which is then translated into a corresponding planning problem $\Pi_{\Xi'}$. $\Pi_{\Xi'}$ is in turn compiled into SAT using one of the methodologies outlined in Section 3 for increasing values of n and the propositional formula generated at each step is fed to a state-of-the-art SAT solver (currently Chaff [27], SIM [18], and SATO [31] are currently supported). As soon as a satisfiable formula is found, the corresponding model is translated back into a partial order plan[11] which is reported to the user.

---

[10] If $m = 0$ then no action is applied and therefore $s. = s.$, i.e. stuttering.
[11] The partial order plan represents attacks on the protocol.

We have run our tool against a selection of (flawed) security protocols drawn from the Clark/Jacob library [10]. For each protocol we have built a corresponding protocol insecurity problem modeling a scenario with a bounded number of sessions in which the involved principals exchange messages on a channel controlled by the most general intruder based on the Dolev-Yao model. Moreover, we assume perfect cryptography (see Section 2) and that all atoms are typed i.e. we exclude type confusion (*strong typing assumption*).[12] Note that since the number of sessions is bounded and strong typing is assumed, then an intruder with finite knowledge is sufficient to find all the possible attacks that can occur in this scenario.

The results of our experiments are reported in Table 1. Each protocol has been analyzed by applying one of the following encoding techniques:[13]

CEA: standard linear encoding with generation of the CEA enabled,
BITWISE: bitwise encoding, and
NoCEA: standard linear encoding with generation of the CEA disabled and therefore combined with the Abstraction Refinement Loop.

For each protocol we give the number of fluents (**F**) and of actions (**Acts**) in the correspondent planning problem, and for each encoding technique we give the smallest value of n at which the attack is found (**N**), the number of propositional variables (**A**) and clauses (**CL**) in the SAT formula (in thousands), the time spent to generate the SAT formula (**EncT**), the time spent by SAT solver to solve the last SAT formula (**Last**), and the total time spent by the SAT solver to solve all the SAT formulae generated for that protocol (**Tot**).[14] In the context of **NoCEA** a comparison between Chaff and SIM has been performed and the solving times are shown together with the number of iterations of the Abstraction Refinement Loop (**#**).

When the **CEA** encoding technique is applied, encoding time largely dominates solving time and this is strictly related to the size of the SAT instances generated. Both the size and the time spent to generate the SAT formulae drop significantly if **BITWISE** is used. This enabled us to encode and find attacks on protocols (e.g. *Andrew*, *KaoChow 2*, *KaoChow 3*, and *Woo-Lam M*) that could not be solved by using the **CEA** encoding technique. However, finding an attack using **BITWISE** can require more iterative-deepening steps (cf. the **N** columns) than those required by applying the standard linear encoding. As a consequence on some instances, such as the *NSCK* and the *SPLICE* protocols, **CEA** performance better than **BITWISE** both in terms of the size the SAT formulae and in encoding time. Moreover, the addition of the bitwise axioms can significantly increase the solving time (see, e.g., *KaoChow 2* and *KaoChow 3* protocols).

As pointed out in [19] type-flaw attacks can be prevented by tagging the fields of a message with information indicating its intended type. Therefore the strong typing assumption is allows us to restrict the search space and focus of the most interesting attacks.

Experiments have been carried out on a PC with a 1.4 GHz CPU and 1 GB of RAM.

Times are measured in seconds.

Table 1. Experiments on protocols from the Clark/Jacob library

Column groups: **CEA**, **BITWISE**, **NoCEA**. Within CEA and BITWISE the sub-columns are N, A(K), CL(K), EncT, Chaff(Last, Tot). Within NoCEA they are N, A(K), CL(K), EncT, Chaff(Last, Tot, #), SIM(Last, Tot, #).

Protocol	F	Acts	CEA N	CEA A(K)	CEA CL(K)	CEA EncT	CEA Chaff Last	CEA Chaff Tot	BIT N	BIT A(K)	BIT CL(K)	BIT EncT	BIT Chaff Last	BIT Chaff Tot	NoCEA N	NoCEA A(K)	NoCEA CL(K)	NoCEA EncT	NoCEA Chaff Last	NoCEA Chaff Tot	NoCEA Chaff #	SIM Last	SIM Tot	SIM #
Andrew	465	15,650	9	145	13,949	7,100	7.6	19.7	11	178	5,348	220.8	4.6	21.6	9	145	2,256	111.4	2.0	12.1	1	1.7	6.2	0
EKE	1,148	10,924	5	62	< 1	0.1	0.0	0.0	5	62	1,630	113.1	1.0	3.1	5	62	783	74.1	0.7	3.7	2	0.0	0.9	0
ISO-CCF-1 U	39	22	4	< 1	6	0.3	0.0	0.1	4	< 1	1	0.1	0.0	0.0	4	< 1	< 1	0.1	0.0	0.0	0	0.0	0.0	0
ISO-CCF-2 M	81	132	4	< 1	2	0.1	0.0	0.0	4	< 1	10	0.5	0.0	0.0	4	< 1	4	0.2	0.0	0.1	0	0.0	0.0	0
ISO-PK-1 U	61	49	4	< 1	17	0.9	0.0	0.0	4	< 1	3	0.2	0.0	0.0	4	< 1	2	0.1	0.1	0.1	0	0.0	0.0	0
ISO-PK-2 M	125	279	4	2	< 1	0.1	0.0	0.0	4	2	22	1.1	0.0	0.0	4	2	10	0.6	0.0	0.0	0	0.1	0.1	0
ISO-SK-1 U	39	25	4	< 1	< 1	0.4	0.0	0.0	4	< 1	1	0.1	0.0	0.0	4	< 1	< 1	0.1	0.0	0.0	0	0.1	0.1	0
ISO-SK-2 M	100	87	4	< 1	3	0.4	0.0	0.0	4	< 1	6	0.5	0.0	0.0	4	< 1	3	0.4	0.0	0.7	1	0.0	0.0	0
KaoChow 1	2,753	2,042	7	36	355	18.4	0.1	0.7	8	41	345	15.8	0.9	1.6	7	36	131	8.0	0.1	0.7	0	0.2	0.6	0
KaoChow 2	32,963	22,344	9	531	35,178	1,494	-	-	11	642	6,244	309.7	180.9	473.1	9	531	1,804	140.4	1.6	15.6	4	9.2	26.4	0
KaoChow 3	49,378	55,727	9	995	-	-	-	-	11	1,206	17,528	1,009.9	303.4	1,063.5	9	995	5,737	585.5	7.5	41.6	5	28.5	127.6	3
NSCK	6,755	5,220	9	115	787	41.3	0.4	1.6	10	127	1,131	46.3	1.3	4.6	9	115	334	17.1	0.3	1.3	0	0.4	1.5	0
NSPK	429	662	7	7	51	2.3	0.1	0.1	7	7	83	3.4	0.1	0.2	7	7	33	1.5	0.1	0.1	0	0.1	0.1	0
NSPK-server	298	604	8	9	212	8.0	0.1	0.2	8	9	115	5.1	0.1	0.2	8	9	54	2.8	0.1	0.2	0	0.1	0.2	0
SPLICE	822	658	9	14	91	4.6	0.1	0.2	11	17	160	6.9	0.2	1.5	9	14	62	3.6	0.1	0.1	1	0.1	0.3	1
Swick 1	496	258	5	4	17	1.0	0.1	0.1	7	6	38	1.7	0.0	0.1	5	4	13	0.8	0.0	0.1	0	0.1	0.1	0
Swick 2	693	450	6	8	59	3.2	0.1	0.1	7	9	65	2.9	0.0	0.1	6	8	29	1.7	0.0	0.1	1	0.1	0.1	0
Swick 3	526	482	6	5	12	0.8	0.1	0.1	6	7	46	2.0	0.0	0.1	6	5	11	0.7	0.0	0.1	0	0.1	0.1	0
Swick 4	1,357	1,283	4	15	64	11.0	0.1	0.2	6	17	162	15.9	0.2	0.7	4	15	57	10.2	0.1	0.3	1	0.3	0.5	0
Stubblebine rep	1,409	2,408	5	15	2,048	82.9	0.3	0.6	3	13	194	10.8	0.1	0.2	5	15	95	6.3	0.1	0.1	1	0.1	0.1	0
Woo-Lam M	45,584	27,051	3	13	-	-	-	-	7	13	-	-	-	-	3	13	-	-	-	-	-	-	-	-
			6	481	-	-	-	-	7	554	5,977	433.6	4.0	12.4	6	481	2,498	304.4	1.8	7.7	1	2.3	5.6	0

(K) indicates that the value in this column are given in thousands;
- means that a memory out has been reached; and
< 1 means that the number of atoms or clauses is less than 1 thousand;

To cope with the above issues it is convenient to apply the **NoCEA** encoding technique that is based on the *Abstraction Refinement Loop* described in Section 4. Of course, by disabling the generation of the conflict exclusion axioms, we are no longer guaranteed that the solutions found are linearizable and hence executable.[15] SATMC therefore looks for conflicting actions in the partial order plan found and extends the previously generated encoding with clauses negating the conflicts (if any). The resulting formula is then fed back to the SAT-solver and the whole procedure is iterated until a solution without conflicts is met or the formula becomes unsatisfiable.

The **NoCEA** encoding technique performs uniformly better on all the analyzed protocols. Another interesting point is that while the time spent by Chaff and SIM to solve the propositional formulae is comparable in most cases (see the columns **Last** in the table **NoCEA**),[16] the number of iterations of the Abstraction Refinement Loop is in most of the cases smaller when SIM is used. This is due to the different strategies implemented by the two solvers. Both solvers incrementally build a satisfying assignment, but while SIM terminates (possibly returning a partial assignment) as soon as all the clauses are subsumed by the current satisfying assignment, Chaff avoids the subsumption check for efficiency reasons and thus halts only when the assignment is total. As a consequence the partial order plans found using Chaff usually contain more actions and thus are likely to have more conflicts. This explains why SIM is in many cases more effective than Chaff.

SATMC is one of the back-ends of the AVISS tool [2]. Using the tool, the user can specify a protocol and the security properties to be checked using a high-level specification language similar to the Alice&Bob notation we used in Section 1 to present our simple authentication protocol. The AVISS tool translates the specification into a rewrite-based declarative Intermediate Format (IF) based on multiset rewriting which is amenable to formal analysis. SATMC can optionally accept protocol specifications in the IF language which are then automatically translated into equivalent planning problems. Thus, by using SATMC in combination with the AVISS tool, the user is relieved from the time-consuming and error-prone activity of providing a detailed specification of the protocol insecurity problem in terms of a planning problem.

6   Related Work

The idea of regarding security protocols as planning problems is not new. An executable planning specification language $\mathcal{AL}_{SP}$ for representing security protocols and checking the possibility of attacks via a model finder for logic programs with stable model semantics has been proposed in [1]. Compared to this approach SATMC performs better (on the available results) and can readily exploit improvements of state-of-the-art SAT solvers.

.. A counterexample in the model checking problem corresponds to a model of the propositional formula generated and, therefore, to solutions of the planning problem.
.. Notice that Chaff performs better on big instances.

Gavin Lowe and his group at the University of Leicester (UK) have analyzed problems from the Clark/Jacob library [10] using Casper/FDR2 [16]. This approach has been very successful for discovering new flaws in protocols. However, first experiments on the search time indicate that SATMC is more effective than Casper/FDR2. Besides that Casper/FDR2 limits the size of messages that are sent in the network and is not able to handle non-atomic keys.

The Murphi model-checker has been used in [26] for analyzing some cryptographic protocols such as the Needham-Schroeder Public-Key. Experimental results indicate that Murphi suffers from state-space explosion. To cope with this problem several restrictions on the model are put in place. For instance, the size of the channel is fixed a priori.

In the context of the AVISS tool we have performed a thorough comparison between the back-ends integrated in it. The On-the-Fly Model-Checker (OFMC) [6] performs uniformly well on all the Clark/Jacob library. However, it is interesting to observe that in many cases the time spent by the SAT solver is equal to or even smaller than the time spent by OFMC for the same protocol. The Constraint-Logic-based model-checker (CL) [9] is able to find type-flaw attacks (as well as OFMC). However, the overall timing of SATMC is better than that of CL. Detailed results about these experiments can be found in [2].

7   Conclusions and Perspectives

The work presented in this paper is part of a larger effort aiming at the construction of an industrial-strength SAT-based model-checker for security protocols. In this paper we have introduced an abstraction/refinement strategy which, by interleaving encoding and solving, allows to halve the size of the propositional encoding for the most complex protocols we have analyzed so far. Since even bigger savings are expected on more complex protocols, the improved procedure presented in this paper paves the way to the analysis of complex protocols (e.g. e-commerce protocols) arising in real-world applications.

As far as the role of the SAT-solver is concerned, our experiments using Chaff and SIM indicate that the ability to return partial models can be as important as the pure solving time. In the future work we plan to tighten the integration with the SAT-solver in such a way that clauses generated by the refinement phase are "learnt" by the solver, thereby directly exploiting the sophisticated search strategies incorporated into the solver. However, since the solving time is still dominated by the encoding time, our current focus is on finding ways to reduce the latter.

Experimental results led with COMPACT [13] indicate that the application of propositional simplification techniques reduces dramatically the dimension of the encodings thereby suggesting that they suffer from some form of redundancy. A close look to the problem reveals that this is due to the fact that linear encodings do not exploit the knowledge about the initial state and therefore they encode protocol behaviors which can be safely neglected. This led us to consider a more sophisticated encoding technique, graphplan-based encoding [21], which leads to even smaller propositional encodings [4]. Notice however that the greater generality of linear encodings is useful when analyzing protocols w.r.t. partially specified initial states.

Another promising approach amounts to treating properties of cryptographic operations as invariants. Currently these properties are modeled as actions (cf. rules (1) and (2) in Section 2) and this has a bad impact on the size of the final encoding. A more natural way to deal with these properties amounts to building them into the encoding but this requires, among other things, a modification of the explanatory frame axioms.

References

[1] L. Carlucci Aiello and F. Massacci. Verifying security protocols as planning in logic programming. *ACM Trans. on Computational Logic*, 2(4):542–580, 2001.

[2] A. Armando, D. Basin, M. Bouallagui, Y. Chevalier, L. Compagna, S. Moedersheim, M. Rusinowitch, M. Turuani, L. Viganò, and L. Vigneron. The AVISS Security Protocols Analysis Tool. In *Proc. CAV'02*. 2002.

[3] A. Armando and L. Compagna. Automatic SAT-Compilation of Protocol Insecurity Problems via Reduction to Planning. In *Intl. Conf. on Formal Techniques for Networked and Distributed Systems (FORTE)*, Houston, Texas, November 2002.

[4] A. Armando, Compagna. L., and P. Ganty. Sat-based model-checking of security protocols. In *Proc.of Formal Methods Europe 2003 (FME03)*, Pisa, Sept. 2003.

[5] David Basin and Grit Denker. Maude versus haskell: an experimental comparison in security protocol analysis. In Kokichi Futatsugi, editor, *Electronic Notes in Theoretical Computer Science*, volume 36. Elsevier Science Publishers, 2001.

[6] David Basin, Sebastian Moedersheim, and Luca Viganò. An On-the-Fly Model-Checker for security protocol analysis. Forthcoming, 2003.

[7] D. Bolignano. Towards the formal verification of electronic commerce protocols. In *Proc.of the IEEE Computer Security Foundations Workshop*, pages 133–146. 1997.

[8] Cervesato, Durgin, Mitchell, Lincoln, and Scedrov. Relating strands and multiset rewriting for security protocol analysis. In *PCSFW: Proceedings of The 13th Computer Security Foundations Workshop*. IEEE Computer Society Press, 2000.

[9] Y. Chevalier and L. Vigneron. A Tool for Lazy Verification of Security Protocols. In *Proceedings of the Automated Software Engineering Conference (ASE'01)*. IEEE Computer Society Press, 2001.

[10] John Clark and Jeremy Jacob. A Survey of Authentication Protocol Literature: Version 1.0, 17. Nov. 1997. URL http://www.cs.york.ac.uk/~jac/papers/drareview.ps.gz.

[11] E. Clarke, A. Gupta, J. Kukula, and O. Strichman. SAT based abstraction-refinement using ILP and machine learning techniques. In *Proc. of Conference on Computer-Aided Verification (CAV'02)*, LNCS, 2002.

[12] Ernie Cohen. TAPS: A first-order verifier for cryptographic protocols. In *Proc. of The 13th Computer Security Foundations Workshop*, 2000.

[13] James M. Crawford and L. D. Anton. Experimental results on the crossover point in satisfiability problems. In Richard Fikes and Wendy Lehnert, editors, *Proceedings of the 11th AAAI Conference*, pages 21–27, California, 1993. AAAI Press.

[14] Grit Denker, Jonathan Millen, and Harald Rueß. The CAPSL Integrated Protocol Environment. Technical Report SRI-CSL-2000-02, SRI International, Menlo Park, CA, October 2000. Available at http://www.csl.sri.com/~millen/capsl/.

[15] Danny Dolev and Andrew Yao. On the security of public-key protocols. *IEEE Transactions on Information Theory*, 2(29), 1983.

[16] B. Donovan, P. Norris, and G. Lowe. Analyzing a library of security protocols using Casper and FDR. In *Proceedings of the Workshop on Formal Methods and Security Protocols*. 1999.

[17] Michael D. Ernst, Todd D. Millstein, and Daniel S. Weld. Automatic SAT-compilation of planning problems. In *Proceedings of the 15th International Joint Conference on Artificial Intelligence (IJCAI-97)*, pages 1169–1177. Morgan Kaufmann Publishers, San Francisco, 1997.

[18] E. Giunchiglia, M. Maratea, A. Tacchella, and D. Zambonin. Evaluating search heuristics and optimization techniques in propositional satisfiability. In *Proceedings of IJCAR'2001*, pages 347–363. Springer-Verlag, 2001.

[19] Heather, Lowe, and Schneider. How to prevent type flaw attacks on security protocols. In *PCSFW: Proceedings of The 13th Computer Security Foundations Workshop*. IEEE Computer Society Press, 2000.

[20] Florent Jacquemard, Michael Rusinowitch, and Laurent Vigneron. Compiling and Verifying Security Protocols. In M. Parigot and A. Voronkov, editors, *Proceedings of LPAR 2000*, LNCS 1955, pages 131–160. Springer-Verlag, Heidelberg, 2000.

[21] Henry Kautz, David McAllester, and Bart Selman. Encoding plans in propositional logic. In *KR'96: Principles of Knowledge Representation and Reasoning*, pages 374–384. Morgan Kaufmann, San Francisco, California, 1996.

[22] G. Lowe. Breaking and fixing the Needham-Shroeder public-key protocol using FDR. In *Proceedings of TACAS'96*, pages 147–166. Springer-Verlag, 1996.

[23] Gawin Lowe. Casper: a compiler for the analysis of security protocols. *Journal of Computer Security*, 6(1):53–84, 1998.

[24] W. Marrero, E. Clarke, and S. Jha. Model checking for security protocols, 1997.

[25] Catherine Meadows. The NRL protocol analyzer: An overview. *Journal of Logic Programming*, 26(2):113–131, 1996.

[26] J.C. Mitchell, M. Mitchell, and U. Stern. Automated analysis of cryptographic protocols using murphi. In *Proc. of IEEE Symposium on Security and Privacy*, pages 141–153. 1997.

[27] Matthew W. Moskewicz, Conor F. Madigan, Ying Zhao, Lintao Zhang, and Sharad Malik. Chaff: Engineering an Efficient SAT Solver. In *Proceedings of the 38th Design Automation Conference (DAC'01)*. 2001.

[28] R. M. (Roger Michael) Needham and Michael D. Schroeder. Using encryption for authentication in large networks of computers. Technical Report CSL-78-4, Xerox Palo Alto Research Center, Palo Alto, CA, USA, 1978. Reprinted June 1982.

[29] L.C. Paulson. The inductive approach to verifying cryptographic protocols. *Journal of Computer Security*, 6(1):85–128, 1998.

[30] D. Song. Athena: A new efficient automatic checker for security protocol analysis. In *Proceedings of the 12th IEEE Computer Security Foundations Workshop (CSFW '99)*, pages 192–202. IEEE Computer Society Press, 1999.

[31] H. Zhang. SATO: An efficient propositional prover. In William McCune, editor, *Proceedings of CADE 14*, LNAI 1249, pages 272–275. Springer-Verlag, 1997.

A Case for Efficient Solution Enumeration

Sarfraz Khurshid, Darko Marinov, Ilya Shlyakhter, and Daniel Jackson

MIT Computer Science and Artificial Intelligence Laboratory
200 Technology Square
Cambridge, MA 02139 USA
{khurshid,marinov,ilya_shl,dnj}@lcs.mit.edu

Abstract. SAT solvers have been ranked primarily by the time they take to find a solution or show that none exists. And indeed, for many problems that are reduced to SAT, finding a single solution is what matters. As a result, much less attention has been paid to the problem of efficiently generating *all* solutions.

This paper explains why such functionality is useful. We outline an approach to automatic test case generation in which an invariant is expressed in a simple relational logic and translated to a propositional formula. Solutions found by a SAT solver are lifted back to the relational domain and reified as test cases. In unit testing of object-oriented programs, for example, the invariant constrains the representation of an object; the test cases are then objects on which to invoke a method under test. Experimental results demonstrate that, despite the lack of attention to this problem, current SAT solvers still provide a feasible solution.

In this context, symmetry breaking plays a significant, but different role from its conventional one. Rather than reducing the time to finding the first solution, it reduces the number of solutions generated, and improves the quality of the test suite.

1   Introduction

Advances in SAT technology have enabled applications of SAT solvers in a variety of domains, e.g., AI planning [15], hardware verification [7], software design analysis [16], and code analysis [27]. These applications typically use a solver to find one solution, e.g., one plan that achieves a desired goal or one counterexample that violates a correctness property. Hence, most modern SAT solvers are optimized for finding one solution, or showing that no solution exists. That is also how the SAT competitions [1] rank solvers.

We have developed an unconventional application of SAT solvers, for software testing [20]. Our application requires a solver, such as mChaff [21] or relsat [4], that can enumerate *all* solutions. We find it surprising that the currently available versions of most modern SAT solvers, including zChaff [21], BerkMin [12], Limmat [6], and Jerusat [22], do not support solution enumeration at all[1], let

---

[*] Support in zChaff is under development and available in an internal version. Support in BerkMin is planned for the next version. There is no plan to add support to Jerusat. (Personal communication with the authors of the solvers.)

E. Giunchiglia and A. Tacchella (Eds.): SAT 2003, LNCS 2919, pp. 272–286, 2004.

alone optimize it. We hope that our application can motivate research in solution enumeration.

Software testing is the most widely used technique for finding bugs in programs. It is conceptually simple: just create a test suite, i.e., a set of test inputs, run them against the program, and check if each output is correct. However, testing is typically a labor intensive process, involving mostly manual generation of test inputs, and accounts for about half the total cost of software development and maintenance [5]. Moreover, inputs for modern programs often have complex structural constraints, which makes manual generation of high quality test suites impractical. Automating testing would not only reduce the cost of producing software but also increase the reliability of modern software.

We have developed the TestEra framework [20] for automated specification-based testing [5] of Java programs. To test a method, the user provides a specification that consists of a precondition (which describes allowed inputs to the method) and a postcondition (which describes the expected outputs). TestEra uses the precondition to automatically generate a test suite of *all* test inputs up to a given *scope*; a test input is within a scope of k if at most k objects of any given class appear in it. TestEra executes the method on each input, and uses the postcondition as a test oracle to check the correctness of each output.

TestEra specifications are first-order logic formulas. As an enabling technology, TestEra uses the Alloy toolset. Alloy [13] is a first-order declarative language based on sets and relations. The Alloy Analyzer [14] is an automatic tool that finds *instances* of Alloy specifications, i.e., assignments of values to the sets and relations in the specification such that its formulas evaluate to true. The analyzer finds an instance by: 1) translating Alloy specification into boolean satisfiability formula, 2) using an off-the-shelf SAT solver to find a solution to the formula, and 3) translating the solution back into sets and relations. The analyzer can enumerate all instances (within a given scope) using a SAT solver that supports enumeration, e.g., mChaff or relsat. The analyzer generates *complete* assignments: if the underlying SAT solver generates a solution with "don't care" bits, the analyzer grounds these bits out.

TestEra translates Alloy instances into test inputs that consist of Java objects; notice that the grounding out of "don't care" bits is necessary to build Java objects with fields that are properly initialized. Some inputs are *isomorphic*, i.e., they only differ in the identity of their objects. For example, consider a singly linked list of nodes that contain elements; two lists that have the same elements (or more precisely, isomorphic elements) in the same order are isomorphic irrespective of the node identities. It is desirable to consider only non-isomorphic inputs; it reduces the time to test the program, without reducing the possibility to detect bugs, because isomorphic test inputs form a "revealing subdomain" [28], i.e., produce identical results. The analyzer has automatic symmetry breaking [23] that eliminates many isomorphic inputs; we discuss this further in Section 3.1.

We initially used TestEra to check several Java programs. TestEra exposed bugs in a naming architecture for dynamic networks [17] and a part of an earlier version of the Alloy Analyzer [20]; these bugs have now been corrected. We have also used TestEra to systematically check methods on Java data structures, such as from the Java Collection Framework [26]. More recently, we have applied TestEra to test a C++ implementation of a fault-tree solver [11].

It is worth emphasizing that our application requires solutions that make complete assignments to primary (independent) variables. The main requirement of TestEra is efficient generation of all these solutions. Since storage of solutions has not (yet) been an issue, generating an implicit representation or a "cover" is not necessary. Moreover, testing necessitates generation of actual solutions and not just their representation. However, a solver that produces only prime implicants can be used as an intermediate step in generating all solutions.

In previous work [20], we presented TestEra as an application of SAT solvers in software testing. This paper makes the following new contributions:

- We describe a compelling application of SAT solvers that suggests that solution enumeration is an important feature that merits research in its own right. To the best of our knowledge, this is the first such application in software testing.
- We provide a set of formulas that can be used to compare different solvers in their enumeration. Our formulas fall into the (satisfiable) "industrial" benchmarks category for SAT competitions [1] and are available online at:

    http://mulsaw.lcs.mit.edu/alloy/sat03/index.html

  We also provide the expected number of solutions, which can help in testing a solver's solution enumeration.
- We provide a performance comparison between mChaff and relsat in enumerating a variety of benchmark data structures.
- We show how TestEra users can completely break symmetries, so that each solution of a boolean formula corresponds to a non-isomorphic test input.

2   Test Generation for Modern Software

Structurally complex data abounds in modern software. A textbook example is a data structure such as red-black trees [9] that implement balanced binary search trees. Another example is intentional names used in the intentional naming systems [2] for dynamic networks; an intentional name describes properties of a service by a hierarchical arrangement of *attributes* and *values*, which enables clients to access services by describing the desired functionality without a priori knowing service locations. Fault trees used in fault tree analysis systems [11] are also complex structures; a fault tree models system failure by relating it to failure of basic events in the system, and a fault tree analyzer computes the likelihood of system failure in the input fault tree over a given period of time. What makes such data complex is not only organization but also their structural constraints. For example, for red-black trees, one such constraint is that the number of black nodes along any path from root to a leaf is the same.

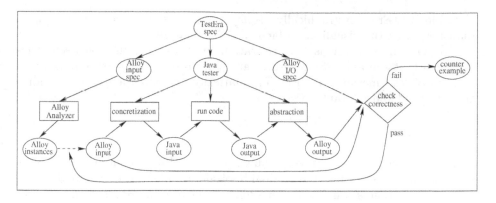

Fig. 1. Basic TestEra framework

Automating generation of input data is the key issue in automating systematic testing of software with structurally complex data. Generation at the representation level by a random assignment of values to object fields is infeasible since the ratio of the number of valid structures and the number of candidate structures tends to zero. A random assignment is very likely to violate (at least) one of the constraints. Generation at an abstract level using a construction sequence (e.g., building a red-black tree with 3 nodes by inserting sequentially 3 elements into an empty tree) is also inefficient: systematic generation of all inputs within a given input size can result in a very large number of construction sequences, most of which generate isomorphic structures. For example, to build all red-black trees with 10 nodes, this approach may require 10! or about 3.6×10^6 sequences, whereas there are only 240 non-isomorphic red-black trees with 10 nodes.

3   TestEra

The TestEra framework [20] provides a novel SAT-based approach for systematic generation of complex structures and automates specification-based testing [5] of Java programs.

Figure 1 illustrates the main components of the TestEra framework. Given a method precondition or an input specification (which describes the constraints that define valid method inputs) in Alloy, TestEra uses the Alloy Analyzer to generate all instances that satisfy the precondition. TestEra automatically *concretizes* these instances to create Java objects that form the test inputs for the method under test. TestEra executes the method on each input and automatically *abstracts* each output to an Alloy instance. TestEra then uses the analyzer to check if the input/output pair satisfy the constraints represented by the postcondition or the input/output specification (which describes the expected outputs). If the postcondition constraints are violated, TestEra reports a concrete counterexample, i.e., an input/output pair that is a witness to the violation. TestEra can graphically display the counterexample, e.g., as a heap snapshot, using the visualization facility of the analyzer.

To perform translations (concretizations and abstractions) between Alloy instances that the analyzer generates and Java objects that are inputs to, or outputs of, the method, TestEra automatically generates a test driver for a given method and relevant classes.

3.1 Symmetry Breaking

The analyzer adapts symmetry-breaking predicates [10] to reduce the total number of instances generated: the original boolean formula that corresponds to the Alloy specification is conjoined with additional clauses in order to generate only a few instances from each isomorphism class [23]. There is a trade-off, however: the more clauses that the analyzer generates, the more symmetries it breaks. But the larger boolean formula can become so large so that solving takes significantly more time, despite the fact that there are fewer instances. The goal of symmetry breaking in the analyzer was to make the analysis faster and not to generate exactly non-isomorphic instances. Therefore, with default symmetry breaking, it can significantly reduce the number of instances, but it is not always optimal, i.e., it may generate more than one instance from some isomorphism classes. (A discussion of algorithms for constructing symmetry-breaking clauses, such as the algorithm implemented in the Alloy Analyzer [23] or a more recent approach [3] is beyond the scope of this paper.)

The analyzer has special support for total orders: for each set $\{a_1, \ldots, a_n\}$ of n elements that is declared to have a total order, the analyzer generates only one order $\{\langle a_1, a_2 \rangle, \langle a_2, a_3 \rangle, \ldots, \langle a_{n-1}, a_n \rangle\}$, out of $n!$ (isomorphic) orders. This support has previously been used for faster analysis. We show in Section 4.3 how TestEra and Alloy users can also use total orders to constrain specifications so that the analyzer generates exactly one instance from each isomorphism class. Conceptually, the idea is to add constraints to ensure that the analyzer generates, from each isomorphism class, only the instance that is the *smallest* with respect to the total orders on the sets whose elements appear in the instance.

4   Non-isomorphic Generation

We next give a simple example that we use throughout this section to explain the fundamental aspects of TestEra's SAT-based generation of non-isomorphic inputs. Consider the following Java code that declares a binary tree and its `removeRoot` method, which removes the root node of its input tree and arranges any remaining nodes as a tree:

```
package testera.example;

class BinaryTree {
    Node root; // root node
    int size;  // number of nodes in the tree
```

```
static class Node {
    Node left;   // left child
    Node right;  // right child
}

void removeRoot() { ... }
}
```

Each object of the class `BinaryTree` represents a binary tree; objects of the inner class `Node` represent nodes of the trees. The method `removeRoot` has one input (i.e., the implicit `this` argument), which is a `BinaryTree`. Let us consider systematic input generation to test this method.

4.1   Specification

For the classes declared in the Java code above, TestEra produces the following Alloy specification:

```
module testera/example/BinaryTree

sig BinaryTree {
  root: option Node,
  size: Integer
}

sig Node {
  left: option Node,
  right: option Node
}
```

The declaration `module` names the specification. The keyword `sig` introduces a *signature* representing a set of indivisible atoms. We use Alloy atoms to model objects of the corresponding classes. Each signature can have *field* declarations that introduce relations between atoms. By default, fields are total functions; `size` is a total function from `BinaryTree` to `Integer`, where `Integer` is a predefined signature. The modifier `option` is used for partial functions (and the modifier `set` for general relations); e.g., `root` is a partial function from `BinaryTree` to `Node`. Partiality is used to model `null`: when the Java field `root` of some object `b` has the value `null`, i.e., points to no object, then the function `root` does not map the atom corresponding to `b` to any other atom.

The method `removeRoot` has only the implicit `this` argument, which is a `BinaryTree`. We consider a simple specification for this method: both precondition and postcondition require only that `this` satisfy the *representation invariant* (also known as a "class invariant") [19] for `BinaryTree`. For `BinaryTree`, the invariant requires that the graph of nodes reachable from `root` indeed be a tree (i.e., have no cycles) and that the `size` be correct; in Alloy, it can be written as follows:

```
fun repOk(t: BinaryTree) {
  all n: t.root.*(left + right) {
    n !in n.^(left + right) // no directed cycle
    sole n.~(left + right)  // at most one parent
    no n.left & n.right }   // distinct children
  t.size = #(t.root.*(left + right)) }   // size is consistent
```

The Alloy *function* repOk records constraints that can be invoked elsewhere in the specification. This function has an argument t, which is a BinaryTree. The function body contains two formulas, implicitly conjoined. The first formula constrains t to be a valid binary tree. The expression left + right denotes the union of relations left and right; the prefix operator '*' is reflexive transitive closure, and the dot operator '.' is relational composition. The expression root.*(left + right) denotes the set of all nodes reachable from root. The quantifier all denotes universal quantification: the formula all n: S { F } holds iff the formula F holds for n bound to each element in the set S. The operators '^', '~', and '&' denote transitive closure, transpose, and intersection, respectively. The formulas sole S and no S hold iff the set S has "at most one" and "no" elements, respectively. If all nodes n are not reachable from itself, have at most one parent, and have distinct children, then the underlying graph is indeed a tree. The second formula constrains the size field of t to match the size of the tree; '#' denotes set cardinality.

4.2   Instance Generation

The Alloy command run repOk for N but 1 BinaryTree instructs the analyzer to find an instance for this specification, i.e., a valuation of signatures (sets) and relations that makes the function repOk evaluate to true. The parameter N needs to be replaced with a specific constant that determines the scope, i.e., the maximum number of atoms in each signature, except those mentioned in the but clause. In our example, N determines the maximum number of Nodes, and the instance has only one BinaryTree. Note that *one* instance has one tree (with several nodes) corresponding to this argument. Enumerating all instances using the Alloy Analyzer generates all trees with up to the given number of nodes.

The following two assignments of sets of *atoms* and *tuples* to signatures and relations in the specification represent binary tree instances for $N=3$:

Instance 1:

```
        BinaryTree = { BT0 }, Node = { N0, N1, N2 }

        root = { (BT0, N1) }, size = { (BT0, 3) }
        left = { (N1, N0) }, right = { (N1, N2) }
```

Instance 2:

```
        BinaryTree = { BT0 }, Node = { N0, N1, N2 }

        root = { (BT0, N0) }, size = { (BT0, 3) }
        left = { (N0, N1) }, right = { (N0, N2) }
```

These instances are isomorphic since we can generate the second instance from the first one by applying to it the permutation that swaps atoms N0 and N1 (i.e., (N0 -> N1, N1 -> N0, N2 -> N2)).

In the sequel, we focus on enumerating instances for test input generation. To compare different ways of enumeration, we consider test inputs of size exactly N. For illustration, consider $N = 5$ in our running example. There are 14 non-isomorphic trees with five nodes [24]. If we use the analyzer without any symmetry breaking, the analyzer generates 1680 instances/trees, i.e., for each of the 14 isomorphism classes, the analyzer generates all 120 distinct trees corresponding to the 5! permutations/labelings of the five nodes. If we use the analyzer with symmetry breaking [23], we can tune how many symmetries to break. With the default symmetry breaking parameter values, the analyzer generates 17 trees with five nodes. If we set these parameter values to break all symmetries, the analyzer generates exactly 14 trees. Notice, however, that doing so can make generation significantly slower since the goal of symmetry breaking in the analyzer was to make analysis faster but not to generate exactly non-isomorphic instances.

4.3   Complete Symmetry-Breaking Using Total Order

We next describe an approach that uses total orders to efficiently break all symmetries in our example. Unlike the built-in symmetry breaking of the analyzer, this approach provides domain specific symmetry breaking and in particular, requires the user to manually add symmetry-breaking predicates.

The analyzer's standard library of models provides a polymorphic signature Ord[t]. Each instantiation of Ord with some set (Alloy signature) t imposes a total order on the elements in t. In consequence, these elements are not indistinguishable any more, and the analyzer does not break any symmetries on that set. However, the analyzer considers only one total order, instead of $(\#t)!$ possible total orders.

In addition to the definition of total order, the analyzer's standard library also provides several Alloy functions for totally-ordered sets. We use two of those functions in the following fact:

```
fact BreakSymmetries {
  all b: BinaryTree {
    all n: b.root.*(left + right) {
    // uses library function to instantiate Ord[Node]
    n.left.*(left + right) in OrdPrevs(n)
    n.right.*(left + right) in OrdNexts(n) } } }
```

The functions OrdPrevs and OrdNexts return the sets of all elements that are smaller and larger than the given element. A *fact* is a formula that expresses (additional) constraints on the instances. The fact BreakSymmetries requires that all trees in the instance (the example instances have only one tree) have nodes in an *in-order* [9]: the nodes in the left (right) subtree of the node n are smaller (larger) than n with respect to the Ord[Node] order. Note that the comparisons are for *node identities*, not for the values in the nodes. (For simplicity of illustration, our example does not even have values.)

We add the above fact to the specification for binary trees, effectively eliminating isomorphic instances. Indeed, the analyzer now generates exactly 14 non-isomorphic trees but it does so faster than using automatically generated symmetry-breaking predicates. In general, the user can break all symmetries by: 1) declaring that each set has a total order and 2) defining a traversal that linearizes the whole instance. The combination of the linearization and the total orders gives a lexicographic order that is used to compare instances. This process enforces generation to produce the canonical structure for each isomorphism class, but it requires the user to manually add symmetry-breaking predicates; this is straightforward to do for data structures, but it can become difficult for large complex specifications.

5 Results

We next present some performance results for solution enumeration obtained with mChaff [21]; we also compare performance of mChaff with that of relsat [4]. To discount the time it takes to write solutions to a file, we slightly modified the standard distributions of mChaff and relsat to disable solution reporting so that the solvers only report the total number of solutions found for each formula. The experiments were performed on a 1.8 GHz Pentium 4 processor with 2 GB of RAM.

5.1 Benchmarks

Table 1 presents the results of mChaff for a set of benchmark formulas that represent structural invariants. Each benchmark is named after the class for which data structures are generated; the structures also contain objects from other classes.

BinaryTree is our running example. LinkedList is the implementation of linked lists in the Java Collections Framework, a part of the standard Java libraries. This implementation uses doubly-linked, circular lists that have a size field and a header node as a sentinel node [9]. (Linked lists also provide methods that allow them to be used as stacks and queues.) TreeMap implements the Map interface using red-black trees [9]. This implementation uses binary trees with parent fields. Each node has a key and a value. HashSet implements the Set interface, backed by a hash table [9]; this implementation builds collision lists for buckets with the same hash code. HeapArray is an array-based implementation of a heap (or a priority queue) [9].

5.2 Performance of mChaff

Table 1 shows results for several input sizes for each benchmark. All scope parameters are set exactly to the given size; e.g., all lists have exactly the given

benchmark	size	#prim	manual symmetry breaking				automatic symmetry breaking			
			#vars	#clauses	#sols	time	#vars	#clauses	#sols	time
BinaryTree	6	86	2120	6686	132	1.05	2333	7018	357	1.50
	7	114	3165	10375	429	6.46	3439	10786	1866	7.45
	8	146	4504	15216	1430	40.46	4831	15682	10286	64.40
	9	182	7775	29618	4862	548.69	8141	30103	60616	1049.93
LinkedList	6	146	2017	6597	203	0.38	2520	7419	5975	3.46
	7	191	2834	9834	877	1.04	3559	11021	52392	68.71
	8	242	3837	14007	4140	4.76	4432	14939	734296	4637.99
	9	299	5852	24411	21147	36.52	6629	25630	—	—
TreeMap	6	203	5203	15162	20	9.10	5542	15668	322	10.85
	7	263	7578	22095	35	110.42	8076	22842	1160	69.09
	8	331	10578	30896	64	254.13	11265	31930	4185	583.62
	9	407	16111	51115	122	741.55	17017	52482	16180	3873.99
HashSet	6	285	5254	19079	462	6.06	5798	19865	693	7.04
	7	373	7540	28881	1716	31.52	8270	29918	3172	30.04
	8	473	10392	41430	6435	151.42	11102	42342	15011	167.30
	9	585	15380	63308	24310	511.51	16277	64441	73519	1587.72
HeapArray	5	56	544	1178	1919	0.55				
	6	72	704	1611	13139	5.10				
	7	90	884	2128	117562	62.62				
	8	110	1084	2735	1005075	1171.64				

Table 1. mChaff performance. All times are in seconds (of total elapsed wall-clock time). For sizes larger than presented, enumeration of solutions for automatically constructed symmetry-breaking predicates takes longer than 1 hour.

number of nodes and the elements come from a set with the given size. For each size, we use mChaff to enumerate solutions for two CNF formulas:

- one with symmetry-breaking predicates generated fully automatically (using the default values of the Alloy Analyzer);
- one with symmetry-breaking predicates added entirely manually to Alloy specifications (as described in Section 4.3).

We tabulate the number of primary variables, the total number of variables, the number of clauses, the number of solutions, and the time it takes to generate all solutions. The time shows the total elapsed time from the start of mChaff with an input being a formula file to the end of generation of all solutions (without writing them in a file). It is worth noting that the time to generate solutions often accounts for more than one-half of the time TestEra takes to test a benchmark data structure implementation [20]; thus, improving efficiency of solution enumeration can significantly improve TestEra's performance.

For BinaryTree, LinkedList, TreeMap, and HashSet, the numbers of non-isomorphic structures are given in the Sloane's On-Line Encyclopedia of Integer Sequences [24]. For all sizes, formulas with manually added symmetry-breaking predicates have as many solutions as the given number of structures, which shows that these predicates eliminate all symmetries. For `HeapArray`, no symmetry-breaking is required: two array-based heaps are isomorphic iff they are identical, since they consist only of integers (i.e., array indices and heap elements) that are not permutable. In TestEra, it is very desirable to generate only non-isomorphic inputs since without breaking isomorphisms it would be impractical to systematically test on all inputs. The factor by which the total number of solutions (including isomorphic solutions) is more than the total number of non-isomorphic solutions, is exponential in the input size. For example, for `TreeMap` and size nine, there are more than 44 million total solutions.

In all cases, formulas with automatic symmetry breaking (using default parameter values) have more solutions than formulas with manual symmetry breaking. Also, in most cases it takes longer to generate the solutions for formulas with automatic symmetry breaking; a simple reason for this is that enumerating a larger number of solutions usually takes a larger amount of time. This is not always the case, however: for `HashSet` and `TreeMap` of size seven, it takes less time to enumerate more solutions. This illustrates the general trade-off in (automatic) symmetry breaking: adding more symmetry-breaking predicates can reduce the number of (isomorphic) solutions, but it makes the boolean formula larger, which can increase the enumeration time. Note that having more variables and clauses (more symmetry-breaking predicates) does not necessarily imply breaking more symmetries. For example, in all examples but `HeapArray`, manual approach generates fewer variables and clauses than the automatic approach, yet manual break more symmetries. The reason for this is that manual approach breaks independent symmetries whereas the automatic approach can break dependent symmetries. In other words, a manual predicate rules out more isomorphic instances per literal of the predicate, so it is "denser". For details, see [23]. The Alloy Analyzer allows users to tune symmetry breaking; we have experimented with different parameter values and the analyzer's default values seem to achieve a sweet spot for our benchmarks.

Note that we do not present numbers for `LinkedList` of size nine with automatic symmetry breaking; for this formula mChaff runs out of memory (2 GB). This suggests that the scheme for clause learning in mChaff [21] may need to be modified when enumerating all solutions. If there is no effective pruning or simplification of clauses added in order to exclude the already found solutions, complete solution enumeration can become infeasible. For all other benchmark formulas, mChaff is able to enumerate all solutions, even when there are more than a million of them. Test inputs that correspond to these solutions, for the sizes from the table, are sufficient to achieve complete code and branch coverage [5] for methods in the respective Java classes.

5.3   Performance Comparison of mChaff with Relsat

Table 2 presents the performance comparison of mChaff with relsat in enumerating *all* solutions for benchmark formulas with manually added symmetry breaking constraints. Enumeration by mChaff seems to be more efficient than that of relsat for the benchmark data structures. The results indicate that the techniques

benchmark	size	# sols	mChaff time	relsat time
	6	132	1.05	4.81
BinaryTree	7	429	6.46	36.28
	8	1430	40.46	268.22
	6	203	0.38	1.21
LinkedList	7	877	1.04	9.08
	8	4140	4.76	78.40
	6	20	9.10	19.22
TreeMap	7	35	110.42	128.27
	8	64	254.13	665.50
	6	462	6.06	52.49
HashSet	7	1716	31.52	475.00
	8	6435	151.42	4100.99
	5	1919	0.55	6.71
HeapArray	6	13139	5.10	77.12
	7	117562	62.62	1073.49

Table 2. Performance comparison of mChaff with relsat in solution enumeration for benchmark formulas with manually added symmetry breaking predicates. All times are in seconds (of total elapsed wall-clock time).

introduced by mChaff for finding the first solution, such as efficient unit propagation, fare reasonably well for solution enumeration. For these benchmarks, it happens that mChaff's default enumeration does not generate any solutions with "don't care" bits. However, we believe mChaff's enumeration technique of obtaining partial solutions with don't-care variables (such that any completion of the solution satisfies the CNF) would also be useful for complete enumeration as grounding out partial solutions with "don't care" bits takes time linear in the number of new solutions generated. Perhaps this technique would also outperform relsat's technique of always producing complete solutions.

5.4 Binary Decision Diagrams

We also conducted some very preliminary experiments using Binary Decision Diagrams (BDDs) in place of SAT solvers. Intuitively, BDDs seem attractive because they make it easier to read off all solutions, once a BDD for a formula has been obtained. Of course, the construction of a BDD itself may be infeasible and can take a long time (and exponential space). We experimented with the CUDD [25] BDD package. We constructed BDDs bottom-up, using automatic variable reordering via sifting [8], from the boolean DAGs from which the CNFs were produced. For all benchmarks, the BDD approach scaled poorly; for non-trivial sizes (over five), the BDD construction led to unmanageably large BDDs (over a million nodes) and did not finish within the alloted time limit of 10 minutes. These results are preliminary and we believe BDD experts can fine tune the performance of BDDs to provide efficient enumeration.

6   Conclusions and Discussion

We have developed an unconventional application of SAT solvers, for software testing. Our application requires a solver that can enumerate all satisfying assignments; each assignment provides a (non-isomorphic) input for the program. In this context, symmetry breaking plays a significant, but different role from its conventional one: rather than reducing the time to finding the first solution, it reduces the number of solutions generated, and improves the quality of the suite of test inputs. The experimental results indicate that it is feasible to use a SAT solver to systematically generate a high quality test suite comprising of structurally complex inputs that would be hard to generate manually.

We envision various other applications of solution enumeration. One natural application is in checking certain classes of logic formulas. For example, consider the formula $\forall x \in D.P(x) \Rightarrow Q(x)$, where D is some (finite) domain, and P and Q are arbitrary predicates. We can simply use a solver to enumerate all x that satisfy $P(x)$ and then for each such x check that $Q(x)$ holds. Alternatively, we can check the validity of the implication (without requiring solution enumeration) by using a solver to directly check satisfiability of the negation: $P(x) \wedge \neg Q(x)$. Usually, the latter approach is preferred because it "opens" Q for the sophisticated optimizations that SAT solvers perform. However, when Q is a very large formula (or a formula that cannot be easily constructed explicitly), the approach with solution enumeration can work better.

Conceptually, TestEra checks that the code under test satisfies the formula $\forall i \in I. \ pre(i) \Rightarrow (\forall o \in O. \ code(i, o) \Rightarrow post(i, o))$, where pre is precondition, I is input domain, O is output domain, $code(i, o)$ denotes execution on input i that results in output o, and $post$ is postcondition. It is possible in some cases to translate (Java) code into a formula $code$ and look for a counter-example using a SAT solver (see e.g. [7, 27, 18]). These translation-based approaches typically build a formula, namely $pre \Rightarrow (code \Rightarrow post)$, which is much bigger than the formula that TestEra builds, namely pre. Therefore, TestEra works better for larger code that does not have many inputs, whereas the traditional approach works better for smaller code that has many possible inputs. Notice that the traditional approach can use any SAT solver, but TestEra requires a solver that can enumerate solutions.

A desirable feature for solvers that can enumerate solutions is to allow users to control the order of enumeration. For example, for testing databases, we would like to get "similar" test cases one after the other so that we can restore the state by using "deltas" and built-in support for rollback, instead of always building the state from scratch. For checking programs, it is desirable to have a solver generate all solutions in the neighborhood (as defined by a given metric) of a particular solution; this would enable testing, for example the entire neighborhood of an execution that gets "close" to a bug.

For testing programs on input sizes for which there exist a very large number of inputs, it is desirable to have a solver that can generate solutions in a random order and thus generate a high quality input sample. Another effective approach for testing on such input sizes is to define a stronger notion of isomorphism, taking into account the domain of application or even the implementation code, and then to enumerate inputs (which are now potentially fewer in number than before). Thus even though it may seem that a solver that specializes on enumeration still must suffer due to the large number of satisfying instances (for a given size), developing such solvers is practically useful.

We hope that our work provides motivation for exploring efficient solution enumeration in modern SAT solvers.

Acknowledgments

This research was funded in part by grant 0086154 from the ITR program of the National Science Foundation and in part by the High Dependability Computing Program from NASA Ames cooperative agreement NCC-2-1298. We thank the anonymous referees and Viktor Kuncak for comments on earlier drafts. We also thank Sharad Malik, Darsh Ranjan, and Zhaohui Fu for discussing and sharing with us the latest implementation of zChaff, and Eugene Goldberg and Alexander Nadel for discussing plans regarding BerkMin and Jerusat.

References

[1] SAT competitions for comparing state-of-the-art solvers. http://www.satlive.org/SATCompetition/.

[2] W. Adjie-Winoto, E. Schwartz, H. Balakrishnan, and J. Lilley. The design and implementation of an intentional naming system. In *Proc. 17th ACM Symposium on Operating Systems Principles (SOSP)*, Kiawah Island, Dec. 1999.

[3] F. A. Aloul, A. Ramani, I. L. Markov, and K. A. Sakallah. Solving difficult SAT instances in the presence of symmetry. In *Proc. 39th Conference on Design Automation*, pages 731–736, 2002.

[4] R. J. Bayardo Jr. and R. C. Schrag. Using CSP look-back techniques to solve real world SAT instances. In *Proc. National Conference on Artificial Intelligence*, pages 203–208, 1997.

[5] B. Beizer. *Software Testing Techniques*. International Thomson Computer Press, 1990.

[6] A. Biere. Limmat satisfiability solver. http://www.inf.ethz.ch/personal/biere/projects/limmat/.

[7] A. Biere, A. Cimatti, E. M. Clarke, M. Fujita, and Y. Zhu. Symbolic model checking using SAT procedures instead of BDDs. In *36thConference on Design Automation (DAC)*, New Orleans, LA, 1999.

[8] K. Brace, R. Rudell, and R. Bryant. Efficient implementation of a BDD package. In *Proc. of the Design Automation Conference (DAC)*, pages 40–45, 1990.

[9] T. H. Cormen, C. E. Leiserson, and R. L. Rivest. *Introduction to Algorithms*. The MIT Press, Cambridge, MA, 1990.

[10] J. Crawford, M. Ginsberg, E. Luks, and A. Roy. Symmetry-breaking predicates for search problems. In *Proc. Fifth International Conference on Principles of Knowledge Representation and Reasoning*, 1996.

[11] J. B. Dugan, K. J. Sullivan, and D. Coppit. Developing a low-cost high-quality software tool for dynamic fault tree analysis. *Transactions on Reliability*, pages 49–59, 1999.

[12] E. Goldberg and Y. Novikov. BerkMin: a fast and robust SAT-solver. In *Proceedings of Design, Automation, and Test in Europe (DATE)*, Mar. 2002.

[13] D. Jackson. Micromodels of software: Modelling and analysis with Alloy, 2001. http://sdg.lcs.mit.edu/alloy/book.pdf.

[14] D. Jackson, I. Schechter, and I. Shlyakhter. ALCOA: The Alloy constraint analyzer. In *Proc. 22nd International Conference on Software Engineering (ICSE)*, Limerick, Ireland, June 2000.

[15] H. Kautz and B. Selman. Planning as satisfiability. In *Proc. European Conference on Artificial Intelligence (ECAI)*, Vienna, Austria, Aug. 1992.

[16] S. Khurshid and D. Jackson. Exploring the design of an intentional naming scheme with an automatic constraint analyzer. In *Proc. 15th IEEE International Conference on Automated Software Engineering (ASE)*, Grenoble, France, Sep 2000.

[17] S. Khurshid and D. Marinov. Checking Java implementation of a naming architecture using TestEra. In S. D. Stoller and W. Visser, editors, *Electronic Notes in Theoretical Computer Science (ENTCS)*, volume 55. Elsevier Science Publishers, 2001.

[18] K. R. M. Leino. A SAT characterization of boolean-program correctness. In *Proc. 10th International SPIN Workshop on Model Checking of Software*, 2003.

[19] B. Liskov and J. Guttag. *Program Development in Java: Abstraction, Specification, and Object-Oriented Design*. Addison-Wesley, 2000.

[20] D. Marinov and S. Khurshid. TestEra: A novel framework for automated testing of Java programs. In *Proc. 16th IEEE International Conference on Automated Software Engineering (ASE)*, San Diego, CA, Nov. 2001.

[21] M. W. Moskewicz, C. F. Madigan, Y. Zhao, L. Zhang, and S. Malik. Chaff: Engineering an efficient SAT solver. In *Proceedings of the 39th Design Automation Conference (DAC)*, June 2001.

[22] A. Nadel. Jerusat SAT solver. http://www.geocities.com/alikn78/.

[23] I. Shlyakhter. Generating effective symmetry-breaking predicates for search problems. In *Proc. Workshop on Theory and Applications of Satisfiability Testing*, June 2001.

[24] N. J. A. Sloane, S. Plouffe, J. M. Borwein, and R. M. Corless. The encyclopedia of integer sequences. *SIAM Review*, 38(2), 1996. http://www.research.att.com/~njas/sequences/Seis.html.

[25] F. Somenzi. CUDD: CU decision diagram package. http://vlsi.colorado.edu/~fabio/CUDD/.

[26] Sun Microsystems. *Java 2 Platform, Standard Edition, v1.3.1 API Specification*. http://java.sun.com/j2se/1.3/docs/api/.

[27] M. Vaziri and D. Jackson. Checking properties of heap-manipulating procedures with a constraint solver. In *Proc. 9th International Conference on Tools and Algorithms for Construction and Analysis of Systems (TACAS)*, Warsaw, Poland, April 2003.

[28] E. J. Weyuker and T. J. Ostrand. Theories of program testing and the the application of revealing subdomains. *IEEE Transactions on Software Engineering*, 6(3), May 1980.

Cache Performance of SAT Solvers: a Case Study for Efficient Implementation of Algorithms

Lintao Zhang[1] and Sharad Malik[2]

[.] Microsoft Research,
1065 La Avenida Ave., Mountain View, CA 94043, USA
[.] Department of Electrical Engineering,
Princeton University, Princeton, NJ 08544, USA
{lintaoz,sharad}@ee.princeton.edu

Abstract. We experimentally evaluate the cache performance of different SAT solvers as a case study for the efficient implementation of SAT algorithms. We evaluate several different Boolean Constraint Propagation (BCP) mechanisms and show their respective run time and cache performances on selected benchmark instances. From the experiments we conclude that a cache friendly data structure is a key element in the efficient implementation of SAT solvers. We also show empirical cache miss rates of several modern SAT solvers based on the Davis-Logemann-Loveland (DLL) algorithm with learning and non-chronological backtracking. We conclude that the recently developed SAT solvers are much more cache friendly in data structures and algorithm implementations compared with their predecessors.

1 Introduction

Boolean Satisfiability (SAT) solvers are beginning to grow out of academic curiosity to become a viable industrial strength reasoning and deduction engine for production tools [3]. As SAT algorithms mature, the implementation issues begin to show their importance as evidenced by the recent development of solvers such as Chaff [12] and BerkMin [6]. Similar to earlier SAT solvers such as GRASP [11], relsat [7] and sato [21], these solvers are based on the Davis-Logemann-Loveland algorithm [5] and incorporate learning and non-chronological backtracking proposed by [11] and [7]. By combining efficient design and implementation together with a careful tuning of the heuristics, these new solvers are often orders of magnitude faster than their predecessors.

Even though it is widely acknowledged that the recent SAT solvers gain considerable speed up due to careful design and implementation, there is little quantitative analysis to support this claim. As the design of SAT solvers becomes more and more sophisticated, more detailed information about the solving process is needed during the implementation of the solver to squeeze the last bit of efficiency out of the solvers because a small speed up of the engine could

E. Giunchiglia and A. Tacchella (Eds.): SAT 2003, LNCS 2919, pp. 287–298, 2004.
© Springer-Verlag Berlin Heidelberg 2004

translate to days of savings for industrial tools in the field. In order to optimize the SAT solvers, it is necessary to have quantitative analysis results on various aspects of the SAT solving process.

In this paper, we analyze one aspect of the SAT solvers, namely the cache performance of the Boolean Constraint Propagation (BCP) operation, to illustrate the importance of careful design of the data structures and algorithms. It is well known that for contemporary computer micro-architectures, the memory hierarchy plays an extremely important role in the performance of the computer system. Algorithms designed with this memory hierarchy in mind, i.e. those that optimize for cache behavior, can gain considerable speedup over non-cache aware implementations as demonstrated by many authors (e.g. [9, 19]). The importance of the cache performance for SAT solvers is first discussed by the authors of SAT solver Chaff [12]. However, no detailed empirical data is provided to support this discussion. In this paper, we use a cache analysis tool to quantitatively illustrate the gains achieved by careful designed algorithms and data structures.

The paper is organized as follows. In Section 2, we describe several mechanisms to perform Boolean Constraint Propagation and how they influence the performance of the SAT solver. We evaluate the cache performance of these mechanisms in a control experiment using the same solver framework to illustrate the importance of cache awareness in algorithm design. In section 3, we evaluate the cache performances of several different SAT solvers that are based on the same basic DLL algorithm with the learning principle. Experiments show that these solvers vary greatly in their cache performance. Finally we draw our conclusions in Section 4.

2   Cache Performance of Different BCP Mechanisms

The Davis-Logemann-Loveland procedure [5] for SAT solving is a search based algorithm. One of the main operations of this procedure is Boolean Constraint Propagation (BCP). After a variable is assigned, the solver needs to propagate the effect of the assignment and determine the unit and conflicting clauses that may occur as a consequence of the variable assignment. Conflicting clauses cause the solver to backtrack from the current search space, while unit clauses result in forced assignments to currently free variables, i.e. these variables are implied. The process of iteratively assigning values to variables implied by unit clauses until no unit clause exists is called Boolean Constraint Propagation (BCP).

In a typical SAT solving process, BCP often takes the most significant part of the total run time. Therefore, BCP algorithms usually dictate the data structure of the clause database and the overall organization of the solver. The implementation of BCP is most essential to the efficiency of the overall implementation of SAT solvers[3]. In this section, we discuss some popular BCP algorithms and empirically show their impact on the overall efficiency of the SAT solver.

---

* In this paper we are only concerned with implementation issues. Improvements in algorithm are often more important, but that is not the subject of this paper.

A simple and intuitive implementation for BCP is to keep counters for each clause. This scheme is attributed to Crawford and Auton [4] by [20]. Similar schemes are subsequently employed in many solvers such as GRASP [11], relsat [7], satz [10] etc. Counter-based BCP has several variations. The simplest counter-based algorithm can be found in solvers such as GRASP [11]. In this scheme, each clause maintains two counters, one for the number of value 1 literals in the clause and one for the number of value 0 literals. Each clause also maintains a count of the total number of literals in it. Each variable has two lists that contain all the clauses where the variable appears as a positive and a negative literal respectively. When a variable is assigned a value, all the clauses that contain the literals of this variable will have their two counters updated. The data structure of this scheme is as follows:

```
struct Clause {
    int literal_array [];
    int num_total_literal;
    int num_value_1_literal_counter;
    int num_value_0_literal_counter;
};
struct Variable {
    int value;
    Clause pos_occurrence_array [];
    Clause neg_occurrence_array [];
};
struct ClauseDatabase {
    int num_clauses;
    Clause clause_array [];
    int num_variables;
    Variable variable_array [];
};
```

After the variable assignment, there are three cases for a clause with an updated counter: if the value 0 count equals the total number of literals, then the clause is a conflicting clause. If the value 0 count is one less than the total number of literals in the clause and the value 1 count is 0, then the clause is a unit clause. Otherwise, the clause is neither conflicting nor unit, we do not need to do anything. If a clause is found to be a unit clause, the solver has to traverse the entire array of the literals in that clause to find the free (unassigned) literal to imply. Given a Boolean formula that has m clauses and n variables, assuming on the average each clause has l literals, then, in the formula each variable on the average occurs lm/n times. Using this BCP mechanism, whenever a variable gets assigned, on the average, lm/n counters need to be updated. Since in this scheme each clause contains two counters, we will call this BCP scheme the *2-Counter Scheme*.

An alternative implementation can use only one counter for each clause. The counter stores the number of non-zero literals in the clause. The data structure for the clauses in this scheme is:

```
struct Clause {
    int num_total_literals;
    int literal_array [];
    int num_value_non_0_literal_counter;
};
```

Notice that the number of total literals in a clause does not change during the search. We can save the storage for it by using some tricks to mark the end of the literal array. For example, we can set the highest bit of the last element in the array to 1 to indicate the end of the array.

In this scheme, when a variable is assigned a value, we only need to update (reduce by 1) all the counters of clauses that contain the value 0 literal of this variable. There is no need to update clauses that contain the value 1 literal of this variable because the counters (number of non-zero literals) do not change. After the assignment, if the non-zero counter for a clause is equal to 0, then the clause is conflicting. If it is 1, the solver needs to traverse the literal array to find the one non-zero literal. There are two cases: if the non-zero literal is a free (unassigned) literal, then the clause is a unit clause and the literal is implied; otherwise it must already evaluate to 1 and the clause is satisfied, so we need to do nothing.

This scheme essentially cuts the number of clauses that need to be updated by half. Whenever a variable gets assigned, on the average, only $lm/2n$ counters need to be updated. However, it does need to traverse clauses more often than the first scheme. If a clause has one literal evaluating to 1 and all the rest evaluate to 0, the previous BCP scheme will know that the clause is already satisfied and no traversal is necessary. In this scheme, the solver still needs to traverse the clause literals because it does not know whether the literal evaluates to 1 or is unassigned. As each clause contains only one counter in this scheme, we will call it the *1-Counter Scheme*.

Notice that in the data structure described above, the counters associated with each clause are stored within the clause structure. This is actually not good from the cache performance point of view. When a variable is assigned, many counters need to be updated. If the counters are located far from each other in memory, i.e. have poor spatial locality, the updates may cause a lot of cache misses. Therefore, to improve the data locality, the counters should be located as close as possible to each other in memory. It is possible to allocate a separate continuous memory space to store the counters. In that case, the data structure becomes:

```
struct Clause {
    int literal_array [];
    int num_total_literal;
};
struct ClauseDatabase {
    int num_clauses;
    Clause clause_array [];
```

```
      int clause_non_0_counter_array [];
      int num_variables;
      Variable variable_array [];
};
```

In this way, we move the counters for the non-zero literals in each clause into a array in the ClauseDatabase structure. We will call this alternative data structure to implement the 1-Counter Scheme as the *Compact 1-Counter Scheme*. From a software engineering point of view, the compact 1-Counter Scheme is not a good idea because it destroys the nice object oriented structure of a clause. However, as we will see in the experimental results, this practice improves the overall efficiency of the solver greatly. For large software projects, these kinds of "hacks" should be avoided in most cases because of maintenance issues. However, for small kernel engines like a SAT solver, such hacks can often be tolerated to improve performance.

Counter-based BCP mechanisms are easy to understand and implement, but they are not the most efficient ones. Modern SAT solvers usually incorporate learning [11, 7] in the search process. Learned clauses often contain many literals. Therefore, the average clause length l is quite large during the solving process, thus making a counter-based BCP engine relatively slow. Moreover, when backtracking from a conflict, the solver needs to undo the counter assignments for the variables unassigned during the backtracking. Each undo for a variable assignment will update lm/n or $lm/2n$ counters on the average depending on the scheme employed.

In [12], the authors proposed a BCP algorithm called *2-literal watching* that tries to address these two problems[4]. For each clause the 2-literal watching scheme keeps track of two special literals called watched literals. Each variable has two lists containing pointers to all the watched literals corresponding to it in either phases. We denote the lists for variable v as pos_watched(v) and neg_watched(v). Initially the watched literals are free. When a variable v is assigned value 1, for each literal p pointed to by a pointer in the list of neg_watched(v), the solver will search for a literal l in the clause containing p that is not 0. There are four cases that may occur during the search:

1. If there exists such a literal l and it is not the other watched literal, then we remove the pointer to p from neg_watched(v), and add the pointer to l to the watched list of the variable of l. We refer to this operation as *moving the watched literal*, because in essence one of the watched pointers is moved from its original position to the position of l.

[.] Another BCP scheme, called head/tail list scheme [20], also claims to have improvements upon counter-based BCP schemes. Unfortunately, there seems to be no *reference implementation* for this scheme. The best solver [6] that employs this scheme is a closed-source solver, therefore, no implementation details are available. Our in-house developed code that uses this scheme cannot validate the claim of it significantly outperforming counter-based schemes. Therefore, we choose not to include the evaluation of this scheme in this paper because our code may not be representative of the best implementation available.

2. If the only such l is the other watched literal and it is free, then the clause is a unit clause, with the other watched literal being the unit literal.
3. If the only such l is the other watched literal and it evaluates to 1, then we need to do nothing since the clause is satisfied.
4. If all literals in the clause are assigned value 0 and no such l exists, then the clause is a conflicting clause.

Unlike counter-based BCP mechanisms, undoing a variable assignment during backtrack in the 2-literal watching scheme takes constant time [12]. As each clause has only two watched literals, whenever a variable is assigned a value, on the average the state of only m/n clauses needs to be updated assuming watched literals are distributed evenly in either phase. However, each update (i.e. moving the watched literal) is much more expensive than updating a counter as in the counter-based scheme.

To evaluate the BCP mechanisms discussed here, we constructed the following controlled experiment. We implemented all of them in the same solver framework based on the SAT solver zchaff [22]. Zchaff already has other data structures optimized for cache performance. Thus, the results of this experiment directly compare only the variations in the different BCP mechanisms with the rest of the environment being constant. Note however that in the original zchaff code, different BCP mechanisms may lead the solver into different search paths because the variables may be implied in different orders by different implication mechanisms. To control this element we modified the zchaff code such that the solver will follow the same search path regardless of the BCP method. This modification incurs a negligible amount of overhead. Also, because of the modification, the search path is different from the default zchaff solver which is evaluated in the next section.

The instances we use for the evaluation are chosen from various application domains with various sizes and run times to avoid biasing the result[5]. They include instances from microprocessor verification [18] (2dlx), bounded model checking (bmc [17], barrel6 [2]), SAT planning [8] (bw_large.d), combinational equivalence checking(c5315), DIMACS benchmarks (hanoi4), FPGA routing [14] (too_large) as well as a random 3-SAT instance with clause-variable ratio of 4.2 (rand)[6]. Table 1 provides statistics on the instances and the solving process common to all of the BCP mechanisms. These statistics include the number of variables, the number of original and learned clauses and literals as well as the ratio of literals to clauses. We also show the number of implications (i.e. variable assignments) needed to solve the problem. All of the experiments are carried out on a Dell PowerEdge 1500sc computer with a 1.13Ghz PIII CPU and 1GB

---

[5] We focus on real world benchmarks and mostly avoid random benchmarks since the real world benchmarks are more relevant for practical SAT solvers which are the focus of this paper. For the same reason we use zchaff instead of a regular (no learning) DLL algorithm because most real world applications use solvers similar to the algorithms used by zchaff.

[6] The random 3-SAT instance has a clause to literal ratio less than 3 because duplicated literals of the same variable in a clause are removed.

Instance Name	Num. Vars	Orig Cls	Orig Lits(k)	Orig Lits/Cls	Lrned Cls	Lrned Lits(k)	Lrned Lits/Cls	Num. Impl(k)
2dlx_cc_mc_ex_bp_f	4583	41704	118	2.83	13756	1180	85.78	3041
barrel6	2306	8931	25	2.76	34416	2489	72.32	12385
bmc-galileo-9	63624	326999	833	2.55	2372	80	33.94	4629
bw_large.d	5886	122412	273	2.23	14899	399	26.75	11110
c5315	5399	15024	35	2.30	83002	7061	85.07	30174
hanoi4	718	4934	12	2.47	4578	259	56.47	364
rand	200	830	2	2.98	25326	500	19.73	1671
too_largefs3w8v262	2946	50416	271	5.38	71772	2847	39.67	9466

Table 1. Statistics of the Instances Used to Evaluate the BCP Mechanisms

main memory. The PIII CPU has a separated level-1 16K data cache and a 16K instruction cache. The level-2 cache is a unified 512K cache. Both L1 and L2 caches have 32 bytes cache lines with 4-way set associativity.

We use `valgrind` [15], an open source cache simulation and memory debugging tool, to perform the cache simulation. The simulated CPU is the same as the CPU we run the SAT solver on. The simulation results are shown in Table 2. In the table, we show the run time (not simulation time) to solve each instance, the number of instructions executed (in billions), the number of data accesses (in billions) and the cache miss rates for data accesses. The cache misses for instructions are usually negligible for SAT solvers and therefore are not shown in the table.

From the table we find that different implementations of the BCP mechanism have significant impact on the runtime of the SAT solver. In comparing the best performing mechanism, i.e. the 2-literal watching scheme, with the worst scheme, i.e. the 2-Counter scheme, we see that for some benchmarks there is a speed difference of almost 20x. Comparing the 2-Counter Scheme and the 1-Counter Scheme, we find that the 1-Counter scheme reduces memory accesses by a small amount but the cache miss rates for the 1-Counter scheme is significantly smaller than that of the 2-Counter scheme. This validates our suspicion that the counter updates are the main sources of cache misses. By reducing the counter updates by half, we reduce the cache miss rate by almost a half. Because the 1-Counter Scheme needs more checks for unit clauses as compared with the 2-Counter scheme, the total memory accesses are not reduced by as much. However, these memory accesses do not cause many cache misses. Comparing the Compact 1-Counter Scheme with the regular 1-Counter scheme, we find that even though the number of the instructions executed and the data accesses are almost the same, the locality of counter accesses improves the level-1 cache miss rates of the Compact 1-Counter Scheme significantly[7]. Therefore, the actual run time is also improved by as much as 2x. This again validates our previous conjecture,

---

* Level-2 cache miss rates do not differ by much, probably because misses in L2 cache are caused by factors other than accessing the variable values, due to the relatively large size of L2 cache.

Instance Name	RunTime (s)	10^9 Instr. Executed	10^9 Data Accesses	L1 Data Miss Rate	L2 Data Miss Rate
2dlx_cc_mc_ex_bp_f	20.09	10.24	4.78	13.54%	16.02%
barrel6	102.15	48.04	22.49	15.09%	16.61%
bmc-galileo-9	16.60	6.83	3.26	5.69%	56.87%
bw_large.d	118.71	28.63	12.82	14.44%	47.16%
c5315	427.47	169.53	81.21	16.48%	23.39%
hanoi4	1.79	1.63	0.78	12.50%	1.16%
rand_1	96.28	39.20	18.36	19.76%	19.09%
too_largefs3w8v262	730.49	171.81	79.58	20.34%	36.73%

(a) 2-Counter Scheme

2dlx_cc_mc_ex_bp_f	14.47	8.43	4.26	8.76%	15.20%
barrel6	67.61	37.78	19.52	9.08%	15.98%
bmc-galileo-9	12.25	6.43	3.17	3.49%	56.57%
bw_large.d	49.09	22.53	11.06	7.22%	28.87%
c5315	268.18	123.82	66.55	11.91%	21.28%
hanoi4	1.47	1.31	0.68	8.32%	1.59%
rand_1	46.73	23.70	13.20	13.22%	13.00%
too_largefs3w8v262	329.31	103.99	57.29	13.45%	32.25%

(b) 1-Counter Scheme

2dlx_cc_mc_ex_bp_f	12.74	8.12	4.31	5.86%	17.48%
barrel6	52.88	36.18	19.77	6.07%	13.69%
bmc-galileo-9	11.55	6.36	3.20	3.22%	55.21%
bw_large.d	39.84	21.97	11.21	5.23%	27.32%
c5315	172.82	116.22	66.92	7.13%	18.10%
hanoi4	1.35	1.25	0.69	4.13%	2.42%
rand_1	22.67	21.81	13.22	6.33%	7.95%
too_largefs3w8v262	141.19	96.29	57.48	8.29%	18.93%

(c) Compact 1-Counter Scheme

2dlx_cc_mc_ex_bp_f	7.26	5.48	2.13	3.83%	14.52%
barrel6	30.35	21.83	8.31	3.79%	12.90%
bmc-galileo-9	8.12	5.90	2.68	2.16%	54.37%
bw_large.d	25.36	17.10	6.97	3.44%	19.82%
c5315	66.70	45.20	17.78	4.27%	14.44%
hanoi4	0.87	0.79	0.31	3.25%	2.75%
rand_1	5.35	3.78	1.58	4.17%	7.66%
too_largefs3w8v262	41.62	25.99	10.28	3.85%	22.83%

(d) 2 Literal Watching Scheme

Table 2. Memory Behavior of Different BCP Mechanisms

i.e. counter updates cause most of the cache misses. The 2-Literal Watching scheme performs best among these schemes, with significantly smaller number of instructions executed as well as the lowest number of data accesses. Moreover, it has the lowest cache miss rate.

3   Cache Performance of Different SAT Solvers

In the previous section, we evaluated several different BCP mechanisms and showed their respective cache performances. In this section, we evaluate the cache behavior of several existing SAT solvers. In the past, SAT solvers were often designed as a validation for various algorithms proposed by academic researchers. Therefore, little emphasis was placed on implementation issues. Recently, as SAT solvers become widely used as a viable deduction engine, increasing emphasis has been placed on implementing them efficiently. Here, we evaluate several SAT solvers that are based on the same principle proposed by [11, 7].

The solvers being evaluated include some of the most widely used SAT solvers such as GRASP [11], relsat [7] and SATO [21]. We also include three winners of the most recent SAT solver competition [16] on structured instances: zchaff [22], BerkMin [6] and limmat [1]. Finally, we also include JeruSAT [13], a SAT solver that was developed very recently. These solvers are very similar in the algorithm employed. For example, they all use unit clause implications in deduction and perform non-chronological backtracking and learning. The major algorithmic difference between them are the decision strategy and clause deletion policy.

Due to different branching heuristics used, these solvers behave very differently on different benchmark instances. Therefore, it is unfair to compare their actual performance by solving a small number of SAT instances. In order to concentrate on the cache behavior of the solvers, we choose some difficult SAT instances and let each solver run under valgrind for 1000 seconds[8]. We make sure that none of the solvers can solve any of the test instances during the run time in order to focus on cache behavior independent of the total run time. Except for relsat, all other solvers use default parameters. We use the option -p0 on relsat to disable the time consuming preprocessing. The benchmarks used include SAT instances generated from bounded model checking [2] (barrel, longmult), microprocessor verification [18] (9vliw, 5pipe), random 3-SAT (rand2) and equivalence checking of xor chains (x1_40). The statistics of the benchmarks we use are shown in the first two columns of Table 3.

The results of the cache miss rates for different solvers are listed in Table 3. From the table we find that earlier SAT solvers such as GRASP, relsat and SATO perform poorly in their memory behavior. The solvers would be around 3x slower than the new contenders such as zchaff and BerkMin solely due to the high cache miss rate. One of the reasons for the high cache miss rates of GRASP and relsat is due to their use of counter-based BCP mechanisms. On

---

[.] Programs running under valgrind are around 30 times slower than on the native machine.

Instance Name	Num. Variables	Num. Clauses	GRASP		SATO		relsat	
			D L1%	D L2%	D L1%	D L2%	D L1%	D L2%
x1_40	118	314	23.72	75.22	1.72	0.00	1.73	0.01
rand_2	400	1680	25.05	60.38	36.79	0.57	11.14	0.12
longmult12	5974	18645	20.86	57.69	19.32	41.68	13.24	19.78
barrel9	8903	36606	21.22	51.73	21.62	55.55	15.73	19.14
9vliw_bp_mc	20093	179492	22.43	88.58	40.49	42.85	14.58	55.34
5pipe	9471	195452	31.14	81.74	36.22	4.22	15.38	33.39

Instance Name	zchaff		BerkMin		Limmat		JeruSAT	
	D L1%	D L2%	D L1%	D L2%	D L1%	D L2%	D L1%	D L2%
x1_40	3.32	26.32	7.17	17.14	6.05	64.45	4.31	2.35
rand_2	6.58	23.46	7.63	11.25	5.83	44.83	5.66	14.90
longmult12	6.46	18.17	7.87	20.80	5.77	40.31	5.61	61.47
barrel9	6.62	33.43	10.98	32.04	7.58	34.39	5.55	54.22
9vliw_bp_mc	9.48	54.37	3.69	9.44	5.68	53.98	7.04	51.95
5pipe	6.14	21.54	5.31	18.51	6.45	46.12	6.49	42.66

Table 3. Memory Behavior of Different SAT Solvers

comparing their cache miss rates with the best that can be achieved by counter-based BCP mechanisms shown in the previous section, we find that they are still under par even considering their BCP mechanisms[9]. SATO uses the same BCP mechanism as BerkMin, but BerkMin seems to be much better optimized in cache performance. One benchmark SATO performs well on is the xor chain verification instance x1_40. The reason is because this is a very small instance and SATO has a very aggressive clause deletion heuristic. Therefore, SATO can fit the clause database into the level-1 cache while other solvers cannot do this due to their more tolerant clause deletion policies. The same reason explains the low cache miss rates for relsat on x1_40 and both SATO and relsat's low L2 miss rates for rand_2. The newer SAT solvers such as zchaff, BerkMin, limmat and JeruSAT all have relatively good cache behavior. Part of the reason for this is due to their better BCP mechanisms, but better design and implementation also contribute to their low cache miss rates.

As a final experiment we show the cache performance of zchaff, BerkMin and Limmat, three state-of-the-art SAT solvers, on two very large and challenging benchmarks. The benchmarks are obtained from Miroslav Velev's microprocessor verification SAT benchmarks [18]. They come from the fvp-unsat-3.0 (pipe_64_16) and SSS-Liveness-SAT-1.1 (2dlx_cc_ex_bp_f_bug3) respectively. These represent the cache performance of the most challenging SAT instances on the best SAT solvers available. We run each instance under `valgrind` for 10000 seconds. The results are shown in Table 4.

From Table 4 we find that as the formulas become very large, both level-1 and level-2 cache miss rates for the solvers increase considerably. Among these three

---

* The benchmarks used are not exactly the same for these two experiments, but the numbers should be representative of their respective behavior.

Instance Name	Num. Variables	Num. Clauses	zchaff		BerkMin		Limmat	
			D L1%	D L2%	D L1%	D L2%	D L1%	D L2%
2dlx	224920	3596385	11.47	91.22	7.38	69.12	6.00	74.01
pipe	64262	2291120	10.49	54.37	8.42	61.95	5.60	51.12

Table 4. Cache Performance of the Best SAT Solvers on Challenging Benchmarks

solvers, zchaff is the oldest one while limmat is the newest. From these limited experimental results it seems that the cache performances of the SAT solvers are still improving incrementally on carefully designed SAT solvers, whether on purpose or coincidentally. In current (and future) generations of microprocessors, the speed difference of main memory (on the mother board) and L2 cache (on die) tends to be large. Therefore, the L2 miss penalty is quite high. From these results we find that there is still a lot of room for improvement in the future for efficient implementation of SAT solvers.

As a final note, we want to point out that cache performance is not the only indication of the overall quality of the SAT solver implementation because different algorithms employed may dictate different memory behavior. Algorithms that have bad memory behavior may be good at pruning search space and reduce the overall runtime of the SAT solver. Even though the solvers compared in this section are all based on very similar algorithms, they do have differences that may bias the results slightly.

4   Conclusion

In this paper, we investigate the memory behavior of several different BCP mechanisms for Boolean Satisfiability Solvers. We find that different implementations of BCP can significantly affect the memory behavior and overall efficiency of the SAT solvers. From the experiments we conclude that a cache friendly data structure is a key element for efficient implementation of SAT solvers. We also show empirical cache miss rate data of several modern SAT solvers based on the Davis-Logemann-Loveland algorithm with learning. We find that recently developed SAT solvers are much more cache friendly compared with their predecessors. We use this as a case study to illustrate the importance of algorithm implementation. Efficient implementation (not necessarily limited to cache behavior) is key to the success of a modern SAT solver and should not be overlooked.

References

[1] Armin Biere. Limmat sat solver. http://www.inf.ethz.ch/personal/biere/projects/limmat/.

[2] Armin Biere, Alessandro Cimatti, Edmund M. Clarke, and Y. Zhu. Symbolic model checking without BDDs. In *Proceedings of Tools and Algorithms for the Analysis and Construction of Systems (TACAS'99)*, 1999.

[3] Per Bjesse, Tim Leonard, and Abdel Mokkedem. Finding bugs in an Alpha microprocessor using satisfiability solvers. In *Proceedings of 13th Conference on Computer-Aided Verification (CAV'01)*, 2001.

[4] J. M. Crawford and L. D. Auton. Experimental results on the crossover point in satisfiability problems. In *Proceedings of the Eleventh National Conference on Artificial Intelligence (AAAI'93)*, pages 21–27, 1993.

[5] Martin Davis, George Logemann, and Donald Loveland. A machine program for theorem proving. *Communications of the ACM*, 5(7):394–397, July 1962.

[6] E. Goldberg and Y. Novikov. BerkMin: A fast and robust SAT-solver. In *Proceedings of the IEEE/ACM Design, Automation, and Test in Europe (DATE)*, 2002.

[7] Roberto J. Bayardo Jr. and Robert C. Schrag. Using CSP look-back techniques to solve real-world SAT instances. In *Proceedings of the National Conference on Artificial Intelligence (AAAI)*, 1997.

[8] H. A. Kautz and B. Selman. Planning as satisfiability. In *Proceedings of the Tenth European Conference on Artificial Intelligence (ECAI'92)*, pages 359–363, 1992.

[9] A. LaMarca and R.E. Ladner. The influence of caches on the performance of heaps. *ACM Journal of Experimental Algorithmics*, 1, 1996.

[10] Chu-Min Li and Anbulagan. Heuristics based on unit propagation for satisfiability problems. In *Proceedings of the Fifteenth International Joint Conference on Artificial Intelligence (IJCAI'97)*, 1997.

[11] João P. Marques-Silva and Karem A. Sakallah. GRASP - a search algorithm for propositional satisfiability. *IEEE Transactions in Computers*, 48(5):506–521, 1999.

[12] Matthew W. Moskewicz, Conor F. Madigan, Ying Zhao, Lintao Zhang, and Sharad Malik. Chaff: Engineering an efficient SAT solver. In *Proceedings of the Design Automation Conference (DAC)*, June 2001.

[13] Alexander Nadel. Jerusat sat solver. http://www.geocities.com/alikn78/, 2002.

[14] G. Nam, K. A. Sakallah, and R. A. Rutenbar. Satisfiability-based layout revisited: Routing complex FPGAs via search-based Boolean SAT. In *Proceedings of International Symposium on FPGAs*, Feburary 1999.

[15] Julian Seward. Valgrind, a open-source memory debugger and cache simulator. http://developer.kde.org/~sewardj/, 2002.

[16] Laurent Simon, Daniel Le Berre, and Edward A. Hirsch. SAT 2002 solver competition report. http://www.satlive.org/SATCompetition/onlinereport.pdf, 2002.

[17] Ofer Strichman. Tuning SAT checkers for bounded model-checking. In *Proceedings of the Conference on Computer-Aided Verification (CAV'00)*, 2000.

[18] M.N. Velev and R.E. Bryant. Effective use of Boolean satisfiability procedures in the formal verification of superscalar and VLIW microprocessors. In *Proceedings of the Design Automation Conference (DAC)*, June 2001.

[19] L. Xiao, X. Zhang, and S. A. Kubricht. Improving memory performance of sorting algorithms. *ACM Journal of Experimental Algorithmics*, 5:1–23, 2000.

[20] H. Zhang and M. E. Stickel. An efficient algorithm for unit propagation. In *Proceedings of the Fourth International Symposium on Artificial Intelligence and Mathematics (AI-MATH'96)*, Fort Lauderdale (Florida USA), 1996.

[21] Hantao Zhang. SATO: an efficient propositional prover. In *Proceedings of the International Conference on Automated Deduction (CADE'97)*, pages 272–275, 1997.

[22] Lintao Zhang. Zchaff SAT solver. http://www.ee.princeton.edu/~chaff/zchaff.php.

Local Consistencies in SAT

Christian Bessière[1], Emmanuel Hebrard[2], and Toby Walsh[2]

[1] LIRMM-CNRS, Université Montpellier II
bssiere@lrmf

[2] Cork Constraint Computation Centre, University College Cork
{e.hbrad,   tw}@4c.ucc.ie

Abstract. We introduce some new mappings of constraint satisfaction problems into propositional satisfiability. These encodings generalize most of the existing encodings. Unit propagation on those encodings is the same as establishing **relatia l ka rc cа isteɴy** on the original problem. They can also be used to establish **(ij)-consɪt ency** on binary constraints. Experiments show that these encodings are an effective method for enforcing such consistencies, that can lead to a reduction in runtimes at the phase transition in most cases. Compared to the more traditional (direct) encoding, the search tree can be greatly pruned.

1   Introduction

Propositional Satisfiability (SAT) and Constraint Satisfaction Problems (CSPs) are two closely related NP-complete combinatorial problems. There has been considerable research in developing algorithms for both problems. Translation from one problem to the other can therefore profit from the algorithmic improvements obtained on the other side. Enforcing a local consistency is one of the most important aspect of systematic search algorithms. For CSPs, in particular, enforcing arc consistency is often the best tradeoff between the amount of pruning and the cost of pruning. The AC encoding [12] has the property that arc consistency in the original CSP is established by unit propagation in the encoding [10]. A complete backtracking algorithm with unit propagation, such as DP [6], therefore explores an equivalent search tree to a CSP algorithm that maintains arc consistency. Likewise, DP on the Direct encoding behaves as the *Forward Checking* algorithm which maintains a weaker form of Arc Consistency [17]. In this paper we show that there is a continuity between direct and support encodings, and following this line, many other consistencies can be simulated by unit propagation in the SAT encoding, for any constraint arity and all in optimal worst case complexity.

The rest of the paper is organized as follows. In section 2 we present the basic concepts used in the rest of the paper. In section 3 we introduce a family of encodings called the *k-AC encodings* where k is a parameter. These encodings enable a large family of consistencies, the so called *relational k-arc-consistency* [8] to be established by unit propagation on the SAT encoding. They work with any constraint arity. Section 4 focuses on binary networks, and shows that

E. Giunchiglia and A. Tacchella (Eds.): SAT 2003, LNCS 2919, pp. 299–314, 2004.
© Springer-Verlag Berlin Heidelberg 2004

these encodings can also be used to establish any *(i,j)-consistency* (another large family of consistencies [9]). We also show that unit propagation on the k-AC encodings can achieve the given level of consistency in optimal time complexity in all cases. Section 5 introduces mixed encodings that combine previous ones to perform a high level of filtering only where it is really needed. And finally, in section 6, we present some experiments, that assess the improvement of these encodings in comparison with the direct encoding. The results also show the ability of this approach to solve large and hard problems by comparing it with the best algorithms for CSPs.

2   Background

2.1   Constraint Satisfaction Problem (CSP)

A *CSP* $\mathcal{P} = (\mathcal{X}, \mathcal{D}, \mathcal{C})$ is a set $\mathcal{X} = \{X_1, \ldots, X_n\}$ of n *variables*, each taking a value from a finite *domain* $D(X_1), \ldots, D(X_n)$ elements of $\mathcal{D}$, and a set $\mathcal{C}$ of e *constraints*, d is the size of the largest domain. A *constraint* C_S, where $S = \{X_{i_1}, \ldots, X_{i_a}\} \subset \mathcal{X}$, is a subset of the Cartesian product of the domains of the variables in S, $C_S \subset D(X_1) \times D(X_2) \times \ldots \times D(X_a)$ that denotes the compatible values for the variables in S. The incompatible tuples are called *nogoods*. We are calling S, the *scope* of C_S and $|S| = a$ its *arity*. An *instantiation* I of a set T of variables is an element of the Cartesian product of the domains of the variables in T. We denote $I[A]$ for the projection of I onto the set of variables A, and $C_S[A]$ the projection of the constraint C_S onto A. An instantiation I is *consistent* if and only if it satisfies all the constraints, that is, $\forall C_S \in \mathcal{C}$ such that $S \subseteq T, I[S] \in C_S$. A *solution* is a consistent instantiation over $\mathcal{X}$.

Let T and S be two distinct sets of variables $T, S \subset \mathcal{X}$, and I an instantiation of T which is consistent. A *support* J of I on S is an instantiation J of S such that $I \cup J$ is consistent. For an instantiation I, if there exists a set S such that I has no support on S, then I doesn't belong to any solution.

2.2   Direct Encoding

The direct encoding [17] is the most commonly used encoding of CSPs into SAT. There is one Boolean variable X_v for each value v of each CSP variable X. $X_v = T$ means the value v is assigned to the variable X. Those variables appear in three sets of clauses :

At-least-one clause : There is one such clause for each variable, and their meaning is that a value from its domain must be given to this variable.
Let X be variable and $D(X) = \{v_1, v_2, \ldots, v_n\}$, then we add the *at-least-one* clause : $Xv_1 \vee Xv_2 \vee \ldots \vee Xv_n$.

At-most-one clause : There is one such clause for each pair of values for each variable, and their meaning is that this variable cannot get more than one value.
Let $v_i, v_j \in D(X), i \neq j$, then we add the *at-most-one* clause : $\neg Xv_i \vee \neg Xv_j$.

Conflict clause : There is one such clause for each nogood of each constraint, and their meaning is that this tuple of values is forbidden.

Let C_{XYZ} be a constraint on the variables X, Y, Z and $[u, v, w] \in D(X) \times D(Y) \times D(Z)$ an instantiation forbidden by C_{XYZ} ($[u, v, w] \notin C_{XYZ}$), then we add the *conflict* clause : $\neg Xu \vee \neg Yv \vee \neg Zw$.

2.3   AC Encoding

The AC encoding [12] enables a SAT procedure to maintain *arc-consistency* during search through *unit propagation*. It encodes not only the structure of the network, but also a consistency algorithm used to solve it. It differs from the direct encoding only on the conflict clauses which are replaced by *support* clauses, the others clauses remain unchanged.

Support clause : Let X, Y be two variables, $v \in D(X)$ a value of X and $\{w_1, \ldots, w_k\}$ the supports of $X = v$ on Y, then we add the *support* clause : $\neg Xv \vee Yw_1 \vee Yw_2 \vee \ldots \vee Yw_k$.

This clause is equivalent to $Xv \rightarrow (Yw_1 \vee Yw_2 \vee \ldots \vee Yw_k)$ which means : as long as Xv holds (i.e, $Xv \neq False$, that is "the value v remains in X's domain"), then at least one of its support must hold. Therefore when all the supports of $X = v$ are falsified Xv is itself falsified.

3   Generalisation of the AC Encoding

The AC encoding can only be applied to binary networks, because support clauses encode the supports of *a single* variable on another *single* variable. Our goal is to encode any kind of support that follows from the definition in section 2.1. The new encoding we introduce here, *k-AC encoding*, represent supports on subsets S of variables of any size, for an instantiation of another subsets T of any size. Since a literal stands for an assignment, an *instantiation* (or a support) of several variables corresponds to a *conjunction* of positive literals. A k-AC clause represents the implication between the instantiation and its supports: if the instantiation holds, one of the supports must hold. Let $[v_1, \ldots, v_p]$ be a support on $X_1, \ldots, X_p$ of a given instantiation on other variables. The conjunction that encodes this support is $(X_1 v_1 \wedge \ldots \wedge X_p v_p)$. To keep the encoding in clausal form, we need then to add an extra variable, say s, for this support and the following equivalence, $s \leftrightarrow (X_1 v_1 \wedge \ldots \wedge X_p v_p)$ which result in the following *equivalence clauses*: $(\neg s \vee X_1 v_1), \ldots, (\neg s \vee X_p v_p)$ and $(\neg X_1 v_1 \vee \ldots \vee \neg X_p v_p \vee s)$. We call s *support-variable*. If the support is unit (say $Y = v$), then there is no need for an extra variable, and the support-variable is the corresponding boolean variable (Yv).

k-AC clause : Let C_S be a constraint, $T = \{X_1, \ldots X_k\} \subset S$ be a set of k variables, $I = [v_1 \in D(X_1), \ldots v_k \in D(X_k)]$ an instantiation of T and $\{s_1, \ldots, s_m\}$ the supports of I on $S - T$, then we add the k-AC clause : $\neg X_1 v_1 \vee \ldots \vee \neg X_k v_k \vee s_1 \vee s_2 \ldots \vee s_m$.

This clause is equivalent to $I \rightarrow (s_1 \vee s_2 \vee \ldots \vee s_m)$ which means : as long as I holds then at least one of its support must hold. Therefore when all the supports of I are falsified I is itself falsified i.e, the k-AC clause is reduced to the conflict clause of length k forbidding I.

In figure 3, we show the four possible k-AC encodings for a ternary constraint. Note that, in the particular cases where the set of support variables is a singleton or the empty set, in other words, $a-k = 1$ or $a-k = 0$, the conjunctions standing for the supports are unit and we do not need to add extra variables.

$\Rightarrow_{encoding}$

X	Y	Z
a	a	b
a	b	b
b	a	a
b	a	b

0-AC encoding	3-AC encoding
$T \rightarrow (S_1 \lor S_2 \lor S_3 \lor S_4)\land$	$((Xa \land Ya \land Za) \rightarrow F)\land$
$(Xa \land Ya \land Zb) \leftrightarrow S_1\land$	$((Xa \land Yb \land Za) \rightarrow F)\land$
$(Xa \land Yb \land Zb) \leftrightarrow S_2\land$	$((Xb \land Yb \land Za) \rightarrow F)\land$
$(Xb \land Ya \land Za) \leftrightarrow S_3\land$	$((Xb \land Yb \land Zb) \rightarrow F)$
$(Xb \land Ya \land Zb) \leftrightarrow S_4$	

2-AC encoding	1-AC encoding
$((Xa \land Ya) \rightarrow Zb)\land$	$(Xa \rightarrow (S_1 \lor S_2)\land$
$((Xa \land Yb) \rightarrow Zb)\land$	$(Xb \rightarrow (S_3 \lor S_1)\land$
$((Xb \land Ya) \rightarrow (Zb \lor Za))\land$	$(Ya \rightarrow (S_4 \lor S_5 \lor S_6))\land$
$((Xb \land Yb) \rightarrow F)\land$	$(Yb \rightarrow S_4)\land$
$((Xa \land Za) \rightarrow F)\land$	$(Za \rightarrow S_7)\land$
$((Xa \land Zb) \rightarrow (Ya \lor Yb))\land$	$(Zb \rightarrow (S_7 \lor S_8 \lor S_9))\land$
$((Xb \land Za) \rightarrow Ya)\land$	$((Ya \land Zb) \leftrightarrow S_1)\land$
$((Xb \land Zb) \rightarrow Ya)\land$	$((Yb \land Zb) \leftrightarrow S_2)\land$
$((Ya \land Za) \rightarrow Xb)\land$	$((Ya \land Za) \leftrightarrow S_3)\land$
$((Ya \land Zb) \rightarrow (Xa \lor Xb))\land$	$((Xa \land Zb) \leftrightarrow S_4)\land$
$((Yb \land Za) \rightarrow F)\land$	$((Xb \land Za) \leftrightarrow S_5)\land$
$((Yb \land Zb) \rightarrow Xa)$	$((Xb \land Zb) \leftrightarrow S_6)\land$
	$((Xb \land Ya) \leftrightarrow S_7)\land$
	$((Xa \land Ya) \leftrightarrow S_8)\land$
	$((Xa \land Yb) \leftrightarrow S_9)$

Table 1. First table: a ternary constraint involving the variables X, Y, Z, the allowed tuples are given. Second table: four possible k-AC encodings of this constraint, T = True and F = False.

The k-AC clauses are a generalisation of support clauses in two different ways. First they capture a larger family of consistencies, *relational k-arc-consistency* (section 3) and (i, j)-*consistency* (section 4). Second they work for any constraint arity. Note that support clauses are 1-AC clauses for binary constraints, and conflict clauses are a-AC clauses for constraints of arity a. For instance, let C_{XYZ} be a constraint on the variables X, Y and Z. If $I = \{X = u, Y = v, Z = w\}$ is an allowed tuple, then the corresponding 3-AC clause is $(Xu \land Yv \land Zw) \rightarrow True$ and is useless. If I is a nogood, then we have $(Xu \land Yv \land Zw) \rightarrow False$, which is a conflict clause $(\neg Xu \lor \neg Yv \lor \neg Zw)$. Direct and support encodings are then particular cases of k-AC encoding.

Recall that in a CSP, a nogood is a forbidden set of assignments, $\neg(X_1 = v_1 \land \ldots \land X_i = v_i)$. And that a Boolean variable correspond to an assignment, the atom $X_i v_i$ represents $X_i = v_i$. For the theorems and proofs below, the word variable will refer to a *CSP* variable, *assignment* to a Boolean variable of the encoding and *support* to a conjunction of assignments in the conclusion of a k-AC clause. An interpretation I is a function that associates a value in $\{0, 1\}$

to the atoms of a set of clauses $\mathcal{B}$. I is a model $(I(\mathcal{B}) = True)$, iff all the clauses in $\mathcal{B}$ are satisfied by I.

Theorem 1 (Correctness and completeness of the k-AC Encoding.)
*I is a model of the set with the **at-least-one**, **at-most-one**, and k-**AC** clauses, iff the assignment such that a variable X take a value v iff $I(Xv) = T$ is a solution of the original constraint network.*

Proof: Suppose that all the assignments of a nogood N are satisfied.

$$N = \neg(X_1 v_1 \wedge X_2 v_2 \wedge \ldots X_n v_n)$$

Let C be the k-AC clause which premiss P is a subset of this nogood

$$C = (X_1 v_1 \wedge X_2 v_2 \wedge \ldots X_k v_k) \rightarrow (s_1 \vee s_2 \vee \ldots s_m), \ s_j = (X_{k+1} j_{k+1} \wedge \ldots X_n j_n)$$

and let S be the rest of this nogood, $S = N - P$. This premiss is satisfied and then the conclusion must be satisfied. Now recall that at-least and at-most clauses ensure that one and only one assignment per (CSP) variable is satisfied. All the supports in C refer to the same variables but are by definition different from S by at least one assignment, (say $X_i v_i$ is the assignment in the nogood, and $X_i j_i$ is the assignment in the support). Since, for this variable, $X_i v_i$ is satisfied, therefore $X_i j_i$ is not, and then the whole conclusion is not satisfied.

Let S be a solution of the original constraint network, and let I be the assignment in which $I(X_i v) = T$ iff, in S, the value v is given to the variable X_i. S gives one and only one value to each variable, the **at-most-one** and **at-least-one** clauses are thus satisfied. Without loss of generality, let C be a k-AC clause which premiss P is an assignment on a set R and conclusion are supports on a set T. If S is a solution then $S[R \cup T]$ is consistent and then $S[T]$ is a support of $S[R]$. Either $P \neq S[R]$ and then C is satisfied since the premiss is falsified, or $S = S[R]$ and then $S[T]$ is one of its support and belongs to C's conclusion. C is then satisfied since both premiss and conclusion are satisfied. □

Unit propagation on the k-AC Clauses corresponds exactly to enforcing relational k-arc-consistency. Relational arc-consistency [8] extends the concept of local consistency, which usually concerns variables, to constraints. A constraint is *relationally arc-consistent* if any instantiation which is allowed on a subset of its variables extends to a consistent instantiation on the whole. *Relational k-arc-consistency* is the restriction of the definition above to sets of variables of cardinality k.

Definition 1 (Relational k-arc-consistency.). *Let $\mathcal{R} = (\mathcal{X}, \mathcal{D}, \mathcal{C})$ be a constraint network, C_S a constraint over the set of variables $S \subset \mathcal{X}$. C_S is relationally k-arc-consistent iff $\forall A \subset S$ such that $|A| = k$ and $\forall I$ a consistent instantiation on A, I can be extented to a consistent instantiation on S in relation to C_S. This means : if $C_S[A]$ is the projection of the relation C_S on A and I is consistent on A, therefore $I \in C_S[A]$.*
A constraint network is relationally k-arc-consistent iff all its constraints are relationally k-arc-consistent.

A k-AC clause is an implication which premiss is a conjunction that stands for the k-instantiation I, and conclusion is a disjunction of supports $s_1 \vee s_2 \vee \ldots \vee s_m$. The k-AC clause for I is $\mathcal{H} = I \rightarrow s_1 \vee s_2 \vee \ldots \vee s_m$. Relational k-arc-consistency ensures that each consistent instantiation of k variables of a constraint can be extented to all the variables of that constraint. In other words, if an instantiation doesn't satisfy this assertion, it is removed from the corresponding constraint, i.e, this tuple is now explicitly forbidden. In the case of the k-AC clauses, when all the supports (which are linked to the conjunction of assignments they represent by equivalence clauses), are falsified, then the premiss must be falsified and this is exactly the nogood corresponding to the k-instantiation, $\mathcal{H} = \neg I$. To prove the equivalence between unit propagation on those encodings, and relational k-arc consistency on the original problem we first recall some definitions given in [1] and slightly modified for our purpose.

A CSP is said to be *empty* if at least one of its variables has an empty domain or at least one of its constraints is empty, i.e, forbids all assignments.

We denote $\mathtt{sat2csp}(\mathcal{P})$ the transformation of a SAT-encoded CSP into a CSP consisting of a variable X_i with a domain $D(X_i) = [v_1, \ldots, v_d]$ for each at-least-one clause $X_i v_1 \vee \ldots \vee X_i v_d$ in $\mathcal{P}$, and a constraint forbidding the nogood $N = (X_1 = v_1 \wedge \ldots \wedge X_k = v_k)$ for each conflict clause $(\neg X_1 v_1 \vee \ldots \vee \neg X_k v_k)$ (or support clause reduced to a conflict clause by unit propagation).

First we show that the relational k-arc consistent closure of a CSP $\mathcal{P}$, written $\mathtt{r\text{-}k\text{-}AC}(\mathcal{P})$ is empty iff the k-AC encoding of $\mathcal{P}$ has an empty image under $\mathtt{sat2csp}$, that is, $\mathtt{sat2csp}(k\text{-}\mathtt{sat}(\mathcal{P}))$ is empty. We ignore the isuue of discovering the emptyness. This is trivial, both in the original problem and in the encoding, when the empty constraint arity is 1, whereas it is not for other arities, though it remains polynomial. Usually, this will be quickly discovered, providing that the empty constraint is small and that the branching heuristic chooses first the variables of this constraint.

Second we prove that, assuming the same branching choices, this equivalence is maintained at each node of the search tree by unit propagation in the encoding. As a corollary, unit propagation on k-AC encoding prunes the search tree equivalently to relational k-rac consistency on the original problem.

Lemma 1 *($\mathcal{P}$) is empty after enforcing relational k-arc consistency iff* $\mathtt{sat2csp}(k\text{-}\mathtt{sat}(\mathcal{P}))$ *is empty.*

Proof: The relational k-arc consistent closure of $\mathcal{P}$ contains all the nogoods of length k forbidding k-instantiations that don't have any support on the rest of the constraint they belong to. By definition, the k-AC clause for such an instantiation is the conflict clause of length k corresponding to the nogood. The later therefore belongs to $\mathtt{sat2csp}(k\text{-}\mathtt{sat}(\mathcal{P}))$. Moreover, all nogoods of length k are added to the relational k-arc consistent closure if and only if they are not supported. Therefore, for any nogood N of length k, $N \in \mathtt{r\text{-}k\text{-}AC}(\mathcal{P})$ iff $N \in \mathtt{sat2csp}(k\text{-}\mathtt{sat}(\mathcal{P}))$

Beside, if $\mathcal{P}$ is emptied by relational k-arc consistency, then the empty constraint arity is always k, since only nogoods of size k are added during the process. $\square$

The proof of Lemmas 1 is based on the fact that the supports of an instantiation are equivalent in the encoding and in the orginal problem. Unit propagation ensures that this is the case as well during search. We consider a relationally k-arc consistent CSP $\mathcal{P}$, an assignment $X = v$ and the induced subproblem $assign(X = v, \mathcal{P})$. In the SAT encoding this corresponds to $assign(Xv = T, k\text{-}\mathtt{sat}(\mathcal{P}))$. We prove that an instantiation looses a support because of an assignment in the CSP if and only if the k-AC clause of this instantiation looses the same support in the encoding by unit propagation of the truth assignment.

Lemma 2 *If an instantiation J, support of another instantiation I in $\mathcal{P}$ is not a support anymore in $assign(X = v, \mathcal{P})$ for relational k-arc consistency, then the support-variable s_J of the corresponding k-AC clause is set to False after unit propagation.*

Proof:
Without loss of generallity, let I be an instantiation on a set T of k variables of a constraint C_S. Let J be a support of I for C_S in $\mathcal{P}$, such that J is not a support of I in $assign(X = v, \mathcal{P})$.

Implicitly, after the assignment $X = v$, all other values in $D(X)$ are removed. If J is not a support, it means that $\exists X \in T - S$, such that $J[X]$ has been removed from its domain $(J[X] \neq v)$.

In the encoding, the assignment $Xv = T$, propagated to the at-most-one clauses yelds the assignments $Xw = F$ for all $w \neq v$. Let s_J be the proposition standing for the support J, then the equivalence clause $(\neg s_J \vee J[X])$ gives the unit clause $\neg s_J$, which is propagated to the k-AC clause. Consequently, the support-variable s_J is set to False (it is not a "support" in the encoding either). At any point of the resolution, a support-variable s_J belongs to the conclusion of a k-AC clause (is not assigned to False) iff its corresponding support J holds in the constraint network. $\square$

Lemmas 1 establishes that if the supports are the same in the original and in the encoded problem, then the problem is empty iff the reformulation is empty. Lemma 2 shows that this is the case during search.

Theorem 2 *Performing full unit propagation on at-least-one, at-most-one and k-AC clauses during search is equivalent to maintain relational k-arc-consistency on the original problem.*

From this follows a strict equivalence between the search trees of an algorithm that maintains relational k-arc consistency in the original problem, and an algorithm that enforces unit propagation on the reformulation.

3.1   Complexity of k-AC Encoding

We assume that n is the number of variables, d is the size of the domains, e is the number of constraints and a denotes their arity. We can ignore the at-most and at-least clauses : there are n at-least clauses each containing d literals, and nd^2 at-most clauses, which are binary. This $O(nd^2)$ space complexity is in all cases lower than the worst space complexity of the k-AC clauses. We therefore focus on the size of the k-AC clauses themselves.

The total number of k-AC clauses is bounded by $e\binom{a}{k}d^k$. We need to cover all the constraints (e). For each constraint, we consider all the subsets of k variables of that constraint $(\binom{a}{k})$. And for each subset, we consider all the instantiations (d^k). The total number of literals for each k-AC clause is bounded by $k + (3(a - k) + 2)(d^{a-k} - 1)$. The premiss contains k literals, and the conclusion at most $d^{a-k} - 1$. Furthermore, if $a - k > 1$, there are also $(d^{a-k} - 1)(3(a - k) + 1)$ additional literals from the equivalence clauses: Each one gives 1 clause of size $a - k + 1$ and $a - k$ clauses of size 2. The space complexity is then $O(ed^k)$ clauses of $O(d^{(a-k)})$ literals, which is still $O(ed^a)$ for any arbitrary constraint and any k. Note that the space complexity of the reformulation and of the original problem are the same. Since unit propagation can be established in linear time, the time complexity is also in $O(ed^a)$, which is optimal worst case time complexity.

4 (i, j)-Consistencies in SAT

In addition to relational k-arc-consistency, k-AC clauses allow us to enforce another very common family of local consistencies (specifically, (i, j)-consistency [9]) by adding the joins of certain constraints and performing the k-AC encoding on this augmented problem.

Definition 2 ((i, j)-Consistency.). *A binary CSP is (i, j)-consistent iff $\forall E_i, E_j$ two sets of i and j distinct variables, any consistent assignment on E_i is a subset of a consistent assignment on $E_i \cup E_j$.*

This family includes many well known consistencies.

– Arc Consistency (AC) corresponds to (1,1)-consistency.
– Path Consistency (PC) corresponds to (2,1)-consistency.
– Path Inverse Consistency (PIC) corresponds to (1,2)-consistency.

If on binary networks, arc consistency is often the best choice, higher level of filtering may sometimes be useful. For instance, path consistency is used in temporal reasoning. However, implementing algorithms to maintain other consistency, and moreover, combining this with improvements like (conflict directed) backjumping and learning requires a lot of work. With our approach, just by setting two parameters, (k and the size of the subsets to consider) and applying a SAT solver to the resulting encoding, we can solve the problem with the chosen consistency and all the other features of the SAT solver.

Definition 3 (Join of Constraints.). *Let C_{S1}, C_{S2} be two constraints, the join $C_{S1} \bowtie C_{S2}$ is the relation on $S1 \cup S2$ containing all tuples t such that $t[S1] \in C_{S1}$ and $t[S2] \in C_{S2}$.*

Theorem 3 *Enforcing (i, j)-consistency is equivalent to enforcing relational i-arc-consistency on the join of all constraints involved in a set of $i + j$ variables, for each of them.*

Proof: Let E_i be a set of i variables, If I, a consistent instantiation on E_i, is (i, j)-inconsistent, then there exists a set E_j of j variables such that $\forall IJ$ a consistent instantiation on $E_i \cup E_j$, $IJ[E_i] \neq I$. Let C be the constraint induced by the join of all the constraints involved in $E_i \cup E_j$. C is the set of all the allowed, i.e, consistent, instantiations on $E_i \cup E_j$, but I is consistent and $I \notin C[E_i]$, therefore C is relationally i-arc-inconsistent (see def 1). Conclusion : if I is (i, j)-inconsistent, then for any set E of $i + j$ variables containing the variables of I, the constraint obtained by joining all constraints which scopes are subsets of E is relationally i-arc-inconsistent. $\square$

The space complexity results of section 3 also apply here, but the number of constraints is equal to the number of subsets of $i + j$ vertices in the constraint graph, i.e, $O(n^{i+j})$, and $a = i + j$. Therefore the worst case space complexity is $O(n^{i+j} d^{i+j})$, and so is the worst case time complexity. This is again optimal.

5   Mixed Encoding

There is a clear relation between the tightness of a constraint and the performance of DP on that constraint encoded with the direct or a k-AC encoding. Consider the binary *not_equal* constraint. It can be encoded by d conflict clauses of size 2 with the direct encoding, whilst $2d$ clauses of size d are required in the AC-encoding even though AC propagation in *not_equal* doesn't achieve much pruning. On the other hand, consider the binary *equal* constraint. This is encoded with $(d-1)^2$ binary clauses in the direct encoding, while you need only $2d$ binary clauses in the AC encoding, and you can expect a lot of AC propagation. The space complexity and the level of propagation is thus linked to the tightness of the constraint. One strategy therefore is to adapt the encoding to the constraint's tightness, i.e, using the direct encoding when the constraint is loose and the AC encoding when it is tight. Moreover we can use, for each constraint, the k-AC clause with the best "adapted" k. The principal issue is to know a *priori* how to pick k. The notion of *m-looseness* [14] give us a way to choose among the different k.

Definition 4 (m-looseness). *A constraint relation R of arity a is called m-loose if, for any variable X_i constrained by R and any instantiation I of the remaining $a - 1$ variables constrained by R, there are at least m supports of I to X_i that satisfy R.*

Theorem 4 (van Beek and Dechter[14]) *A constraint network with domains that are of size at most d and relations that are m-loose is relationally $(k, (\lceil \frac{d}{d-m} \rceil - 1))$-consistent for all k.*

Proof: See [14]. $\square$

We can restrict this to relational $(k,1)$-consistency (that is relational k-arc-consistency) and then we have the relation $\lceil\frac{d}{d-m}\rceil - 1 \geq 1$ which is reduced to : $m \geq \frac{d}{2}$. This means that, given a subset of variables, if all the relations that constrain these variables are $\frac{d}{2}$-loose or more (every instantiations of this subset minus one variable have at least $\frac{d}{2}$ supports on this variable) then these constraints are relationally k-arc-consistent for any k. Therefore enforcing relational k-arc-consistency will not give any pruning, at least initially. In addition, the direct encoding would be more compact for such constraints. This suggests to use support clauses whenever the number of supports is lower than $\frac{d}{2}$ and conflict clauses otherwise. Moreover, for a given constraint arity a, the choice is now extented to any k-AC clause with k between 1 and a. To make a choice, we associate a treshold T_k on the number of supports above which we choose $(k+1)$-AC clauses rather than k-AC to encode a particular instantiation. To compute the mixed encoding of a given constraint we use the following algorithm:

First we consider all the instantiations of size 1 (all the values of all the variables), and for each of them we count the number of supports (of size $a-1$), if this number is less than T_1 then we add the corresponding 1-AC clause. In a second step, we consider all the instantiations of size 2 containing a non-yet-encoded instantiation of size 1, if the number of supports of this instantiation is less than T_2 we encode it with a 2-AC clause, and so on for a steps.

We propose $T_{a-1} = \frac{d}{2}$ whilst we don't have yet any sound value for T_k with k less than $a-1$.

Theorem 5 (Correctness and completeness of the mixed encoding.)
*I is a model of the set containing the **at-least-one**, **at-most-one**, and any k-**AC** clauses, according to the rules above, iff I is a solution of the constraint network.*

Proof: By definition, all the nogoods have at least one k-AC clause which premiss is one of its subsets, then theorem 1 (correctness) can be applied.

Theorem 1 says that k-AC encoding is complete for any k, therefore, k-**AC** clauses are satisfied by I, and so is any combination of them. $\square$

5.1 Complexity

Theorem 6 *The mixed encoding ($k = [1,2]$) requires less than $\frac{3}{2}d^2$ literals to encode a binary constraint, This limit can be asymptoticly reached.*

Proof: Let us consider the Boolean matrix of the constraint. Let r (resp c) be the number of *rows* (resp *colomns* encoded with support clauses, $0 \leq r, c \leq d$. These clauses have each less than $d/2$ literals, so we have $(r+c)\frac{d}{2}$ literals for the supports clauses. There are $(d-r)(d-c)$ elements of the matrix wich are not covered by the support clauses. Besides, there are $r(d-c)$ and $l(d-r)$ elements which are covered by only one support clause. On each row/colomn containing these elements, there are at most $\frac{d}{2}$ 0, so at most, half of them are 0. For each 0 we need a conflict clause of 2 literals, then, to encode these elements, we need

$(r + c)\frac{d}{2} - rc + d^2$ literals. If $c > \frac{d}{2}$, then this number of literals increase when r decrease, and if $c < \frac{d}{2}$, then it increase with r. This number is maximized when r is max and l min or vice versa. The worst case is then $r = d$ and $c = 0$, in that case there are $\frac{3}{2}d^2$ literals.

Let C be a constraint on variables with odd domains and relation matrix as given in the margin (that is a checkerboard of 0 and 1, plus a full column of 0 and a full but one row of 1). The rows are all but one encoded with support clauses, half of this clauses are of size $\lfloor\frac{d}{2}\rfloor$ and the other half are of size $\lceil\frac{d}{2}\rceil$, the last row is encoded with a nogood. All the colomns but one are encoded with $\lfloor\frac{d}{2}\rfloor$ conflict clauses, the remaining one with a unary support clause. That is $3 + \lfloor\frac{d}{2}\rfloor^{\cdot} + \lfloor\frac{d}{2}\rfloor \times \lceil\frac{d}{2}\rceil + \lfloor\frac{d}{2}\rfloor \times 2 \times (d-1)$ literals, which is asymptoticly equal to $\frac{\cdot}{\cdot}d^{\cdot}$ $\square$.

1	0	1	...	0	0
0	1	0	...	1	0
1	0	1	...	0	0
.	.	.	...	.	0
.	.	.	...	.	0
0	1	0	...	1	0
1	1	1	...	1	0

6   Experimental Results

We have performed a set of experiments to compare the different encodings. Section 6 and 1 give a concrete idea of the improvement, in term of pruning and cpu time, in comparison with direct encoding. In section 1 we also show that mixed encodings are an even better way to encode heterogeneous or structured problems. And finally, in section 1 we compare this approach with the state of the art in CSP. For all the random instances, we used Bessiere and Frost's random generator. The CSPs are defined by 5 parameters, the number of variables, the size of the domains, the density (i.e, the number of constraints), their arity and their tightness (i.e, the number of nogoods per constraint). The four first parameters are fixed and the tightness is given on x axis, the y axis giving the cpu time or the number of backtracks. We generally focused on results at the phase transition, when the number of satisfiables instances is the closest to 50%. We used Berkmin SAT solver [11] on generated cnf files.

k-AC Encodings. This experiment involves 1-AC, 2-AC and direct encoding on the following class of ternary networks, 30 variables, 10 values, 60 constraints. 1-AC and 2-AC encodings need both 5 times less backtracks than direct encoding[3] at the phase transition. But only 2-AC encoding translate this greater filtering into a cpu time reduction (again a factor 5). We can explain this by the amount of propagations needed to perform the same filtering in 1-AC encoding, because of the extra variables.

(i,j)-Encodings. This experiment involves PIC encoding, AC encoding and direct encoding on two classes of networks. A sparse class, 150 variables, 15 values,

---

[*] experiments with Chaff showed an even greater difference, about a factor 10 for backtracks, and 15 for cpu time.

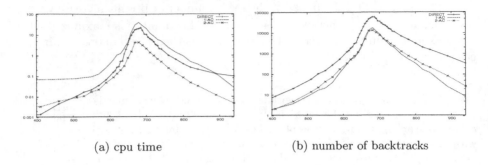

(a) cpu time   (b) number of backtracks

Fig. 1. Cpu time and number of backtracks of BerkMin on GAC (1-AC), 2-AC and Direct (3-AC) encoding.

350 binary constraints, and a dense one, 70 variables, 10 values, 310 binary constraints. According to the theory, PIC prunes even more the search tree than AC, (i.e, the backtracks are less numerous). However, on dense networks, where the gain in pruning is more evident, the amount of extra variables, as previously for 1-AC encoding, slow down the resolution.

Mixed Encoding. To emphasize the benefits of the mixed encoding on more structured problem, we used the Instruction Scheduling Problem, introduced in [15], The problem is to find a minimum length instruction schedule for a basic block of instructions (a straight-line sequence of code with a single entry point) subject to precedence, latency, and resource constraints. Basic blocks are represented as DAG (Direct Acyclic Graph). To model this problem, van Beek used one variable for each instruction, its domain represents the possible positions in the total order we have to find. The constraints are: $instruction_i < instruction_j + k$ for each arc ij labelled with k in the DAG ($instruction_j$ must wait at least k cycles after $instruction_i$), and an AllDiff constraint on all the variables. The domains are initiallized with a lower bound on the number of cycles required, and the instance is solved, if no solution is found they are incremented and the instance is solved again. The first solution encountered is the optimal solution. Each point of the figure 1 represents the runtime Berkmin needed to find the optimal solution on the mixed encoding (x axis) and AC encoding (y axis). (all instances have the same parameters : 20 instruction, 40 constraints of latency and a latency between 1 and 3 inclusive). Figure 1 compares the mixed and direct encodings. The mixed encoding is almost always better in cpu-time, compared to the direct or the AC encoding. The number of backtracks is nearly the same as in AC encoding, while the space complexity is greatly reduced (mostly because of the alldiff constraint).

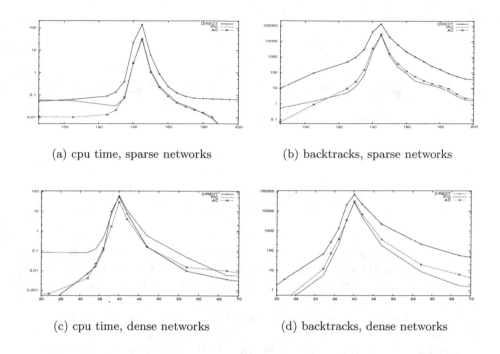

(a) cpu time, sparse networks   (b) backtracks, sparse networks

(c) cpu time, dense networks   (d) backtracks, dense networks

Fig. 2. Cpu time and number of backtracks of BerkMin on Direct, AC and PIC encoding for two classes of networks.

Comparison with the State of the Art in CSP. We also measured the efficiency of this approach in comparison with the state of the art for CSP solvers. We have done these comparisons on the following classes:

- (a) binary sparse : ¡180 variables, 15 values, 450 constraints, 147 nogoods¿.
- (b) binary dense : ¡90 variables, 10 values, 400 constraints, 38 nogoods¿.
- (c) ternary dense : ¡10 variables, 10 values, 100 constraints, 208 nogoods¿.
- (d) ternary medium : ¡30 variables, 6 values, 75 constraints, 109 nogoods¿.
- (e) ternary sparse : ¡50 variables, 10 values, 70 constraints, 790 nogoods¿.

For binary classes, 100 instances were generated and solved by MAC[4] , (Maintain Arc Consistency) with AC2001 algorithm [5], and the dynamic variable ordering (dvo) H1_DD_x [2], which outperforms the well known dom/deg heuristic. The same instances were translated in SAT problems with AC and PIC encoding, and then solved by BerkMin561.

For ternary classes 100 instances were also generated and solved by NFCx [3], where x is 0 or 5, using GAC2001 [5], and dom/deg dvo [4] without singleton

---

* For MAC the number of backtracks can be slightly overestimated, since the value given is in fact the number of visited nodes.

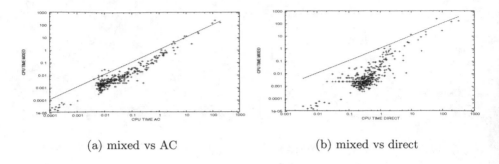

(a) mixed vs AC   (b) mixed vs direct

Fig. 3. 20 instructions, 40 latency constraints, max latency 3. cpu time for Berk-Min on Mixed encoding (y axis) and AC encoding or Direct encoding (x axis).

propagation [13]. Here again, the same instances were translated with 1-AC, 2-AC, 3-AC, mixed(1) and mixed(2) encodings, and solved by BerkMin561 [5]. The results of our approach take also into account the translation duration, which include the time spent on reading the csp file and writing the cnf. Note that this duration is insignificant when the problem is really hard, and can be dramaticly reduced by not creating a temporary file[6]. The first observation is that the performance of BerkMin on high filtering k-AC encodings (all but direct) is better on sparse than on dense networks. There are at least two reasons for that behaviour : firstly, for dense networks, at the cross-over point, the constraints are loose, and then there is not much propagation. Moreover, recall that k-AC clauses encode the *supports*, and they are more numerous when the constraints are loose. A 1-AC clause (and its equivalence clauses) for a ternary constraint can have between 1 and $3d^2$ literals, according to the number of supports, that can therefore make a great difference for the SAT solver. However, The results below show that this approach can really handle large and hard problems. The best algorithm should probably always be to solve the original problem rather than its reformulation, but when good algorithms are hard to make, reformulation is a good alternative. For example NFC is certainly more distant from the "best possible algorithm" than MAC is, and then BerkMin on the right k-AC encoding is very close, and sometimes better, than NFC. In the same way, there are very few good PIC [7] or "Maintain Relational K-Arc Consistency " algorithms.

7   Conclusion

We presented a new family of mappings of constraint problems into satisfaction problems, and proved the optimality in space and time complexity of these encodings. We also proved that performing full unit propagation on k-AC encoding

---

[.] all Christian Bessiere's algorithms ran on a 1.6 GHz pentium, whereas BerkMin ran on a 1.8 GHz one, BerkMin's results are then corrected by a factor 1.8/1.6.

[.] most of this time is spent on i/o

class (a)	MAC2001	AC + BM	PIC + BM
#backtracks	55559	66749	62006
total time	**39.6**	165	178
translation	N/A	1.37	1.2

class (b)	MAC2001	AC + BM	PIC + BM
#backtracks	56718	136139	103173
total time	**17.0**	354	373
translation	N/A	0.5	0.3

		NFC	1-AC	2-AC	3-AC	mix(1)	mix(2)
(c)	time	**0.1**	6.5	5.5	2.1	5.5	5.4
(c)	trans	N/A	0.87	1.8	1.8	1.9	2
(d)	time	**0.37**	3.26	0.84	1.14	0.84	0.86
(d)	trans	N/A	0.37	0.42	.72	0.45	0.46
(e)	time	18.40	59.8	15.5	85.8	11.4	**9.5**
(e)	trans	N/A	2.4	1.6	2.4	1.6	1.6

(a) results of MAC2001 and BerkMin on AC and PIC encodings.

(b) Results of NFC and BerkMin on different encodings on 3 classes of ternary networks.

Fig. 4. total time is cpu time for MAC and BerkMin's cpu time + translation duration, all in seconds.

is the same as enforcing relational k-arc-consistency on the original problem, or used in a slightly different way, (i,j)-consistency. We showed how to mix the different encodings to take advantage of their best individual features. And finally we demonstrated preliminary experimental results of the efficiency of the introduced encodings.

From a constraint programming perspective, these new encodings are a very easy way to implement and test algorithms for enforcing a wide range of filterings, all in optimal worst case time complexity.[7] Such encodings also profit from the sophisticated branching heuristics and other algorithmic features of the SAT solver (like non-chronological backtracking and nogood learning). Given the recent rapid advances in SAT solvers, they offer an alternative way to solve hard problem instances. From the satisfiability perspective, these encodings are useful for modelling, since many real life problems are likely to have straightforward representations as CSPs whereas SAT models are often not as easy to make. Modelling is also far more understood for CSPs than for SAT. These encodings allow the SAT research community to take advantage of CSP modelling results.

Acknowledgements

We thank Ian P. Gent for helpful discussions about SAT encoding. We also thank Eugene Goldberg for his giving us a slightly modified version of Berkmin561. This work was supported by Science Foundation Ireland.

References

[1] F. Bacchus, X. Chen, P. van Beek, and T. Walsh. Binary vs. non-binary constraints. *Artificial Intelligence*, 140(1-2):1–37, 2002.

---

* this goal was also pursued in [16], though the approach was completly different

[2] C. Bessière, A. Chmeiss, and L. Saïs. Neighborhood-based variable ordering heuristics for the constraint satisfaction problem. In *Proceedings CP'01*, pages 565–569, 2001. Short paper.

[3] C. Bessière, P. Meseguer, E.C. Freuder, and J. Larrosa. On forward checking for non-binary constraint satisfaction. *Artificial Intelligence*, 141:205–224, 2002.

[4] C. Bessière and J.C. Régin. MAC and combined heuristics: two reasons to forsake FC (and CBJ?) on hard problems. In *Proceedings CP'96*, pages 61–75, 1996.

[5] C. Bessière and J.C. Régin. Refining the basic constraint propagation algorithm. In *Proceedings IJCAI'01*, pages 309–315, 2001.

[6] M. Davis, G. Logemann, and D. Loveland. A machine program for theorem proving. *Communications of the ACM*, 5:394–397, 1962.

[7] Romuald Debruyne. A property of path inverse consistency leading to an optimal PIC algorithm. In *Proceedings ECAI'00*, pages 88–92, 2000.

[8] R. Dechter and P. van Beek. Local and global relational consistency. *Theoretical Computer Science*, 173(1):283–308, 1997.

[9] E.C. Freuder. A sufficient condition for backtrack-bounded search. *Journal of the ACM*, 32:755–761, 1985.

[10] I.P. Gent. Arc consistency in SAT. In *Proceedings ECAI'02*, 2002.

[11] E. Golberg and Y. Novikov. Berkmin: a fast and robust sat-solver. In *Proceeding DATE'02*, pages 142–149, 2002.

[12] S. Kasif. On the parallel complexity of discrete relaxation in constraint satisfaction networks. *Artificial Intelligence*, 45:275–286, 1990.

[13] B.A. Nadel. Constraint satisfaction algorithms. *Computational Intelligence*, 5:188–224, 1989.

[14] P. van Beek and R. Dechter. Constraint tightness and looseness versus local and global consistency. *Journal of the ACM*, 44:549–566, 1997.

[15] Peter van Beek and Kent Wilken. Fast optimal instruction scheduling for single-issue processors with arbitrary latencies. *Lecture Notes in Computer Science*, 2239:625–639, 2001.

[16] G. Verfaillie, D. Martinez, and C. Bessière. A generic customizable framework for inverse local consistency. In *Proceeding AAAI'99*, pages 169–174, 1999.

[17] T. Walsh. SAT v CSP. In *Proceedings CP'00*, pages 441–456, 2000.

Guiding SAT Diagnosis with Tree Decompositions

Per Bjesse, James Kukula, Robert Damiano, Ted Stanion, and Yunshan Zhu

Advanced Technology Group, Synopsys Inc.

Abstract. A tree decomposition of a hypergraph is a construction that captures the graph's topological structure. Every tree decomposition has an associated tree width, which can be viewed as a measure of how tree-like the original hypergraph is. Tree decomposition has proven to be a very useful theoretical vehicle for generating polynomial algorithms for subclasses of problems whose general solution is NP-complete. As a rule, this is done by designing the algorithms so that their runtime is bounded by some polynomial times a function of the tree width of a tree decomposition of the original problem. Problem instances that have bounded tree width can thus be solved by the resulting algorithms in polynomial time. A variety of methods are known for deciding satisfiability of Boolean formulas whose hypergraph representations have tree decompositions of small width. However, satisfiability methods based on tree decomposition has yet to make an large impact. In this paper, we report on our effort to learn whether the theoretical applicability of tree decomposition to SAT can be made to work in practice. We discuss how we generate tree decompositions, and how we make use of them to guide variable selection and conflict clause generation. We also present experimental results demonstrating that the method we propose can decrease the number of necessary decisions by one or more orders of magnitude.

1   Introduction

Tree decomposition [6] is a graph theoretic concept which abstractly captures topological structure in a variety of problems [4] such as constraint satisfaction [12], Gaussian elimination [21], database query processing [9], and image computation [13].

The topological structure of a Conjunctive Normal Form (CNF) formula can be represented as a hypergraph, where the vertices of the hypergraph correspond to the variables of the CNF and the hyperedges correspond to the clauses. Given a small treewidth tree decomposition for a hypergraph of a CNF formula, a variety of methods are known for deciding its satisfiability [12, 10, 3].

In this paper we report on our effort to learn whether satisfiability solving guided by tree decomposition can be made to work in practice. To do this, we attempt to find tree decompositions of small treewidth for significant problems, and to incorporate methods based on tree decomposition into a state-of-the-art SAT solver.

E. Giunchiglia and A. Tacchella (Eds.): SAT 2003, LNCS 2919, pp. 315–329, 2004.
© Springer-Verlag Berlin Heidelberg 2004

The end result of the work presented in this paper is a satisfiability checking method that given a bounded width tree decomposition of a problem instance will be guaranteed to run in quadratic time. We present the methods we use for generating tree decompositions and show how we make use of the tree decomposition to guide diagnosis and conflict clause generation. We also present experimental results that demonstrate that there are real-life SAT instances with small tree width, where our tree-sat method decreases the number of necessary decisions by one or more orders of magnitude.

2   Preliminaries

In the remainder of this paper, we will focus on augmenting GRASP-like [16] implementations of the Davis-Putnam-Loveland-Logemann (DPLL) method [11] with tree decomposition guidance. We refer readers unfamiliar with decision, deduction, and diagnosis components of such algorithms, including conflict graphs and backjumping, to [23].

3   Tree Decomposition

Given a hypergraph $G = (V, E)$, where V is a set of vertices and E a set of hyperedges with $e \subseteq V$ for each $e \in E$, a *tree decomposition* of G is a triple (N, F, χ) where

1. N is a set of nodes,
2. $F \subset N \times N$ is a set of arcs such that (N, F) forms an unrooted tree,
3. $\chi : N \rightarrow 2^V$ associates a subset of vertices with each tree node,
4. for every hyperedge $e \in E$, there is some node $n \in N$ such that $e \subseteq \chi(n)$,
5. for every $n_1, n_2, n_3 \in N$, if n_2 lies on the path in (N, F) between n_1 and n_3, then $\chi(n_1) \cap \chi(n_3) \subseteq \chi(n_2)$. This means that, for each vertex $v \in V$, the set of nodes that contain v form a subtree of (N, F).

Informally, this means that a tree decomposition of a CNF formula will be an unrooted tree, whose nodes contain subsets of the variables in the formula. This tree needs to fulfill the two properties that (1) the set of variables in each clause in the CNF needs to be a subset of some node, and (2) the set of nodes that contain a variable v from the original CNF must form a subtree of the tree decomposition.

The *treewidth* of a tree decomposition is $\max_{n \in N} |\chi(n)| - 1$. The treewidth of a hypergraph is the smallest treewidth of any of its possible tree decompositions.

Figure 1 shows an example of a CNF formula and an associated tree decomposition of treewidth 3.

4   Tree Decomposition and DPLL-SAT

We will now relate each of the core algorithmic components of modern DPLL-based SAT solvers—decision, deduction, and diagnosis—to tree decomposition.

CNF

(1 -2 4)
(1 2 8)
(2 3 -8)
(-0 1 2)
(0 1 -3)
(-2 4 7)
(0 -1 6)
(2 -3 5)

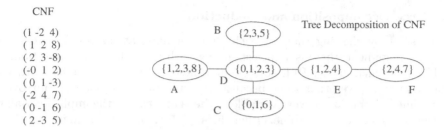

Fig. 1. A CNF formula and its tree decomposition

First, we show that a tree decomposition of a CNF formula will necessarily include paths corresponding to all possible chains of deduction resulting from assigning one or more variables. Next, we describe a set of conditions under which all chains from decision variables to an unsatisfied clause will pass through some common tree node, so that a legitimate conflict clause can be constructed from the variables in that node. Finally, we outline constraints on decision variable selection that insure these conditions are met.

Before we proceed further, however, we would like to note that a SAT procedure that restricts itself to constructing conflict clauses contained in tree nodes will be polynomial in the problem size for any class of problems with treewidth bounded by w. To see this, we can think of a SAT procedure as a series of conflict clause constructions. The number of conflict clauses constructed will be limited to $O(n3^w)$, where n is the size of the CNF formula, since any tree decomposition can be trivially reduced to have $O(n)$ nodes and each node with w variables can only generate 3^w clauses. When the i'th conflict clause is constructed, the CNF formula will have grown to size $O(n + i)$ from the addition of the prior conflict clauses. The time required to build the i'th conflict clause is linear in this size, since a chain of implications could propagate through a large fraction of the clauses. Thus the total SAT procedure will be bounded by

$$\sum_{i=0}^{n3^w}(n + i) = O(n^2 9^w)$$

There exists SAT algorithms for bounded treewidth problems that are linear rather than quadratic in the problem size [12]. However, these algorithms do not make use of the strength of DPLL solvers and as far as we know, none of these algorithms have managed to scale in practice. As our goal is to explore techniques that are effective for typical industrial problem instances, and DPLL solvers have proved to be very competitive in this context, our focus in this paper will thus be on modifying a DPLL solver to generate bounded conflict clauses even though the resulting complexity is superlinear.

4.1   Tree Decomposition and Deduction

Let us start by showing that for each chain of implications resulting from a variable assignment, there is a corresponding path in the tree decomposition.

The construction of DPLL solvers guarantees that every variable with an implied value is given that value because of some antecedent clause. Moreover, at the time of implication every variable in the clause except the implied variable must have been given some value, either through decision or through implication. Let us consider an arbitrary chain of implications $v_0, v_1, \ldots, v_k$, where v_0 is a decision variable, v_i was given an implied value before v_j whenever $0 < i < j$, and for each $i < k$, v_i appears in the antecedent c_{i+1} of v_{i+1}.

Lemma 1. *For any chain of implications there is path in the tree decomposition $n_0, n_1, \ldots, n_l$ together with a mapping $p : [0..l] \rightarrow [0..k]$, satisfying*

- *Either $n_i = n_{i+1}$, or n_i and n_{i+1} are adjacent in the tree decomposition.*
- $v_{p(i)} \in \chi(n_i)$
- $p(0) = 0$
- $p(l) = k$
- *If $p(i) = j$, then either $p(i+1) = j$ or $p(i+1) = j+1$.*
- *If $p(i) \neq p(i+1)$, then $n_i = n_{i+1}$.*

This node sequence is built up from paths in the tree between nodes containing successive antecedent clauses. To see this, focus on some particular subsequence v_i, v_{i+1}, $i > 0$, together with the antecedents c_i of v_i and c_{i+1} of v_{i+1}. By rule 4 of tree decomposition, clause c_i must be contained by some tree node n_{c_i}, and similarly clause c_{i+1} must be contained by some $n_{c_{i+1}}$ Variable v_i must be in clause c_i since c_i is the antecedent of v_i, and so v_i must be in node n_{c_i}. Variable v_i must also be in clause c_{i+1} since v_i, v_{i+1} is part of an implication chain, so v_i must also be in node $n_{c_{i+1}}$. Since variable $v+i$ is in both nodes n_{c_i} and $n_{c_{i+1}}$, by rule 5 of tree decomposition it must be in every tree node on the path between them. In this way we can build up the complete path in the tree decomposition by appending the paths joining each pair of successive antecendent clauses c_i and c_{i+1}.

As an example, consider the CNF formula in Figure 1. Assume that the variables v_1 and v_2 in the implication chain $7, 2, 8$ has the antecendent clauses $c_1 = (-2\,4\,7)$ and $c_2 = (1\,2\,8)$. The following is then a path and mapping corresponding to the chain:

i	0	1	2	3	4	5
n_i	F	F	E	D	A	A
$p(i)$	0	1	1	1	1	2
$v_{p(i)}$	7	2	2	2	2	8

4.2   Tree Decomposition and Diagnosis

Next we define a set of conditions under which a conflict clause can be constructed from variables within a single tree node.

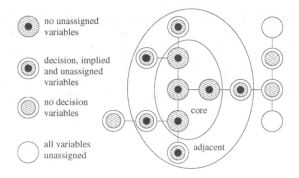

Fig. 2. Core and adjacent nodes

Definition 1. *Given a CNF formula, a tree decomposition of it, and a partial assignment composed of decision and deduction assignments, a core subtree N_C is a nonempty maximal subtree that satisfies:*

- *For each decision variable v in the partial assignment, some node in N_C or some node adjacent to N_C contains v.*
- *Every variable in every node of N_C is assigned.*

Definition 2. *Given a non-conflicting partial assignment, a further decision assignment is* admissible *relative to a tree decomposition if the decision variable is chosen from among the nodes adjacent to a core of the starting partial assignment.*

In Figure 2, the core subtree is made up from the four nodes in the middle that are fully assigned. Any variable that is contained in the five nodes adjacent to the core are admissible for assignment.

Suppose a admissible decision assignment to variable v results in a conflict. The unsatisfied clause cannot be contained in any of the nodes of the core N_C, because the starting partial assignment which was non-conflicting had already assigned values to all the variables in the nodes of N_C. $N \setminus N_C$ in general forms a forest of subtrees, one subtree of which contains n_D, the node adjacent to N_C which contains v. The unsatisfied clause must exist in this subtree, because no decision in n_D can cause any implication in any other subtree, since any implication path would have to traverse nodes in N_C but all variables in N_C have already been assigned. This leads to:

Lemma 2. *If a conflict is deduced from a admissible decision assignment, a conflict clause can be constructed from the variables in a single tree node, in particular from the variables in n_D.*

This can be seen by considering the chains of implications from the decision variables to the unsatisfied clause. Every implication chain to the unsatisfied clause starts at a node in $N_C \cup \{n_D\}$ and ends at a node containing the conflict in the subtree which contains n_D. N_C touches this subtree at n_D, therefore every implication chain from a decision variable to the unsatisfied clause must pass through n_D. A conflict clause must include enough variables assigned by decision or deduction to generate the conflict. Since every implication chain includes a variable in n_D, a conflict clause can be built from these variables.

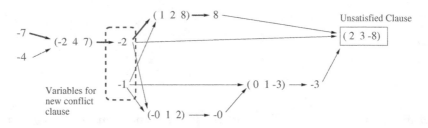

Fig. 3. An implication graph

As an example, consider the implication graph in Figure 3 where a conflict arises from the admissible decision to assign variable 1 the value 0 in the CNF from Figure 1. To see that this decision is admissible, note that earlier decisions has assigned variables 4 and 7 the value 0, which has led to an implied value for variable 2. Node F hence forms a core, which means that all variables in the adjacent node E are admissible. As the current conflict stems from an admissible decision, we can construct a conflict clause from the variables in a single node— the variables 1 and 2 in node E.

One concern regarding conflict clause construction is that the tree decomposition for the original CNF formula might not be a valid decomposition for the new formula that includes the constructed conflict clause. But since each constructed clause is contained within an original tree node, the original tree decomposition does remain valid as conflict clauses are added.

4.3   Tree Decomposition and Decision

A SAT procedure needs a way to select a decision variable whenever unassigned variables remain. We have seen that admissible decision assignments let us build conflict clauses from variables contained in a single tree node. However, admissible decision assignments are not available before a core subtree has been formed. We now describe an complete method to select decision variables that permits us to maintain our constraint on sets of variables in conflict clauses.

Definition 3. *Given a CNF formula and a partial assignment consisting of decision assignments and deductions, an unassigned variable is a compatible decision candidate relative to a tree decomposition of the CNF formula if it satisfies the following criteria:*

- *If no decision assignments have been made yet, then any unassigned variable is a compatible candidate.*
- *If no core subtree exists, then a compatible candidate must be contained in a node that contains all current decision variables.*
- *If a core subtree exists for the current partial assignment, a compatible candidate must be contained in a node adjacent to a core.*

That selecting decision variables from among compatible candidates is an effective strategy is established by the following two lemmas.

Lemma 3. *If each decision in a partial assignment is selected from candidates compatible with a tree decomposition, and unassigned variables remain, then a compatible candidate exists.*

The argument for this is inductive. As long as no core exists, the strategy calls for selecting variables from some common node. Once no unassigned variables are left in that common node, then that node becomes the seed for a core subtree, guaranteeing the existence of a core. A core is defined to be maximal, so nodes adjacent to a core must have unassigned variables. As long as one picks compatible candidates, further compatible candidates will exist until the assignment is complete.

Lemma 4. *If a decision assignment to a compatible candidate results in a conflict, then a conflict clause can be built from variables in a single tree node.*

To see this, we consider two cases. As long as a core does not exist, all decision variables come from some common node. A conflict clause can be constructed from the variables of this node, e.g. from the decision variables themselves. When a core exists, the compatible candidates are just those which give admissible assignments, so by Lemma 2 a conflict clause can be built from some single node.

With a decision variable selection strategy that supports conflict clause construction from single tree nodes, we can conclude that:

Theorem 1. *Given a tree decomposition for a CNF formula, decision and diagnosis can be performed so that for each conflict clause constructed, some node contains all the variables in the clause.*

5   Constructing Tree Decompositions

The effectiveness of the conflict clause construction method we have described relies on first constructing, for a given CNF instance, a tree decomposition of small width. Clearly some CNF instances will not have any small width decompositions, and for these the methods we have described will not provide a useful bound on the sizes of conflict clauses or their number. But we expect that many practical problems will have small width. For example, it has been observed [18] that digital circuits tend to have small cutwidth. Small cutwidth implies small treewidth, so we expect our method to be effective on a large fraction of digital circuits. Moreover, the reverse is not necessarily true, so our method has the potential of being effective even on some classes of problems that have an intractable cutwidth.

Finding a minimal width tree decomposition for an arbitrary graph is NP-complete [5], so an efficient general algorithm is unlikely to exist. For fixed k, checking if a graph has treewidth k and constructing a width k tree decomposition if it does can be done in time linear in the size of the graph [7]. Unfortunately, current algorithms grow in cost very rapidly with k, and are only practical for very small k, roughly $k \leq 4$ [19]. Much more efficient algorithms have been developed for approximating k with bounded error [2], but even these appear to be too costly for industrial problems.

Due to the limitations of the direct approaches to computing a tree decomposition, we have taken a different approach. We rely on the facts that tree decompositions can be derived from orderings of hypergraph vertices [8]—CNF variables in our case—and that a plethora of robust and scalable CNF variable ordering methods is available. Given an ordering of variables, we use the following algorithm to compute a decomposition of a hypergraph representation G:

1. Let v be the next unprocessed variable in the variable order.
2. Add a new tree node n to the tree decomposition D with $\chi(n) = \bigcup_{e \ni v} e$
3. Update the hypergraph G by deleting all hyperedges containing v and adding a single hyperedge $e_n = \chi(n) \setminus \{v\}$.
4. Add arcs to the tree decomposition so that n is connected to every node n' whose hyperedge $e_{n'}$ just was deleted from G in step 3.
5. If unprocessed variables exists, goto 1, else we are done.

As an example, in Figure 4 we illustrate the generation of the tree decomposition from Figure 1 using the variable order $8, 5, 6, 0, 1, 3, 7, 2, 4$. The first variable in the order, 8, is contained in the clauses $(1\,2\,8)$ and $(2\,3\,-8)$, so when the node A is built, $\chi(A)$ is set to $\{1, 2, 3, 8\}$. The new hyperedge $e_A = (1\,2\,3)$ then replaces the two clauses with variable 8. Variable 5 generates $\chi(B) = \{2, 3, 5\}$ and $e_B = (2\,3)$. Variable 6 generates $\chi(C) = \{0, 1, 6\}$ and $e_C = (0\,1)$. The next variable in the order, 0, is contained in the clauses $(-0\,1\,2)$, $(0\,1\,-3)$, and $(0\,-1\,6)$, as well as the added hyperedge e_C. So $\chi(D)$ will be $\{0, 1, 2, 3\}$. Since node D incorporates hyperedge C, an arc is added between nodes C and D. Separate nodes could be built to reflect the elimination of variables 1 and 3, but since node D already includes all the variables involved, one can compact the tree by just letting node D serve for the elimination of all three variables $0, 1$, and 3. To reflect this, arcs are also added between node D and nodes A and B. The new hyperedge $e_D = (2)$ then replaces the clauses and earlier added hyperedges containing the eliminated variables. The creation of nodes E and F then continues in the same pattern.

The algorithm we use to build tree decompositions reduces the problem of finding a good tree decomposition to the problem of finding a good variable order. We have explored the use of two different methods for constructing variable

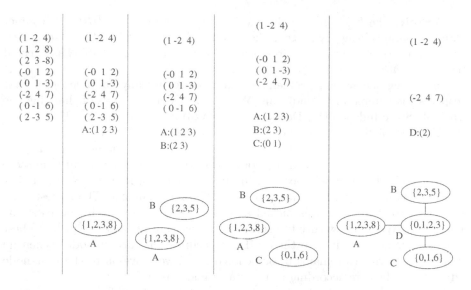

Fig. 4. Constructing a tree decomposition

orders for industrial-sized problems; a simple greedy method and a method based on linear arrangement. The simple greedy method we tried was the min-degree heuristic [2, 20] which is fast and known to give reasonable results. Each next variable in the order is determined on the fly from the reduced hypergraph from which earlier variables have been eliminated. The variable chosen to be next in the order is the variable for which $|\bigcup_{e \ni v} e|$ is smallest, being the size of the tree node to be constructed to eliminate the variable.

The second heuristic we explored is built using the MINCE linear placement engine [1]. The objects placed by MINCE are the hyperedges of the graph, corresponding to the CNF clauses. MINCE then generates a linear order of the clauses, attempting to reduce the cutwidth, the largest number of variables appearing in clauses on both sides of a cut. We then convert the clause order to a variable order by placing v_1 before v_2 if the last clause containing v_1 occurs before the last clause containing v_2. Since MINCE orders the clauses in a way to keep clauses with common variables near each other, our hope is that the tree decompositions generated from the MINCE clause order will have small width.

6   Implementing a Tree-Based DPLL Solver

Let us consider the practicalities of integrating the tree decomposition approach to satisfiability solving into a modern Chaff-like [17] DPLL engine. In order to make use of a tree decomposition, we need to (1) modify the conflict clause generation; and (2) control the selection of decision variables so that we only use compatible candidates.

We solve the first problem by modifying the standard *1-UIP* [22] conflict clause generation slightly so that the conflict clauses that it returns are forced to contain variables exclusively from the last node a decision variable was selected from (the *choice node*)

In solving the second problem, we are free to use any variable order that respects the compability conditions. We will consider two ways of changing the Variable State Independent Decaying Sum (VSIDS) variable order [17].

The first of these approaches is the *static node* VSIDS order: Given a tree decomposition of a CNF problem, we generate a order on the nodes in the tree decomposition once and for all by computing the average initial variable score in each node. Each node's position in this order is static in the sense that it will not change during the search for a satisfying assignment. The largest node according to this measure is picked as the initial choice node in the tree (the *root node*). Whenever we need to select a new decision variable, we pick the best scored variable according to the VSIDS measure from current node. When no unassigned variables remains in our choice node, we move on to the best node adjacent to the core according to the static node order.

The second approach, the *dynamic node* variable order, differs from the static order in that we do not necessarily exhaust a choice node before we move on to the next node (with the exception of the initial root node, as this is required for compatibility). Instead, we pick a new choice node for each decision, by selecting that node adjacent to the core that will allow us to pick the highest scored variable according to VSIDS. In the dynamic node order, we pick the root node to be the node with the highest average of the smallest 10% of the nodes. The rationale for this is that we want to find a balance between being locked into a root node for the smallest number of decisions possible, and still make decisions using strong variables. In contrast to the static node order, we pick a new root node every time the proof search is restarted.

7   Experimental Results

In this section, we present the experimental performance of the tree-based SAT using the dynamic and the static variable order. Our objective is to show that satisfiability solving based on tree decomposition has potential in the sense that there are classes of formulas where it can be used to decrease the number of decisions significantly compared to a standard DPLL solver.

Our benchmark problems are a mixture of industrial and public benchmarks: The Dubois problems are a series of random benchmarks generated by the *gensathard* program that is included in the DIMACS benchmark distribution [15]. The remaining problems are inhouse-generated equivalence checking problems. In particular, the *addm_x_y* examples are equivalence checks between different ways of implementing the addition of x y-bit numbers. The two implementations in each benchmark differ in the order they process individual bits from the different words. Note that the different *addm* benchmarks of varying size are not related in the sense that any one is a simple extension of the other—no adder trees in any of the problems have substructures that are even remotely similar.

Benchmark	Size (v/c)	Tree width	Static tree-sat # dec	Dyn. tree-sat # dec	Traditional sat # dec
dubois_50	150/400	4	**101**	101	2 647
dubois_500	1500/4000	4	**1 226**	2 002	58 316
dubois_1000	3000/8000	4	**3 351**	6 576	242 616
dubois_2000	6000/16000	4	**10 301**	18 223	712 153
addm_4_3	253/842	18	**938**	965	1 410
addm_4_4	433/1548	29	**2 653**	5 313	3 684
addm_4_5	661/2242	37	**6 141**	17 311	12 545
addm_4_6	937/3194	42	**23 735**	31 370	32 796
addm_4_7	1261/4314	51	1 480 277	**83 533**	134 426
addm_5_3	406/1367	29	**3 716**	12 060	8 344
addm_5_4	701/2382	41	**37 842**	64 132	42 486
addm_5_5	1076/3677	50	651 478	1 109 646	**166 847**
97686	4566/13987	170	**1914**	6228	9 485

Table 1. Experimental results

We found that tree decompositions constructed using the MINCE-based heuristic generally gave significantly smaller treewidths than those constructed using the simple greedy heuristic, therefore the SAT results we report here used the MINCE-based heuristic. Since underlying MINCE engine, is randomized, different tree decompositions are generated for each run. We report the average value of ten SAT runs for the tree-based solvers.

To illustrate the different behavior of the two tree decomposition heuristics, we gathered information on the distribution of node sizes for each when run on the *addm_5_5* problem.

Percentile	Greedy	MINCE
20th	6	21
median	7	33
80th	12	41
max	95	46

The largest node in the MINCE-based decomposition, with 46 variables, is less than half that of the greedy decomposition. The greedy heuristic generates many small nodes and just a few large ones, while the MINCE-based heuristic generates a much more uniform distribution. Since the number of conflict clauses that can be constructed is exponential in the size of the nodes, the disadvantage of a few large nodes more than outweighs the advantage of many smaller ones. The computational cost of the MINCE-based heuristic is dominated by that of the underlying MINCE engine, making the greedy heuristic considerably faster. For example, on the *addm_5_5* problem the greedy heuristic was about 35x faster than the MINCE heuristic.

Our core DPLL solver is on par with Berkmin and ZChaff in terms of speed, but the additions for doing tree decomposition and handling the core are un-optimized. We therefore focus on comparing the proof methods in terms of the necessary number of decisions. This has the added benefit that it provides an implementation and platform independent measure of the potential of the methods.

As Table 1 indicates, the Dubois problems seem to have the interesting characteristic that their tree decompositions widths is held constant at 4 even when problem size increases. This means that they are easy for our tree-based solvers in the sense that very short conflict clauses will be needed to solve them. The experimental data confirms that both the static and dynamic tree-based DPLL solver needs orders of magnitude fewer decisions than our reference standard DPLL solver.

In contrast, the generated tree decompositions for the *addm_x_y* examples increase with the size and number of operands. The tree width for the larger examples ranges from 18 up to over 50. Still, the tree decomposition seems to be helpful, especially using the static variable order. However, for the very largest example (*addm_5_5*), the tree-based methods do many times worse than the plain solver. One of the potential reasons for this is that it becomes harder for our current tree decomposition engine to find a high quality decomposition as the problem size increases.

The industrial circuit 97686 has the largest tree width of all the benchmarks. Interestingly, it can still be solved using relatively few decisions by the static variable ordering algorithm. Even a tree decomposition with a high width can thus sometimes can be helpful.

As can be seen from the table, the static variable order seems to be better in almost all cases than the dynamic variable order. The results hence indicate that for these examples it is advantageous to keep the variables that are related by the nodes together in the variable order, rather than to try to emulate the VSIDS order as closely as possible.

Additional insight into the behavior of the conflict clause construction method we have described can be gained from the distribution of conflict clauses constructed. We gathered data for the *addm_5_5* problem:

Percentile	Static tree-sat	Traditional sat
20th	40	35
median	42	105
80th	44	141
max	45	351

This data shows that with conventional methods for decision and diagnosis, most conflict clauses constructed are longer than even the longest clauses constructed with our method based on tree decomposition.

We would like to note that although our results show consistent improvements in the number of decisions, we presently do not fare as well in terms of runtime. This is partly due to an unoptimized implementation of tree-based DPLL and partly due to the overhead of our unsophisticated tree decomposer. For example, the tree-based DPLL engines are a factor three to ten times slower per decision in terms of the pure DPLL engine on the *addm* examples, and we incur between three seconds and four minutes of overhead from generating the tree decompositions. However, some recent developments that we discuss in our conclusions in Section 9 suggest that a more efficient implementation of our method will have the potential to scale well also in terms of runtime.

8 Related Work

There have been other attempts to construct efficient decision methods for formulas with low treewidth. One such approach is Darwiche's Decomposable Negation Normal Form (DNNF) [10]. Formulas that have bounded treewidth can be checked for satisfiability in linear time by first translating them into DNNF, and then applying a simple test. We expect there to be examples where satisfiability solving based on DNNF translation is superior to tree-based DPLL, since our theoretical complexity bound is quadratic. However, one appealing aspect of our approach is that the underlying SAT solving machinery that we are making use of is mature and is known to perform well in practice on industrial examples. Moreover, our light-weight integration makes it possible for us to interface to new solvers as soon as the state-of-the-art for DPLL solving advances.

The oracle we use for generating tree decompositions, MINCE, has previously been applied to the generation of variable orders both for BDDs and for SAT-solvers [1]. In this context, MINCE is used as a preprocessing step that generates an initial order. In contrast, we use MINCE to construct a tree decomposition that not only guides the variable ordering process in its entirety, but also guides the construction of conflict clauses.

There are strong parallels between our static node variable order and Amir and McIlrath's heuristic for generating a DPLL variable order from a decomposition of a propositional theory [3]. In this work the partitions are first ordered based on their *constrainedness*—the ratio of clauses contained in a given partition to the number of partition variables. The propositional variables are then ordered in a way that respects this partition order. A significant difference between our use of a given tree decomposition and Amir and McIlrath's is that we guarantee that all generated conflict clauses have length bounded by the treewidth of the decomposition.

9 Conclusions and Future Work

There has been a lot of research into tree decomposition, and there exists a rich theory about how tree decompositions can be used to solve NP-complete problems. However, the prior work in this field has not focused primarily on attempting to leverage tree decomposition to achieve speed ups on large real-life satisfiability problems. There has also been very little research that has aimed to combine the strengths of state-of-the-art satisfiability solvers with a tree decomposition generator that is practical for realistic problems.

The work that we have presented here represents a first step in this direction, and we hope that the results we have shown will stimulate further research. The simple approach we have presented already shows promise in the sense that it can decrease the number of necessary decisions for solving problems significantly, as witnessed by the order of magnitude improvements for the Dubois problems. Moreover, our work demonstrates that there exists heuristics that can process problems containing many thousands of variables and clauses in reasonable time, and still provide results that can help improve SAT solving efficiency. Finally, we have shown that there are structured problems from real-life, such as the *addm* problems, that have reasonable tree widths and where even an unrefined procedure that attempts to leverage tree decompositions can improve the decision count substantially.

We believe that tree decomposition can be a valuable tool for SAT. However, there is still work that remains to be done. For example on the benchmark problems, the best runs of MINCE often provide as few as half as many decisions than the average values that we have reported. As future work, we would therefore like to study how we can improve the tree decomposition engine, and tune our tree-based DPLL solver.

It is already clear that there is a lot more to be gained: In a parallel development to our original conference paper in SAT 2003, Huang and Darwiche have introduced a variable ordering heuristic that is a continuation of Darwiche's work on DNNF [14]. They guide decision variable selection in a DPLL solver using *DTrees*—the subclass of tree decompositions that correspond to full binary trees. Huang and Darwiche's variable selection heuristic uses the DTree to compute an order on the tree nodes that never changes during the execution of the search. However, just like in our case, the SAT engine is free to chose any decision variable within the current node until it is full. The most important difference between Darwiche and Huang's work, and the work presented in this paper is that we not only use the tree decomposition to guide decision, but use the structural information to enforce bounded conflict clause construction. Other, less significant, differences are that (1) we use a different oracle for generating tree decompositions, (2) we consider full tree decompositions rather than DTrees, and (3) our node order may change during the execution at the price of more runtime overhead. The DTree-based heuristic extends our experimental results by demonstrating that even in the case where conflict clauses are not bounded, runtimes can be improved significantly on a number of structured unsatisfiable benchmarks by navigating a subclass of tree decompositions in a completely static way. We are very excited about the results achieved by Huang and Darwiche, and we are eager to investigate how much further we can get by combining the benefits of their lower overhead approach to decision guiding with the power of our bounded conflict clause generation.

References

[1] F. Aloul, I. Markov, and K. Sakallah. Faster SAT and Smaller BDDs via Common Function Structure. In *Proc. Intl. Conf. on Computer-Aided Design*, pages 443–448, 2001.

[2] E. Amir. Efficient Approximation for Triangulation of Minimum Treewidth. In
     Proc. Conf. on Uncertainty in Artificial Intelligence, 2001.
[3] E. Amir and S. McIlraith. Solving satisfiability using decomposition and the
     most constrained subproblem. In *Proc. Workshop on Theory and Applications of
     Satisfiability Testing*, 2001.
[4] S. Arnborg. Efficient Algorithms for Combinatorial Problems on Graphs with
     Bounded Decomposability - A Survey. *BIT*, 25:2–23, 1985.
[5] S. Arnborg, D. G. Corneil, and A. Proskurowski. Complexity of finding embed-
     dings in a k-tree. *SIAM Journal of Algebraic and Discrete Methods*, (8), 1987.
[6] H. Bodlaender. A Tourist Guide through Treewidth. *Acta Cybernetica*, 11, 1993.
[7] H. Bodlaender. A linear time algorithm for finding tree-decompositions of small
     treewidth. In *Proc. ACM Symposium on the Theory of Computing*, 1993.
[8] H. Bodlaender, J. Gilbert, H. Hafsteinsson, and T. Kloks. Approximating
     Treewidth, Pathwidth, Frontsize, and Shortest Elimination Tree. *Journal of Al-
     gorithms*, 18:238–155, 1995.
[9] C. Chekuri and A. Rajaraman. Conjunctive query containment revisited. In *Proc.
     Int'l Conf. on Database Theory*, volume LNCS 1186, pages 56–70, 1997.
[10] A. Darwiche. Compiling knowledge into decomposable negation normal form. In
     Proc. Intl. Joint Conf. on Artificial Intelligence, 1999.
[11] M. Davis, G. Logeman, and D. Loveland. A machine program for theorem-proving.
     Communications of the ACM, 5(394–397), 1962.
[12] R. Dechter and J. Pearl. Network-based heuristics for constraint-satisfaction prob-
     lems. *Artificial Intelligence*, 34(1):1–34, 1988.
[13] A. Gupta, Z. Yang, P. Ashar, L. Zhang, and S. Malik. Partition-Based Decision
     Heuristics for Image Computation using SAT and BDDs. In *Proc. Intl. Conf. on
     Computer-Aided Design*, pages 286–292, 2001.
[14] J. Huang and A. Darwiche. A structure-based variable ordering heuristic for SAT.
     In *Proc. Intl. Joint Conf. on Artificial Intelligence*, 2003.
[15] D. Johnson and M. Trick, editors. *The Second DIMACS Implementation Chal-
     lenge*. DIMACS series in Discrete Mathematics and Theoretical Computer Sci-
     ence. American Mathematical Society, 1993.
     (see http://dimacs.rutgers.edu/challenges/).
[16] J. P. Marques Silva and K. A. Sakallah. GRASP—a new search algorithm for
     satisfiability. In *Proc. Intl. Conf. on Computer-Aided Design*, pages 220–227,
     1996.
[17] M. Moskewicz, C. Madigan, Y. Zhao, L. Zhang, and S. Malik. Chaff: Engineering
     an efficient SAT-solver. In *Proc. of the Design Automation Conf.*, 2001.
[18] M. Prasad, P. Chong, and K. Keutzer. Why is ATPG easy? In *Proc. of the Design
     Automation Conf.*, 1999.
[19] H. Roehrig. *Tree Decomposition: A Feasibility Study*. M.S. Thesis, Max-Planck-
     Instit. Inform. Saarbruecken, 1998.
[20] D. Rose. Triangulated Graphs and the Elimination Process. *J. of Discrete Math-
     ematics*, 7:317–322, 1974.
[21] D. Rose and R. Tarjan. Algorithmic Aspects of Vertex Elimination on Directed
     Graphs. *SIAM J. Appl. Math.*, 34(1):176–197, 1978.
[22] L. Zhang, C. Madigan, M. Moskewicz, and S. Malik. Efficient conflict driven
     learning in a boolean satisfiability solver. In *Proc. Intl. Conf. on Computer-Aided
     Design*, 2001.
[23] L. Zhang and S. Malik. The quest for efficient boolean satisfiability solvers. In
     Proc. of the Computer Aided Verification Conf., 2002.

On Computing k-CNF Formula Properties

Ryan Williams[*]

Computer Science Department
Carnegie Mellon University
Pittsburgh, PA 15213
ryanw@cs.cmu.edu

Abstract. The latest generation of SAT solvers (e.g. [10,7]) generally have three
key features: randomization of variable selection, backtracking search, and some
form of clause learning. We present a simple algorithm with these three features
and prove that for instances with constant Δ (where Δ is the clause-to-variable
ratio) the algorithm indeed has good worst-case performance, not only for computing SAT/UNSAT but more general properties as well, such as maximum satisfiability and counting the number of satisfying assignments. In general, the algorithm can determine any property that is computable via *self-reductions* on the
formula.

One corollary of our findings is that for all fixed Δ and $k \geq 2$, *Max-k-SAT* is
solvable in $O(c^n)$ expected time for some $c < 2$, partially resolving a long-
standing open problem in improved exponential time algorithms. For example,
when $\Delta = 4.2$ and $k = 3$, *Max-k-SAT* is solvable in $O(1.8932^n)$ worst-case
expected time. We also improve the known time bounds for exact solution of
$\#2SAT$ and $\#3SAT$, and the bounds for k-SAT when $k \geq 5$.

1 Introduction/Background

Exponential time algorithms for SAT with improved performance have been theoretically studied for over 20 years. Beginning in 1979, Monien and Speckenmeyer [8] gave
a $\tilde{O}(1.618^n)$ worst-case time algorithm for 3-SAT [8].[1] Reviewing the literature, it appears that studies in improved worst-case time bounds for SAT were mostly dormant
for many years, until a resurgence in the late 1990s (e.g. [11,4,1]). The first improvements used DPLL-style variants, where variables were repeatedly chosen in some way,
and the algorithm recursed on both possible values for the variables. The improved
time bounds came about due to clever case analysis about the number of variables or
the number of clauses removed from consideration in each of these recursive branches.
In 1999, Schöning [14] gave a $\tilde{O}(1.3333^n)$ algorithm for 3-SAT that is essentially the
WalkSAT algorithm [15]; this was followed by a $\tilde{O}(1.3303^n)$ improvement a couple of
years later [6].

The work on *Max-k-SAT* has been less successful than that for k-SAT: it has been
open whether or not *Max-k-SAT* can be solved in c^n steps for $c < 2$. In this work, we

---

[*] Supported in part by an NSF Graduate Research Fellowship and the NSF ALADDIN Center
(http://www.aladdin.cs.cmu.edu/).

[1] Note: All time bounds in this paper will be *worst-case*.

E. Giunchiglia and A. Tacchella (Eds.): SAT 2003, LNCS 2919, pp. 330–340, 2004.
© Springer-Verlag Berlin Heidelberg 2004

will resolve the question in the affirmative, when the clause density is constant. Further, there has been strong recent progress in counting satisfying assignments [2]: #2SAT and #3SAT are solvable in 1.3247^n and 1.6894^n time, respectively. Our approach supplants these bounds, having 1.2923^n and 1.4461^n expected time. Also, for large k, our algorithm outperforms any other in the SAT case. For example, for $k = 20$, Schöning's random walk algorithm runs in $\tilde{O}(1.9^n)$ whereas ours runs in $\tilde{O}(1.8054^n)$. This bound improvement occurs for all $k \geq 5$. It is important to stress that our randomized method is of the Las Vegas variety and thus *complete*, unlike the previous randomized algorithms for these problems [11,14,6] which are Monte Carlo (with one-sided error).

One disadvantage of some improved exponential time algorithms is their limited applicability: often, an improved algorithm for one variant of SAT yields little or no insight about other SAT variants. Here, our strategy can in general be applied to determine most interesting hard-to-compute properties of an arbitrary k-CNF formula that have been considered, under conditions that we will formally specify. We deliberately make our approach as abstract as possible, so that perhaps its ideas may be useful in other areas as well.

2   Notation

Let $T(n)$ be super-polynomial and $p(n)$ be a polynomial. We will express runtime bounds of the form $T(n) \cdot p(n)$ as $\tilde{O}(T(n))$, the tilde meaning that we are suppressing polynomial factors.

Boolean variables will be denoted as $x_i \in \{true, false\}$. Literals (negated or non-negated variables) will be denoted by $l_i \in \{x_i, \overline{x_i}\}$. F will denote a Boolean formula in conjunctive normal form over variables $x_1, \ldots, x_n$. We represent F as a family of subsets over $\{x_1, \overline{x_1}, \ldots, x_n, \overline{x_n}\}$. The sets of F are called clauses. We will implicitly assume that F has no trivial clauses containing both x_i and $\overline{x_i}$ for some i. The number of clauses in F is denoted by $m(F)$, the number of variables is $n(F)$, and the density of F is $\Delta(F) = m(F)/n(F)$. Typically we will just call these n, m, and Δ when the formula F under consideration is clear. Two special kinds of formulas are $\top$ and $\bot$. $\top$ is the empty formula $\varnothing$, or trivially true formula. $\bot := \{\varnothing\}$, the formula with a single, empty constraint, a trivially false formula.

The formula $F[x_i = v]$ is the formula that results when value $v \in \{true, false\}$ is substituted for variable x_i in F.

3   Self-reducible Properties

Let us formalize the sort of formula properties that are computable by the algorithm we will describe. Intuitively, they are those properties that may be described due to *self-reducibility*. For example, satisfiability of a formula F is a self-reducible property, since satisfiability of F may be deduced by testing satisfiability on the smaller formulas $F[x = true]$ and $F[x = false]$.

Definition 1. *Let f be a function from k-CNF formulas and natural numbers to a set V. f computes a* **feasibly self-reducible property** *iff:*

(1) $\forall i \in \mathbb{N}$, $f(\top, i)$ and $f(\bot, i)$ are polytime computable.
(2) There exists a polytime computable function g such that

$$f(F, n) = g(x, f(F[x = true], n - 1), f(F[x = false], n - 1)),$$

for all formulas F and variables x.

In English, this means we can easily compute f on F using g, provided we are given f's values when some variable is true, and when it is false.

To motivate this definition, we demonstrate that interesting (e.g. NP and $\#P$ complete) properties normally determined of SAT instances are feasibly self-reducible, provided we begin our computation of $f(F, n)$ with $n = n(F)$. The following table shows some of the properties that fall under our framework, given g and f's definition on the trivially true and trivially false formula. We also provide our algorithm's expected time bounds when $k = 2$ and $k = 3$ for these various properties. For the first two rows of the table, the v_i are truth values; for the second two rows, they are natural numbers.

f	$g(x, v_1, v_2)$	$f(\top, i)$	$f(\bot, i)$	$k = 2$	$k = 3$
SAT	$v_1 \vee v_2$	true	false	trivial	1.4461^n
UNSAT	$v_1 \wedge v_2$	false	true	trivial	1.4461^n
Max-SAT	$\max\{o(x_1, F) + v_1, o(\overline{x_1}, F) + v_2\}$	0	0	c^n ($c < 2$) if $\Delta = O(1)$	
#SAT	$v_1 + v_2$	2^i	0	1.2923^n	1.4461^n

(We define $o(l, F)$ to be the number of occurrences of literal l in F.) $\#SAT(F)$ is the number of satisfying assignments. *Max-SAT*(F) is the maximum number of clauses satisfied by any assignment. (A simple modification of the algorithm will be able to extract an assignment satisfying the maximum number, with no increase in asymptotic runtime.) We remark that *Max-SAT* is the only function above that uses the variable x in the specification for g.

4   Algorithm

We now present a way to compute any feasibly self-reducible f on k-CNF formulas with density Δ. The methodology is quite similar in nature to previous improved exponential time algorithms using dynamic programming [13,16]. The three major differences here are the use of randomness, the manner in which dynamic programming is employed, and the tighter analysis that results from analyzing k-CNF formulas.

Roughly speaking, the algorithm first chooses a random subset of δn variables, then does a standard depth-first branching on these variables, for some calculated $\delta \in (0, 1)$. After depth δn has been reached, the algorithm continues branching, but saves the computed f-values of all formulas considered after this point. The major point is that for suitable δ (depending on k and Δ), the space usage necessary is small, and the expected runtime is greatly reduced asymptotically.

4.1   Preliminary Initialization

Before the main portion of the algorithm is executed, a few preliminary steps are taken to set up the relevant data structures.

0. Let $\Delta = m/n$, and δ be the smallest root of the polynomial $\Delta\delta^k + \delta - \Delta$ over the interval $(0, 1)$. (Existence of such a root will be proven later.) Since k is constant, one can numerically compute this root to a suitable precision in polynomial time.

1. Let F be a k-CNF formula with n variables. Choose a random permutation σ : $[n] \rightarrow [n]$. (We will naturally think of σ as a permutation on the variables of F.) Define $F_{cover} \subseteq F$ as:

$$F_{cover} := \{c \in F \mid \forall l_i \in c. \ l_i \in \{x_{\sigma(1)}, \overline{x_{\sigma(1)}}, x_{\sigma(2)}, \overline{x_{\sigma(2)}}, \ldots, x_{\sigma(n)}, \overline{x_{\sigma(n)}}\}\}.$$

That is, F_{cover} is the subset of clauses c such that all k variables of c are contained in $\{x_{\sigma(1)}, \ldots, x_{\sigma(\delta n)}\}$. Define $F_{rem} := F - F_{cover}$. (F_{cover} is the set of clauses "covered" by the first δn variables of σ, and F_{rem} is the set that might possibly "remain", when the first δn variables of σ are set to values.)

2. Define $\leq_c$ to be a lexicographic (total) ordering on the clauses of F_{rem}, where the ordering is obtained from the variable indices. For instance, given $i_1 < j_1 < k_1$ and $i_2 < j_2 < k_2$, $\{x_{i_1}, x_{j_1}, x_{k_1}\} \leq_c \{x_{i_2}, x_{j_2}, x_{k_2}\}$ iff either $i_1 < i_2$ or ($i_1 = i_2$ and $j_1 < j_2$) or ($i_1 = i_2$ and $j_1 = j_2$ and $k_1 \leq k_2$). Define c_i to be the ith clause w.r.t. the ordering $\leq_c$.

3. Let f be some feasibly self-reducible function we wish to compute for F. Let V be the co-domain of f. (Typically, V is either $\{true, false\}$ or $\mathbb{N}$.) Initialize the set $Learned \subseteq \{0, 1\}^{m(F_{rem})} \times \{1, \ldots, n\} \times V$ of learned f-values as empty.

4.2 Search

The search portion of the algorithm recurses on a formula F_r and integer i, which are initially F and n, respectively.

Compute-f(F_r, i):

1. [If $i = 0$ then either $F_r = \bot$ or $F_r = \top$; take step 2.]

2. If $F_r = \bot$ or $F_r = \top$, return $f(\top, i)$ or $f(\bot, i)$, respectively.

3. (*Branching phase*) If $i \geq n - \delta n$, then return:
$g(x_{\sigma(n-i+1)},$ Compute-f($F_r[x_{\sigma(n-i+1)} = true], i - 1$),
$\qquad\qquad\quad$ Compute-f($F_r[x_{\sigma(n-i+1)} = false], i - 1$)).

4. (*Learned values phase*)

Else, let $F_r^k \subseteq F$ be the set of original k-clauses in F that correspond to the clauses that *remain* (possibly $< k$-)clauses in F_r. (We say a clause of F *remains* in F_r if it (a) has not been satisfied by the partial assignment that reduced F to F_r, and (b) has also not been *falsified*; that is, at least one of its literals has not been set false.) It follows that $F_r^k \subseteq F_{rem}$; see the analysis in the following subsection.

Represent F_r as a pair $(b(F_r^k), i)$, where $b(F_r^k)$ is a vector of $m(F_{rem})$ bits: for $j = 1, \ldots, m(F_{rem})$,

$$b(F_r^k)[j] := 1 \iff c_j \in F_{rem} \text{ (the jth clause in $\leq_c$) remains in F_r.}$$

(The analysis subsection will further explain why this uniquely represents F_r.[2])

---

[2] To illustrate with an example, suppose $F = \{\{a, z\}, \{\overline{a}, b\}, \{\overline{a}, y\}, \{x, y\}\}$ and $F_r = F[a = true, y = false] = \{\{b\}, \{x\}\}$. Then $F_r^k = \{\{\overline{a}, b\}, \{x, y\}\}$: $\{\overline{a}, y\}$ is not included because

5. If $(b(F_r), i, v) \in$ *Learned* for some v, then return v.

6. Else, let b_t and b_f be the bit vector representations of $F[x_{\sigma(n-i+1)} = true]$ and $F[x_{\sigma(n-i+1)} = false]$, respectively.

Set $v_t :=$ Compute-f$(f(F[x_{\sigma(n-i+1)} = true]), i - 1)$ and
$v_f :=$ Compute-f$(f(F[x_{\sigma(n-i+1)} = false]), i - 1)$.

Set $v := g(x_{\sigma(n-i+1)}, v_t, v_f)$.

Update *Learned* $:=$ *Learned* $\cup \{(b(F_r), i, v)\}$, and return v.

4.3   Details of Analysis

Sketch of correctness   Here, we assume the choice of δ is suitable and defer its justification until later. We consider each step in the above algorithm one by one.

- Steps 1 and 2, the base cases, are clear. Step 3 is obvious assuming Compute-f$(F_r[x_{\sigma(i)} = true], i - 1)$ and Compute-f$(F_r[x_{\sigma(i)} = false], i - 1)$ return correct answers.

- i always equals the number of variables that have not been set to values by the algorithm; the proof is a simple induction. Hence if $i < n - \delta n$, then the first δn variables have all been set in F_r, so letting $F_r^k \subseteq F$ be the set of original k-clauses in F that correspond to the clauses of F_r, $F_r^k \subseteq F_{rem}$ follows from the definition of F_{rem}: any clause $c \in F_r^k$ *cannot* be in F_{cover} (if $i < n - \delta n$, then the first δn variables have been set, hence by definition by F_{cover}, every literal in $c \in F_{cover}$ has been set, so $c \notin F_r^k$), hence $c \in F_{rem}$.

- In Steps 4 and 5, notice the representation $(b(F_r^k), i)$ tells us two things: (a) which clauses of F_{rem} have been either satisfied or falsified (and which have not) to yield F_r, and (b) which variables have been set to values in F_r (those variables that have been set are just those $x_{\sigma(j)}$ where $j < i$).

Thus, if literals of these variables appear in the (un-satisfied and un-falsified) clauses specified by $b(F_r^k)$, we may infer that these literals are *false*, as in the example of the footnote on the preceding page. Therefore we can reconstruct F_r given the pair $(b(F_r^k), i)$: $b(F_r^k)$ tells us F_r^k, which is the set of clauses in F that remain in F_r, and i tells us which literals in those clauses of F_r^k do not appear in F_r (*i.e.* are set to *false*). Hence the map $F_r \mapsto (b(F_r), i)$ is 1-1, and it is semantically correct to return v for $f(F_r)$ if $(b(F_r), i, v) \in$ *Learned* in Step 5.

For every f-value computed, it is stored in *Learned* and search for it before recomputing. The *Learned* set used in step 5 can be implemented using a binary search tree, where the keys are pairs containing (a) the $m(F_{rem})$ bit vector representations of the F_rs and (b) the variable index i. The relevant operations (insert and find) take only polynomial time.

---

it's falsified, and $\{a, z\}$ is not included because it's satisfied. If the ordering $\leq_c$ is given by $\{a, z\} \leq_c \{\overline{a}, b\} \leq_c \{\overline{a}, y\} \leq_c \{x, y\}$, then $b(F_r^k) = [0\ 1\ 0\ 1]$. Observe that, given $b(F_r^k)$ and the knowledge that a and y have been set to some values, this is enough to reconstruct F_r: the presence of y in $\{x, y\} \in F_r$ implies that y was set *false*, and the presence of a in $\{\overline{a}, b\} \in F_r$ implies that a was set *true*.

Runtime analysis We claim the algorithm devotes $\tilde{O}(2^{\delta n})$ time for the branching phase (when $i \leq \delta n$) and a separate count of $\tilde{O}(2^{E[m(F_{rem})]})$ expected time for the learned values phase, where $E[m(F_{rem})]$ is the expected number of clauses in F_{rem} over the choice of random σ. Hence in total, the expected runtime is $\tilde{O}(2^{E[m(F_{rem})]} + 2^{\delta n})$, and the optimal choice of δ to minimize the this will make $E[m(F_{rem})] = \delta n$.

To simplify the analysis, we consider an "unnatural" procedure, for which our algorithm has runtime no worse than it. The procedure will perform the phases of the algorithm described above, but in the opposite order. It will *first* (a) construct the *Learned* set of f-values recursively, saving each discovered value as it goes along. Then it will (b) run the branching phase until depth δn, in which case it simply refers to the corresponding stored value in *Learned*.

It is clear that if the runtime of (a) is bounded by T, then the runtime of this procedure is $\tilde{O}(2^{\delta n} + T)$, as looking up an f-value in *Learned* takes only polynomial time. Thus it suffices for us to prove that (a) takes $\tilde{O}(2^{E[m(F_{rem})]})$ expected worst-case time. Each $(b(F_r), i)$ pair's f-value in *Learned* is computed at most once, and is determined in polynomial time using g and assuming the f-values for smaller F_r are given. (We defer the cost of computing the f-values for smaller F_r to those smaller formulas). Moreover, the base cases $f(\top, i)$ and $f(\bot, i)$ are polytime computable by self-reducibility.

Thus the total time used by the learned formula phase will be at most

$poly(n) \cdot$ [number of possible $(b(F_r), i)$ pairs] $= \tilde{O}(2^{E[m(F_{rem})]})$,

since the total number of pairs possible in *Learned* is at most $n \cdot 2^{m(F_{rem})}$.

Let us specify the procedure for constructing *Learned* more formally. Start with $(\bot, i)$ and $(\top, i)$ for every $i = 1, \ldots, n$, and put $(\bot, i, f(\bot, i))$ and $(\top, i, f(\top, i))$ in *Learned*.

0. Initialize $j := n - 1$.

1. Repeat steps 2-3 until $j = \delta n$:

2. Set $\mathcal{F} := \{F_r \cup \{c \in F | x_{\sigma(j)} \in c \vee \overline{x_{\sigma(j)}} \in c\} \mid \exists v. (b(F_r), j + 1, v) \in$ *Learned*$\}$.

3. For all $F_r \in \mathcal{F}$,

Find v_t and v_f such that $(b(F_r[x_{\sigma(j)} = true]), i + 1, v_t)$ and $(b(F_r[x_{\sigma(i)} = false]), i + 1, v_t)$ in *Learned*, using a search tree.

Put $(b(F_r), i, g(x_{\sigma(i)}, v_t, v_f))$ in *Learned*, and set $i := i - 1$.

Notice we are always placing a value for a new pair in *Learned*. Hence we place at most $n2^{m(F_{rem})}$ values in *Learned*, in total. Each iteration of the for-loop for a fixed F_r takes polynomial time. The number of possible F_r in $\mathcal{F}$ is at most $2^{m(F_{rem})}$ (though it will be much less in most cases). There are at most $n - \delta n$ repetitions of the repeat loop, hence this procedure takes $\tilde{O}(2^{E[m(F_{rem})]})$ expected worst-case time.

Theorem 1. *For every k and Δ, there exists a constant $c < 2$ such that any feasibly self-reducible f on k-CNF Boolean formulas with density Δ is computable in $O(c^n)$ expected (worst-case) time.*

Proof. It suffices to show that the optimal choice of δ is always less than 1. Let c_i be a k-CNF clause. For a randomly chosen σ, the probability that a particular variable v is among the first δn variables is δ. Hence the probability that every variable in c_i is among the first δn variables designated by σ is at least $\delta^k[1 - o(1)]$. (So the probability that $c_i \in F_{cover}$ is this quantity.) More precisely, the probability is

$$\prod_{i=0}^{k-1} \frac{\delta n - i}{n - i} \geq \delta^k \prod_{i=0}^{k-1}\left(1 - \frac{i}{n}\right) \geq \delta^k\left(1 - \frac{d}{n}\right),$$

for some constant $d > 0$. Thus the probability that $c_i \in F_{rem} = F - F_{cover}$ is *at most* $1 - \delta^k[1 - o(1)]$. For each clause $c_i \in F$, define an indicator variable X_i that is 1 if $c_i \in F_{rem}$, and 0 otherwise. Then the expected number of clauses in F_{rem} is

$$E[m(F_{rem})] = \sum_{i=1}^{m} E[X_i] \leq m \cdot [1 - \delta^k(1 - d/n)],$$

by linearity of expectation. Hence the expected time for the learned value phase is (modulo polynomial factors)

$$2^{E[m(F_{rem})]} \leq 2^{[1-\delta^k](1-d/n)\Delta n} = 2^{[1-\delta^k]\Delta n - d \cdot \Delta \cdot [1-\delta^k]} \in \tilde{O}(2^{[1-\delta^k]\Delta n}),$$

and the optimal choice of δ satisfies the equation

$$\delta = (1 - \delta^k)\Delta \implies \Delta\delta^k + \delta - \Delta = 0.$$

Notice that the variance in $m(F_{rem})$ will be small in general (more precisely, susceptible to Chernoff bounds), thus our expectation is not a mathematical misnomer; we will not analyze it in detail here.

We now show that for $k > 0$ and $\Delta > 0$, the polynomial $p(x) = \Delta x^k + x - \Delta$ has at least one root $x_0 \in (0, 1)$; the theorem will follow. First, $p(x)$ has at least one real root r. Note $p(1) = 1$ for all k and Δ, while $p(0) = -\Delta$. Since $p(1) > 0$ and $p(0) < 0$, it follows that there is an $r \in (0, 1)$ satisfying $p(r) = 0$.   □

We have empirically observed that as either Δ or k increase, the relevant root of $p(x)$ approaches 1. Thus, if either k or Δ are unbounded functions in terms of n, we cannot (using the above analysis) guarantee any $\delta < 1$ such that the aforementioned procedure takes at most $2^{\delta n}$ time.

4.4 Max-k-SAT Solution

Ever since Monien and Speckenmeyer [9] showed in 1980 that there exists an algorithm for *Max-3-SAT* running in $\tilde{O}(2^{m/3})$, it has been a well-studied open problem as to whether *Max-k-SAT* could actually be solved in $O(c^n)$ time for $c < 2$. All previous exact algorithms for this problem have runtimes of the form $O(c^m)$, with c decreasing slowly over time (e.g. [9,1,4]). One prior algorithm was very close to a c^n time bound: a $(1 - \epsilon)$ approximation scheme was given by Hirsch [5] that runs in $(c_\epsilon)^n$ for some c_ϵ; however, c_ϵ approaches 2 as ϵ approaches 0.

A corollary of our above theorem is that when k and the clause density Δ are constant, there exists a less-than-2^n algorithm. While this is probably the more relevant situation for applications, it remains open whether *Max-k-SAT* can be solved when Δ is an unbounded function of n.

Corollary 1. *For every constant k and Δ, there exists a constant $c < 2$ such that Max-k-SAT on formulas of density Δ is solvable in $\tilde{O}(c^n)$ expected time.*

4.5 Improvements on Counting and SAT for High k

If the property we seek is some function on the satisfying assignments of F, then a better runtime bound can be achieved; we will outline our modified approach here. For instance, if we wish to count the number of satisfying assignments or determine satisfiability, then we can use the unit clause rule in branching. The unit clause rule has been used since [3] for reducing SAT instances.

Rule 1 *(Unit clause) If $\{l_j\} \in F$ then set $F := F[l_j = true]$.*

For feasibly self-reducible f on satisfying assignments, we incorporate the unit clause rule into the previous algorithm, between Steps 2 and 3. Now we observe that, in order to say that a clause $c \in F$ is not in F_{rem}, rather than requiring *all* k variables of c to be assigned values in the first δn variables, only $k - 1$ of the variables need to be assigned: if one of them makes c true, c is no longer present; if $k - 1$ literals are *false* in c then the unit clause rule applies.

This gives us a slightly better equation for δ, namely

$$\delta = (1 - \delta^k - k\delta^{k-1})\Delta,$$

since the probability that at least $k - 1$ variables of any clause c appear in the first δn variables of σ is at least $1 - \delta^k - k\delta^{k-1}$, the third term coming from the fact that there are k ways to choose $k - 1$ of the variables in c that appear. As might be expected, this equation yields better time bounds. There is no longer a strict dependence on Δ, and we obtain bounds such as the following:

Theorem 2. *#3SAT is solvable in $\tilde{O}(1.4461^n)$ expected time, for any constant Δ.*

Proof. (Sketch) Assume Δ is fixed. We wish to compute the largest possible $\delta \in (0, 1)$ s.t. $\delta = (1 - \delta^3 - 3\delta^2)\Delta$, *i.e.*

$$\Delta\delta^3 + 3\Delta\delta^2 + \delta - \Delta = 0.$$

This cubic equation has three solutions; the only one that is non-negative for $\Delta > 0$ is:

$$\delta(\Delta) = -\frac{F^{1/3}}{12\Delta} + \frac{3\Delta - 1}{F^{1/3}} - 1 + i\frac{\sqrt{3}}{2}\left(\frac{F^{1/3}}{6\Delta} + \frac{2 - 6\Delta}{F^{1/3}}\right),$$

where

$$F = \left(-108\Delta + 108 + 12\sqrt{\frac{12}{\Delta} - 27 + 162\Delta - 243\Delta^2}\right)\Delta^2.$$

When $\Delta = 1$, it is straightforward to calculate that $\delta(1) = \sqrt{2} - 1 \approx 0.414\ldots$.

For $\Delta > 1$, the solution for $\delta(\Delta) > \sqrt{2}-1$. F is dominated by $c_1 \cdot i \cdot \Delta^3 - c_2 \cdot \Delta^3$ for some constants c_1, c_2. Note that every term in the expression for δ involving F cancels out these Δ^3 terms (by taking a cube root and then dividing by Δ). Hence expression for δ approaches a certain constant as Δ increases.

Our numerical experiments show that $\delta(\Delta) \to 0.53208\ldots$ as $\Delta \to \infty$. Therefore $2^{0.53208n} \leq 1.4461^n$ is the bound. $\qquad\square$

For $k \geq 5$, even an improvement in SAT (over previous algorithms) is observed, using similar reasoning. The best known algorithm in that case has been that of Paturi, Pudlak, Saks, and Zane [12], which has the bounds 1.5681^n and 1.6370^n for $k = 5$ and 6. We have found through numerical experiments that our algorithm does strictly better for $k \geq 5$. An example:

Corollary 2. *5-SAT and #5-SAT are solvable in $\tilde{O}(1.5678^n)$ expected time for any constant Δ; 6-SAT and #6-SAT are solvable in $\tilde{O}(1.6065^n)$ for any constant Δ.*

Proof. (Sketch) Numerical solution of the equations $\delta^5 + 5\Delta\delta^4 + \delta - 1 = 0$ and $\delta^6 + 6\Delta\delta^5 + \delta - 1 = 0$ show that we can upper bound δ by $\delta \leq 0.6486\ldots$ and $\delta \leq 0.6839\ldots$, respectively. These δ-values yield the time bounds of the corollary. $\qquad\square$

A sharper improvement can be given for #2SAT, since for large Δ, single variable branches can remove many variables due to the unit clause rule. Specifically, in the worst case, one variable is assigned to a value in one branch, while at least 2Δ variables are assigned in another.

Theorem 3. *#2SAT is solvable in $\tilde{O}(1.2923^n)$ expected time.*

Proof. (Sketch) Let c be a constant to be fixed later. We consider the following algorithm for computing #2SAT.

Given a k-CNF F,

(0) If $\{x\} \in F$ (resp. $\{\overline{x}\} \in F$), recursively compute the number of SAT assignments A for $F[x = true]$ (resp. $F[x = false]$), and return A.

(1) If $\Delta \geq c$, then we claim the average number of occurrences per variable in the 2-CNF F is at least $2c$. (Let o be the average; then $o \cdot n = 2m$, where m is the number of clauses.) Therefore there is at least one variable x that occurs in $2c$ clauses. Recursively compute the number of SAT assignments $A_{x=t}$ for $F[x = true]$. Then compute the number of SAT assignments $A_{x=f}$ for $F[x = false]$. Return $A_{x=t} + A_{x=f}$.

(2) If $\Delta < c$, then return Compute-f(F, n) for $f = \#SAT$, i.e. the dynamic programming algorithm for counting satisfying assignments.

Now we analyze the algorithm. Case (0) is just the unit clause rule.

Case (1) analysis: It can be shown that the worst case is when x appears either all positively or all negatively. In this case, if $x = true$ then only one variable (x) is removed from F in one recursive branch. When $x = false$, all of the variables in clauses with x are set to true by case (0); hence at least $2c + 1$ variables are removed from F in the other branch.

The analysis of Case (2) is simply that from previous sections.

This gives an upper bound on the time recurrence in terms of n:

$$T(n) \leq \max\{T(n-1) + T(n-2c),\ 2^{\delta n}\},$$

where $\delta \in (0,1)$ is the smallest value such that $\delta < (1 - \delta^2 - 2\delta)c$. (The first term in the max corresponds to Case (1), the second term to Case (2).)

We want an integer c such that the runtime is minimized (so the two terms in the max are roughly equal). To do this, let $\lambda(A, B)$ be $r \in (1, 2)$ satisfying $1 - 1/r^A - 1/r^B = 0$. (Note that r is unique, and $O(\lambda(A, B)^n)$ is an upper bound on the recurrence $T(n) = T(n - A) + T(n - B)$, $T(1) = 1$.)

Observe that as c increases, $\lambda(1, 2c)$ decreases while $\delta \in (0, 1)$ such that $\delta = (1 - \delta^2 - 2\delta)c$ increase. Thus, we wish to choose c such that both $\lambda(1, 2c)^n$ and $2^{\delta n}$ s.t. $\delta = (1 - \delta^2 - 2\delta)c$ are minimized.

Choosing $c = 3$, $\lambda(1, 2c) \approx 1.2555$ and $\delta \approx 0.36993$, so $2^{\delta} \approx 1.2923$. Hence the runtime of the procedure is upper bounded by $\tilde{O}(1.2923^n)$.   □

5   Conclusion

We have shown, in a very general manner, how various hard properties of k-CNF properties may be determined in less than 2^n steps. It is interesting to formulate our main result in the language of *strong backdoors* [17]. Informally, for a function f computing a property of F, a strong backdoor is a subset S of variables in F, defined with respect to some "subsolver" A that runs in polynomial time and solves special cases of the function f. A strong backdoor S has the property that systematically setting all possible assignments to the variables of S allows one to compute f on F, by using A to solve each reduced F that results from a variable assignment. Since each assignment takes only polynomial time to evaluate with A, the overall runtime is bounded by $2^{|S|} poly(n)$.

Our main result may be stated in terms of randomly chosen backdoors.

Theorem 4. *For any feasibly self-reducible f and k-CNF formula F with constant clause density Δ, there exists a $\delta \in (0, 1)$ s.t. a random subset S of δn variables is a strong backdoor for f with probability at least $1/2$, with respect to a subsolver A that runs with $2^{\delta n}$ preprocessing (and polynomial time on each assignment to variables in S).*

While the dynamic programming scheme used in this paper is very applicable, one obvious caveat is that the procedure requires exponential space in order to achieve this. Therefore one open problem is to find algorithms that can compute self-reducible formula properties in *polynomial* space. Another question (which we believe to be not so difficult) is how to derandomize the algorithm– i.e. convert it a deterministic one, without much loss in efficiency. A further direction is to use some clever properties of *Max-k-SAT* when $\Delta = \omega(1)$ to get an less-than-2^n algorithm for general *Max-k-SAT*.

Finally, it is worth exploring what other useful properties of CNF formulas can be expressed via our definition of self-reducible functions, to determine the full scope of the method we have described. One hard problem that probably *cannot* be computed with it is solving quantified Boolean formulas; this is because in QBFs, it seems crucial to maintain the fixed variable ordering given by the quantifiers. On the other hand, if we assume the number of quantifier alternations is small, this may permit one to use a variable-reordering approach of the form we have described.

6   Acknowledgements

Many thanks to the anonymous referees for their very helpful comments.

References

1. N. Bansal and V. Raman. *Upper bounds for MaxSat: Further improved.* Proc. of the 10th ISAAC, 1999.
2. V. Dahllöf, P. Jonsson, Magnus Wahlstrm. *Counting Satisfying Assignments in 2-SAT and 3-SAT.* Proc. of COCOON, 535-543, 2002.
3. M. Davis and H. Putnam, *A computing procedure for quantification theory.* Journal of the ACM, 7(1):201-215, 1960.
4. E. A. Hirsch. *New worst-case upper bounds for SAT.* Journal of Automated Reasoning, Special Issue II on Satisfiability in Year 2000, 2000. A preliminary version appeared in Proceedings of SODA 98.
5. E. A. Hirsch. *Worst-case Time Bounds for MAX-k-SAT with respect to the Number of Variables Using Local Search.* ICALP Workshop on Approximation and Randomized Algorithms in Communication Networks, 69-76, 2000.
6. T. Hofmeister, U. Schoening, R. Schuler, O. Watanabe. *A probabilistic 3-SAT algorithm further improved.* Proc. of STACS, 192202, 2002.
7. I. Lynce and J. Marques-Silva. *Complete unrestricted backtracking algorithms for satisfiability.* In Fifth International Symposium on the Theory and Applications of Satisfiability Testing, 2002.
8. B. Monien, E. Speckenmeyer, *3-satisfiability is testable in $O(1.62^r)$ steps*, Bericht Nr. 3/1979, Reihe Theoretische Informatik, Universität-Gesamthochschule-Paderborn.
9. B. Monien and E. Speckenmeyer. *Upper bounds for covering problems.* Bericht Nr. 7/1980, Reihe Theoretische Informatik, Universität-Gesamthochschule-Paderborn.
10. M. Moskewicz, C. Madigan, Y. Zhao, L. Zhang, and S. Malik. *Chaff: Engineering an Efficient SAT Solver.* Proc. DAC-01, 2001.
11. R. Paturi, P. Pudlak, and F. Zane. *Satisfiability Coding Lemma.* Proc. of the 38th IEEE FOCS, 566-574, 1997.
12. R. Paturi, P. Pudlak, M. E. Saks, and F. Zane. *An improved exponential-time algorithm for k-SAT.* Proc. of the 39th IEEE FOCS, 628-637, 1998.
13. M. Robson. *Algorithms for maximum independent sets.* Journal of Algorithms, 7(3):425-440, 1986
14. U. Schoening. *A probabilistic algorithm for k-SAT and constraint satisfaction problems.* Proc. of the 40th IEEE FOCS, 410-414, 1999.
15. B. Selman, H. Kautz, and B. Cohen. *Local Search Strategies for Satisfiability Testing.* Cliques, Coloring, and Satisfiability: Second DIMACS Implementation Challenge, 1993.
16. R. Williams. *Algorithms for quantified Boolean formulas.* Proc. ACM-SIAM SODA, 299-307, 2002.
17. R. Williams, C. Gomes, and B. Selman. *Backdoors To Typical Case Complexity.* To appear in Proc. of IJCAI, 2003.

Effective Preprocessing with Hyper-Resolution and Equality Reduction

Fahiem Bacchus and Jonathan Winter

Department of Computer Science, University Of Toronto,[*]
Toronto, Ontario, Canada
[fbacchus|winter]@cs.toronto.edu

Abstract. HypBinRes, a particular form of hyper-resolution, was first employed in the SAT solver 2CLS+EQ. In 2CLS+EQ, HypBinRes and equality reduction are used at every node of a DPLL search tree, pruning much of the search tree. This allowed 2CLS+EQ to display the best all-around performance in the 2002 SAT solver competition. In particular, it was the only solver to qualify for the second round of the competition in all three benchmark categories. In this paper we investigate the use of HypBinRes and equality reduction in a preprocessor that can be used to simplify a CNF formula prior to SAT solving. We present empirical evidence demonstrating that such a preprocessor can be extremely effective on large structured problems, making some previously unsolvable problems solvable. The preprocessor is also able to solve a number of non-trivial instances by itself. Since the preprocessor does not have to worry about undoing changes on backtrack, nor about keeping track of reasons for intelligent backtracking, we are able to develop a new algorithm for applying HypBinRes that can be orders of magnitude more efficient than the algorithm employed inside of 2CLS+EQ. The net result is a technique that improves our ability to solve hard problems SAT problems.

1 Introduction

In this paper we investigate the use of a particular hyper-resolution rule, HypBinRes, along with equality reduction to preprocess CNF encoded SAT theories. HypBinRes is an inference rule that attempts to discover new binary clauses. These binary clauses are in turn used to detect that a literal is either forced or must be equivalent to other literals. In either case the input formula can then be reduced to one not containing that literal.

The HypBinRes rule was developed as part of the SAT solver 2CLS+EQ [1]. This solver was designed to further investigate the use of additional reasoning at every node of a DPLL search tree to prune the search tree [2, 3]. In the 2002 SAT competition 2CLS+EQ displayed the best all around performance, being the only solver to qualify for the second round of the competition in all three benchmark categories: industrial, handmade, and random problems. Furthermore, 2CLS+EQ was the top contributor to the SOTA (state of the art) solver: it solved 18 problems that were not solved by any other solver, (second was zchaff which was the sole solver of 15 problems) [4]. This

---

[*] This research was supported by the Canadian Government through their NSERC program.

E. Giunchiglia and A. Tacchella (Eds.): SAT 2003, LNCS 2919, pp. 341–355, 2004.

performance demonstrated that the right kind of additional reasoning can be very effective. Furthermore, in [2] empirical evidence was presented demonstrating that it is the specific use of HypBinRes and equality reduction that is key to 2CLS+EQ's performance.

The competition results demonstrated two other things about the use of HypBinRes. First, it can be quite expensive to utilize inside of the DPLL search, often resulting in a significant slow down in the per-node search rate of the solver. On some problems, the pruning produced is so dramatic that there is a significant net improvement in solution times. However, on many problems the overheads are such that state of the art DPLL SAT solvers, like zchaff [5], can solve the problem faster, even though they search many more nodes. Second, there are a number of problems on which HypBinRes and equality reduction is so effective that 2CLS+EQ can solve the problem without doing any search.

These two observations lead us to investigate the use of HypBinRes and equality reduction as a preprocessor for simplifying SAT problems prior to invoking a DPLL search. First, much of the expense in the implementation of HypBinRes comes from the fact that it was being used dynamically inside of a DPPL search. Using HypBinRes dynamically means that it must maintain sufficient information to allow all of the changes it makes to the theory to be undone on backtrack. Furthermore, because 2CLS+EQ utilizes intelligent backtracking, information also has to be maintained so that the reasons for failures can be computed. Since at each node HypBinRes and equality reduction can produce large changes to the theory, computing and maintaining all of this information becomes quite expensive. All of that extra work can be avoided in a preprocessor. Second, that HypBinRes with equality reduction was actually able to solve some hard problems prior to search, gave us reason to believe that it could usefully simplify other problems even if it is not able to solve them completely.

In this paper we report on the results of our investigation into the use of HypBinRes and equality reduction as a preprocessor. A short summary being that such a preprocessor is often extremely effective in improving net solution times, in contrast with the mixed results about preprocessing reported in [6]. In the sequel we will first describe HypBinRes and equality reduction in more detail. Then we will sketch a new algorithm suitable for implementing it in a preprocessor. Empirical results from an implementation of this algorithm are presented next, followed by some conclusions.

2   HypBinRes+eq

HypBinRes is a rule of inference involving a hyper-resolution step (i.e., a resolution step that involves more than two input clauses). It takes as input a single n-ary clause ($n \geq 2$) $(l_1, l_2, ..., l_n)$ and $n - 1$ binary clauses each of the form $(\bar{l}_i, \ell)$ ($i = 1, ..., n - 1$). It produces as output the new binary clause (ℓ, l_n). For example, using HypBinRes hyper-resolution on the inputs (a, b, c, d), $(h, \bar{a})$, $(h, \bar{c})$, and $(h, \bar{d})$, produces the new binary clause (h, b).

HypBinRes is equivalent to a sequence of ordinary resolution steps (i.e., resolution steps involving only two clauses). However, such a sequence would generate clauses of intermediate length while HypBinRes side-steps this, only generating the final binary clause. In a SAT solver it is generally counter productive to add all of these intermediate clauses to the theory.[1] However, can be very useful to add the final binary clause.

It should also be noted that if the input n-ary clause is itself binary, HypBinRes reduces to the simple resolution of binary clauses. For example, HypBinRes on the "n-ary" clause (a, b) and the clause $(h, \bar{a})$ yields the new binary clause (h, b).

HypBinRes could also be used to generate unit clauses if we allow it to consider one more binary clause. For example, (a, b, c, d), $(h, \bar{a})$, $(h, \bar{b})$, $(h, \bar{c})$, and $(h, \bar{d})$, when hyper resolved together produces the unit clause (h). Equivalently, one can do as we do in our implementation. We can apply HypBinRes as specified above and then a separate single step of ordinary resolution of binary clauses. In our example, the HypBinRes step uses only the first 3 binary clauses would produce (h, d), then an ordinary resolution step with clause $(h, \bar{d})$ produces (h).

Once binary clauses are available equality reduction can performed. If the theory F contains $(\bar{a}, b)$ as well as $(a, \bar{b})$ (i.e., $a \Rightarrow b$ as well as $b \Rightarrow a$), then we can generate a new formula EqReduce(F) by equality reduction. Equality reduction involves (a) replacing all instances of b in F by a, (b) removing all clauses which now contain both a and $\bar{a}$, (c) removing all duplicate instances of a (or $\bar{a}$) from all clauses. This process might generate new binary clauses.

For example, EqReduce($\{(a, \bar{b})$, $(\bar{a}, b)$, $(a, \bar{b}, c)$, $(b, \bar{d})$, $(a, b, d)\}$) $=$ $\{(a, \bar{d})$, $(a, d)\}$. Clearly EqReduce(F) will have a satisfying truth assignment if and only if F does. Furthermore, any truth assignment for EqReduce(F) can be extended to one for F by assigning b the same value as a.

Finally, we can apply the standard reduction of unit clauses. If we have a unit clause (ℓ) in the theory, we can remove all clauses containing ℓ, and then remove $\bar{\ell}$ from all remaining clauses. We use UR(ℓ) to denote such an application of this inference rule. The iterative application of UR until no more unit clauses remain is commonly known as unit propagation UP.

We can apply unit reduction (UR), HypBinRes, and equality reduction to a CNF theory until no more new inferences can be made with these rules. We call the resultant theory the *HypBinRes+eq closure*. A theory in which these three inference rules can infer nothing new is called HypBinRes+eq closed. Interestingly, a Church-Rosser result holds for this collection of inference rules.

Theorem 1. *The HypBinRes+eq closure of a CNF theory $\mathcal{F}$ is unique up to renaming. That is, the order in which the inference rules are applied is irrelevant, as long as we continue until we cannot apply them anymore.*

Proof. We show that these inference rules satisfy the Church-Rosser property. Namely, if a CNF theory T can be reduced to T_1 or T_2 by zero or more applications of the above three inference rules, then there exists another expression that both T_1 and T_2 can be reduced to (up to renaming[2]). From this we immediately obtain the theorem.

First, we show that for any sequence of two rule applications r_1 and r_2, there is some other sequence of rule applications $\pi(r_1, r_2)$ such that $r_2[r_1[T]]$ is equivalent to $\pi(r_1, r_2)[r_2[T]]$ up to renaming. This is shown by exhaustive case analysis. Each case

---

[1] These clauses are not like conflict clauses. Adding conflict clauses does appear to be useful.

[2] Renaming might be necessary because the EqReduce rules might be applied in different ways.

is easy, but there are many of cases. Hence, we only give a couple of examples. Say that r_1 is $\{(a, x), (a, y), (\bar{x}, \bar{y}, z)\} \vdash (a, z)$, and r_2 is $\{(a, z), (c, z), (\bar{c}, \bar{a}, \bar{s})\} \vdash (s, z)$. r_2 depends of r_1, and thus might not be applicable to T. In this case we view $r_2[T]$ as being a null operation, i.e., $r_2[T] = T$. Hence, $r_2[r_1[T]]$ is equivalent to $r_2[r_1[r_2[T]]]$. Another case is with the same r_1 but with r_2 being UR(x). Now $r_2[r_1[T]]]$ is equivalent to $r_1'[r_2[T]]$ where r_1' is $\{(a, y), (\bar{y}, z)\} \vdash (a, z)$. From this result it follows that for any theory $r[\pi[T]]$ where π is a sequence of rule applications, there exists an alternate sequence π' such that $r[\pi[T]]$ is equivalent to $\pi'[r[T]]$: we simply push r in one step at a time.

Second, we observe that for any two theories T_1 and T_2 equivalent up to renaming, and any inference rule r_1, Church-Rosser holds for $r_1[T_1]$ and T_2. We simply rename the literals in r_1 according to the renaming function between T_1 and T_2 and apply the renamed r_1 to T_2: $r_1'[T_2]$. The result is clearly equivalent to $r_1[T_1]$. That is, the empty sequence and r_1' transform $r_1[T_1]$ and T_2 to equivalent theories.

Finally, we consider two theories, T_1 and T_2 for which Church-Rosser holds. We show that for any rule application r Church-Rosser still holds for $r[T_1]$ and T_2. Church-Rosser means that there exists two sequences π_1 and π_2 such that $\pi_1[T_1]$ is equivalent to $\pi_2[T_2]$. Hence, $r[\pi_1[T_1]]$ is equivalent to $r'[\pi_2[T_2]]$ where r' is r appropriately renamed. By our first result, there are sequences π_1' and π_2' such that $\pi_1'[r[T_1]]$ is equivalent to $r[\pi_1[T_1]]$, and $\pi_2'[r'[T_2]]$ is equivalent to $r'[\pi_2[T_2]]$. Thus π_1 applied to $r[T_1]$ and $\pi_2'; r'$ applied to T_2 make these two theories equivalent. With the previous base case and this inductive step we have shown that Church-Rosser holds.

Now we can conclude that since no rules can be applied in the HypBinRes+eq closure it must be the case that we that any two sequences of rules reaching closure must yield the same theory (to renaming). Church-Rosser holds between the two theories obtained by these two sequences. Hence, if these theories were different, there would be a non-empty sequence of rules applicable to at least one of them (to move them both to an equivalent theory). That is, the theories could not both be HypBinRes+eq closed.
∎

The practical significance of Theorem 1 is that we are free to apply these inference rules in any order; we are guaranteed to reach the same final result. We now turn our attention to an algorithm for computing the HypBinRes+eq closure. Our new algorithm does not implement HypBinRes directly, rather it exploits the close relation between HypBinRes and unit propagation.

2.1   UP and HypBinRes+eq

Unit propagation is the iterative procedure of applying all unit reduction rules until no more unit clauses remain. Unit propagation can also be done on a trial basis. That is we can choose a literal to set to be true and then perform unit propagation. We call this *unit propagating a literal*, and denote it UP(a), where a is the literal that has been initially set to true. When UP(a) causes another literal ℓ to become true we use the notation UP$(a) \vdash \ell$. If UP$(a) \vdash \ell$ as well as UP$(a) \vdash \bar{\ell}$, we have detected that a is a *failed literal*, and it must be the case that the original theory $\mathcal{F} \vdash \bar{a}$. We can then reduce $\mathcal{F}$ by performing UP$(\bar{a})$.

In the sequel we will generally suppress mention of the underlying CNF theory, $\mathcal{F}$, upon which the various the reasoning processes are being run.

Theorem 2. *UP is more powerful than a single HypBinRes resolution step, but not as powerful as a sequence of HypBinRes resolution steps. More precisely:*

1. *If (a, b) can be produced by a single HypBinRes step, then either $UP(\bar{a}) \vdash b$ or $UP(\bar{b}) \vdash a$.*
2. *There are theories from which a binary clauses (a, b) can be produced from a sequence of HypBinRes steps, but neither $UP(\bar{a}) \vdash b$, nor $UP(\bar{b}) \vdash a$.*
3. *In a theory with no unit clauses (we can remove all units by an initial unit propagation phase), if $UP(\bar{a}) \vdash b$ then there is a sequence of HypBinRes steps that produce (a, b).*

Proof. (1) Any HypBinRes step is of the form $\{(l_1, l_2, ..., l_n), (\ell, \bar{l}_1), \ldots, (\ell, \bar{l}_{n-1})\}$, $\vdash (\ell, l_n)$, and $UP(\bar{\ell}) \vdash l_n$. Note that it need not be the case that $UP(\bar{l}_n) \vdash \ell$.

(2) An example is $\{(a, x), (a, y), (\bar{x}, \bar{y}, c), (\bar{c}, h), (\bar{c}, i), (\bar{i}, \bar{h}, q), (b, m), (b, o), (\bar{m}, \bar{o}, \bar{q})\}$. The binary clauses (a, c), $(\bar{c}, q)$, and $(\bar{q}, b)$ can be produced by 3 HypBinRes steps, after which two more resolution steps produce (a, b). However, $UP(\bar{a}) \vdash \{x, y, c, h, i, q\}$, but not b, and $UP(\bar{b}) \vdash \{m, o, \bar{q}\}$ but not a.

(3) We prove this result by induction. First we define an ordering on the literals entailed by $UP(\bar{a})$. Stage 0 of $UP(\bar{a})$ involves reducing the theory by removing a from all clauses, and removing all clauses containing $\bar{a}$. All literals appearing in unit clauses of the reduced theory are said to be entailed at step one. At stage 1 the one step literals are used to further reduce the clauses of the theory, and all literals appearing in new unit clauses are said to be entailed at step 2. In general, the literals entailed at step i are those appearing in unit clauses of the reduced theory produced at stage $i - 1$. We prove the theorem by induction on the stage at which b is produced.

If b is a step 1 literal then the clause (a, b) must have appeared in the initial theory: i.e., a zero length sequence of HypBinRes steps suffices.

Say b is entailed at step i, and that it was entailed by the clause $(l_1, \ldots, l_k, b)$ becoming unit. Hence, the negation of each of the l_i was entailed at earlier steps, and by induction for each there is a sequence of HypBinRes steps producing the binary clause $(a, \bar{l}_j)$ for each $j \in \{1, \ldots, k\}$. Hence, one more step of HypBinRes suffices to produce (a, b). ∎

2.2 Achieving HypBinRes+eq Closure with UP

Achieving HypBinRes+eq closure involves repeatedly applying HypBinRes, UR, and equality reduction until nothing new can be inferred. Theorem 2 shows that we can achieve HypBinRes closure by repeatedly applying UP on the literals of the theory.

More precisely, we first reduce the theory by unit propagating all unit clauses it might contain. Then for each remaining literal ℓ we can perform $UP(\ell)$, adding to the theory a new binary clause $(\bar{\ell}, a)$ for every literal a such that $UP(\ell) \vdash a$. By (1) above, one pass over all of the literals ensures that we find all binary clauses that can be inferred by one HypBinRes step. Adding the entailed binary clauses then ensures that the second pass can find all binary clauses inferable by two HypBinRes steps. By (2) we must add the entailed binary clauses found in the first pass, else UP would not be powerful enough. Adding these clauses makes all of the inputs to the second HypBinRes step available in the theory, and by (1) allows UP to capture the second HypBinRes step. These passes are continued until we find no new binary clauses; clearly at this stage we have achieved HypBinRes closure: there is no instance of the HypBinRes rule that can be applied.

Equality reduction and unit propagation can now added to compute the HypBinRes+eq closure. One obvious way to see this is to consider the iterative process were we wait until HypBinRes closure is achieved, then perform all equality reduction and unit propagations, iterating these two steps until we find nothing new. By Theorem 1 this particular sequence of operations will compute the HypBinRes+eq closure. In practice, however, the flexibility ensured by Theorem 1 is very important for efficiency. For example, it is always a good idea to perform UR immediately whenever we find a unit clause.

Part (3) of Theorem 2 tells us that we do not achieve anything greater than HypBinRes closure using multiple applications of UP: UP cannot infer anything more than HypBinRes. Hence the process just described computes precisely the HypBinRes+eq closure.

2.3   A Real Algorithm

The process described in the previous section would make a hopelessly inefficient algorithm. However, we can develop an efficient algorithm by using UP in a more refined manner that tries to avoid consuming too much space and wasted work. The basic idea is that we can often tell when $\mathrm{UP}(\ell)$, for some literal ℓ, will not yield anything new. A good example of this is when $\mathrm{UP}(a) \vdash \ell$ and $\mathrm{UP}(a)$ yields nothing new—$\mathrm{UP}(\ell)$ cannot either. Space is also an issue with the previous process. If we add a binary clause $(\bar{\ell}, a)$ for every a such that $\mathrm{UP}(\ell) \vdash a$, we could end up storing the transitive closure of the binary subtheory, which can be quadratic in the number of literals. This would make it impossible to deal with the large CNF theories that are now commonplace.

Our algorithm utilizes the implicit implication graph represented by a set of binary clauses [7]: the nodes are all of the literals in the theory and each binary clause (a, b) represents the two edges $\bar{a} \Rightarrow b$ and $\bar{b} \Rightarrow a$. In the following discussion we will interchangeably refer to a set of binary clauses as an implication graph and vice versa. Our implementation actually works with sets of binary clauses, performing operations on the implication graph (like traversing it) by corresponding operations on the binary clauses.

First we remove all unit clauses from the input CNF by doing an initial unit propagation. Then all of the input binary clauses are collected, and used to represent an implication graph. The aim of the algorithm is to generate an augmented implication graph (new set of binary clauses) that satisfies the following property: if (a, b) is present in the HypBinRes+eq closure, then in the implication graph b is reachable from $\bar{a}$ and a is reachable from $\bar{b}$. In particular, the clause (a, b) need not be in the final set of binary clauses, but it must be derivable by resolution steps involving only the computed set

Table 1 Graph Search Algorithm for computing HypBinRes closure

```
Visit(ℓ)
1.   if ℓ is MARKED return
2.   CurrentImplicants := {}
3.   foreach l s.t. (ℓ̄,l) is in the implication graph
4.       if l ∈ CurrentImplicants
5.           delete (ℓ̄,l) from the implication graph.
6.       else
7.           Visit(l)
8.           CurrentImplicants ∪= DescendantsOf(l)
9.   UPImplicants := {l s.t. UP(ℓ) ⊢ l}
10.  NewImplicants := UPImplicants - CurrentImplicants
11.  foreach l ∈ NewImplicants
12.      if l ∈ CurrentImplicants
13.          continue
14.      else
15.          add (ℓ̄,l) to implication graph.
16.          Visit(l)
17.          CurrentImplicants ∪= DescendantsOf(l)
18.  MARK ℓ
19.  return
```

of binary clauses. Thus we avoid materializing the transitive closure of the implication graph.[3]

The basic algorithm is presented Table 3. It is based on a depth first post-order traversal of the implication graph $\mathcal{G}$. The traversal is started at the set of literals (nodes) that have no parents in $\mathcal{G}$, i.e., they do not appear in any binary clauses, only their negations do. When the search completes its visit of a literal that literal is marked. The mark indicates that unit propagating that literal will not be able to discover anything new (at least for now).

First the algorithm visits all current children of the literal ℓ, recursively achieving a marked status for each child. As the children are visited the set of literals currently reachable from ℓ are accumulated (line 8). It can be that a new edge is added to the graph when recursively solving a child, and that new edge might reach some other child l of ℓ. This means that there is now another path to l from ℓ rather than the direct edge $(\bar{\ell}, l)$. To keep the graph small, the direct edge can be deleted (line 5). After all of the current children are marked, ℓ is unit propagated and all of the entailed literals accumulated (line 9). Note that ℓ entails all of its children *simultaneously*, so its set of unit implicants will in general be larger than the union of its children's unit implicants.

Any unit implicant not in the set of currently reachable literals then needs to be added to the graph (in order to converge on the HypBinRes closure). This is done at lines 11–17. Each of these new children can then be visited to ensure that they are

---

[3] The original 2CLS+EQ algorithm did explicitly represent the transitive closure of the implication graph as do the two other preprocessors that reason with the binary clauses, 2SIMPLIFY [8] and 2CL_SIMP [3]. All of these algorithms have difficulty dealing with larger formulas. The preprocessor we report on here follows "leaner" approach suggested in [9, 10].

properly marked. Again, the algorithm tries to minimize the number of new children added, by skipping those that are already reachable from some previously processed child (line 13).

Note that the algorithm makes no attempt to minimize the size of the implication graph (although this is an option if space is very tight). For example, it does not go back to check whether previous children might be reachable from children processed later—this would be too expensive.

Dealing with equivalence reduction and unit propagation of forced literals is conceptually straightforward. Equivalent literals can be detected with a depth-first search using Tarjan's strongly connected components (SCC) algorithm [11] (or more modern improvements). And forced literals (detected during the unit propagations of the algorithm) can be unit propagated. The complexity lies with finding ways to perform these operations incrementally and in more efficient orders. For example, we want to interrupt the graph search to unit propagate forced literals as soon as they are detected.

To make equivalent literal detection incremental we do an initial SCC and equality reduction prior to invoking **Visit**. Subsequently, we restrict the SCC computation so that it is only invoked when a new binary clause has been added. Furthermore we only search in the neighborhood of the new edges for newly created components (any new SCC must include one of the new edges). In this way we avoid examining the entire graph for every small change.

For forced literals, whenever a failed literal is detected, we interrupt the graph search and immediately unit propagate the negation of the failed literal. This marks some of the nodes in the graph as true or false, and can also generate new binary clauses that are immediately added to the graph. The graph search is then continued. To deal with the changes in the graph we make it backtrack immediately from any true or false node, otherwise it continues as before. Hence, it will only traverse the new edges in that part of the graph it has not yet searched. We utilize the literal marks to ensure that it eventually goes back to consider the new edges in the part it has already searched

All of the inference rules utilize the literal marks to inform the graph search of any incremental work it still needs to do. The mark represents the condition that unit propagating the literal will not yield any new edges in the graph. So whenever one of the rules is activated, it unmarks those literals that might now generate something new under unit propagation. There are only two cases:

(1) A new binary clause (a, b) is added to the theory. The new clause represents the new edges $\bar{a} \Rightarrow b$ and $\bar{b} \Rightarrow a$. This means that $\bar{a}$ and $\bar{b}$ along with anything upstream of them could now potentially generate new unit implicants. Hence, all these literals are unmarked so that the graph search can reconsider them.

(2) An n-ary clause $(l_1, \ldots, l_{k+1})$ is reduced in size to the clause $(l_1, \ldots, l_k)$. This could happen from a unit propagation forcing $\bar{l}_{k+1}$ or from $\bar{l}_{k+1}$ becoming equivalent to one of the other l_i. For any literal ℓ such that $UP(\ell) \vdash \bar{l}_i$ for any $i \in \{1, \ldots, k\}$ it could that $UP(\ell)$ now makes this clause unit. For example, it could be that $UP(\ell) \vdash \{\bar{l}_1, \ldots, \bar{l}_{k-1}\}$. Previously, this would have only reduced the clause to binary, but now that the clause has been reduced in size we get that $UP(\ell) \vdash l_k$. Hence, we must unmark ℓ. In general, we unmark all literals upstream of any of the $\bar{l}_i$. Note that (2) is simply a generalization of (1).

Table 2 The Hypre Preprocessing Algorithm

Hypre
```
1.  Unit Propagate all unit clauses.
2.  Find all SCC and perform all EqReduce steps
3.  UNMARK all nodes in implication graph
4.  while there is an UNMARKED node
5.    foreach ℓ s.t. ℓ is marked, and has no parents
6.        Visit(ℓ)
7.        Perform Incremental SCC
9.        UNMARK all nodes according to the two cases above.
10. end.
```

Visit(ℓ)
```
1.   if ℓ is MARKED return
2.   CurrentImplicants := {}
3.   foreach l s.t. (ℓ̄,l) is in the implication graph
4.     if l ∈ CurrentImplicants
5.         delete (ℓ̄,l) from the implication graph.
6.     else
7.         Visit(l)
7.5        if ℓ is MARKED return
8.         CurrentImplicants ∪= DescendantsOf(l)
9.   UPImplicants := {l s.t. UP(ℓ) ⊢ l}
9.1  if contradiction detected
9.2      UP(ℓ̄)
9.3      MARK all literals whose truth value is set
9.4      UNMARK all nodes according to the two cases above.
9.5      return
10.  NewImplicants := UPImplicants - CurrentImplicants
11.  foreach l ∈ NewImplicants
12.    if l ∈ CurrentImplicants
13.        continue
14.    else
14.5       Note that these new edges do not cause unmarking
15.        add (ℓ̄,l) to implication graph.
15.5       Note that these visits cannot detect a contradiction
15.6       since line 9 didn't.
16.        Visit(l)
17.        CurrentImplicants ∪= DescendantsOf(l)
18.  MARK ℓ
19.  return
```

With this unmarking process, we run the graph search until there is no unmarked literal (or a contradition is detected, or all clauses become satisfied). Unit propagations of forced literals are done immediately, and strongly connected component detection and equivalent literal reduction performed after the graph search is completed. Both the unit propagations as well as the equality reductions might remove node marks, in which case we may have to perform another iteration. Once no more changes can be made we have achieved HypBinRes+eq closure.

Table 3 Performance on two "hard" MITERS problems.

Problem	HyPre	Berkmin	2SIMPLIFY +Berkmin	2CL_SIMP +Berkmin	Zchaff	2SIMPLIFY +Zchaff
c6288-s	1.05	> 604,800	> 30,000	> 604,800	> 604,800	> 604,800.0
c6288	1.05	> 604,800	> 30,000	> 604,800	> 604,800	> 604,800.0

The final algorithm is presented in Table 3. The changes to **Visit** are indicated by fractional line numbers.

3   Empirical Results

Perhaps the most dramatic demonstration of the power of HypBinRes+eq closure comes from the two problems c6288-s, and c6288 from João Marques-Silva's MITERS test suite. Both of these problems are detected to be UNSAT by the preprocessor. Table 3 shows the time required by the preprocessor and by the SAT solvers BerkMin 5.61[4], and zchaff. All times reported are in CPU seconds on a 3GB, 2.4 GHx Pentium-IV. These two solvers were used in our tests as they are probably the most powerful current SAT solvers.

It can be seen that although the preprocessor solves both problems is about a second, the unsimplified problem is unsolvable by either of these state of the art SAT solvers. We ran each problems for a week before aborting the run. We also tried two other preprocessors (discussed in more detail in Section 4) 2SIMPLIFY [8] and 2CL_SIMP [3]. Both of these preprocessors do some form of binary clause reasoning, but as can be seen from the table, neither are effective on this problem.

Similar results are achieved on the BMC2 test suite from the 2002 competition, Table 4. Again all of these problems were solved by the preprocessor (all are SAT). These problems were also solved by the original 2CLS+EQ solver without search. However, as noted above the implementation of HypBinRes+eq closure in 2CLS+EQ is much less efficient. This is verified by the results of the table.

Table 4 Performance on the BMC2 suite

Problem	HyPre	2CLS+EQ	Berkmin	Zchaff	Problem	HyPre	2CLS+EQ	Berkmin	Zchaff
BMC2-b1	0.01	0.03	0.00	0.02	BMC2-b4	2.33	7.44	10.56	36.81
BMC2-b2	0.05	0.16	0.05	0.13	BMC2-b5	28.31	321.60	520.92	3492.23
BMC2-b3	0.31	0.84	0.62	1.89	BMC2-b6	214	11,193	3,426	>20,000

---

[4] The most recent public release.

Table 5 Performance on various families of problems. Time in CPU seconds. Bracketed numbers indicate number of failures for that family. Numbers in **bold** face indicate family/solver combinations where preprocessing reduced the net solution times (i.e. preprocessing plus solving), or allowed more problems to be solved.

Family (#probs)	HyPre	Berkmin-Orig	Berkmin-Pre	Zchaff-Orig	Zchaff-Pre
BMC (76)	18,819 (53)	41,751 (1)	29,254 (1)	48,225 (1)	30,009 (1)
BMCTA (2)	1,210 (2)	7,835 (0)	**6,015** (0)	34,741 (1)	**24,977** (1)
Cache (5)	23,620 (5)	61,771 (2)	42,327 (2)	30,725 (1)	43,932 (2)
w10 (4)	1,377 (4)	516 (0)	65 (0)	1,603 (0)	615 (0)
Checker (4)	8 (4)	4,092 (0)	**3,191** (0)	1,763 (0)	2,318 (0)
Comb (3)	352 (3)	22,017 (1)	**21,196** (1)	60,000 (3)	**36,845** (1)
IBM-Easy (2)	1,454 (2)	1,092 (0)	61 (0)	678 (0)	2 (0)
IBM-Med (2)	47,880 (2)	5,075 (0)	6 (0)	13,342 (0)	6 (0)
IBM-Hard (3)	1,767 (3)	38,997 (1)	**3,560** (0)	47,581 (2)	**28,920** (0)
Lisa (29)	16 (29)	107,379 (2)	135,612 (3)	185,091 (7)	111,482 (2)
f2clk (3)	674 (3)	22,663 (1)	**19,451** (0)	40,662 (2)	**28,687** (1)
fifo (4)	875 (3)	56,014 (2)	**1,015** (0)	29,040 (1)	**8,929** (0)
ip (4)	735 (4)	1,466 (0)	**145** (0)	30,667 (1)	**1,687** (0)
w08 (3)	8,672 (2)	8,542 (0)	151 (0)	27,670 (1)	**309** (0)
rule_03 (20)	4,409 (14)	134,258 (6)	**10,501** (0)		
rule_06 (20	15,509 (5)	113,360 (4)	**5,587.62** (0)		

Table 5 shows some additional results summed over families of problems. Most of these families came from the 2002 SAT competition. The families 03_rule and 06_rule come from the IBM Formal Verification Benchmark suite (we did not run Zchaff on these families). The number in brackets after the family name is the number of problems in the family. The time given is the total time to "solve" all problems in the family in CPU seconds. In these experiments a single problem time-out of 20,000 CPU seconds was imposed, and 20,000 was added to the total time to "solve" the family for each failed problem. The number in brackets after the total time is the number of problems the solver failed to solve. For the preprocessor "failure to solve" means that the problem was not resolved at preprocessing time. The first set of times is the time required by the preprocessor, then for each of the two SAT solvers we give the time require to solve the original problems, then the time require to solve the preprocessed problems (i.e., those that had not be resolved by the preprocessor).

The data shows that HypBinRes+eq closure almost always makes the problems easier to solve. Only for Cache with Zchaff, and Lisa with Berkmin does the time to process the family increase. The preprocessor also allows Berkmin to solve one more problem from the IBM-Hard family, one more from the f2clk, two more from the fifo family, 6 more from the 03_rule family, and 4 more from the 06_rule family. It improves Zchaff even more, allowing Zchaff to solve 2 more from Comb, 2 more from IBM-Hard, five more from Lisa, one more from f2clk, fifo, ip, and w08. Preprocessing on rare occasion makes a problem harder as in the case for one problem in the Lisa family for Berkmin, and one problem in the Cache family for Zchaff. Interestingly, for the Lisa family and Berkmin, the preprocessing allowed Berkmin to solve one problem it could not solve before, and stopped it from solving one that it could solve before.

Frequently, especially for Berkmin, once we add the time to perform the preprocessing the gains in total solution time are minimal, or even negative. The net gains for Zchaff are better. Nevertheless, the preprocessor almost always makes the problem easier, so only in the case of IBM-Med does it cause a serious slow down in the total solution time. We feel that this is acceptable given that it also allows some problems to be solved that previously could not be solved.

Table 6 provides some data about typical reductions in size in the CNF formulas produced by the preprocessor (these are of course formulas that are unresolved by preprocessing). The data shows that for the most part the preprocessor is able to achieve a significant reduction in the number of variables in the theory. Furthermore, it is clear that for problems like IBM-Med-b1, which contains more than 2 million binary clauses, it would be impossible to maintain the transitive closure of the binary clauses. (In fact, 2CLS+EQ, whose algorithm for achieving HypBinRes+eq closure involves realizing the transitive closure, is unable to complete the preprocessing of the larger problems.) The implicit representation of these clauses in an implication graph used in our implementation avoids this blowup. In fact, we see that generally the final theory has a reduced number of binary clauses, even though the outputed set of clauses contains within its transitive closure all implications of HypBinRes+eq.

On a number of families, e.g., Bart, Homer, GridMN, Matrix, Polynomial, sha, and fpga-routing-2 the preprocessor had no effect. In these problems there are no useful binary clauses in the original theory.

It can be noted however, that at least for the Bart family using HypBinRes+eq dynamically during the search had dramatic benefits even though it was useless as a preprocessor. In particular, from data gathered during the SAT-2002 competition 2CLS+EQ was able to solve all 21 instances of the Bart family in 1.1 seconds with an average of only 25 nodes expanded during the search. Berkmin on the other hand took 77.6 seconds to solve the family and zchaff required 39,708 seconds and still was only to solve 5 of the 21 problems.

4   Previous Work

Some of the processing done in our preprocessor is similar to techniques used in previous preprocessors, e.g., COMPACT [12], 2SIMPLIFY [8], 2CL_SIMP [3], and COMPRESSLITE[14].

COMPACT, and COMPRESSLITE both try to detect failed literals, with COMPRESSLITE doing a more elaborate test including finding literals forced by both ℓ and $\bar{\ell}$. Both of these preprocessors use unit propagation to do their work. However, the inferences performed are not as powerful. In particular, COMPACT employs only the failed literal rule: if $UP(a) \vdash$ FALSE then force $\bar{a}$. By Theorem 2, HypBinRes will also detect all failed literals. COMPRESSLITE on the other hand employs two inference rules

1. If $UP(\ell) \vdash a$ and $UP(\bar{\ell}) \vdash a$, then force a.
2. If $UP(\ell) \vdash a$ and $UP(\bar{\ell}) \vdash \bar{a}$ then perform EqReduce replacing a by ℓ.

Table 6 Size reduction on various problems instances. #Vars is the number of variables, #N-ary is the number of non binary clauses, and #Bin is the number of binary clauses

Problem	Original			After Processing		
	#Vars	#N-ary	#Bin	#Vars	#N-ary	#Bin
BMC-b10	42,405	42,327	98,804	10,229	14,978	35,263
BMC-b74	209,211	208,117	501,283	181,057	182,006	472,656
BMCTA-b1	64,909	87,909	108,066	40,705	73,866	158,966
BMCTA-b2	87,029	118,197	144,802	55,871	100,978	215,793
Cache-b1-s1-0	113,080	96,120	326,995	63,872	70,160	191,079
Cache-b2-s2-0	227,210	195,260	666,835	117,991	131,310	377,647
w10-b3	32,745	21,123	82,290	12,059	13,995	46,922
w10-b4	36,291	23,499	91,621	13,657	15,856	53,158
Checker-b3	1,155	33,740	5,528	1,029	30,158	4,903
Checker-b4	1,188	34,732	5,688	1,062	31,150	5,069
Comb-b2	31,933	21,364	91,097	15,707	17,517	84,952
Comb-b3	4,774	3,342	12,988	2,405	2,806	9,552
IBM-Easy-b2	29,605	35,046	115,506	15,276	29,297	59,389
IBM-Easy-b3	48,109	58,023	156,958	17,967	34,014	94,779
IBM-Med-b1	212,091	207,243	2,313,020	64,779	133,805	1,204,190
IBM-Med-b2	125,646	122,454	1,378,757	29,165	69,545	339,252
IBM-Hard-b2	22,984	27,508	90,298	11,788	21,803	49,480
IBM-Hard-b3	33,385	41,040	121,986	14,757	26,583	72,340
Lisa-b28	1,453	6,954	968	1,347	6,494	1,439
Lisa-b21	2,125	10,464	1,576	1,447	6,724	1,260
f2clk-b2	27,568	20,352	58,761	10,395	11,478	56,342
f2clk	34,678	25,662	74,001	14,895	16,388	84,901
fifo-b3	194,762	118,508	407,666	29,724	52,026	87,117
fifo-b4	259,762	158,108	543,766	41,574	72,826	122,374
ip-b3	49,967	36,407	124,732	14,042	15,592	48,462
ip-b4	66,131	48,275	165,208	19,226	21,352	65,791
w08-b2	120,367	81,080	344,201	34,861	40,515	211,796
w08-b3	132,555	89,489	379,993	40,856	47,091	243,426

HypBinRes+eq captures both rules. For (1) HypBinRes will conclude $(\bar{\ell}, a)$ and (ℓ, a) from which a can be inferred. For (2) HypBinRes will conclude $(\bar{\ell}, a)$ and $(\ell, \bar{a})$, from which EqReduce can make the equality reduction. HypBinRes+eq closure is in fact more powerful that these two rules. For example, COMPRESSLITE is only able to remove 3.3% of the literals from the two Miters problems c6288-s and c6288, whereas our HypBinRes+eq preprocessor can completely solve these problems. Judging from the times quoted in [14] COMPRESSLITE is also currently much less efficient than our implementation.

2SIMPLIFY and 2CL_SIMP are much closer to our work in that both try to do extensive reasoning with binary clauses. 2CL_SIMP does not employ a HypBinRes rule, but it does more extensive subsumption reasoning. 2SIMPLIFY on the other hand employs the implication graph and also implements a rule that on closer inspection turns out to be equivalent to HypBinRes. This rule is called the derive shared implications [8]. The rule is as follows: from a n-ary clause $(l_1, \ldots, l_n)$ the set of literals reachable from l_i (in the implication graph) is computed for all $i \in \{1, \ldots, n\}$. All literals in the intersection of these sets is detected to be forced. For a literal a to be in each of these sets, it means that $l_i \Rightarrow a$ for all i. That is, we have the n binary clauses $(\bar{l}_1, a), \ldots, (\bar{l}_n, a)$, and the n-ary clause $(l_1, \ldots, l_n)$. This is clearly a restricted case of HypBinRes. In fact, 2SIMPLIFY searches in a very limited way for literals in all but one of these sets, and from them learns new binary clauses (this version is precisely HypBinRes). 2SIMPLIFY also performs equality reduction.

However, 2SIMPLIFY does not compute the closure (in fact it does not even detect all single applications of HypBinRes). Nor was any theoretical analysis of the HypBinRes rule provided. Furthermore, like 2CL_SIMP it computes the transitive closure of the implication graph (binary subtheory). Thus on many examples these two programs fail to run. For example, on the BMC family of 76 problems, 2SIMPLIFY aborts on 68 of the problems, and 2CL_SIMP on 45 problems. Even when these programs run they seem to be ineffective on industrial benchmarks, or at least much less effective than the HypBinRes+eq preprocessor. Table 3 provided some evidence of this. Table 7 provides some more.

Table 7 2SIMPLIFY and 2CL_SIMP Performance on the BMC2 suite. Total solution time (preprocessor + solver) shown.

Problem	2CL_SIMP + Berkmin	2SIMPLIFY + Berkmin	Berkmin	Problem	2CL_SIMP + Berkmin	2SIMPLIFY + Berkmin	Berkmin
BMC2-b1	0.01	0.01	0.00	BMC2-b4	16.58	0.47	10.56
BMC2-b2	0.08	0.05	0.05	BMC2-b5	226.88	251.55	520.92
BMC2-b3	0.83	0.40	0.62	BMC2-b6	3,751.03	3,849.22	3,426.58

5 Conclusion

We have presented a new preprocessor for CNF encoded SAT problems. The empirical evidence demonstrates that although it can sometimes take a lot of time to do its work, it generally produces a significant simplification in the theory. This simplification is sometimes the difference between being able to solve the problem and not being to solve the problem. Hence, it is fair to say that it improves our ability to solve hard SAT problems.

6   Acknowledgements

We thank the referees for very useful comments and for pointing out that the binary clause reasoning employed in [3] had also been implemented as a preprocessor.

References

[1] Bacchus, F.: Enhancing davis putnam with extended binary clause reasoning. In: Proceedings of the AAAI National Conference. (2002) 613–619

[2] Bacchus, F.: Exploring the computational tradeoff of more reasoning and less searching. In: Fifth International Symposium on Theory and Applications of Satisfiability Testing (SAT-2002). (2002) 7–16 Available from www.cs.toronto.edu/~fbacchus/2clseq.html.

[3] Van Gelder, A., Tsuji, Y.K.: Satisfiability testing with more reasoning and less guessing. In Johnson, D., Trick, M., eds.: Cliques, Coloring and Satisfiability. Volume 26 of DIMACS Series in Discrete Mathematics and Theoretical Computer Science. American Mathematical Society (1996) 559–586

[4] Simon, L., Berre, D.L., Hirsch, E.A.:   The sat2002 competition.   Technical report, www.satlive.org (2002) available on line at www.satlive.org/SATCompetition/.

[5] Moskewicz, M., Madigan, C., Zhao, Y., Zhang, L., Malik, S.: Chaff: Engineering an efficient sat solver. In: Proc. of the Design Automation Conference (DAC). (2001)

[6] Lynce, I., Marques-Silva, J.P.:   The puzzling role of simplification in propositional satisfiability.   In: EPIA'01 Workshop on Constraint Satisfaction and Operational Research Techniques for Problem Solving (EPIA-CSOR). (2001) available on line at sat.inesc.pt/~jpms/research/publications.html.

[7] Aspvall, B., Plass, M., Tarjan, R.: A linear-time algorithms for testing the truth of certain quantified boolean formulas. Information Processing Letters **8** (1979) 121–123

[8] Brafman, R.I.: A simplifier for propositional formulas with many binary clauses. In: Proceedings of the International Joint Conference on Artifical Intelligence (IJCAI). (2001) 515–522

[9] Morrisette, T.: Incremental reasoning in less time and space. submitted manuscript (2002) available from the author e-mail threesat2000@yahoo.com.

[10] Van Gelder, A.:   Toward leaner binary-clause reasoning in a satisfiability solver.   In: Fifth International Symposium on the Theory and Applications of Satisfiability Testing (SAT 2002).   (2002) on line pre-prints available at gauss.ececs.uc.edu/Conferences/SAT2002/sat2002list.html.

[11] Tarjan, R.: Depth first search and linear graph algorithms. SIAM Journal on Computing **1** (1972) 146–160

[12] Crawford, J.M., Auton, L.D.: Experimental results on the crossover point in random 3-sat. Artificial Intelligence **81** (1996) 31–57

[13] Li, C.M., Anbulagan: Heuristics based on unit propagation for satisfiability problems. In: Proceedings of the International Joint Conference on Artifical Intelligence (IJCAI). (1997) 366–371

[14] Berre, D.L.: Exploiting the real power of unit propagation lookahead. In: LICS Workshop on Theory and Applications of Satisfiablity Testing. (2001)

Read-Once Unit Resolution

Hans Kleine Büning[1] and Xishun Zhao[2]

. Department of Computer Science,
Universität Paderborn
33095 Paderborn (Germany)
kbcsl@upb.de
. Institute of Logic and Cognition,
Zhongshan University
510275, Guangzhou, (P.R. China)
hsdp08@zsu.edu.cn

Abstract. Read-once resolution is the resolution calculus with the restriction that any clause can be used at most once in the derivation. We show that the problem of deciding whether a propositional *CNF*-formula has a read-once unit resolution refutation is *NP*-complete. In contrast, we prove that the problem of deciding whether a formula can be refuted by read-once unit resolution while every proper subformula has no read-once unit resolution refutation is solvable in quadratic time.

Keywords: Propositional formulas, tree resolution, unit resolution, read-once resolution

1 Introduction

For propositional formulas in conjunctive normal form (*CNF*) the problem of deciding whether a resolution proof exists is *coNP*-complete. Some restricted resolution calculi have been proposed and investigated in the literature. For example, for $k \geq 1$ a k-resolution step is a resolution step in which one of the two parent clauses consists of at most k literals. A k-resolution proof is a resolution proof in which each step is a k-resolution step. The problem of deciding whether a k-resolution proof exists can be solved in polynomial time for any fixed k. Another restriction is the input-resolution which demands that one of the parent clauses is always an input clause. Whether an input-resolution proof exists can be solved in linear time [3]. The input resolution is equivalent to the unit resolution with respect to refutations.

Please note that in a resolution proof a clause may be used several times, and as a result, a resolution proof may have super-polynomial steps in the size of the input formula. There do exist formulas [2,4,10] for which any resolution proof requires super-polynomial steps. Iwama and Miyano [5] restricted the occurrences of each clause in a resolution proof to guarantee that the size of the proof is polynomial. For example, they introduced the notion of Read-Once Resolution [5,6,8] which is the resolution calculus with the restriction that any initial clause and

E. Giunchiglia and A. Tacchella (Eds.): SAT 2003, LNCS 2919, pp. 356–369, 2004.
© Springer-Verlag Berlin Heidelberg 2004

any generated clause can be used at most once in a resolution derivation. That means, in a read-once resolution step we have to remove the parent clauses and to add the resolvent to the current set of clauses. The class of *CNF*-formulas for which a read-once resolution proof exists is denoted as *ROR*. Since every read-once proof has polynomial size, *ROR* is in *NP* and shown to be *NP*-complete [5,6]. In [6], Kleine Büning and Zhao studied the restriction of *ROR* to k-resolution. It has been shown in [6] that the problem of determining whether a *CNF* formula is refutable by read-once k-resolution remains *NP*-complete [6] for any fixed $k \geq 2$.

In this paper, we are interested in the question whether the read-once restriction has some influence on the power of unit resolution (i.e., 1-resolution). We denote the class of formulas for which a read-once unit-resolution refutation exists as *ROU*. Although whether a unit resolution refutation for a CNF formula exists can be decided in linear time, we will show that *ROU* is *NP*-complete. This result and results in [6] show that the restriction of read-once can sometimes make a polynomial-time solvable resolution problem intractable. However, input-resolution is an exception. We shall show that a *CNF* formula has an input resolution refutation if and only if it is refutable by read-once input-resolution. In other words, the read-once property has no effect on the power of the input resolution. This is quite interesting because input resolution and unit resolution have the same refutation ability and the read-once restriction influences unit resolution greatly.

Usually, for a *NP*-complete or *coNP*-complete class C of *CNF*-formulas, MC, the class of formulas in C such that their proper subformulas are not in C, is in D^P and intractable. Here D^P is the class of problems which can be described as the difference of two *NP*-problems. As an example, the unsatisfiability problem is *coNP*-complete while minimal unsatisfiability is D^P-complete [7]. However, *ROU* is quite different, although the class *ROU* is *NP*-complete, we present an algorithm deciding *MROU* in time $O(n^2)$. That means, the problem of whether a formula has a read-once unit resolution proof whereas any proper subformula has no such refutation is tractable.

The rest of this paper is organized as follows. In section 2, we recall some notions and terminologies that will be used later. In section 3, we first prove that the existence of read-once unit resolution proof is *NP*-complete. Then we prove that the read-once restriction has no effect on input resolution. Therefore, the existence of read-once input resolution proof can be solved in linear time. The first part of section 4 is devoted to investigate the properties of *MROU*. It is proved that every resolution tree of a *MROU* formula is coherent (see Definition 2 in section 4) and that if a formula has a coherent read-once unit resolution tree then it is in *MROU*. This leads to an algorithm for determining whether a *CNF* formula is in *MROU*. The computational complexity of this algorithm is analyzed in the second part of section 4.

2   Notation

Literals are propositional variables or their negations. A clause is a disjunction of literals. We use L for literals, $x, y, \cdots$ for variables, and $f, g, h, \cdots$ for clauses. Sometimes we write $(x_1, \cdots, x_k \rightarrow L)$ for a clause $(\neg x_1 \vee \cdots \vee \neg x_k \vee L)$. A unit clause consists of exactly one literal. A *CNF*-formula is a conjunction of clauses. In this paper we regard *CNF*-formulas as multi-sets of clauses. That means, multiple occurrences of a clause will be considered as different clauses. We use the capital letters $F, G, H, ..$ for formulas. $var(F)$ is the set of variables occurring in F. Given two formulas F and G, $F + G$ is the multi–set union (or, conjunction) of F and G. The number of positive (negative respectively) occurrences of a variable in a formula F is denoted as $pos(x, F)$ ($neg(x, F)$ respectively).

A *CNF*-formula is called minimal unsatisfiable if and only if F is unsatisfiable and every proper subset of clauses of F is satisfiable. $MU(1)$ is the set of minimal unsatisfiable formulas F with $n + 1$ clauses where n is the number of variables of F. *SAT* is the set of satisfiable *CNF*-formulas.

3   Read-Once Unit Resolution

A read-once unit resolution step is a unit resolution step, where both parent clauses cannot be used as parent clauses within other resolution steps. We use the abbreviation ROU-resolution for read–once unit resolution. A ROU-resolution refutation is a sequence of ROU-resolution steps leading to the empty clause.

Please keep in mind that *CNF*-formulas are given as multi-sets of clauses. For example the formula $F = x \wedge (\neg x \vee y) \wedge (\neg x \vee \neg y)$ has no ROU-resolution refutation, because after the first resolution step we have the formula $y \wedge (\neg x \vee \neg y)$, and after the second step the formula $\neg x$. However, the formula $F \wedge x$ has a ROU-resolution refutation.

ROU is the class of formulas for which a ROU-resolution refutation exists.

Theorem 1. *The class ROU is NP-complete.*

Proof. Obviously, the class ROU is in *NP*. The completeness will be shown by a reduction from the satisfiability problem *3-SAT* for propositional formulas in *3-CNF*. It is well-known that *3-SAT* remains *NP*-complete, even if we demand that any literal occurs at most twice in the formula. Let $F = f_1 \wedge ... \wedge f_m$ be a formula over the variables $x_1, \cdots, x_n$ with clauses $f_i = (L_{i_1} \vee L_{i_2} \vee L_{i_3})$ in which each literal occurs at most twice.

We introduce new pairwise different variables $a_j, b_j, d_j, e_j, w_j^1, w_j^2, x_j^+, x_j^-$ for $1 \leq j \leq n$, and c_i for $(1 \leq i \leq m)$.

We associate to each variable x_j the formula

$$V_j := w_j^1 \wedge w_j^2 \wedge a_j \wedge b_j \wedge d_j \wedge e_j \wedge$$
$$(w_j^1, a_j, d_j \rightarrow x_j^+) \wedge (w_j^2, b_j, e_j \rightarrow x_j^+) \wedge$$
$$(w_j^1, d_j, b_j \rightarrow x_j^-) \wedge (w_j^2, e_j, a_j \rightarrow x_j^-)$$

x_j^+ respectively x_j^- can be derived from V_j twice by the ROU-resolution using distinct clause sets. For example for x_j^+ take the derivation sequences

$$w_j^1, (w_j^1, a_j, d_j \to x_j^+) \vdash_{\overline{RO\text{-}Unit\text{-}Res}} (a_j, d_j \to x_j^+), a_j \vdash_{\overline{RO\text{-}Unit\text{-}Res}} (d_j \to x_j^+), d_j$$
$$\vdash_{\overline{RO\text{-}Unit\text{-}Res}} x_j^+, \text{ and,}$$

$$w_j^2, (w_j^2, b_j, e_j \to x_j^+) \vdash_{\overline{RO\text{-}Unit\text{-}Res}} (b_j, e_j \to x_j^+), b_j \vdash_{\overline{RO\text{-}Unit\text{-}Res}} (e_j \to x_j^+), e_j$$
$$\vdash_{\overline{RO\text{-}Unit\text{-}Res}} x_j^+.$$

But there is no derivation $V_j \vdash_{\overline{RO\text{-}Unit\text{-}Res}} x_j^+ \wedge x_j^-$.

Suppose the contrary. Since w_j^1 and w_j^2 occur exactly once as unit-clauses, w.l.o.g. we assume that the clause $(w_j^1, a_j, d_j \to x_j^+)$ respectively $(w_j^2, e_j, a_j \to x_j^-)$ is used for the derivation of x_j^+ respectively x_j^-. That means, the unit clause a_j must be applied twice, but a_j occurs only once positively in the formula V_j. Hence, it is not possible to derive x_j^+ and x_j^- simultaneously by means of ROU-resolution.

For any literal L over a variable x we define

$$L^* = \begin{cases} x^+, & \text{if } L = x \\ x^-, & \text{otherwise} \end{cases}$$

We associate to each clause $f_i = (L_{i_1} \vee L_{i_2} \vee L_{i_3})$ the formula

$$C_i := (L_{i_1}^* \to c_i) \wedge (L_{i_2}^* \to c_i) \wedge (L_{i_3}^* \to c_i)$$

and we define

$$H = \bigwedge_{1 \le i \le n} V_j \wedge \bigwedge_{1 \le i \le m} C_i \wedge (\neg c_1 \vee \cdots \vee \neg c_m)$$

Next we prove $H \in ROU$ if and only if $F \in SAT$.

For the direction from right to left we suppose F is satisfiable. W.l.o.g. let $x_1 = \cdots = x_s = 1$ and $x_{s+1} = \cdots = x_n = 0$ be a satisfying truth assignment for F. Please note that any literal occurs at most twice in F. For $x_j (1 \le j \le s)$ respectively $x_j (s + 1 \le j \le n)$ we derive using V_j two occurrences of x_j^+ respectively x_j^-. Since any clause f_i contains x_j for some $(1 \le j \le s)$ or $\neg x_j$ for some $(s + 1 \le j \le n)$, the formula C_i contains a clause $(x_j^+ \to c_i)$ for some $j \le s$ or $(x_j^- \to c_i)$ for some $s + 1 \le j$. Now we apply a ROU-resolution step to one of the previously derived unit clauses x_j^+ respectively x_j^- and this clause. We obtain the unit clause c_i. Finally, we iteratively apply ROU-resolution to $c_1 \wedge \cdots \wedge c_m$ and $(\neg c_1 \vee \ldots \vee \neg c_m)$ obtaining the empty clause. Hence, there is a ROU-resolution refutation for H.

For the reverse direction we assume that there is a ROU-resolution refutation for H. Since $(\neg c_1 \vee \ldots \vee \neg c_m)$ is the only negative clause, and H is a Horn formula, every unit clause c_i must be derived. That means for any $i(1 \le i \le m)$ a clause $(L_{i_{j_i}}^* \to c_i)$ must be used for some j_i. Therefore, we have a derivation for $L_{i_{j_i}}^*$

for each i $(1 \leq i \leq m)$. W.l.o.g. we assume that $\{L^*_{i_{j_i}} \mid 1 \leq i \leq m\} \subseteq \{x^+_r \mid 1 \leq r \leq s\} \cup \{x^-_r \mid s+1 \leq r \leq n\}$ for some s. Please note, that for a variable x only one of x^+ and x^- can be derived in H. Obviously the truth assignment $x_r = 1(1 \leq r \leq s)$ and $x_r = 0(s+1 \leq r \leq n)$ satisfies the formula F. Altogether we have shown $H \in ROU$ if and only if $F \in SAT$.

It is well known that a formula has a unit resolution refutation if and only if it has a input-resolution refutation. However, if we add the restriction of read-once the situation changes since the next theorem shows that the read-once property has no influence to input-resolution.

Theorem 2. *For any formula F, F is refutable by input-resolution if and only if F is refutable by read-once input-resolution.*

Proof. We prove the non-trivial direction from left to right by an induction on n, the number of variables. For $n = 1$ the formula has the form $x \wedge \neg x$ for which obviously a read-once input-resolution refutation exists. Now we suppose that F is a formula over $n+1$ variables for which an input-resolution refutation exists. In the final step of the refutation one of the parent clauses must be a unit clause L, which belongs to the clauses of F. We define $F[L = 1]$ to be the formula obtained by deleting any clause containing L and removing any occurrence of $\neg L$. That means, we assign $L = 1$ and simplify the formula. $F[L = 1]$ must be unsatisfiable and have an input-resolution refutation. By the induction hypothesis we obtain a read-once resolution refutation for $F[L = 1]$. Now we add the removed occurrences of $\neg L$ to the clauses. The formula is denoted as $F[L = 1](\neg L)$ which is obviously a subformula of F. Then there is a read-once input-resolution derivation for $\neg L$. Since L is a unit clause in F and not used previously we obtain a read-once input-resolution refutation for F.

4   Minimal Formulas with ROU-refutation

The class $MROU$ is the set of ROU-formulas for which any proper subformula is not in ROU. Formally, we define

$$MROU := \{F \in ROU \mid \forall f \in F, F - \{f\} \notin ROU\}$$

An example of a formula in ROU but not in $MROU$ is the formula $F = x \wedge (\neg x \vee y) \wedge (x \vee \neg y) \wedge \neg x$. The example is of interest, because there is a ROU-resolution refutation using all clauses of F and leading to the empty clause by resolving the clauses in the given order. The shortest derivation uses only the first and the last clause.

Examples of $MROU$-formulas are minimal unsatisfiable Horn formulas in which no literal can be removed without violating the unsatisfiability.

Our algorithm for deciding $MROU$ is based on the following properties for $MROU$-formulas:

1. For any variable the number of positive occurrences equals the number of negative occurrences.

2. Any order of applications of ROU-resolution steps leads to the empty clause.
3. Replacing multiple occurrence of a variable by introducing new variables which occur exactly once results in a minimal unsatisfiable formula.
4. In every ROU-resolution tree there is no unit clause L occuring in a branch more than once.
5. In every ROU-resolution tree, for any variable x, x and $\neg x$ can not occur as unit clauses in different branches if one of them has height greater than 1.

4.1 A Characterization of *MROU*

In this subsection we shall use coherent resolution trees to characterize formulas in *MROU*. At first we collect some simple properties of *MROU*. For a formula F and literals $L_1, \cdots, L_k$ we use

$$F \vdash_{RO\text{-}Unit\text{-}Res} L_1 \wedge \cdots \wedge L_k$$

to denote the following statement: Each L_i is derivable from F by the ROU-resolution and the derivations of L_i and L_j ($i \neq j$) use disjoint sets of clauses from F.

For a clause $f = L_1 \vee \cdots \vee L_k$, $\overline{f}$ is the conjunction $\neg L_1 \wedge \cdots \wedge \neg L_k$.

Proposition 1. *Suppose F has a ROU-resolution refutation in which every clause of F appears. Then for any $f \in F$, $F - \{f\} \vdash_{RO\text{-}Unit\text{-}Res} \overline{f}$.*

Proof. We prove the proposition by induction on the number of clauses. Suppose F contains only two clauses. Since each clause occurs in the resolution proof for F, F is the formula $\{x, \neg x\}$. Clearly the assertion holds. Suppose F has more clauses and the first step in the resolution proof is

$$\frac{L \quad \neg L \vee g}{g}.$$

Then the formula $F' = F - \{L, \neg L \vee g\} + \{g\}$ meets the assumption of the proposition. By the induction hypothesis, we have $F' - \{g\} \vdash_{RO\text{-}Unit\text{-}Res} \overline{g}$. Hence,

$$F - \{\neg L \vee g\} \vdash_{RO\text{-}Unit\text{-}Res} L \wedge \overline{g},$$
$$F - \{L\} \vdash_{RO\text{-}Unit\text{-}Res} \neg L$$

Let f be an arbitrary clause in $F - \{L, \neg L \vee g\}$. By the induction hypothesis we get $F' - \{f\} \vdash_{RO\text{-}Unit\text{-}Res} \overline{f}$. Since we can replace the clause g by the resolution step $\frac{L \quad \neg L \vee g}{g}$ we obtain $F - \{f\} \vdash_{RO\text{-}Unit\text{-}Res} \overline{f}$.

Lemma 1. *If $F \in MROU$, then*

1. *$pos(x, F) = neg(x, F)$ for any variable $x \in var(F)$.*
2. *For every unit clause $L \in F$ and for every clause $\neg L \vee g \in F$,*
   $F' := (F - \{L, \neg L \vee g\}) + \{g\}$ is in MROU.

Proof. For Part (1) we only need the precondition that F has a ROU-resolution refutation in which every clause of F appears. Please note that each ROU-resolution step removes a unit clause and an occurrence of the complementary literal. Furthermore, after the first ROU-resolution step the remaining steps still form a ROU-resolution refutation of the resulting formula. Then we can prove (1) by induction.

For part (2) we consider the unit clause L and an arbitrary clause $\neg L \vee g$ in F. Since $F \in MROU$ there is some ROU-resolution tree, say T, for F. If T contains the resolution step

$$\frac{L \quad \neg L \vee g}{g}$$

then removing L and $\neg L \vee g$ we obtain a ROU-resolution tree for F'. Then F' is in $MROU$, because $F \in MROU$ and any ROU-resolution tree for F' can be extended to a ROU-resolution tree for F by replacing g by $\frac{L \quad \neg L \vee g}{g}$.

Now we suppose that the resolution step $\frac{L \quad \neg L \vee g}{g}$ does not occur in the tree T. Since $F \in MROU$, any clause in F must occur in the tree. By Proposition 1, we have $F - \{\neg L \vee g\} \vdash_{\overline{RO\text{-}Unit\text{-}Res}} L \wedge \overline{g}$. Suppose $g = (L_1 \vee \cdots \vee L_m)$. Then for $L, \neg L_1, \cdots, \neg L_m$ there are ROU-derivations with disjoint clause sets. If in the derivation of L, L is not an initial unit clause, then the initial unit clause L has been used in one of the trees for $\neg L_i$. We can exchange the derivation and the initial unit clause. The new derivation is again a ROU-derivation for $L, \neg L_1, \cdots, \neg L_m$. Now we can resolve the initial unit clause L with $\neg L \vee g$ and finally resolve g with $\neg L_1, \cdots \neg L_m$ generating the empty clause. Altogether we have a ROU-resolution refutation where L and $\neg L \vee g$ have been resolved. That is the case we discussed above. Hence, we obtain $F' \in MROU$.

For our algorithm we need some properties of ROU-resolution trees. At first we define an ordering of nodes which are labelled by unit clauses.

Definition 1. *Let T be a ROU-resolution tree for a formula F. For literals L_1 and L_2, we write $L_1 <_T L_2$ if and only if T has a branch in which nodes labelled by the unit clause L_1 and L_2 occur and the height of the node L_1 is less than the height of the node L_2.*

$<_T^*$ *denotes the transitive closure of $<_T$.*

Please note that in case $L_1 <_T L_2 <_T L_3$ there may be two different branches, one containing L_1 and L_2, and the other containing L_2 and L_3.

Definition 2. *Let T be a ROU-resolution tree for a formula F. We say T is coherent if*

1. *every clause of F occurs in the tree T.*
2. *There is no literal L such that $L <_T^* L$.*
3. *There is no variable x such that x and $\neg x$ occur as unit clauses in different branches and one of them has height (i.e., the distance from the root) greater than 1.*

4. *For all literals L and K, if $L <_{T_L} K <^*_{T_L} \neg L$, then there is no occurrence of K outside T_L, where T_L is the subtree whose root is the node labelled by L.*

Lemma 2. *If $F \in MROU$ then every ROU-resolution tree T for F is coherent.*

Proof. Suppose $F \in MROU$. Let T be an arbitrary ROU-resolution tree for F. We prove that T satisfies the four items given in definition 2.

Ad 1. Since $F \in MROU$ every clause must be used in a ROU-derivation. That means, every clause occurs in the tree.

Ad 2. Suppose there is some L with $L <^*_T L$, say $L <_T L_1 <_T \cdots <_T L_m <_T L$. If the nodes with labels $L, L_1, \cdots, L_m, L$ lie in the same branch, then we can reduce the tree. Let T^1_L be the subtree with the lower node labelled by L as its root and T^2_L the subtree with the higher node labelled by L as root. Clearly, T^2_L is a proper subtree of T^1_L. We replace the tree T^1_L by T^2_L. The resulting tree is a ROU-resolution tree for F with fewer clauses, this contradicts the assumption of $F \in MROU$.

Suppose some of the nodes lie in different branches, say, L_{i-1} and L_{i+1}. Then we have a path from L_{i-1} to L_i in one branch and in a different branch a path from L_i to L_{i+1}. Let $T^1_{L_i}$ and $T^2_{L_i}$ be the corresponding subtrees with the respective two nodes labelled by L_i as their roots. Now we exchange the subtree $T^1_{L_i}$ by $T^2_{L_i}$ and vice versa. The resulting tree T' is a ROU-resolution tree for F in which L_{i-1}, L_i and L_{i+1} lie in one branch. We can perform the transformation until all the nodes lie in one branch. Then we obtain the previous case and therefore a contradiction to $F \in MROU$.

Ad 3. Suppose there is some x for which a node with label x and a node with label $\neg x$ in different branches exist. Let T_x resp. $T_{\neg x}$ be the associated subtrees of x and $\neg x$ in T. If one of the nodes, say x, has height greater than 1, then we can construct a new ROU-resolution tree with fewer clauses. Take the two trees T_x and $T_{\neg x}$ and resolve x and $\neg x$. That contradicts the fact that $F \in MROU$.

Ad 4. Suppose there are some literals L and K for which $L <_{T_L} K <^*_{T_L} \neg L$ and K has an occurrence outside T_L. Without loss of generality we can assume that $K <_{T_L} \neg L$ (otherwise, we can perform the transformation described in Ad 2 within T_L to get a new tree in which K and $\neg L$ occur in one branch). Let T_K be the subtree with root label K, where the node lies outside T_L. Further, let $T_{K,L}$ be the subtree in T_L with root K. Now we exchange $T_{K,L}$ and T_K. We obtain a ROU-resolution tree, in which L and $\neg L$ occur as node labels in different branches and the node with label $\neg L$ has height greater than 1. That is a contradiction to $F \in MROU$ (see also case 3).

Next we will prove a technical lemma on transformations of coherent trees.

Lemma 3. *Let T be a coherent ROU-resolution tree for a formula F and $L, \neg L \vee g$ clauses of F. Then T can be transformed into a coherent ROU-resolution tree T' for F in which the resolution step $\frac{L, \ \neg L \vee g}{g}$ occurs.*

Proof. Suppose T is a coherent ROU-resolution tree for F. At first we prove that for an arbitrary fixed clause $f \in F$, T can be transformed into a coherent resolution tree T^* with the following properties. (1) For each literal L of f, T^* contains a derivation $T^*_{\neg L}$ of the unit clause $\neg L$, and (2) the empty clause is obtained by iteratively applying unit resolution steps which resolves $\neg L$ with f.

Suppose the last resolution step in T is to resolve y and $\neg y$ to produce the empty clause. Without loss of generality suppose y occurs in the branch b from f to the empty clause. Suppose y comes from f. Then any clause in b above y contains the literal y and therefore its resolution partner must be a unit clause. If one clause in the path does not contain y, then the occurrence of y in f (if it contains y) has been resolved somewhere. This would imply that occurrences of y in lower levels come from other clauses containing y, contradicting the assumption that y comes from f. Since T is a ROU-resolution tree, T itself is as required.

Suppose y does not come from f, then it comes from another clause, say $y \vee h \in F$. Again, y occurs in every clause in the path from $y \vee h$ to y. Now remove all occurrences of y in the path from $y \vee h$ to y, and add the resolution step which resolves $y \vee h$ and $\neg y$ from $T_{\neg y}$ which is the derivation of $\neg y$ in T. The resulting tree is denoted as T'. This transformation is shown in Fig 1. Next we show T' is coherent. Obviously, every clause in F occurs in the new tree.

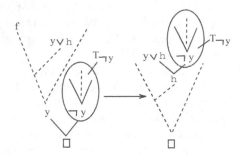

Fig. 1.

Please notice that during the transformation only one unit clause was generated and it lies in the first level of the new tree, that is, it is one parent clause of the empty clause.

First we show condition 2 in definition 2 holds for T'. Suppose on the contrary $L <^*_{T'} L$ for some literal.

Suppose L is the generated unit clause. Because each unit clause K such that $\neg L <_{T'_{\neg L}} K <^*_{T'_{\neg L}} L$ occurs only in $T'_{\neg L}$ (which equals to $T_{\neg L}$) and since it cannot be that $L <^*_{T'_{\neg L}} L$, there is another occurrence of the unit clause L not in the subtree $T'_{\neg L}$. However, this will imply that T is not coherent since L and $\neg L$ occur in different branches of T, a contradiction.

Suppose L is not the generated literal. The generated unit clause is denoted as L'. Since L' is the only generated unit clause and it lies in the first level, so it has no contribution to $L <^*_{T'} L$. Thus, $L <^*_T L$. This is impossible since T is coherent.

Next we show that the condition 3 holds for T'. Suppose there are two nodes which are labelled by L and $\neg L$ and one of them has height greater than one. Since T is coherent, one of L and $\neg L$ must be the generated unit clause. Without lose of generality we suppose L is the generated unit clause. This will imply that $\neg L$ occurs twice in one branch. Thus, the condition 2 of Definition 2 is false for T', a contradiction.

Finally we show condition 4 holds for T'. Suppose $L <_{T'_L} K <^*_{T'_L} \neg L$ and K has an occurrence outside T'_L. Since T is coherent, L must be the generated unit clause. That is, K occurs in $T'_{\neg L}$. This would imply that $\neg L <^*_{T'} \neg L$, a contradiction since we have proved that condition 2 holds for T'.

Altogether, T' is coherent.

Now in the new tree T', the branch from f to the empty clause becomes shorter. Applying iteratively the above procedure until one of the unit clauses in the lowest level comes from f, we can obtain our desired tree, denoted as T^*.

Please note that f is an arbitrary fixed clause of F. Thus we can take f as $\neg L \vee g$. Further, without loss of generality we just simply assume that T is T^*. Suppose, in T, L is not directly obtained from the initial unit clause L. Since T is coherent, the initial unit clause L has been used in one of the subtrees for other unit clauses. Then we exchange the derivation T_L and the initial L. If in the resulting tree the resolution step concerning L is $\frac{L \quad \neg L \vee g'}{g'}$, then we just remove all occurrences of $\neg L$ in the path from $\neg L \vee g$ to $\neg L \vee g'$ and finally add the step $\frac{L \quad \neg L \vee g}{g}$. The transformations are shown in Fig 2.

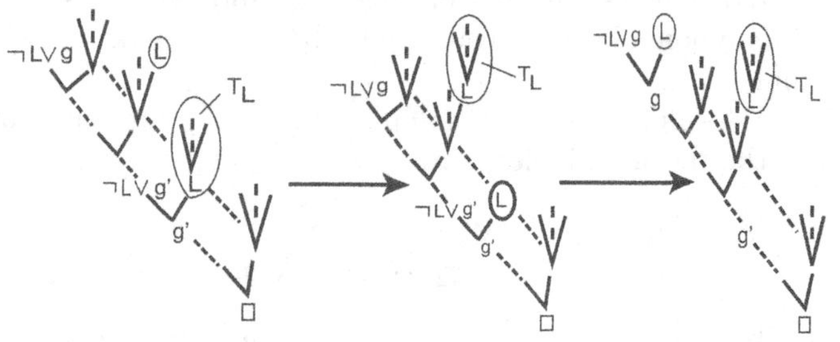

Fig. 2.

It is not hard to see that the transformation from the first tree to the second tree in Fig 2 preserves the coherence. If g' is the empty clause in the second tree of Fig 2, i.e. the last step is to resolve $\neg L$ and the circled L, then by the previous proof we know the final tree is coherent. Suppose g' is not empty. The path from $\neg L \vee g'$ to $\neg L \vee g$ in the second tree of Fig 2 contains no unit clauses and in the path from g' to $\neg L \vee g$ in the final tree of Fig 2 only g' could probably be a unit clause. That means the second transformation in Fig 2 does not generate new unit clauses. Thus, clauses above g' in the two mentioned paths have no contribution to the coherence. Consequently, the final tree is a coherent.

Therefore, the final tree is as required.

Lemma 4. *If F has a coherent ROU-resolution tree T then $F \in MROU$.*

Proof. (Induction on the number of clauses.) If F has only two clauses then clearly the lemma is true. Now suppose F has more than two clauses. Suppose on the contrary $F \notin MROU$. Then there is a proper subformula G of F such that $G \in MROU$. Now let L and $\neg L \vee g$ be two clauses in G. We see that $F' = (F - \{L, \neg L \vee g\}) + \{g\}$ is not in $MROU$ since $F' - G' = F - G$ is nonempty and $G' \in MROU$ by Lemma 1, where $G' = G - \{L, \neg L \vee g\} + \{g\}$. By lemma 3 w.l.o.g we can assume that in T the resolution step $\frac{L, \ \neg L \vee g}{g}$ occurs. We remove L and $\neg L \vee g$. We get a read-once resolution tree for F' which is clearly coherent since T is coherent. By the induction hypothesis, F' is in $MROU$. This contradicts the fact that $F' \notin MROU$.

For deciding whether a formula F has a coherent ROU-resolution tree, we make use of some properties of minimal unsatisfiable formulas.

Proposition 2. *Let F be a minimal unsatisfiable formulas such that $pos(x, F) = neg(x, F) = 1$ for each $x \in var(F)$. Then*

1. *F contains at least two unit clauses.*
2. *F is in $MU(1)$.*
3. *$F \in MROU$.*

The first two statements are well known and quite simple to prove. That the formula F is in $MROU$ follows immediately from (1) and (2).

In proving the polynomial time solvability of $MROU$ we make use of a transformation of formulas.

Suppose F is a formula with $pos(x, F) = neg(x, F)$ for any variable x, $\{x_1, x_2, \cdots, x_n\}$ is the set of variables, and

$$x_i \vee f_1, \quad x_i \vee f_2, \cdots, \quad x_i \vee f_k \\ \neg x_i \vee g_1, \neg x_i \vee g_2, \cdots, \neg x_i \vee g_k \qquad (1)$$

are the clauses containing x_i or $\neg x_i$ in an arbitrary but fixed order.

For each clause $x_i \vee f_j$, we introduce a new variable x_{ij}. We replace $x_i \vee f_j$ by $x_{ij} \vee f_j$ and $\neg x_i \vee g_j$ by $\neg x_{ij} \vee g_j$.

We perform the procedure iteratively for $i = 1, 2, \cdots, n$. Finally, we obtain a formula, called $S(F)$, in which any literal occurs exactly once.

We will show that $S(F) \in MU(1)$ if $F \in MROU$, where $MU(1)$ is the class of the minimal unsatisfiable formulas with $n + 1$ clauses and n variables. Please note that $S(F)$ depends on the enumeration of the clauses in F.

Lemma 5. *If $F \in MROU$ then $S(F) \in MU(1)$ for any $S(F)$.*

Proof. We prove the lemma by induction on the number of clauses.

If F has only two clauses then obviously the lemma holds.

Suppose $F \in MROU$ has more than two clauses. We pick a unit clause, say x_i with $pos(x_i, F) = k$. Without loss of generality, we assume that x_i corresponds to x_{ik} in $S(F)$. Let $\neg x_i \vee g$ be the clause which corresponds to $\neg x_{ik} \vee g^*$ in $S(F)$. By Lemma 1, the formula $F' := F - \{x_i, \neg x_i \vee g\} + \{g\}$ is in $MROU$ and by the induction hypothesis, $S(F')$ is in $MU(1)$. Now we add the literal $\neg x_{ik}$ to the clause in $S(F')$ which is obtained from g during the transformation process and add x_{ik} as a unit clause to $S(F')$. Obviously, the resulting formula is $S(F)$ (up to renaming). Since $S(F') \in MU(1)$, we get $S(F) \in MU(1)$.

Suppose $S(F) \in MU(1)$. Then each literal in $S(F)$ occurs exactly once. By Proposition 2, $S(F) \in MROU$. Again because $S(F) \in MU(1)$, we can construct in linear time a ROU-resolution tree for $S(F)$. Now in the tree replacing the introduced variables by the initial ones we obtain a ROU-resolution tree for F in which each clauses of F occurs.

The next theorem presents a characterization of $MROU$, which will be used for our algorithm.

Theorem 3. *Let F be a formula with $pos(x, F) = neg(x, F)$ for every variable x. For an arbitrary fixed $S(F)$ and for an arbitrary fixed ROU-resolution refutation tree T for F obtained from $S(F)$, $F \in MROU$ if and only if $S(F) \in MU(1)$ and T is coherent.*

Proof. The direction from left to right follows from Lemma 2 and Lemma 5. The inverse direction follows from Lemma 4.

4.2   Complexity of MROU

In this subsection we show that $MROU$ can be decided in quadratic time. The main task of the decision procedure given below is to test whether $S(F)$ is in $MU(1)$ and to test whether the associated tree is coherent

Proposition 3. *Let T be a ROU-resolution tree for a formula F. Whether T is coherent can be decided in time $O(|F|^2)$.*

Proof. We present a procedure deciding whether the four conditions given in Definition 2 are satisfied.

1. To test whether all clauses of F have been used in the tree T, that means in the refutation, costs not more than quadratic time.

2. The next task is to decide whether $L <^*_T L$ for some literal L. According to the definition of $<^*_T$, we can use the following approach to decide condition 2 of the coherency of T. Try to find for each literal L all literals L' such that $L <^*_T L'$, and finally to see whether L is among these literals. However, this algorithm would cost cubic time. To design a more efficient algorithm we transform the resolution tree T into a directed graph with direction from the root to the leafs. We are only interested in the nodes labelled by unit clauses. Therefore, at first

we transform T into a tree T' as follows: Suppose in T we have two nodes n_1 and n_2 labelled by the literal L_1 and the literal L_2, and there is a directed path from n_1 to n_2. If no node on the path different from n_1 and n_2 has a unit clause label, then add an arc from n_1 to n_2 to T. We perform these steps as long as possible. Finally we remove all nodes labelled by a non-unit clause. The resulting tree is called T'. Obviously, we have $L <_T^* L$ if and only if $L <_{T'}^* L$.

Now we transform T' into a directed graph $G(T)$ as follows: If in T' two different nodes with the same label occur then we join them together as one node. Then after several steps there are no two different nodes with the same label. Finally, we delete the node labelled by the empty clause. We denote the directed graph as $G(T)$. The reduction from T to $G(T)$ costs linear time in the size of T assuming random access.

Claim. There is no literal L with $L <_T^* L$ if and only if $G(T)$ is acyclic.

The claim follows from the following observations. Suppose $L_1 <_T L_2$, that means, L_1 and L_2 occurs in one branch of T. From the reduction from T to $G(T)$ we can see that there is a path from L_1 to L_2 in $G(T)$. Conversely, if there is an edge from L_1 to L_2 then we have $L_1 <_T L_2$ by the reduction from T to $G(T)$.

The strongly connected components of a directed graph can be computed in linear time [9]. $G(T)$ is acyclic if and only if $G(T)$ has only strongly connected component consisting of a single node. Hence condition 2 of Definition 2 (coherence) can be decided in quadratic time.

3. To test condition 3, we employ the following procedure. For each variable x, we first go through the tree T to check whether x occurs, then delete all branches containing a node labelled by x, then check whether $\neg x$ occurs as a unit clause in the remaining part, finally check whether $\neg x$ or x has height greater than 1. Clearly, for any fixed x, the above procedure costs linear time. Thus, checking condition 3 of the coherency costs not more than quadratic time.

4. For the last condition of the coherency, we check for each node labelled by a unit clause L whether $\neg L$ occurs in T_L, and check in T_L whether there is a K such that $K <_{T_L}^* \neg L$ which also has an occurrence outside T_L.

Please note that in $G(T_L)$ we can compute in linear time all vertices from which there is a path to $\neg L$. That means, to compute all K such that $L <_{T_L} K <_{T_L}^* \neg L$ costs not more than linear time. To see whether a K occurs outside T_L needs linear time. Therefore, to check the last condition needs quadratic time.

Altogether we have seen that the coherency can be decided in quadratic time.

Theorem 4. *MROU can be solved in time $O(n^2)$ where n is the length of the formula.*

Proof. According to the previous results we can employ the following procedure to check whether a given formula F is in *MROU*.

Input: A *CNF*-formula F
Output: yes if $F \in MROU$; no if otherwise
    if there is some $x \in var(F)$ such that $pos(x, F) \neq neg(x, F)$ then return no
    transform F into a $S(F)$; if $S(F) \notin MU(1)$ then return no

construct from $S(F)$ a ROU-resolution tree T for F
if T is coherent then return yes else return no

To check whether $pos(x, F) = neg(x, F)$ for each $x \in var(F)$ costs linear time in the size of F depending on an appropriate data structure. The transformation costs linear time, and the size of $S(F)$ is the same as that of F. To test if $S(F) \in MU(1)$ costs no more than linear time. This can be seen as follows. According to Proposition 2, to test whether $S(F) \in MU(1)$ we just need to apply the unit resolution iteratively, and to see whether the final formula is $\{\sqcup\}$. To construct a resolution tree from $S(F)$ needs linear time. To decide whether the tree is coherent needs quadratic time(see Proposition 3). Altogether, the algorithm requires no more than $O(|F|^2)$ steps.

5   Conclusion

Our procedure for deciding *MROU* has running time $O(|F|^2)$ where checking the coherency for the generated tree is the time consuming part. Whether the coherence can be solved much faster is not known. It remains an open question whether there is a linear time algorithm for *MROU*.

References

1. Aharoni, R., Linial, N.: Minimal Non-Two-Colorable Hypergraphs and Minimal Unsatisfiable Formulas, *Journal of Combinatorial Theory* **43** (1986), 196–204.
2. Chvatal V., Szemeredi T.: Many Hard Examples for Resolution, *Journal of ACM*, **35** (1988), 759–768.
3. Dowling, W.J., Gallier, J.H.: Linear-Time Algorithms for Testing the Satisfiability of Propositional Horn Formulas, *Journal of Logic Programming* **1** (1984), 267–284.
4. Haken, A.: The Intractability of Resolution, *Theoretical Computer Science* **39** (1985) 297–308.
5. Iwama, K., Miyano, E.: Intractability of Read-Once Resolution, *Proceedings Structure in Complexity Theory, 10th Annual Conference*, IEEE, 1995, 29–36.
6. Kleine Büning, H., Zhao, X.: The Complexity of Read-Once Resolution, *Annals of Mathematics and Artificial Intelligence*, **36** (2002), 419–435.
7. Papadimitriou, C.H., Wolfe, D.: The Complexity of Facets Resolved, *Journal of Computer and System Science* **37** (1988) 2–13.
8. Szeider, S.: NP–Completeness of Refutability by Literal-Once Resolution, *LNAI 2083, Automated Reasoning* (2001), 168–181.
9. Tarjan, R.E.: Depth-First Search and Linear Graph Algorithms, *SIAM Journal on Computing*, **1** (1972), 146–160.
10. Urquhart A.: Hard Examples for Resolution, *Journal of ACM*, **34** (1987), 209–219.

The Interaction Between Inference and Branching Heuristics

Lyndon Drake and Alan Frisch

Artificial Intelligence Group, Department of Computer Science
University of York, York YO10 5DD, United Kingdom
{lyndon,frisch}@cs.york.ac.uk,
http://www.cs.york.ac.uk/~\{lyndon,frisch\}

Abstract. We present a preprocessing algorithm for SAT, based on the HypBinRes inference rule, and show that it does not improve the performance of a DPLL-based SAT solver with conflict recording. We also present evidence that the ineffectiveness of the preprocessing algorithm is the result of interaction with the branching heuristic used by the solver.

1 Introduction

Propositional satisfiability (SAT) is the archetypal NP-complete problem [3]. It is possible to solve a wide range of problems from classes such as constraint satisfaction [18], quasigroup completion [19], and model checking [2], by encoding them into SAT and solving the SAT representation of the problem. In this paper we assume that SAT problem instances are given in conjunctive normal form (CNF), where a formula is the conjunction of a set of clauses. Each clause consists of a disjunction of literals, where a literal is either an atom or its negation.

The fastest available complete SAT solvers (such as Chaff [15]) use variants of the DPLL [5, 4] search algorithm for SAT. These solvers typically augment the search procedure with two fast inference techniques: unit propagation (included in the original DPLL algorithm) and conflict clause recording [14].

In this paper we present a preprocessor for complete SAT solvers, based on a binary clause inference rule called HypBinRes [1]. We show that this preprocessor does not improve the performance of Chaff, and give evidence that this is due to interaction with the branching heuristic used by Chaff.

2 HypBinRes Preprocessing

HypBinRes (or hyper-binary resolution) is a rule for inferring binary clauses. These binary clauses are resolvents of the original formula, but applying the HypBinRes rule can discover these binary consequences more efficiently than resolution. We first define the HypBinRes rule and discuss its implementation as a preprocessor for a SAT solver, and then give experimental results that demonstrate that HypBinRes preprocessing is ineffective at improving the performance of Chaff on a wide range of SAT instances.

E. Giunchiglia and A. Tacchella (Eds.): SAT 2003, LNCS 2919, pp. 370–382, 2004.
© Springer-Verlag Berlin Heidelberg 2004

2.1   The HypBinRes Rule

Given the literals $a_1, \ldots, a_n$ and h, where $\bar{a}_i$ is the complement of a_i, an example of applying HypBinRes is:

$$
\begin{array}{c}
a_1 \vee a_2 \vee a_3 \\
\bar{a}_1 \vee h \\
\bar{a}_2 \vee h \\
\hline
a_3 \vee h
\end{array}
$$

More formally, given a clause of the form $(a_1 \vee \cdots \vee a_n)$ and a set of $n-1$ satellite clauses of the form $(\bar{a}_i \vee h)$ each containing a distinct $a_i \in \{a_1, \ldots, a_n\}$, we can infer the clause $(a_x \vee h)$ where $a_x \in \{a_1, \ldots, a_n\}$ and $a_x \neq a_i$ for all $i \neq x$. This inference rule, called HypBinRes, corresponds to a series of binary resolution [16] steps resulting in a binary clause, and was defined by Bacchus [1]. It is more efficient to use HypBinRes to infer a set of binary clauses than to use resolution to find the same set of clauses, because resolution may take several steps and generate intermediate resolvent clauses.

De Kleer [6] describes an almost identical resolution rule called H5-3 which also infers binary clauses. The H5-3 rule differs by restricting operation to inference on negative satellite clauses, and by not requiring binary satellite clauses in the resolution.

2CLS+EQ [1] is a complete DPLL-based SAT solver that uses HypBinRes during search. 2CLS+EQ is unusual in its use of complex reasoning during search, and is the only publically available solver that uses non-unit resolution during search to achieve comparable performance to the fastest conventional SAT solvers.[1]

2.2   Using HypBinRes in a Preprocessor

While 2CLS+EQ is competitive, it is still outperformed by the best conventional SAT solvers. We decided to examine the effect of HypBinRes as a preprocessor for such a conventional SAT solver. Adding HypBinRes-implied clauses to SAT instances could result in the search space being pruned, and as a result improve runtime performance.

The effect of preprocessing using HypBinRes may vary from problem class to problem class. The constraint satisfaction problem (CSP) is another NP-complete problem, for which two main SAT encodings exist: the direct encoding [18], and the support encoding [10]. The support encoding of a CSP instance generally results in better SAT solver runtime performance than the direct encoding, as DPLL on the support encoding enforces arc consistency [11] on the formula. Furthermore, we have previously shown [8] that HypBinRes, when applied to the direct encoding will generate support clauses for the functional

---

[*]   Van Gelder's 2clsVER [17] uses other reasoning during search, but its performance is not competitive with the fastest solvers.

constraints in the instance. As a consequence, we might expect improved performance on SAT-encoded CSP instances after preprocessing with HypBinRes.

Experiments with 2CLS+EQ showed that for many other problem classes, a reasonable number of HypBinRes clauses can be generated during preprocessing. In order to make 2CLS+EQ perform well, its data structures and heuristics have been designed to interact efficiently with HypBinRes resolution during search. As a result, it is not a particularly efficient solver when used with HypBinRes disabled during search, but the number of inferences made during preprocessing suggested to us the hypothesis that a separate HypBinRes preprocessor would improve the performance of a more conventional SAT solver.

2.3   Results

In order to test this hypothesis, we wrote a preprocessor (based on Swan, our SAT solver) to apply HypBinRes. The preprocessor takes as input a file in DIMACS CNF format [7], and produces as output a CNF file containing the HypBinRes resolvents of the formula in addition to the original clauses. Our implementation of HypBinRes is not the most efficient possible, but the preprocessor performs fast enough that the runtime required by the preprocessor is neglible compared to the solver runtime on most instances.

Fig. 1 shows the results of comparing Chaff runtime performance at solving 100 hard random CSP instances using the direct encoding,[2] before and after applying HypBinRes as a preprocessor. The instances included a mix of satisfiable and unsatisfiable problems. Points below the line indicate improved performance resulting from the preprocessing, while those above the line show decreased performance. All times are given in seconds, and all the experiments in this section were carried out on a Pentium III 650 machine with 192MB of RAM.

The overall impact of using HypBinRes as a preprocessor for Chaff is negligible: the geometric mean runtime for the original problem instances is 1.74 seconds, while the corresponding mean for the preprocessed instances is 1.73 seconds (the medians were identical). However, performance on individual problem instances varies greatly, with some problems displaying much better and others much worse performance (with an interquartile range of 2.9s for the original instances and 3.1s for the preprocessed instances).

These results for HypBinRes preprocessing were not confined to SAT-encoded CSP instances, and in fact all problem classes in the SATLIB library [12] display substantially similar behaviour (though there are minor variations). For example, Fig. 2 shows Chaff's performance on 22 quasigroup (Latin squares) problems. The quasigroup problems contain a mix of satisfiable and unsatisfiable instances.

---

[*] While these CSP instances are random, the SAT instances derived using the direct encoding contain some structure as a side-effect of the encoding. That structure can be exploited by HypBinRes [8]. In our experiments, Chaff (noted for its high performance on structured SAT instances) often outperformed SATZ (which performs best on random SAT instances) by one or two orders of magnitude on the SAT-encoded CSP instances.

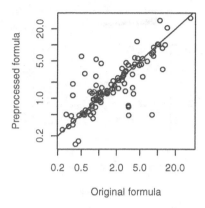

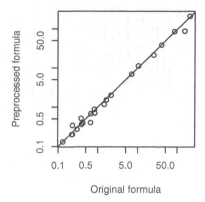

Fig. 1. Chaff runtimes for CSP instances before and after HypBinRes preprocessing (log-log scale, times in seconds)

Fig. 2. Chaff runtimes for SATLIB QG instances before and after HypBinRes preprocessing (log-log scale, times in seconds)

In all the instances we examined, the change in runtime was matched by an approximately proportional change in the number of search nodes explored. This suggests that the runtimes were not significantly affected by the increased size of the preprocessed formulae, but were instead determined by the size of the search space explored.

2.4   Analysis

Intuitively, while making the inferences required to generate implied clauses must take time and the extra clauses will add some overhead to the search procedure, we would expect the additional clauses to result in a smaller search space. Instead, what we observed was that Chaff sometimes explored a larger search space on a preprocessed formula than on the corresponding original formula. Two possible explanations for this observed behaviour are that the additional clauses interact poorly with the conflict clause generation, and that they interact poorly with the heuristic. Interaction with the conflict clauses is difficult to study, but interaction with the heuristic is more practical to investigate so we examined that first (see Section 4 for some results on conflict analysis).

The heuristic used by Chaff is, like many SAT heuristics, based on counting literals. Adding implied clauses is likely to change the relative literal counts by increasing the occurrences of some literals relative to others. If the original heuristic ordering was nearly optimal, this change in relative literal counts may result in degraded performance on the preprocessed formula.

Two quasigroup instances which demonstrate the variability in performance are qg3-09, which took 201 seconds to solve before preprocessing and 212 seconds

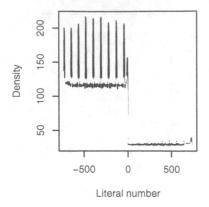

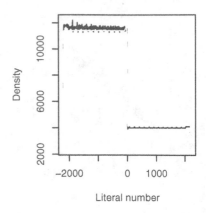

Fig. 3. Density plot of literals in qg3-09. The grey line shows the original formula, and the black line the preprocessed formula. Chaff performed worse on the preprocessed formula

Fig. 4. Density plot of literals in qg5-13. The grey line shows the original formula, and the black line the preprocessed formula. Chaff performed much better on the preprocessed formula

afterwards, and qg5-13, which took 149 seconds to solve before preprocessing and 88 seconds afterwards. If these instances were satisfiable the faster times might simply be explained by the branching heuristic making a fortunate set of early decisions in order to find a single satisfying assignment. However, both these instances are unsatisfiable, so if the branching order is the main factor controlling performance then the heuristic must make choices that cover a larger proportion of the search space.

Figures 3 and 4 show density plots of the literal counts for these two instances. There are many hundreds of literals in these instances – too many for a bar plot of the literal counts. A density plot shows a smoothed version of the counts, while still displaying interesting features. Differences between the two lines in each plot correspond to differences in the literal count between the original and preprocessed formulae. In both cases it is clear that in each instance the relative literal counts have been altered by the additional clauses, and this effect is proportionally much greater in the case of qg3-09 (which Chaff performed worse on after preprocessing).

The differences between the two plots suggest that poor performance may be the result of the heuristic making poor decisions on the preprocessed formula, due to altered literal counts. It is not possible to be confident on this point based purely on counting literals in the formula, as the actual heuristic decisions in Chaff depend on the conflict clauses generated during search. However, the differences between these two instances and others we studied were enough to motivate further investigation of the interaction between implied clauses and the heuristic.

3   Isolating the Heuristic

One method for investigating the effect of implied clauses on a heuristic is to prevent the heuristic considering the additional clauses. In other words, if we ensure that only the clauses in the original formula are visible to the heuristic, the additional clauses will not significantly change the decision order. The additional clauses might directly terminate a branch early, or result in new unit propagations, but if a static heuristic is used this will not result in a larger search tree. There is potentially a small negative impact on runtime performance from handling a larger formula.

To achieve this in the context of Chaff, we would have to not only hide the additional clauses from the heuristic, but prevent them from being involved in conflict clause generation, as the heuristic used in Chaff is deliberately biased towards literals appearing in recent conflict clauses. We decided instead to implement implied clause hiding in our own SAT solver Swan (which is easy to modify and does not use conflict recording), by annotating the preprocessed formula with hints identifying the additional implied clauses, and altering the heuristic to stop it counting literals in implied clauses.

3.1   Hint Syntax

In order to separate the marking of implied clauses from the particular preprocessor or solver implementation being used, we decided to extend the DIMACS CNF format [7]. We did this by giving implied clause "hints" as a comment line in the preamble, of the form c !! hint implied-clause 2 45 47 0, where 2, 45 and 47 are clause numbers. Clauses are assumed to be numbered in the order they appear in the formula, starting at 1. The format is intended to be easy for people to read and modify, and for software to parse.

Because hints are placed in a comment line, they will be ignored by solvers that have not been modified to interpret the hints. As a consequence of this, a preprocessor that adds implied clauses can include a hint line in its output, and the resulting file still conforms to the DIMACS CNF format specification. Existing SAT solvers, such as Chaff, can be run on the preprocessed formula without modification (the implied clauses will still be used by the solver), while only a relatively simple parsing change is required in order to use the hints.

3.2   Results

We tested implied clause hiding on the same set of quasigroup instances as before, using our modified version of Swan, and two static heuristics (one a static version of the VSIDS heuristic used in Chaff). The two heuristics behaved similarly with respect to the visibility of the implied clauses (though different problem classes performed better with one or the other heuristic).

All times are given in seconds, and all the experiments in this section were carried out on a dual processor Pentium III 750 machine with 1GB of RAM. A timeout of 5 minutes was used for the single solution counting experiments

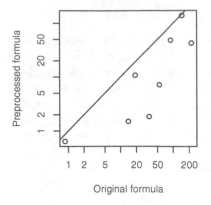

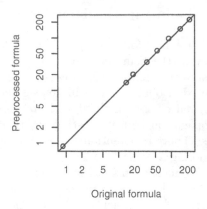

Fig. 5. Swan runtimes for SATLIB QG instances before and after preprocessing (log-log scale, times in seconds)

Fig. 6. Swan runtimes for SATLIB QG instances before and after preprocessing, hiding the implied clauses (log-log scale, times in seconds)

(Figures 5 and 6), and a timeout of 20 minutes when we configured Swan to count all solutions of the satisfiable instances. Swan failed to solve a number of the instances within the time limit, though in some cases it solved the preprocessed version of those instances.

There are two main points to note from these results: on average, preprocessing a formula with HypBinRes reduced the runtime by more than 50%; and using hints to hide the additional clauses from the heuristic resulted in no performance improvement after preprocessing. We obtained similar results when we configured Swan to count all solutions on the satisfiable instances. When implied clauses were hidden from the heuristic, the search space on the preprocessed problem was no larger than that on the original problem, as expected (though there was occasionally a small runtime penalty). We also ran experiments on the SAT-encoded CSP instances, but Swan was unable to solve many of the instances within the time limit. However, on the instances that Swan was able to solve we observed similar behaviour to the results for the quasigroup instances.

3.3   Analysis

HypBinRes generates only binary clauses, which are most likely to contribute to a proof if they are involved in unit propagation. When the implied binary clauses are visible, the heuristic is more likely to make decisions resulting in unit propagation on those clauses, as the literals in the implied clauses will receive higher weightings.

The heuristic used by SATZ [13] is based on the idea of choosing to assign the literal which will result in the most unit propagations. Bacchus points out that while this heuristic is expensive to compute, given a literal l counting the number of binary clauses containing the negation of l gives the number of unit propagations resulting from assigning l true [1, Observation 6]. The effect on a literal counting heuristic (like those used in Chaff and Swan) of adding HypBin-Res clauses will be to give higher priority to assigning those variables which will result in unit propagations (though the value ordering for a particular variable will not be optimal).

These results go some way towards explaining the performance results from Chaff. The heuristic used by Chaff gives precedence to literals appearing in recently generated conflict clauses, so as the search space is explored literals from clauses in the input formula (including those added by HypBinRes) will contribute relatively little to the literal counts. If the search proceeds for long enough, the clauses in the input formula will be effectively ignored in preference to those generated by conflict recording.

However, this explanation cannot completely account for the observed behaviour of Chaff, as on some of the SAT-encoded CSP instances the size of the search space explored by Chaff increased. These larger search spaces must be due either to poor branching heuristic choices early in the search space because of the altered literal counts, or to the generation of different conflict clauses because of the different decisions and unit propagations.

4   Further Results

In order to determine how much effect the implied clauses were having on the heuristic, we made two further modifications to Swan: first adding a form of conflict analysis, and then quantifying the size of branches pruned by implied clauses. Both of these modifications extend the work presented in the previous section by using the hints to identify the implied clauses in the problem instances.

4.1   Conflict Analysis

The conflict analysis uses the implication graph (e.g. Fig. 7) to identify the clauses involved in a conflict. Each vertex in the implication graph represents an assignment, and the incident edges to a vertex are labelled with the clause that causes the assignment at that vertex. Conflict analysis operates in much the same way as the conflict analysis engine in a solver with conflict clause recording [14, 20], with two differences: instead of stopping at the first unique implication point as Chaff and other similar solvers do we trace all the way back to the decision variables, and we do not generate a conflict clause. If an implied clause is involved in a conflict, either directly by the current assignment making the clause false, or indirectly by being involved in the sequence of unit propagations that leads to another clause becoming false, then that clause is considered to have pruned a branch.

If implied clauses are never involved in conflicts, then those clauses will have no impact on the conflict clauses generated by the conflict recording in Chaff (and

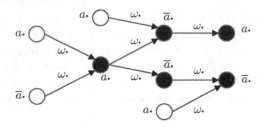

Fig. 7. Implication graph

other similar solvers). By examining how often implied clauses are involved in conflicts, we can estimate the effect of the additional clauses on conflict recording.

4.2   Branch Pruning Measurement

Quantifying the size of the branches pruned by an implied clause is important because not all pruning will significantly reduce the search space. Given the overhead of generating implied clauses, such clauses are unlikely to significantly improve runtime performance unless they prune large portions of the search space. Fig. 8 shows the start of a search space that might be explored by a DPLL solver, where ⊗ denotes a failure node in the search space, and ⊘ a node which only fails when implied clauses are added to the original formula. If an implied clause becomes false under the assignment at node a, then the implied clause is involved in the conflict but does not contribute to pruning the search space because node a was already a failure node. Even if an implied clause becomes false under the assignment at node b, the clause does not result in any significant pruning because the successors of b, nodes c and d, were already failure nodes. However, a significant portion of the search space would be pruned if node e became a failure node because of an implied clause.

By measuring the amount of the search space pruned by a particular conflict, we can find out if a few large subtrees have been pruned, or a large number of smaller subtrees. The conflict analysis engine can be used to identify the implied clauses responsible for pruning a particular branch. We know that there is a relationship between the branching order of DPLL and the ordered resolutions that will prune the search space explored by backtracking search [9]. If we do find instances where large subtrees have been pruned, it should be possible to show the relationship between the branching heuristic ordering and the literals being resolved on.

Our current implementation is not completely automated, and operates by printing the assignments and unit propagations involved while exploring the search space on the original and preprocessed instances. We have formatted the output so that the Unix `diff` program can be used to compare the two sets of output, but the output from `diff` has to be examined by a human (we intend to automate this process in the near, as the search spaces are often very large and take some time to examine).

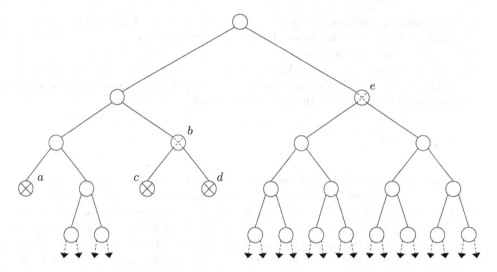

Fig. 8. Search space pruning by implied clauses (⊗ denotes a failure node, and ⊙ a node which only fails when implied clauses are added to the original formula)

Table 1. Results from running Swan on the SAT-encoded CSP instance `direct0006`

Implied clauses	Visible to heuristic	Time (s)	Branches	Unit propagations	Backtracks
No	–	278.00	582234	21857718	291034
Yes	No	215.90	457883	14338664	228860
Yes	Yes	3.45	5460	220985	2623

4.3   Results

Our experiments suggest that the implied clauses added by HypBinRes preprocessing are rarely involved in conflicts. For example, these additional clauses are never involved in conflicts on the quasigroup instances we have examined. As a result, on many instances there will be little or no pruning of the search space as a result of adding HypBinRes clauses.

Furthermore, it appears that when the added clauses do participate in conflicts, each new conflict prunes a relatively small portion of the search space. For example, consider the SAT-encoded CSP instance `direct0006`. Table 1 shows the results of running Swan on the original instance, the preprocessed instance but with the additional clauses hidden from the heuristic, and the preprocessed instance with the additional clauses visible to the heuristic. It is clear that even with the branching order fixed, the number of branches was significantly reduced by adding the implied clauses, which suggests that the additional clauses are involved in pruning branches. However, examining the differences between the search spaces explored shows that many small subtrees were pruned, rather than a few large ones. Relaxing the restriction on the branching order by making the implied clauses visible to the heuristic is what produces the really significant improvements to the size of the search space and the runtime.

5   Related Work

Our current preprocessor implementation differs from 2CLS+EQ in one important respect: it does not perform the EqReduce operation described in [1]. While we could have modified 2CLS+EQ to operate as a preprocessor, including the EqReduce operation, doing so is not trivial and would limit the possibility of experiments examining the effect of adding EqReduce to HypBinRes. However, Bacchus has since written a preprocessor called HYPRE that combines HypBinRes and EqReduce. HYPRE can take a long time to run, but substantially improves solver performance on many problems. It can also solve some hard problems without search.

6   Future Work

There is little existing work which analyses the reasons for the performance impact of a preprocessing algorithm, so we plan to investigate the interactions between inference and branching heuristics for other inference rules such as length bounded resolution and neighbour resolution. Other popular heuristics may exhibit different behaviour in the presence of additional clauses, so we will also study other heuristics.

Based on the density plots (Figures 3 and 4) it may be possible to automatically determine how many implied clauses to add based on the relative alterations to the literal counts in the formula. If this technique is effective, it may be useful as an automatic measure for the likely effect on relative solver performance resulting from other preprocessing techniques that alter the formula.

Reasoning about the search space explored by a solver with conflict clause recording is difficult, but it would be possible to add clause recording to Swan, and then show which implied clauses contribute to the conflict clause generation. This would make it possible to study the interaction between the clauses added during preprocessing by HypBinRes and those added during search by conflict recording.

7   Conclusion

The effect of adding implied clauses on SAT solver performance can strongly interact with the branching heuristic used. In particular, using the HypBinRes rule to add clauses can result in Chaff searching a larger search space, with a corresponding increase in the solution time.

By isolating the effect of the additional clauses on the heuristic, we have shown that a DPLL solver without conflict recording such as Swan performs best when the implied clauses are visible to the heuristic. While the variable performance exhibited by Chaff on HypBinRes preprocessed instances could be partially due to interaction with conflict clauses, we have also shown that much of the poor performance on these instances can be explained by the Chaff heuristic ignoring the clauses added by HypBinRes in favour of those generated by conflict clause recording.

8   Acknowledgements

We are grateful to Ian Gent for making available the CSP instances used in his paper, to Fahiem Bacchus and the authors of Chaff for making their SAT solvers publically available, to Dan Sheridan for his collaboration in developing the hints format, to Oliver Kullman for his suggestions, and to Toby Walsh and other members of the APES research group for their advice and encouragement. The first author is supported by EPSRC Grant GR/N16129 (for more information see http://www.cs.york.ac.uk/aig/projects/implied).

References

[1] Fahiem Bacchus. Extending Davis Putnam with extended binary clause reasoning. In *Proceedings of the National Conference on Artificial Intelligence (AAAI-2002)*, 2002 (to appear).

[2] Armin Biere, Alessandro Cimatti, Edmund Clarke, and Yunshan Zhu. Symbolic model checking without BDDs. In W.R. Cleaveland, editor, *Tools and Algorithms for the Construction and Analysis of Systems. 5th International Conference, TACAS'99*, volume 1579 of *Lecture Notes in Computer Science*. Springer-Verlag Inc., July 1999.

[3] S.A. Cook. The complexity of theorem-proving procedures. In *Proceedings of the Third ACM Symposium on Theory of Computing*, pages 151–158, 1971.

[4] Martin Davis, George Logemann, and Donald Loveland. A machine program for theorem-proving. *Communications of the ACM*, 5:394–397, 1962.

[5] Martin Davis and Hilary Putnam. A computing procedure for quantification theory. *Journal of the ACM*, 7:201–215, 1960.

[6] Johan de Kleer. A comparison of ATMS and CSP techniques. In *Proceedings of the 11th Joint International Conference on Artificial Intelligence (IJCAI-89)*, pages 290–296, 1989.

[7] DIMACS. Suggested satisfiability format. Available from ftp://dimacs.rutgers.edu/pub/challenge/satisfiability/doc/satformat.tex, May 1993.

[8] Lyndon Drake, Alan Frisch, Ian Gent, and Toby Walsh. Automatically reformulating SAT-encoded CSPs. In *Proceedings of the International Workshop on Reformulating Constraint Satisfaction Problems (at CP-2002)*, Cornell, 2002.

[9] Alan M. Frisch. Solving constraint satisfaction problems with NB-resolution. *Electronic Transactions on Artificial Intelligence*, 3:105–120, 1999.

[10] Ian P. Gent. Arc consistency in SAT. In *ECAI 2002: the 15th European Conference on Artificial Intelligence*, 2002.

[11] Pascal Van Hentenryck, Y. Deville, and C.M. Teng. A generic arc consistency algorithm and its specializations. *Artificial Intelligence*, 57(2–3):291–321, 1992.

[12] Holger H. Hoos and Thomas Stützle. SATLIB: An online resource for research on SAT. In Ian Gent, Hans van Maaren, and Toby Walsh, editors, *SAT2000: Highlights of Satisfiability Research in the Year 2000*, volume 63 of *Frontiers in Artificial Intelligence and Applications*, pages 283–292. IOS Press, 2000.

[13] Chu Min Li and Anbulagan. Look-ahead versus look-back for satisfiability problems. In *Proceedings of the International Conference on Principles and Practice of Constraint Programming*, 1997.

[14] João P. Marques Silva and Karem A. Sakallah. Conflict analysis in search algorithms for satisfiability. In *Proceedings of the IEEE International Conference on Tools with Artificial Intelligence*, Nov 1996.

[15] M. Moskewicz, C. Madigan, Y. Zhao, L. Zhang, and S. Malik. Chaff: Engineering an efficient SAT solver. In *39th Design Automation Conference*, Las Vegas, June 2001.

[16] J. A. Robinson. A machine-oriented logic based on the resolution principle. *Journal of the Association for Computing Machinery*, 12(1):23–41, 1965.

[17] Allen van Gelder. Combining preorder and postorder resolution in a satisfiability solver. In Henry Kautz and Bart Selman, editors, *Electronic Notes in Discrete Mathematics*, volume 9. Elsevier Science Publishers, 2001.

[18] Toby Walsh. SAT v CSP. In *Proceedings of CP-2000*, LNCS-1894, pages 441–456. Springer-Verlag, 2000.

[19] Hantao Zhang and Jieh Hsiang. Solving open quasigroup problems by propositional reasoning. In *Proceedings of International Computer Symposium*, 1994.

[20] Lintao Zhang, Conor F. Madigan, Matthew H. Moskewicz, and Sharad Malik. Efficient conflict driven learning in a Boolean satisfiability solver. In *Proceedings of ICCAD*, 2001.

Hypergraph Reductions and Satisfiability Problems

Daniele Pretolani

Università di Camerino
Dipartimento di Matematica e Informatica
Via Madonna delle Carceri
62032 Camerino (MC) Italy
daniele.pretolani@unicam.it

Abstract. Although several tractable classes of SAT are known, none of these turns out to be easy for *optimization, probabilistic* and *counting* versions of SAT. These problems can be solved efficiently for formulas with bounded *treewidth*. However, the resulting time bounds are exponential in the treewidth, which means exponential in the size of the largest clause.

In this paper we show that solution methods for formulas with treewidth two can be combined with specialized techniques for dealing with "long" clauses, thus obtaining time bounds independent of the treewidth. This leads to the definition of a new class of tractable instances for SAT and its extensions. This class is related to a particular class of *reducible* hypergraphs, that extends partial 2-trees and hypertrees.

1   Introduction

Several classes of linearly solvable SAT instances have been proposed in the literature, such as *SLUR* (that includes *balanced* and *Extended Horn* formulas [10]); *Q-Horn* (that includes renamable Horn and 2-SAT formulas, see [11] for a comparison to other classes); *nested* and *co-nested* formulas [6,7].

A quite different picture arises for *optimization versions* of SAT, namely *Max SAT* and *Min SAT* (the latter often referred to as *Min-ONEs* or DISTANCESAT). These problems are NP-hard also for Horn 2-SAT formulas; linear algorithms are known only for nested and co-nested formulas. Similar observations hold for the *probabilistic versions* of SAT, namely *PSAT* [8] and *CPA* [3]; the relations between optimization and probabilistic versions of SAT are discussed in [9]. Linear time algorithms are known for CPA on balanced formulas [9] and PSAT on *hypertrees* [1].

The problem of *counting* satisfying assignments for a CNF formula, i.e. *#SAT*, is #P-complete (in decision version) and completeness holds true also for 2-SAT formulas. We are not aware of nontrivial classes of formulas where #SAT can be solved in polynomial time. We refer the interested reader to [2] for an overview of recent and less recent results on #SAT and related problems.

E. Giunchiglia and A. Tacchella (Eds.): SAT 2003, LNCS 2919, pp. 383–397, 2004.
© Springer-Verlag Berlin Heidelberg 2004

Efficient solution methods exist for satisfiability problems on *bounded tree-width* formulas. A formula σ (with n variables) has *treewidth* k if the *co-occurrence graph* representing σ (see [5]) has treewidth k, i.e. it is a *partial* k-*tree* [4]. In this case, optimization SAT can be solved in $O(3^k n)$ time [5], while #SAT can be solved in $O(n 2^{O(k)})$ time (see [2]). Probabilistic SAT reduces to a linear system of $O(3^k n)$ variables and constraints [1].

Note that the above bounds are linear in n for any fixed k, but they are *exponential* in k, where k may be up to $n - 1$. Furthermore, clause length has a direct impact on the treewidth, indeed, one has $k \geq L - 1$, where L is the size of the largest clause.

In this paper we introduce some new techniques that allow processing "long" clauses efficiently, at least under particular circumstances. Combining these new techniques with known methods for partial 2-trees we define a new class of tractable formulas, namely, the class *R-SAT* of *reducible formulas*. For this class, we obtain complexity bounds that are linear in the length of the formula *and* in the treewidth. Notably, R-SAT is defined in terms of a corresponding class of *reducible hypergraphs*, which is an extension of partial 2-trees as well as hyper-trees.

The structure of the paper is as follows. In the next section, we introduce notation and define the class of reducible hypergraphs; moreover, we show that R-SAT is incomparable with the tractable classes mentioned above. Section 3 describes algorithms for optimization, probabilistic and counting version of SAT. The last section suggests some directions for further research. Examples of application of our algorithms are given in the Appendix.

2 Definitions

We denote a CNF formula by the pair $\sigma = (\mathcal{X}, \mathcal{C})$, where $\mathcal{X}$ is the set of variables (propositions) and $\mathcal{C}$ is the set of clauses. A clause $C \in \mathcal{C}$ is a conjunction of $|C|$ *literals*, where a literal is either a variable $x \in \mathcal{X}$ or a *negated* variable $\neg x$. We assume that a clause does not contain both literals x and $\neg x$. We also assume that σ does not contain *unit clauses*, that is, $|C| > 1$ for each $C \in \mathcal{C}$; as we shall see in the next section, this is not a limiting assumption. The *length* $\mathcal{L}(\sigma)$ is the total number of literals in σ, that is $\mathcal{L}(\sigma) = \sum_{C \in \mathcal{C}} |C|$.

A *hypergraph* $\mathcal{H}$ is a pair $(\mathcal{V}, \mathcal{E})$ where $\mathcal{V}$ is the set of *nodes* and $\mathcal{E}$ is the set of *hyperarcs*; a hyperarc is a nonempty subset $e \subseteq \mathcal{V}$. We assume that each hyperarc e contains at least two nodes, i.e. $|e| \geq 2$. An *edge* is a hyperarc e with $|e| = 2$; a *hyperedge* is a hyperarc h such that $|h| > 2$. The size of $\mathcal{H}$ is the sum of the sizes of its hyperarcs.

To a given CNF formula σ we associate a hypergraph $\mathcal{H}_\sigma = (\mathcal{V}_\sigma, \mathcal{E}_\sigma)$, where $\mathcal{V}_\sigma = \mathcal{X}$ and $\mathcal{E}_\sigma$ contains a hyperarc $e(C)$ for each clause $C \in \mathcal{C}$. Hyperarc $e(C)$ contains the nodes corresponding to the variables in C, that is, $e(C) = \{x \in \mathcal{V}_\sigma : x \in C \vee \neg x \in C\}$. Note that $\mathcal{H}_\sigma$ may contain *parallel* hyperarcs, that is, distinct hyperarcs containing the same nodes. Note also that $\mathcal{H}_\sigma$ provides an incomplete representation of σ, since it does not distinguish negated and non-

negated variables. A complete representation of CNF formulas is provided by
directed hypergraphs, see e.g. [1].

The hypergraph $\mathcal{H} = (\mathcal{V}, \mathcal{E})$ in Figure 1 is associated with the formula containing the clauses: $a = (\neg u_5 \vee u_4 \vee u_6)$, $b = (\neg u_6 \vee u_4)$, $c = (\neg u_3 \vee u_5)$,
$d = (\neg u_2 \vee u_1 \vee u_3)$, $e = (\neg u_7 \vee \neg u_4)$, $f = (\neg u_7 \vee u_3)$. Here, $\mathcal{V} = \{u_1, u_2, \ldots, u_7\}$
and $\mathcal{E} = \{a, b, \ldots, f\}$. Nodes (variables) and hyperarcs (clauses) are represented
by circles and triangles, respectively; each triangle is connected to the nodes in
the corresponding hyperarc. We have e.g. a hyperedge $a = \{u_4, u_5, u_6\}$ and an
edge $b = \{u_4, u_6\}$. Observe that each triangle points towards the non-negated
variables in the corresponding clause. Indeed, we adopt the standard representation of directed hypergraphs, which provides a complete visual description of
the underlying formula.

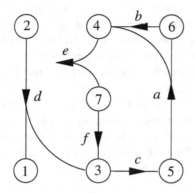

Fig. 1. A hypergraph $\mathcal{H} = (\mathcal{V}, \mathcal{E})$

Given a hypergraph $\mathcal{H} = (\mathcal{V}, \mathcal{E})$ we define for each node $u \in \mathcal{V}$ the sets $\mathcal{E}_E(u)$
and $\mathcal{E}_H(u)$ of edges respectively hyperedges *incident with* (i.e. containing) u:

$$\mathcal{E}_E(u) = \{e \in \mathcal{E} : |e| = 2, \ u \in e\}; \quad \mathcal{E}_H(u) = \{e \in \mathcal{E} : |e| > 2, \ u \in e\};$$

moreover, we define $\mathcal{E}(u) = \mathcal{E}_E(u) \cup \mathcal{E}_H(u)$. Given a hyperarc $h \in \mathcal{E}$ we define
the set of *neighbour edges* of h:

$$N_E(h) = \bigcup_{u \in h} \mathcal{E}_E(u).$$

For the hypergraph in Figure 1 we have $\mathcal{E}_E(u_4) = \{e, b\}$ and $\mathcal{E}_H(u_4) = \{a\}$. For
hyperedge $d = \{u_1, u_2, u_3\}$, we have $N_E(d) = \{f, c\}$.

2.1   Hypergraph Reduction

Here we introduce several *hypergraph reduction* operations, that consist in deleting hyperarcs and nodes, possibly adding new hyperarcs. To begin with, we

describe the hypergraph counterparts of some well known *graph reduction* operations.

series reduction $series(e_1, u, e_2)$: given the edges $e_1 = \{v_1, u\}$ and $e_2 = \{v_2, u\}$, such that $\mathcal{E}(u) = \{e_1, e_2\}$, delete e_1, e_2 and u; add to $\mathcal{E}$ an edge $e = \{v_1, v_2\}$.

parallel reduction $parallel(e_1, e_2)$: given the edges $e_1 = \{u, v\}$ and $e_2 = \{u, v\}$, delete edge e_2.

tail reduction $tail(a, u)$: given hyperarc a and $u \in a$, with $\mathcal{E}(v) = \{a\}$ for each $v \in a \setminus \{u\}$, delete hyperarc a and each node in $a \setminus \{u\}$.

Recall that (undirected) graphs of treewidth 2 can be reduced to a set of isolated nodes by a sequence of series, parallel and tail reductions (see [4], Theorem 33). Moreover, the *hypertrees* defined in [1] are reduced by tail operations only. Note that a tail reduction can delete hyperedges of arbitrary size.

Our approach consists in combining classical *edge-reducing* operations, such as *series* and *parallel*, with operations that delete hyperedges, such as *tail*. In particular, we introduce operations that reduce a hyperedge to an edge, that may be later deleted by edge-reducing operations.

To this aim, a quite simple extension of the *tail* reduction may be defined: replace a hyperedge h by an edge $e = \{u, v\}$ if $u, v \in h$ and $\mathcal{E}(y) = \{h\}$ for each $y \in h \setminus \{u, v\}$. That is, we allow *two* nodes in h (instead of one) to belong to other hyperarcs besides h. A more powerful operation can be obtained if we also accept edges connecting nodes u, v to other nodes in h. In formal terms, instead of requiring $\mathcal{E}(y) = \{h\}$ for each $y \in h \setminus \{u, v\}$, we impose the following (weaker) condition:

$$\mathcal{E}_H(y) = \{h\}, \ \mathcal{E}_E(y) \subseteq \mathcal{E}_E(u) \cup \mathcal{E}_E(v) \qquad \qquad \forall y \in h \setminus \{u, v\}. \qquad (1)$$

Note that an edge $e = \{x, y\} \subseteq h \setminus \{u, v\}$ is not allowed according to (1). A further extension is obtained if we do not require u and v to belong to h, that is, we do not assume $\{u, v\} \subseteq h$ in (1).

We thus define the following reduction operation, where we assume that edges connecting nodes u and v to other nodes in h are deleted together with h.

hyperedge reduction $hyperedge(h, u, v)$: given a hyperedge h and two nodes u and v satisfying (1), delete h; for each node $y \in h \setminus \{u, v\}$ delete y and $\mathcal{E}_E(y)$; add to $\mathcal{E}$ an edge $\{u, v\}$.

The hypergraph on the left in Figure 2 shows the effect on the hypergraph in Figure 1 of reduction $hyperedge(a, u_3, u_4)$. Note that $u_4 \in a$, while $u_3 \notin a$; both nodes u_5 and u_6 have incident edges, namely: $\mathcal{E}_E(u_5) = \{c\} \subset \mathcal{E}_E(u_3)$, and $\mathcal{E}_E(u_6) = \{b\} \subset \mathcal{E}_E(u_4)$. Here, hyperedge a and edges b and c are replaced by a single edge $g = \{u_3, u_4\}$. Observe that edges generated by reduction operations are represented by lines (omitting triangles) in Figure 2.

We say that a hypergraph $\mathcal{H}$ is *reducible* if it can be reduced to a set of isolated nodes by means of a sequence of *series, parallel, tail* and *hyperedge* operations. In particular, we shall assume that we are given a reducible $\mathcal{H} = (\mathcal{V}, \mathcal{E})$, and a *reduction sequence* $R = \{r_1, r_2, \ldots, r_q\}$ of operations that reduce $\mathcal{H}$ to a single

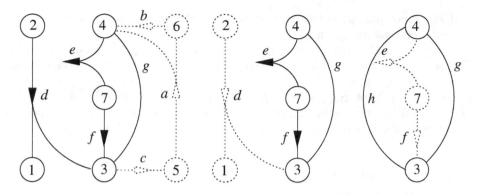

Fig. 2. A reduction sequence for hypergraph $\mathcal{H}$ in Figure 1

node. Let $|\mathcal{V}| = n$ and $|\mathcal{E}| = m$. Each reduction operation decreases the number of hyperarcs by at least one, therefore, at most m operations are needed. Moreover, there are at most $n-1$ *series*, *tail* and *hyperedge* operations, since these operations delete at least one node. A reduction sequence for the hypergraph $\mathcal{H}$ in Figure 1 is $R = \{hyperedge(a, u_3, u_4), tail(d, u_3), series(e, u_7, f), parallel(g, h), tail(g)\}$. The second and third hypergraph from the left in Figure 2 show the effect of operations $tail(d, u_3)$ and $series(e, u_7, f)$, respectively.

2.2   Comparison to Other Classes

We say that a CNF formula σ is *reducible* if the corresponding $\mathcal{H}_\sigma$ is reducible. We denote by *R-SAT* the class of reducible formulas. Here we show that R-SAT is incomparable to some well known tractable classes of formulas.

The *co-occurrence graph* $G_\sigma = (V_\sigma, E_\sigma)$ representing $\sigma = (\mathcal{X}, \mathcal{C})$ contains one node for each variable $x \in \mathcal{X}$, and one arc (x, x') for each pair of variables that appear in the same clause. A clause of length l in σ induces a clique of l nodes in G_σ. Therefore (see [4]) G_σ has treewidth $k \geq L - 1$, where L is the maximum clause length in σ. In particular, G_σ is a partial 2-tree only if $L \leq 3$. On the contrary, a reducible formula may contain clauses of length up to n. Moreover, if L is the maximum clause length in a reducible formula σ, G_σ may contain a clique of size $L+2$, resulting in a treewidth $k \geq L+1$. In particular, this happens if $\mathcal{H}_\sigma$ contains a hyperedge h of size L, an edge $\{u, v\}$ with $u, v \notin h$, and the edges $\{u, y\}$ and $\{v, y\}$ for each $y \in h$. Note that a reduction $hyperedge(h, u, v)$ can be applied.

Consider now the representation exploited in [7]. Given $\sigma = (\mathcal{X}, \mathcal{C})$, define a bipartite graph $B_\sigma = (O_\sigma, D_\sigma, E_\sigma)$ where $O_\sigma = \mathcal{X}$, $D_\sigma = \mathcal{C}$, and E_σ contains an arc (x, c) if either x or $\neg x$ appears in clause c. Observe that Figures 1 and 3 can be considered as a drawing of some graphs B_σ, where nodes in O_σ and D_σ are represented by circles and triangles, respectively. It is not difficult (see e.g. the above example of treewidth $k = L + 1$) to find examples where σ is reducible and B_σ is not planar; in this case, σ is not *nested* nor *co-nested*, according to the characterizations given in [7].

Next we show that the formula σ represented by the hypergraph on the left in Figure 3 cannot be solved by Algorithm SLUR [10]. Here σ contains the clauses $a = (\neg u_1 \vee u_2 \vee u_3)$, $b = (\neg u_1 \vee \neg u_4 \vee \neg u_5)$, $c = (\neg u_2 \vee u_3)$, $d = (\neg u_3 \vee u_4)$, and $e = (\neg u_3 \vee u_5)$. Suppose to assign to u_1 the value *true*; no contradictions are detected by unit resolution, however, the remaining formula is unsatisfiable, Observe that σ is satisfiable, and we have $u_1 = \textit{false}$ in each truth assignment satisfying σ. This shows that algorithm SLUR may fail on σ. Recall that balanced formulas can be solved by SLUR.

Finally, the reducible hypergraph on the right in Figure 3 corresponds to a formula that is not Q-Horn.

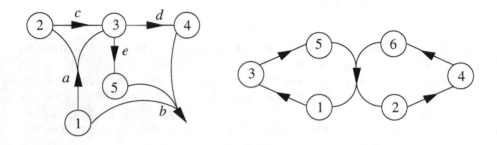

Fig. 3. Reducible formulas not in SLUR and not in Q-Horn

3   Algorithms

In this section we shall provide linear time algorithms for optimization and counting versions of SAT, and describe (omitting details) the reduction of probabilistic SAT to solving a system of linear inequalities. All the proposed methods share a common structure. We are given a reducible formula σ, the associated hypergraph $\mathcal{H} = (\mathcal{V}, \mathcal{E})$, and a reduction sequence $R = \{r_1, r_2, \ldots, r_q\}$ for $\mathcal{H}$. When a reduction operation is applied to $\mathcal{H}$, the informations associated with the deleted portion of the hypergraph are *projected* onto the remaining portion. In particular, we associate suitable values (weights, probabilities,...) to edges and nodes in the hypergraph; *series*, *parallel* and *hyperedge* reductions set or modify edge values, while *tail* deals with node values.

A *(partial) truth assignment* on σ is an assignment of values *true*, *false* to (a subset of) the variables in σ. Given a set $\bar{\mathcal{X}} \subseteq \mathcal{X}$ of k propositional variables, a partial truth assignment on $\bar{\mathcal{X}}$ can be seen as a set of k literals, containing either x or $\neg x$ for each $x \in \bar{\mathcal{X}}$. The set of 2^k partial truth assignments for $\bar{\mathcal{X}}$ is denoted by $W_{\bar{\mathcal{X}}}$. We shall denote by x and $\bar{x}$ the literals x or $\neg x$, respectively. Moreover, we shall identify nodes in $\mathcal{H}$ and variables in σ. Therefore, if $e = \{u, v\} \in \mathcal{E}$,

$W_e = W_{\{u,v\}}$ denotes the set of partial truth assignments to variables u and v. Finally, we denote by $C(a)$ the clause represented by $a \in \mathcal{E}$.

3.1 Optimization Satisfiability

We consider a general optimization version of SAT, where weights are associated with both nodes and clauses. Let $w(C) \in \mathbb{R}$ be the weight of clause $C \in \mathcal{C}$, and let $c(x) \in \mathbb{R}$ be the weight of variable $x \in \mathcal{X}$. Note that positive as well as negative weights are allowed. The goal is to maximize the sum of the weights of satisfied clauses and variables set to *true*. It is easy to show that both Max SAT and Min SAT (as well as SAT) arise as particular cases of the above problem. We safely assume that σ contains no unit clauses, since adding a unit clause $\{x\}$ or $\{\neg x\}$ with weight w is equivalent to adding to $c(x)$ a constant w or $-w$, respectively.

For each edge $e = \{u, v\}$ in $\mathcal{H}$ (including those generated by reductions) we introduce four values f_W^e, for $W \in W_e$, namely f_{uv}^e, $f_{\bar{u}v}^e$, $f_{u\bar{v}}^e$ and $f_{\bar{u}\bar{v}}^e$. Moreover, for each node u in $\mathcal{H}$ we maintain two values f_u and $f_{\bar{u}}$.

At the beginning, the values f represent the weights of binary clauses and variables. More precisely, given $W \in W_e$, we have $f_W = w(C(e))$ if W satisfies $C(e)$, and $f_W = 0$ otherwise. Moreover, we have $f_u = c(u)$ and $f_{\bar{u}} = 0$. At each reduction, the values f are set or updated in order to keep track of the maximum weight that can be obtained from the deleted part of $\mathcal{H}$. Let us consider in detail the values f computed by each reduction operation. Note that edges and hyperedges are treated in different ways in a *tail* reduction.

Series Let $e_1 = \{x, u\}$ and $e_2 = \{u, y\}$ in $series(e_1, u, e_2)$; $e = \{x, y\}$ is the returned edge. Let

$$f_{xy}^e := \max\left\{ (f_{xu}^{e_1} + f_{uy}^{e_2} + f_u), (f_{x\bar{u}}^{e_1} + f_{\bar{u}y}^{e_2} + f_{\bar{u}}) \right\};$$

$f_{x\bar{y}}^e$, $f_{\bar{x}y}^e$ and $f_{\bar{x}\bar{y}}^e$ are defined in a similar way.

Parallel Let $e_1 = \{x, y\}$ in $parallel(e_1, e_2)$; Let

$$f_{xy}^{e_1} := f_{xy}^{e_1} + f_{xy}^{e_2}$$

and update $f_{x\bar{y}}^{e_1}$, $f_{\bar{x}y}^{e_1}$ and $f_{\bar{x}\bar{y}}^{e_1}$ in a similar way.

Tail (edge) Let $e = \{u, v\}$ in $tail(e, u)$. Let

$$f_u := f_u + \max\left\{ (f_{uv}^e + f_v), (f_{u,\bar{v}}^e + f_{\bar{v}}) \right\};$$

and update $f_{\bar{u}}$ in a similar way.

Tail (hyperedge) Let $V = h \setminus \{u\}$ and $w_h = w(C(h))$ in $tail(h, u)$. Let $\overline{W}$ be the partial truth assignment on V yielding the maximum weight M:

$$M = \sum_{v \in V} \max\{f_v, f_{\bar{v}}\}.$$

Assume without loss of generality that setting $u = true$ satisfies $C(h)$: accordingly, let $F_u = M + w_h$ be the best weight obtained from h when $u = true$. In order to define $F_{\bar{u}}$, two cases must be considered. If $\overline{W}$ satisfies $C(h)$, we have $F_{\bar{u}} = F_u = M + w_h$. Otherwise, a truth assignment satisfying $C(h)$ can be obtained from $\overline{W}$ by "flipping" the assignment for a node $x \in V$ yielding a minimum weight decrease

$$\delta = \min_{v \in V} |f_v - f_{\bar{v}}| = |f_x - f_{\bar{x}}|;$$

in this case, $F_{\bar{u}}$ is the best of two options, namely, $C(h)$ satisfied or not:

$$F_{\bar{u}} = \max\left\{(M - \delta + w_h), M\right\}.$$

Set $f_u = f_u + F_u$ and $f_{\bar{u}} = f_{\bar{u}} + F_{\bar{u}}$.

Hyperedge: Let $e = \{u, v\}$ be the edge added by $hyperedge(h, u, v)$; let $V = h \setminus \{u, v\}$ and $w_h = w(C(h))$. We process each $W \in W_{\{u,v\}}$ separately; let us consider $u = v = true$. First we compute the best assignment for each node $x \in V$, taking into account the edges in $\mathcal{E}_E(x)$ incident with u or v. Define the weight F_x obtained setting $x = true$:

$$F_x = f_x + \sum_{a=\{u,x\}\in\mathcal{E}} f^a_{xu} + \sum_{a=\{v,x\}\in\mathcal{E}} f^a_{vx}$$

and define $F_{\bar{x}}$ in a similar way. Let $\overline{W}$ be the best assignment to nodes in V, that is, the one yielding an overall weight

$$M_{uv} = \sum_{x \in V} \max\left\{F_x, F_{\bar{x}}\right\}.$$

Suppose that clause $C(h)$ is satisfied by $\overline{W}$, or by setting $u = v = true$: in this case, set $f^e_{uv} = M_{uv} + w_h$. Otherwise, a truth assignment satisfying $C(h)$ can be obtained from $\overline{W}$ by flipping the assignment for a node in V at minimum loss

$$\delta = \min_{x \in V} |F_x - F_{\bar{x}}|;$$

f^e_{uv} is the best of two options, namely, $C(h)$ satisfied or not:

$$f^e_{uv} = \max\left\{(M_{uv} - \delta + w_h), M_{uv}\right\}.$$

At the end of the reduction sequence, if u is the remaining node, the optimal solution is given by $\max\{f_u, f_{\bar{u}}\}$. It is easy to see that each edge-reducing operation requires constant time, while tail and hyperedge reductions require $O(|h|)$ time. We can therefore state the following theorem.

Theorem 1. *SAT, Weighted Max SAT and Weighted Min SAT for a reducible formula σ can be solved in linear $O(\mathcal{L}(\sigma))$ time.*

3.2 Probabilistic Satisfiability

Probabilistic versions of SAT are based on the following principle: *consistent* probabilities of variables and clauses are the result of a *probability distribution* over the 2^n *possible worlds*, that is, truth assignments. The probability of a clause (a variable) is the sum of the probabilities of the possible worlds where the clause is satisfied (the variable is set to *true*, respectively). Given an assignment $\pi \in [0,1]^m$ of probabilities to clauses, the *probabilistic satisfiability* problem PSAT [8] consists in checking whether π is consistent. The *check of coherence* problem CPA [3] consists in checking whether an assignment $\eta \in [0,1]^n$ of probabilities to variables is consistent, with the further constraint that clauses have probability one, that is, only possible worlds satisfying the formula can be considered.

A solution method for formulas represented by partial 2-trees has been devised in [1]; this method is based on *probability variables*. Given a partial truth assignment $W \in W_{\bar{\mathcal{X}}}$, the probability variable p_W denotes the sum of the probability values assigned to the possible worlds where the variables in $\bar{\mathcal{X}}$ are set according to W. By introducing suitable probability variables, as well as linear constraints relating these variables to each other, one obtains a linear system S which has a solution if and only if the probability assignment is consistent. Moreover, the number of linear constraints and (nonnegative) variables is linear in the length of the formula.

As far as edges are concerned, the method described in [1] can be applied to reducible hypergraphs. We do not describe this method here. Instead, we show that the probability variables technique can be extended in order to deal with hyperedges.

Consider a hyperedge h where $|h| = k$; without loss of generality, assume that $C(h)$ contains nonnegated variables. Our goal is to relate the probability π_h assigned to $C(h)$ to the probabilities of variables in $C(h)$. For each node $v \in h$, the probability variable p_v denotes the probability of v being *true*. Clearly, $1 - p_v$ is the probability of v being *false*.

We introduce the following two conditions:

$$1 - \pi_h \leq \min_{v \in h} 1 - p_v; \tag{2}$$

$$\pi_h \leq \sum_{v \in h} p_v; \tag{3}$$

note that (2) maps into k linear inequalities $1 - \pi_h \leq 1 - p_v$, $v \in h$.

Assume that there are no constraints relating the probability variables p_v to each other; in other words, probabilities can be distributed arbitrarily over the 2^k partial truth assignments in W_h. In this situation, Conditions (2) and (3) are *necessary and sufficient* to conclude that π_h and the p_v are consistent, that is, they are the result of the same probability distribution.

Due to space limits, we cannot provide a formal proof of correctness for the above claim. An intuitive explanation can be given as follows. First, we must assign an overall probability $1 - \pi_h$ to the partial truth assignment where $C(h)$ is not satisfied, that is, all the variables in $C(h)$ are set to *false*. Clearly, this is possible if and only if Condition (2) holds. Moreover, we must assign an overall probability π_h to the possible worlds where at least one variable in $C(h)$ is set to *true*. This is possible if and only if (3) holds. To see why this is true, consider the extreme situation where (3) holds as an equality, i.e. $\pi_h = \sum_{v \in h} p_v$. In this case, for each $v \in h$ we assign probability p_v to the partial truth assignment where v is set to *true* and the other variables in $C(h)$ are set to *false*. Clearly, if $\pi_h < \sum_{v \in h} p_v$, we consider partial truth assignments where more than one variable is set to *true*.

In a *tail* reduction of a hyperedge h we add Conditions (2) and (3) to the linear system S. Indeed, there are no edges connecting nodes in h, which implies that no constraints relate the probability variables p_v, $v \in h$.

Consider now a reduction *hyperedge*(h, u, v). Here, there are no edges connecting nodes in $V = h \setminus \{u, v\}$, however, the variables p_x for $x \in V$ are related to the probability variables in $W_{\{u,v\}}$ due to the edges in $\mathcal{E}_E(u) \cup \mathcal{E}_E(v)$. Nevertheless, once one partial truth assignment W for u and v is fixed, the variables p_x are not related to each other. Therefore, it suffices to consider the four truth assignments in $W_{\{u,v\}}$ separately.

For each $W \in W_{\{u,v\}}$ we introduce a variable π_W, representing the probability of $C(h)$ *given that* W is chosen; note that π_W is *not* a probability variable. Moreover, for each $x \in V$ we introduce the probability variable p_{Wx} of x being *true* given W. Then, we adapt Conditions (2) and (3) to relate π_W to the variables p_{Wx}. Clearly, the equation $\pi_h = \pi_{uv} + \pi_{u\bar{v}} + \pi_{\bar{u}v} + \pi_{\bar{u}\bar{v}}$ must be added, together with the constraints (see [1]) relating each p_{Wx} to the probability variable p_W.

Each *tail* and *hyperedge* operations on a hyperedge h of size k require $O(|k|)$ new variables and constraints. Combining methods for edges and hyperedges, we obtain a system S with $O(\text{size}(\mathcal{H}))$ variables and constraints. We therefore state our second main result.

Proposition 1. *PSAT and CPA on a reducible formula σ can be reduced to solving a system of $O(\mathcal{L}(\sigma))$ non-negative variables and constraints.*

3.3   Counting Satisfying Assignments

We assume that σ does not contain unit clauses, otherwise, unit resolution may be applied. For each edge $e = \{u, v\}$ in $\mathcal{H}$ (including those generated by reductions) we introduce four *counting variables* N_W^e, for $W \in W_e$, namely N_{uv}^e, $N_{\bar{u}v}^e$, $N_{u\bar{v}}^e$ and $N_{\bar{u}\bar{v}}^e$. Moreover, for each node u in $\mathcal{H}$ we maintain two values N_u and $N_{\bar{u}}$.

At the beginning, the values N represent the satisfying assignments for clauses. More precisely, given $W \in W_e$, we have $N_W^e = 1$ if W satisfies $C(e)$, and $N_W = 0$ otherwise. Moreover, we have $N_u = N_{\bar{u}} = 1$. At each reduction, the values N are set or updated in order to keep track of the partial truth assignments satisfying the deleted part of $\mathcal{H}$. Let us consider in detail the operations performed for each reduction operation.

Series Let $e_1 = \{x, u\}$ and $e_2 = \{u, y\}$ in $series(e_1, u, e_2)$; $e = \{x, y\}$ is the returned edge. Let

$$N_{xy}^e := (N_{xu}^{e_1} \times N_{uy}^{e_2} \times N_u) + (N_{x\bar{u}}^{e_1} \times N_{\bar{u}y}^{e_2} \times N_{\bar{u}});$$

$N_{x\bar{y}}^e$, $N_{\bar{x}y}^e$ and $N_{\bar{x}\bar{y}}^e$ are defined in a similar way.

Parallel Let $e_1 = \{x, y\}$ in $parallel(e_1, e_2)$; Let

$$N_{xy}^{e_1} := N_{xy}^{e_1} \times N_{xy}^{e_2}$$

and update $N_{x\bar{y}}^{e_1}$, $N_{\bar{x}y}^{e_1}$ and $N_{\bar{x}\bar{y}}^{e_1}$ in a similar way.

Tail (edge) Let $e = \{u, v\}$ in $tail(e, u)$. Let

$$N_u := N_u \times \left(N_{uv}^e \times N_v + N_{u\bar{v}}^e \times N_{\bar{v}} \right)$$

and update $N_{\bar{u}}$ in a similar way.

Tail (hyperedge) Let $V = h \setminus \{u\}$ and $w_h = w(C(h))$ in $tail(h, u)$. Assume without loss of generality that $C(h)$ contains nonnegated variables. First we compute the total number N_V of assignments obtained from nodes in V; N_V is the sum over all $W \in \mathcal{W}_V$ of the number of satisfying assignments, counted so far, where variables in V are fixed according to W; more precisely:

$$N_V = \sum_{W \in \mathcal{W}_V} \prod_{l \in W} N_l = \prod_{v \in V} (N_v + N_{\bar{v}}).$$

Since we assumed that setting $u = true$ satisfies $C(h)$, we let $N_u = N_V$. In order to update $N_{\bar{u}}$, we must compute the number N_h of truth assignments in $\mathcal{W}_V$ that falsify $C(h)$ when $u = false$:

$$N_h = \prod_{v \in V} N_{\bar{v}}.$$

Then we let $N_{\bar{u}} = (N_V - N_h)$.

Hyperedge: Let $e = \{u, v\}$ be the edge added by $hyperedge(h, u, v)$, $V = h \setminus \{u, v\}$ and $w_h = w(C(h))$. We process each $W \in \mathcal{W}_{\{u,v\}}$ separately; let us consider the case $u = v = true$. First, we consider each node $x \in V$, taking into account the edges in $\mathcal{E}_E(x)$ incident with u or v. The total number of satisfying assignments, counted so far, where $u = v = x = true$ is given by

$$D_x = N_x \times \prod_{a=\{u,x\} \in \mathcal{E}} N_{ux}^a \times \prod_{a=\{v,x\} \in \mathcal{E}} N_{vx}^a;$$

the number $D_{\bar{x}}$ is computed in a similar way. Then we count the total number N_V of satisfying assignments for nodes in V:

$$N_V = \sum_{W \in \mathcal{W}_V} \prod_{l \in W} D_l = \prod_{x \in V} (D_x + D_{\bar{x}}).$$

Suppose that setting $u = v = true$ satisfies clause $C(h)$; in this case, we set $N_{uv}^e = N_V$. Otherwise, in order to compute N_{uv}^e we must count the number N_h of assignments where $C(h)$ is false. Assume without loss of generality that $C(h)$ is falsified by setting $x = true$ for each $x \in V$; we have:

$$N_h = \prod_{v \in V} D_v.$$

Finally we let $N_{uv} = (N_V - N_h)$.

At the end of the reduction sequence, if u is the remaining node, the total number of satisfying assignments is given by $N_u + N_{\bar{u}}$. It is easy to see that each edge-reducing operation requires constant time, while tail and hyperedge reductions require $O(|h|)$ time. We can therefore state our last result.

Theorem 2. *#SAT for a reducible formula σ can be solved in linear $O(\mathcal{L}(\sigma))$ time.*

4   Further Work

In this paper we defined a class of CNF formulas that are easy for several propositional satisfiability problems. The solution methods proposed here are based on a suitable *elimination sequence* for variables and clauses; similar techniques have been exploited quite often in the SAT literature, see e.g. [1,5,6]. Our contribution consists in treating "long" clauses in a particular way, which makes the computational cost of each elimination linear (instead of exponential) in the clause length. As a result, the computational complexity of solving SAT, as well as optimization and counting versions, becomes linear both in the length and in the treewidth of the formula. For probabilistic versions of SAT, a linear bound on the size of a particular linear system can be obtained.

Though not addressed here, a formal definition of reducible hypergraphs can be given by means of *edge expansion* operations, or equivalently, in terms of *hypergraph composition*. We are thus faced with the *recognition problem* for this class, that is: find a reduction sequence for a given hypergraph, or prove that none exists. We leave it as an open question whether a fast (e.g. linear time) recognition algorithm for reducible hypergraphs can be devised.

In this paper, we combined known reduction techniques for partial 2-trees and special techniques for reducing hyperedge to edges. Possibly, a wider class of tractable formulas may be defined by more complex reduction rules. The goal is to define a general class of "partial 2-hypertrees", that is, a sort of hypergraph counterpart of partial 2-trees. A further possibility is to adopt reduction techniques for partial k-trees, for $k > 2$. In the long term, this may lead to a hierarchy of tractable formulas corresponding to "partial k-hypertrees". This direction of research is closely related to *fixed parameter tractable* classes of SAT problems, see e.g. [12].

The concept of treewidth, which is at the core of our approach, turns out to be crucial also in relation with Bayesian inference; see [2] for a detailed discussion.

It is an interesting open question whether our techniques may have an impact in this broader context.

One final comment: as for many other tractable subclasses, the impact of reducible formulas on the practice of SAT solving might be questionable. Two ideas, however, may be worth investigating: applying reduction operations within SAT solvers, and finding relevant benchmarks, such as SAT encodings of real life problems, that are close enough to a reducible formula.

Acknowledgements

I wish to thank the Referees for their many helpful comments, and for pointing out reference [2].

References

1. K. A. Andersen and D. Pretolani, "Easy Cases of Probabilistic Satisfiability", *Annals of Mathematics and A.I.* **33** No. 1 (2001) 69–91.
2. F. Bacchus, S. Dalmao, T. Pitassi, "DPLL with Caching: A new algorithm for #SAT and Bayesian Inference", ECCC Report TR03-003 (2003). Available at `http://eccc.uni-trier.de/eccc-local/Lists/TR-2003.html`
3. M. Baioletti, A. Capotorti, S. Tulipani and B. Vantaggi, "Elimination of Boolean variables for Probabilistic Coherence", *Soft Computing* **4** No. 2 (2000) 81-88.
4. H. L. Bodlaender, "A partial k-arboretum of graphs with bounded treewidth", *Theoretical Computer Science* **209** (1998) 1–46.
5. Y. Crama, P. Hansen and B. Jaumard, "The Basic Algorithm for pseudo-Boolean programming revisited", *Discrete Applied Mathematics* **29** (1990) 171–185.
6. D. Knuth, "Nested Satisfiability", *Acta Informatica* **28** (1990) 1–6.
7. J. Kratochvíl and M. Křivánek, "Satisfiability of co-nested formulas", *Acta Informatica* **30** (1993) 397-403.
8. N. J. Nilsson, "Probabilistic Logic", *Artificial Intelligence* **28** (1986) 71–87.
9. D. Pretolani, "Probability Logic and Optimization SAT: the PSAT and CPA Models", presented at SAT 2002, Cincinnati (Ohio) May 2002. Accepted for publication in *Annals of Mathematics and A.I.*.
10. J. S. Schlipf, F. S. Annexstein, J. V. Franco and R. P. Swaminathan, "On finding solutions for extended Horn formulas", *Information Processing Letters* **54** (1995) 133-137.
11. H. van Maaren, "A Short Note on Some Tractable Cases of the Satisfiability Problem", *Information and Computation* **158** (2000) 125–130.
12. S. Szeider, "On fixed-parameter tractable parametrizations of SAT", proceedings of the Sixth Int. Conf. on Theory and Applications of Satisfiability Testing, S. Margherita Ligure (Italy) May 5–8, 2003, 108–116.

Appendix: Examples

Let us consider the formula represented by the reducible hypergraph in Figure 1. We show the effect of the operations in the reduction sequence. For the sake of simplicity, we denote by i and $\bar{i}$ the literals u_i and $\bar{u}_i$.

Optimization Satisfiability

We assign a weight 1 to each node, and a weight $|C|$ to each clause C.

(1) $hyperedge(a, u_3, u_4)$: $V = \{u_5, u_6\}$; the values F for u_5 and u_6 are determined by clauses $c = (\neg u_3 \vee u_5)$ and $b = (\neg u_6 \vee u_4)$, respectively; $C(a)$ is satisfied by $u_4 = true$, while for $u_4 = false$, $\overline{W}$ falsifies $C(a)$, and δ must be computed;
  - $u_3 = u_4 = true$: $F_5 = f_{35}^c + f_5 = 2 + 1 = 3$, $F_{\bar{5}} = f_{35}^c + f_{\bar{5}} = 0 + 0 = 0$;
    $F_6 = f_{46}^b + f_6 = 2 + 1 = 3$, $F_{\bar{6}} = f_{46}^b + f_{\bar{6}} = 2 + 0 = 2$;
    $f_{34}^g = w_a + M_{34} = 3 + \max\{3, 0\} + \max\{3, 2\} = 9$;
  - $u_3 = true, u_4 = false$: $F_5 = f_{35}^c + f_5 = 2 + 1 = 3$, $F_{\bar{5}} = f_{35}^c + f_{\bar{5}} = 0 + 0 = 0$;
    $F_6 = f_{46}^b + f_6 = 0 + 1 = 1$, $F_{\bar{6}} = f_{46}^b + f_{\bar{6}} = 2 + 0 = 2$;
    $M_{3\bar{4}} = \max\{3, 0\} + \max\{1, 2\} = 5 = F_5 + F_{\bar{6}}$; $\delta = \min\{3 - 0, 2 - 1\} = 1$;
    $f_{34}^g = \max\{(M_{3\bar{4}} - \delta + w_a), M_{3\bar{4}}\} = \max\{7, 5\} = 7$;
  - $u_3 = false, u_4 = true$: $F_5 = f_{35}^c + f_5 = 2 + 1 = 3$, $F_{\bar{5}} = f_{35}^c + f_{\bar{5}} = 2 + 0 = 2$;
    $F_6 = f_{46}^b + f_6 = 2 + 1 = 3$, $F_{\bar{6}} = f_{46}^b + f_{\bar{6}} = 2 + 0 = 2$;
    $f_{34}^g = w_a + M_{34} = 3 + \max\{3, 2\} + \max\{3, 2\} = 9$;
  - $u_3 = false, u_4 = false$:
    $F_5 = f_{35}^c + f_5 = 2 + 1 = 3$, $F_{\bar{5}} = f_{35}^c + f_{\bar{5}} = 2 + 0 = 2$;
    $F_6 = f_{46}^b + f_6 = 0 + 1 = 1$, $F_{\bar{6}} = f_{46}^b + f_{\bar{6}} = 2 + 0 = 2$;
    $M_{3\bar{4}} = \max\{3, 2\} + \max\{1, 2\} = 5 = F_5 + F_{\bar{6}}$; $\delta = \min\{3 - 2, 2 - 1\} = 1$;
    $f_{34}^g = \max\{(M_{3\bar{4}} - \delta + w_a), M_{3\bar{4}}\} = \max\{7, 5\} = 7$.

(2) $tail(d, u_3)$: $V = \{u_1, u_2\}$; $C(d)$ is satisfied by $u_3 = true$; $M = f_1 + f_2 = 2$; $\overline{W}$ is $u_1 = u_2 = true$, which satisfies $C(d)$;
  - $u_3 = true$: $f_3 := f_3 + M + w_d = 1 + 2 + 3 = 6$;
  - $u_3 = false$: $f_{\bar{3}} := f_{\bar{3}} + M + w_d = 0 + 2 + 3 = 5$;

(3) $series(e, u_7, f)$: $h = \{u_3, u_4\}$ is the added edge:
  - $u_3 = u_4 = true$: $f_{34}^h = \max\{(f_{47}^e + f_{37}^f + f_7), (f_{4\bar{7}}^e + f_{3\bar{7}}^f + f_{\bar{7}})\} = \max\{(0 + 2 + 1), (2 + 2 + 0)\} = 4$;
  - $u_3 = true, u_4 = false$: $f_{3\bar{4}}^h = \max\{(f_{\bar{4}7}^e + f_{37}^f + f_7), (f_{\bar{4}\bar{7}}^e + f_{3\bar{7}}^f + f_{\bar{7}})\} = \max\{(2 + 2 + 1), (2 + 2 + 0)\} = 5$;
  - $u_3 = false, u_4 = true$: $f_{\bar{3}4}^h = \max\{(f_{47}^e + f_{\bar{3}7}^f + f_7), (f_{4\bar{7}}^e + f_{\bar{3}\bar{7}}^f + f_{\bar{7}})\} = \max\{(0 + 0 + 1), (2 + 2 + 0)\} = 4$;
  - $u_3 = false, u_4 = false$: $f_{\bar{3}\bar{4}}^h = \max\{(f_{\bar{4}7}^e + f_{\bar{3}7}^f + f_7), (f_{\bar{4}\bar{7}}^e + f_{\bar{3}\bar{7}}^f + f_{\bar{7}})\} = \max\{(2 + 0 + 1), (2 + 2 + 0)\} = 4$;

(4) $parallel(g, h)$:
  - $f_{34}^g = f_{34}^g + f_{34}^h = 9 + 4 = 13$;

- $f_{34}^g = f_{34}^g + f_{34}^h = 7 + 5 = 12$;
- $f_{\overline{34}}^g = f_{\overline{34}}^g + f_{\overline{34}}^h = 9 + 4 = 13$;
- $f_{3\overline{4}}^g = f_{3\overline{4}}^g + f_{3\overline{4}}^h = 7 + 4 = 11$;

(5) $tail(g, u_3)$:
- $f_3 := f_3 + \max\{(f_{34}^g + f_4), (f_{3\overline{4}}^g + f_{\overline{4}})\} = 6 + \max\{(13 + 1), (12 + 0)\} = 20$;
- $f_{\overline{3}} := f_{\overline{3}} + \max\{(f_{\overline{34}}^g + f_4), (f_{\overline{34}}^g + f_{\overline{4}})\} = 5 + \max\{(13 + 1), (11 + 0)\} = 19$;

The maximum weight is $\max\{f_3, f_{\overline{3}}\} = 20$.

Counting Satisfying Assignments

(1) $hyperedge(a, u_3, u_4)$: $V = \{u_5, u_6\}$; the values D for u_5 and u_6 are determined by clauses $c = (\neg u_3 \vee u_5)$ and $b = (\neg u_6 \vee u_4)$, respectively; $C(a)$ is satisfied by $u_4 = true$, while for $u_4 = false$ we have $N_a = D_5 \times D_{\overline{6}}$ assignments falsifying $C(a)$;
- $u_3 = u_4 = true$: $D_5 = 1$, $D_{\overline{5}} = 0$, $D_6 = D_{\overline{6}} = 1$; $N_{34}^g = N_V = 1 \times 2 = 2$;
- $u_3 = true$, $u_4 = false$: $D_5 = 1$, $D_{\overline{5}} = 0$, $D_6 = 0$, $D_{\overline{6}} = 1$;
  $N_{3\overline{4}}^g = N_V - N_a = (1 \times 1) - D_5 \times D_{\overline{6}} = 0$;
- $u_3 = false$, $u_4 = true$: $D_5 = D_{\overline{5}} = D_6 = D_{\overline{6}} = 1$; $N_{\overline{3}4}^g = N_V = (2 \times 2) = 4$;
- $u_3 = false$, $u_4 = false$: $D_5 = D_{\overline{5}} = 1$, $D_6 = 0$, $D_{\overline{6}} = 1$;
  $N_{\overline{34}}^g = N_V - N_a = (2 \times 1) - D_5 \times D_{\overline{6}} = 1$.

(2) $tail(d, u_3)$: $V = \{u_1, u_2\}$; $C(d)$ is satisfied by $u_3 = true$; for $u_3 = false$, $C(d)$ is falsified by $u_1 = false$ and $u_2 = true$;
- $u_3 = true$: $N_3 = N_3 \times N_V = 1 \times 2 \times 2 = 4$;
- $u_3 = false$: $N_{\overline{3}} = N_{\overline{3}} \times (N_V - N_h) = 4 - (N_{\overline{1}} \times N_2) = 3$;

(3) $series(e, u_7, f)$: $h = \{u_3, u_4\}$ is the added edge;
- $u_3 = u_4 = true$: $N_{34}^h = N_{47}^e N_{37}^f N_7 + N_{4\overline{7}}^e N_{3\overline{7}}^f N_{\overline{7}} = 0 + 1 = 1$;
- $u_3 = true$, $u_4 = false$: $N_{3\overline{4}}^h = N_{\overline{4}7}^e N_{37}^f N_7 + N_{\overline{4}\overline{7}}^e N_{3\overline{7}}^f N_{\overline{7}} = 1 + 1 = 2$;
- $u_3 = false$, $u_4 = true$: $N_{\overline{3}4}^h = N_{47}^e N_{\overline{3}7}^f N_7 + N_{4\overline{7}}^e N_{\overline{3}\overline{7}}^f N_{\overline{7}} = 0 + 1 = 1$;
- $u_3 = u_4 = false$: $N_{\overline{34}}^h = N_{\overline{4}7}^e N_{\overline{3}7}^f N_7 + N_{\overline{4}\overline{7}}^e N_{\overline{3}\overline{7}}^f N_{\overline{7}} = 0 + 1 = 1$;

(4) $parallel(g, h)$:
- $N_{34}^g = N_{34}^g \times N_{34}^h = 2 \times 1 = 2$
- $N_{3\overline{4}}^g = N_{3\overline{4}}^g \times N_{3\overline{4}}^h = 0 \times 2 = 0$;
- $N_{\overline{3}4}^g = N_{\overline{3}4}^g \times N_{\overline{3}4}^h = 4 \times 1 = 4$;
- $N_{\overline{34}}^g = N_{\overline{34}}^g \times N_{\overline{34}}^h = 1 \times 1 = 1$;

(5) $tail(g, u_3)$:
- $N_3 = N_3 \times (N_{34}^g N_4 + N_{3\overline{4}}^g N_{\overline{4}}) = 4 \times (2 + 0) = 8$;
- $N_{\overline{3}} = N_{\overline{3}} \times (N_{\overline{3}4}^g N_4 + N_{\overline{34}}^g N_{\overline{4}}) = 3 \times (4 + 1) = 15$.

The number of satisfying assignments is $N_3 + N_{\overline{3}} = 23$.

SBSAT: a State-Based, BDD-Based Satisfiability Solver

John Franco[1], Michal Kouril[1], John Schlipf[1], Jeffrey Ward[1], Sean Weaver[1], Michael Dransfield[2], and W. Mark Vanfleet[2]

[1] The University of Cincinnati, Cincinnati 45221-0030, USA
John.Schlipf@UC.Edu
[2] Department of Defense, United States of America

Abstract. We present a new approach to SAT solvers, supporting efficient implementation of highly sophisticated search heuristics over a range of propositional inputs, including CNF formulas, but particularly sets of arbitrary boolean constraints, represented as BDDs. The approach preprocesses the BDDs into state machines to allow for fast inferences based upon any single input constraint. It also simplifies the set of constraints, using a tool set similar to standard BDD engines. And it memoizes search information during an extensive preprocessing phase, allowing for a new form of lookahead, called *local-function-complete* lookahead. This approach has been incorporated, along with existing tools such as lemmas, to build a SAT tool we call SBSAT.

Because of its memoization of constraint-by-constraint lookahead information, besides incorporation of standard lemmas, SBSAT also provides a convenient platform for experimentation with search heuristics. This experimentation is ongoing.

We show the feasibility of SBSAT by comparing it to zChaff on several of the benchmarks. We also show an interesting dependence of some standard benchmarks upon simply the independent/dependent variable distinction.

Keywords: Satisfiability, State Machine, Binary Decision Diagram, DAG.

1 Introduction

Recent development of technologies for solving the propositional Satisfiability (PSAT) problem has been so successful it has captured the attention of people working in areas as diverse as theoretical physics and computer engineering. Significant among those technologies are the use of conflict-resolution or lemmas to turn a tree search into a DAG search, the development of advanced "lemma heuristics" for choosing the "best" lemmas, partial lookahead for information that can be used to enhance "search heuristics", non-chronological backtracking, and advanced data structures [LMS02, ZMMM01]. Newer technologies, *e.g.* based on autarkies [Kul98] and symmetry [Gol02, GN02], show great promise.

But many PSAT problems still are difficult, and many of those do not naturally appear as CNF problems. One can translate them to CNF and apply a

E. Giunchiglia and A. Tacchella (Eds.): SAT 2003, LNCS 2919, pp. 398–410, 2004.

CNF solver. This translation need not expand the formula by more than a constant factor [Sch89], but new variables must be added to achieve this. Moreover, some information may be hidden by the translation, such as clustering of dependencies or distinctions between so-called independent and dependent variables. Exploiting this information may speed up search.

Simply to emphasize that we are allowing non-CNF input, we shall refer to our problems as *PSAT* rather than just SAT. One standard representation for complex boolean functions is the *Reduced Ordered Binary Decision Diagram* (*BDD*) [Bry86]. A BDD is a canonical DAG representation of a boolean function in terms of constants 0 and 1 and the if-then-else operator *ite*. For example, the tree representation of $ite(x_1, ite(x_2, 1, 0), 0)$, with the two 0 nodes are merged, is the BDD expressing $x_1 \wedge x_2$.

Many problems in microprocessor design, for example, such as in design verification and interconnect synthesis, are more naturally expressed with BDD constraints rather than CNF constraints. Standard logical operations are easily implemented on BDDs, and BDDs have been used successfully in many cases over the past 10 years. However, as the number of variables grow, BDDs can grow exponentially, limiting pure BDD methods.

An obvious next step is to develop a hybrid algorithm, combining BDD tools and Davis-Putnam-Loveland-Logemann (*DPLL*) search [DLL62].[3] We propose here a new variety of hybrid. Typically, the input will be a PSAT problem expressed in terms of BDD's. We also take full advantage of the huge memory capacity now routinely available on general purpose, low-cost computers, to precompute (compute once, before starting the brancher) as much as feasible. (Preprocessing in SAT solvers is not new, but our automata (see Sect. 3) provide new ways to use memoization.)

1. We do as much BDD-type preprocessing as is feasible. We created a new BDD operation, called *strengthening*, for simplifying a collection of BDDs while avoiding size explosion (Sect. 2).
2. Before applying a DPLL-style search procedure, we precompute as much static information as possible, to speed backtracking. (Sect. 3.)
3. We propose a new search heuristic for choosing the next variable to assign a value to. Since BDDs encode more complex interrelationships than do CNF formulas, we can precompute *complete* lookahead information on individual BDDs, considering all partial truth assignments for a single input BDD. We can then combine that local (single BDD at a time) lookahead information during the search. We call this *local-function-complete* lookahead. (See

---

[.] For other approaches, see, *e.g.* [PG96, GA98, PK00, KZCH00].

Sect. 3.) This is critical because, since we do not break inputs into CNF, there is more information in a *single* BDD than in a *single* clause.

2   BDD Preprocessing

SBSAT first preprocesses the input formulas, typically BDDs, before preparing for the DPLL-type search. We borrow and modify techniques from BDD solvers, avoiding techniques that will explode BDD size.

If input is given in CNF (as in some benchmarks), we first cluster the clauses into BDD's as far as is feasible. Because of processing limitations (see Sect. 3), we limit our clusters to at most 17 variables except for a few special types of functions. On the `dlx` problem set [Vel00], which was made available to us in both CNF and CMU's less processed *trace* format, we found we were able to recapture most of the structure present in the trace format but lost in the CNF.

The first operation is quite primitive: Individual BDDs not broken into CNF formulas may force some variables to be true or false or force some equivalences $x_i \leftrightarrow x_j$ or $x_i \oplus x_j$ (here $\oplus$ is exclusive or). All such inferences from a single BDD may easily be found by a depth-first search of the BDD. By contrast, if we translated first to CNF, a few such inferences would appear as unit clauses, but most would be spread across several different clauses and thus no longer be obvious. We identify such inferences and simplify the input. Moreover, after every simplification we repeat this process, since one simplification of the constraints may make others apparent.

If a variable appears in only one BDD — in only b_1 among $b_1, \ldots, b_m$ below — we may use Boolean *existential quantification*:

$$\exists x_i(b_1 \wedge \cdots \wedge b_k) \text{ is logically equivalent to } \exists x_i(b_1) \wedge b_2 \wedge \cdots \wedge b_m.$$

We now search for a satisfying assignment for $\exists x_i(b_1) \wedge b_2 \wedge \cdots \wedge b_m$.. Since the latter problem has fewer variables than the original, it is *usually* easier to search. If the smaller problem is unsatisfiable, so is the larger. And any satisfying assignment I to the smaller problem can be expanded to a satisfying assignment to the larger — just choose any value x_i to satisfy b_i (along with the assignments in I). Existential quantification is used by BDD solvers in the same way.

A simple BDD solver, given BDDs $b_1, b_2, \ldots, b_k$, may just conjoin them, at the cost of potential explosive size growth. We made a simple adaptation of that, which we call *strengthening*: Given two BDDs, b_1 and b_2, conjoin b_1 with the *projection* of b_2 onto the variables of b_1 — $b_1 \wedge \exists \boldsymbol{x} b_2$, where $\boldsymbol{x}$ is the set of variables appearing in b_2 but not in b_1. Strengthening each b_i against all other b_js sometimes lets us infer additional literals or equivalences as above.

At other times it is useful to decouple conjunctions. (Of course, there is no unique way to do this.) We use the *restrict* function of [CM90, CM91] to "prune out" duplicated logic, removing from f all branches that contradict the constraint c.[4]

There appear to be two gains to restricting. It can make our state machines (see Sect. 3) smaller. And it often *appears*, by avoiding duplicated information, to make our local-function-complete lookahead heuristic's evidence combination rule work better. (However, it can, in odd cases, also lose local information.) Sometimes it also reveals some of the inferences that strengthening would.

Since most BDD-type preprocessing is sometimes useful and sometimes not, we provide user control for how much to do, presuming the user can learn what works well on what classes of problems.

3   State Machines

We normally preprocess boolean constraints into acyclic Mealy machines, called SMURFs (for "State Machine Used to Represent Functions"). We may assume each constraint implies no literals, since those would have been trapped during preprocessing. The inputs to SMURFs are literals assumed or inferred to be true; the outputs are sets of literals that are forced to be true (analogous to unit resolution in CNF); and the states correspond to what "portion" — or *residual* — of the constraint remains to be satisfied. SMURFs are described in Fig. 3. For a set of constraint BDDs, we compute the SMURFs for each of the separate BDDs and merge states with equal residual functions, maintaining one pointer into the resultant automaton for the current state of each constraint.

The SMURF structure described, for a boolean formula with n variables, can have, in the worst case, close to 3^n states. Thus an Achilles' heel of SBSAT can be handling long input functions. In most benchmarks, that has been little practical problem: all individual constraint are reasonably short — *except*[5] for a small special group of functions: long clauses, long exclusive disjunctions, and "assignments" $\lambda_0 = \lambda_1 \wedge \cdots \wedge \lambda_k$ and $\lambda_0 = \lambda_1 \vee \cdots \vee \lambda_k$ (where the λ_i's are literals). To solve the space problem for these special functions, we create special data structures; these take little space and can simulate the SMURFs for the functions exactly with little time loss. For a long clause we store only (i) whether the clause is already satisfied, and (ii) how many literals are currently not assigned truth values. Storing exclusive disjuncts is similar. For the assignments, we store both the value (0,1, or undefined) of the left-hand-side literal and the number of right-hand-side literals with undefined truth values.

· *Restrict* is similar to the function called generalized cofactor (gcf) or constrain [Bra90, CM91]. Both $restrict(f, c)$ and $gcf(f, c)$ agree with f *on interpretations where c is satisfied*, but are generally somehow simpler than f. Both are highly dependent upon variable ordering, so both might be considered "non-logical." We choose *restrict* because the BDDs produced tend to be smaller.

· as In, for example, dlx benchmark suite made available by Miroslaw Velev [Vel00].

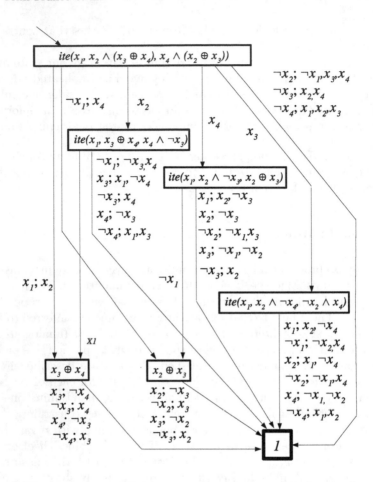

Fig. 1. We preprocess BDDs into deterministic Mealy machines called "SMURFs." This example explains construction. *ite* denotes if-then-else and ⊕ denotes exclusive or.

The SMURF represents $ite(x_1, x_2 \wedge (x_3 \oplus x_4), x_4 \wedge (x_2 \oplus x_3))$. It represents, in part, BDDs for the function under all possible variable orderings — since we cannot know in what order the brancher considers the variables.

The start state (upper left) represents the original function. On the left is a transition from the start state labeled "$x_1; x_2$"; this means that, from that state, on input x_1, the automaton makes a transition and outputs $\{x_2\}$. If the brancher guesses, or infers, that x_1 is true, it will "tell" the automaton to branch on x_1. The output of $\{x_2\}$ tells the brancher that x_2 must also be true — the analogue of unit inference in CNF. This transition goes to a state labeled $x_3 \oplus x_4$, meaning that, after x_1, x_2 are set to 1, what remains to be satisfied — the *residual function* — is $x_3 \oplus x_4$. On the upper right are three transitions shown with one arrow. The first is from the start state on input $\neg x_2$; it outputs $\{\neg x_1, x_3, x_4\}$ and goes to state 1 — meaning the original BDD is now satisfied, *i.e.*, that there is no residual constraint to satisfy.

4   The LSGB Heuristic

We also precompute information for choosing variables to branch on. The *weight* of a transition counts the number of literals forced on the transition, plus the expected number of literals forced below that state, where a forced literal after m additional choices is weighted $1/K^m$. (K, set experimentally, is currently 3.) In Fig. 3, the transition out of the start state on $\neg x_1$ has weight

$$1 + (\frac{1}{K} + \frac{1}{K} + \frac{1}{K} + \frac{1}{K})/4;$$

the transition out on x_4,

$$0 + (\frac{1}{K^2} + \frac{2}{K} + \frac{1}{K} + \frac{2}{K} + \frac{2}{K} + \frac{1}{K})/6.$$

At brancher time we need only look up these individual weights in a table.

For the special data structures defined above, the calculation above is simulated. If a disjunction $\lambda_1 \vee \cdots \vee \lambda_m$ with k still unassigned variables were represented as a SMURF, the weight of λ_i is 0 (since the clause immediately becomes satisfied, nothing more can be forced), and the weight of $\neg\lambda_i$ is $1/(2K)^{k-1}$. We directly code this in our simulated SMURF. Exclusive disjunctions are similar. Assignments are similar but break into cases; one recurrence relation is hard to solve, so we just precompute the weights as a function of the number of unassigned λ_i's and look them up at branch time.

Our "locally-skewed, globally-balanced" ($LSGB$) search heuristic is similar to the "Johnson heuristic" on CNF formulas where $K = 2$. The intuition is to branch toward forced inferences as quickly as possible to narrow the search space (or get lemmas fast). To pick the next variable to branch on: For each variable x_i, compute (i) the sum S_i^+ of the *weights* of transitions on x_i out of all current SMURF states and (ii) the sum S_i^- of the *weights* of transitions on $\neg x_i$. A high sum represents a high "payoff." For an ideal branching variable x_i, both x_i and $\neg x_i$ force many literals, so we branch on the variable x_i where $S_i^+ \cdot S_i^-$ is maximum. For that variable, branch first toward the larger of S_i^+, S_i^-.[6]

There are circumstances where other search heuristics are known to work well. LSGB was intended for applications where not much is known about — or easily determined about — the given PSAT problem. We show it performs well there. If the problem is known to have a lot of exploitable structure, it may be better to specify a different heuristic. We allow the experienced user some choice. The SMURF structure admits such heuristics as well; on a simple heuristic, it may not be needed, but (except for preprocessing time) it does not hinder either. Work is needed on hybrid heuristics.

In Sect. 6, we present benchmark problems where zChaff substantially outperforms SBSAT with the LSGB heuristic, and benchmarks where SBSAT with LSGB substantially outperforms zChaff. And we conjecture what features of a problem may make it easy or difficult for zChaff.

---

[*] We borrow the idea of taking the product from Freeman [Fre95].

5   Lemmas

Except for data structures and search heuristics, SBSAT generalizes standard
DPLL-type searches. Having SMURFs output forced literals allows generalizing
unit clause propagation. SBSAT also makes extensive use of backjumping, recent
advanced data structures, and lemmas.

SBSAT creates clause lemmas, not BDD lemmas, for efficiency. It creates
lemmas lazily — during branching rather than precomputation, memoizing them
in a lemma cache to avoid regeneration. SBSAT creates a lemma when a literal
is forced and resolves lemmas during backtracking. Lemmas that seem useful
are cached. During the search, if a partial assignment negates all but one literal
of a clause, that last literal is inferred true. SBSAT uses a modified Chaff-type
data structures for the cache [LMS02]. Chaff versions restart when they fills its
lemma cache; SBSAT continues, deleting the lemma least recently used. More
work is needed here.

6   Computational Results

Our primary interest has been to provide a platform for sophisticated search
heuristics. We report work on three sets of benchmark problems: a set of bounded
model checking (BMC) problems, our own set of random BMC-type problems,
and the dlx benchmark suite.[7] We compared SBSAT to zChaff [MMZZM01], the
most successful CNF solver to date on several of the dlx problem sets.

6.1   Bounded Model Checking (BMC)

Model checking involves checking temporal logic properties on a finite Kripke
model (basically, finite state automaton, with states representing possible states
of the device, plus a function mapping states to proposition letters true in those
states). The transition function represents which states can move to which other
states in unit time. The transition function might be intrinsically nondeterministic, or there might be nondeterminism reflecting inputs from outside the system
which are not represented in the description. *Linear time temporal logic (LTL)*
expresses properties of a single possible time-sequence of states (as opposed to
considering branching time possibilities). Typical statements include $p \lor q$ (p or
q is true in the current time state), Xp (p is true in the next time state), Gp
(p is true in all future states), and Fp (p is true in some future state). Connectives, both boolean and temporal, can be nested. Statements are used to assert
that some/every sequence of legal steps satisfies some key properties. In *bounded
model checking* [BCCZ], the temporal conditions are checked for only a fixed
number of time steps.

If only k time steps are to be traced, BMC essentially makes k copies of
the automaton and translates temporal statements as boolean combinations of

---

[*] Made available by Miroslaw Velev [Vel00].

statements about individual time steps. The states can of course also be encoded with vectors of boolean variables: for m variables per state and k time steps, there are km variables for state-time. The transition function can be represented as boolean conditions between adjacent states. Thus the BMC translation of whether there is a "run" of the system satisfying the constraints becomes a boolean satisfiability problem.

With benchmark BMC problems, SBSAT has suffered, as often happens, from working with CNF input files, since then it must reconstruct what a likely raw form input might have been. (We have no way to estimate how much faster SBSAT would have been had we had input in more complicated BDDs.) Our results have been mixed. Our best results, in Table 6, came from an encoding of verifying a multiplier.[8] The final number, 04, 08, *etc.*, refers to the k above — the number of generations of execution traced; to save space we include only examples for k divisible by 4. The zChaff times below are from our own experiments; we were unable to match reported times — we do not know whether this is due to the version of chaff used or to some additional processing they used.

Table 1. SBSAT and zChaff times on bounded model checking benchmarks.

Benchmark	SBSAT:				zChaff	
	choices	total time (seconds)	preprocss (seconds)	branch (seconds)	choices	total time (seconds)
longmult.04	534	1.280	0.580	0.610	536	0.170
longmult.08	12697	39.030	1.940	36.910	27003	35.140
longmult.12	34601	166.400	4.210	161.930	127305	362.780
longmult.16	36581	243.370	7.560	235.490	101410	231.270
longmult.20	36857	326.360	11.930	314.010	154020	554.660
longmult.24	25300	252.600	17.410	234.710	141858	451.540
longmult.28	24941	298.500	23.800	274.150	210849	876.520
longmult.32	25650	359.020	32.130	326.240	203553	645.030
longmult.36	23639	374.520	41.110	332.690	173858	669.370
longmult.40	11780	249.870	50.450	198.620	195874	700.690

Note that when the number of time steps is large, SBSAT is substantially faster than zChaff. We conjecture that this is partly due to a limitation of the chaff heuristic: An oversimplification is that the chaff family of solvers branch on variables that have occurred most frequently in the lemmas occurring so far. Thus they rank the variables, identifying variables that have proved important in the past. In effect, they seem to be narrowing in on some sort of "backdoor" variable set: roughly, a relatively small set of key variables, where, given one — or all — assignments to these key variables, solving the rest of the problem is relatively easy.

---

[*] Available from CMU at http://www-2.cs.cmu.edu/ modelcheck/bmc.html

When the number of generations in BMC gets substantially larger than the number of states in the machine, there is a great deal of symmetry between the variables representing the same state at different times. In that case, it seems the historical heuristic values for all these variables should be similar — and that to choose a branching variable effectively requires careful examination of the *current* point in the search space. This is what the LSGB heuristic does.

We also randomly generated a series of test problems which we refer to as "sliding window problems," They take our conjectured key property of BMC problems above to the extreme. The automaton is reduced to two states — one boolean variable. Since there is only one variable per state, except for a random factor noted below, there is relatively little difference between any two variables. We have a group of assertions below asserting some properties of each state, its immediate predecessor state, and states potentially several time steps in the future, corresponding, e.g., to an assertion

$$p \vee XXXXXXp \vee XXXXXXXXXXXXXXXXp.$$

For m time steps, pick two random boolean functions, $f(v_1, v_{i_1}, \ldots, v_{i_k}, v_{m/2})$, $g(v_1, v_{j_1}, \ldots, v_{j_l}, v_{m/2})$ (with variables explicitly listed, in order of subscript, and k, l are relatively small). The constraint set is

$$\{f(v_{1+h}, v_{i_1+h}, \ldots, v_{i_k+h}, v_{\frac{m}{2}+h}) : 0 \leq h \leq \tfrac{m}{2}\}$$
$$\cup$$
$$\{g(v_{1+h}, v_{j_1+h}, \ldots, v_{j_k+h}, v_{\frac{m}{2}+h}) = o_h : 0 \leq h \leq \tfrac{m}{2}\},$$

where each o_h is randomly chosen to be 0 or 1. Here, f represents a simple constraint, and g represents a constraint that changes upon unpredictable input from outside the system. (Note that this differs from BMC problem representations where the existence of random inputs is shown by nondeterminism. Our g's break symmetries.)

Experimental results are shown in Table 6.

On the small examples, SBSAT is slower than zChaff; this is due entirely to preprocessing time. Otherwise, SBSAT is far faster, and the difference increases with time. This seems to confirm our conjecture about why SBSAT is faster than zChaff on the BMC problems.

6.2   The dlx Problem Set

The dlx problem sets, arising from microprocessor-verification work at Carnegie Mellon, seemed almost prohibitively difficult before the Chaff solvers. Now Chaff versions can solve them all easily. We show below that, with suitable tuning, SBSAT can be relatively competitive with zChaff. But more interesting, this example reveals what may be good ways to incorporate user-supplied heuristics into a solver. "Suitable tuning" involved two steps. First, the LSGB heuristic, which is tailored for circumstances where lemmas are not a sufficient tool for

Table 2. Times for zZChaff and SBSAT on sliding windows problems. SBSAT lemmas were disabled.

#Variables	Satisfiable?	zChaff (seconds)	SBSAT (seconds)
60	sat	0.15	0.76
60	unsat	1.74	1.05
80	sat	1.00	1.38
80	unsat	149.53	9.98
100	sat	8.92	1.47
100	unsat	2288.11	153.70
120	sat	> 10000	89.90
120	unsat	> 10000	4259.74

speeding up searches, was replaced with a simplification, "nChaff" (near Chaff), of Chaff's heuristic.

Second, on these problem sets, much of what zChaff seems to exploit is the difference between independent and dependent variables. Here a *dependent variable* is one which the user *defines*, in the trace format versions of the dlx set, in terms of an assignment, such as $v_0 = ite(v_1, v_2, v_3)$ or $v_0 = v_1 \wedge \cdots \wedge v_{37}$. On the dlx benchmarks, telling SBSAT always to branch on dependent variables before independent variables can speed up the search massively.[9] There are other problems where branching on independent variables first significantly speeds up the search. (This suggests the possibility of dovetailing choices, alternating between branching on independent and dependent variables.) Importantly, by staying in the user domain rather than CNF, SBSAT can easily separate variables which the user describes as dependent.

Times reported below include preprocessing times. SBSAT input was in CMU's trace format, not CNF, allowing for automatic detection of dependent variables. Measurements were taken on a 2GHz Pentium 4/Linux v. 2.4.7 platform with 2GB RAM. SBSAT's lemma cache size was set to 20000.

Now zChaff still is faster; *e.g.*, on `dlx2_cc` it runs over four times as fast as the best SBSAT run. This is in part due to its simplicity and optimization — and thus fast execution: zChaff makes 24,305 decisions, whereas, with dependent-

---

[*] This is in conflict with a frequent intuition that we should branch first on independent variables, since they are forced anyway. There has indeed been a fair amount of discussion about independent variables and dependent variables and whether to branch on one type before the other [KMS97, CGPRST01, Sht00, GS99]. In the case of the dlx problems we gain immensely by branching first on dependent variables. We do not know whether this effect is intrinsic to the logic of these circuit design problems or to the way the designer thought about them. Of course, we also do not know how zChaff would perform if the user were allowed to input domain knowledge — in this case that it is wise to branch on dependent variables first.

Table 3. Times (in seconds) for zChaff and SBSAT on some dlx problems.

Benchmark	zChaff	SBSAT /LSGB	SBSAT /LSGB depndt-1st	SBSAT nChaff	SBSAT nChaff depndt-1st
Satisfiable:					
dlx2_cc_bug01	1.00	347.92	12.31	6.22	5.25
dlx2_cc_bug08	1.13	3.12	2.99	3.64	3.01
dlx2_cc_bug40	0.94	272.38	10.01	14.78	4.34
Unsatisfiable:					
dlx2_dlx2_aa	0.14	1.98	0.99	2.34	0.92
dlx2_dlx2_cc	1.43	1361.02	18.39	14.76	6.18

first variable choice, SBSAT makes 25,921 backtracks under LSGB and 15,079 backtracks under nChaff. But SBSAT is becoming competitive.[10]

Recall also that SBSAT gives the user a choice of heuristics, letting the user gain experience with what kinds of heuristic help on various classes of problems. We believe that the success of the dependent-variables-first heuristic illustrates the utility of providing that choice.

As noted earlier, regardless of whether the LSGB or nChaff search heuristic is employed, great performance for SBSAT seems to depend more on choosing dependent variables first than on the search heuristic. From the way the zChaff results follow the SBSAT results and considering that SBSAT does not take into account any weighting of lemmas as zChaff does when choosing a variable for branching, one might suppose that the strength of zChaff on the dlx benchmarks is more due to zChaff's heuristic happening to choose dependent variables before independent variables than to restarts, data structures, or any other feature of zChaff.[11] We leave investigation of this remark to a future paper.

Our goal was partly to build a solver for PSAT problems not handled well by other solvers. BMC and "sliding window" problems turn out to be hard for the current lemma-driven methods, we conjecture because of the uniformity of their variables. We find that a stronger lookahead, coupled with extensive preprocessing to allow fast lookaheads, speeds up such searches.

We are indebted to the anonymous referees for many helpful suggestions and a corrected reference to the literature.

This research was partially supported by U.S. Department of Defense grant MDA 904-02-C-1162.

[**] Also, zChaff is more heavily optimized than SBSAT. Of course, we do not know how much SBSAT can be optimized.

[**] Indeed, there are some trade-offs between speed and extensive processing. For example, in its current form, zChaff is somewhat handicapped by its clever data structures for lemma handling, which limit the search heuristics it can use effectively.

References

[BCCZ] Biere, A., A. Cimatti, E. Clarke, and Y. Zhu. Symbolic Model Checking without BDDs. TACAS 1999. *Lecture Notes in Computer Science* 1579 (1999).

[Bra90] Brace, K.S., R.L. Rudell, and R.E. Bryant. Efficient Implementation of a BDD Package. *ACM Proceedings of the 27th ACM/IEEE Design Automation Conference* (1990) 40–45.

[Bry86] Bryant, R.E.: Graph-Based Algorithms for Boolean Function Manipulation. *IEEE Transactions on Computers* C-35 (1986) 677–691.

[CGPRST01] Cimatti, A., E. Giunchiglia, M. Pistore, M. Roveri, R. Sebastiani, and A. Tacchella.: NuSMV Version 2: BDD-Based + SAT-Based Symbolic Model Checking. Available from http://sra.itc.it/people/roveri/papers/IJCAR01.ps.gz.

[CM90] Coudert, O., and J.C. Madre: A Unified Framework for the Formal Verification of Sequential Circuits, *Proc. of Int'l Conf. on Computer-Aided Design*, 1990, pp. 126-129.

[CM91] Coudert, O., and J.C. Madre: Symbolic Computation of the Valid States of a Sequential Machine: Algorithms and Discussion. *Proc. of Int'l Worshop on Formal Methods in VLSI Design*, Miami FL, USA, January 1991.

[DLL62] Davis, M., G. Logemann, and D. Loveland.: A Machine Program for Theorem Proving. *Communications of the Association of Computing Machinery* 5 (1962) 394–397.

[Fre95] Freeman, J.: Improvements to Propositional Satisfiability Search Algorithms. Ph.D. Dissertation in Computer and Information Science, University of Pennsylvania, 1995.

[GS99] Giunchiglia, E., and R. Sebastiani.: Applying the Davis-Putnam Procedure to Non-Clausal Formulas. *Proceedings of AI*IA '99, Lecture Notes in Artificial Intelligence, #1792*, Springer Verlag, 1999.

[Gol02] Goldberg, E.: Testing Satisfiability of CNF Formulas by Computing a Stable Set of Points. *Proceedings of the Fifth International Symposium on the Theory and Applications of Satisfiability Testing* (2002) 54–69. (Available from: http://gauss.ececs.uc.edu/Conferences/SAT2002/Abstracts/goldberg.ps Expanded version submitted for publication.)

[GN02] Goldberg, E., and Y. Novikov.: BerkMin: A Fast and Robust Sat-Solver Design. *Proceedings Design, Automation, and Test in Europe* 2002 142–149.

[GA98] Gupta, A., and P. Ashar.: Integrating a Boolean Satisfiability Checker and BDDs for Combinational Equivalence Checking. *Proceedings 11th IEEE International Conference on VLSI Design: VLSI for Signal Processing* (1998) 222–225.

[KZCH00] Kalla, P., Z. Zeng, M.J. Ciesielski, and C. Huang.: A BDD-based Satisfiability Infrastructure Using the Unate Recursive Paradigm. *Proceedings Design, Automation, and Test in Europe* 2000 232–236.

[KMS97] Kautz, H., D. McAllester, and B. Selman.: Exploiting Variable Dependency in Local Search. Available from http://www.cs.washington.edu/homes/kautz/papers/dagsat.ps.

[Kul98] Kullmann, O. Heuristics for SAT Algorithms: Searching for Some Foundations. Submitted for publication. (Available from http://cs-svr1.swan.ac.uk/~csoliver/heur2letter.ps.gz).

[LMS02] Lynce, I., and J. Marques-Silva.: Efficient Data Structures for Backtrack Search SAT Solvers. *Proceedings of the Fifth International Symposium on the Theory and Applications of Satisfiability Testing* (2002) 308–315. (Available from: http://gauss.ececs.uc.edu/Conferences/SAT2002/Abstracts/lynce.ps.

[MMZZM01] Moskewicz, M.W., C. Madigan, Y. Zhao, L. Zhang, and S. Malik.: Engineering a (Super?) Efficient SAT Solver. *Proceedings of the 38th ACM/IEEE Design Automation Conference* (2001).

[PK00] Paruthi, V., and A. Kuehlmann.: Equivalence Checking Combining a Structural SAT-Solver, BDDs, and Simulation. *Proceedings of the IEEE International Conference on Computer Design: VLSI in Computers and Processors* 2000, 459–464.

[PG96] Puri, R., J. Gu.: A BDD SAT Solver for Satisfiability Testing: An Industrial Case Study. *Annals of Mathematics and Artificial Intelligence*, 17 (1996) 315–337.

[SAFS95] Schlipf, J.S., F. Annexstein, J. Franco, and R. Swaminathan.: On finding solutions for extended Horn formulas. *Information Processing Letters*, 54 (1995) 133–137.

[Sch89] Schöning, U.: *Logic for Computer Scientists*. Springer Verlag (1980) 22.

[Sht00] Shtrichman, O.: Tuning SAT Checkers for Bounded Model Checking. *Proceedings of the 12th International Computer Aided Verification Conference* 2000.

[Sta94] Stålmarck, G.: A System for Determining Propositional Logic Theorems by Applying Values and Rules to Triplets that are Generated from a Formula. Swedish Patent No. 467,076 (approved 1992), U.S. Patent No. 5,276,897 (1994), European Patent No. 0403,454 (1995).

[Vel00] Velev, M.N.: Superscalar Suite 1.0. Available from: http://www.ece.cmu.edu/~mvelev.

[ZMMM01] Zhang, L., C. Madigan, M. Moskewicz, and S. Malik.: Efficient Conflict Driven Learning in a Boolean Satisfiability Solver. *Proceedings of ICCAD 2001*.

Computing Vertex Eccentricity in Exponentially

Large Graphs: QBF Formulation and Solution

Maher Mneimneh, Karem Sakallah

Advanced Computer Architecture Laboratory
University of Michigan
{maherm,karem}@umich.edu

Abstract. We formulate eccentricity computation for exponentially large graphs as a decision problem for Quantified Boolean Formulas (QBFs.) and demonstrate how the notion of eccentricity arises in the area of formal hardware verification. In practice, the QBFs obtained from the above formulation are difficult to solve because they span a huge Boolean space. We address this problem by proposing an eccentricity-preserving graph transformation that drastically reduces the Boolean search space by decreasing the number of variables in the generated formulas. Still, experimental analysis shows that the reduced formulas are unsolvable by state-of-the-art QBF solvers. Thus, we propose a novel SAT-based decision procedure optimized for these formulas. Despite exponential worst-case behavior of this procedure, we present encouraging experimental evidence showing its superiority to other public-domain solvers.

1 Introduction

Finding the eccentricity of a vertex in a graph, the radius of a graph and its diameter have numerous practical applications. An example is the location problem [5] where one is interested in positioning a facility in a location such that a certain objective is met (e.g., maximum distance traveled to the facility is minimized.) The problem can be solved by finding a vertex with minimum eccentricity; such a vertex is a possible candidate for the location of the facility. To compute the eccentricity of a vertex, one can find the shortest path (i.e., the path corresponding to the shortest distance) between the vertex and all other vertices in the graph; the maximum distance obtained determines the eccentricity. Although efficient linear implementations for computing the single-source shortest path problem exist [4], such algorithms can not handle graphs with an exponential number of nodes; any linear-time algorithm would still require exponential time to finish.

State-transition graphs (STGs) of finite-state machines (FSMs) are examples of exponentially large graphs arising in practice. These STGs usually arise from sequential logic circuits. A sequential circuit with n memory elements has 2^n states; the corresponding STG has 2^n vertices. The sequential circuit can be viewed as a succinct representation of the STG's adjacency matrix [8, 13]. For a given present state and input, the sequential circuit computes the next state and output. The operation of the sequential circuit starts from a known state called the initial state. We are interested in computing

E. Giunchiglia and A. Tacchella (Eds.): SAT 2003, LNCS 2919, pp. 411–425, 2004.

the eccentricity of the initial state. This problem has several practical applications. Next, we illustrate its significance to the area of formal hardware verification.

Design verification - ensuring that a digital system performs its desired functionality - has become a challenging if not a showstopping endeavor for present-day designs. For digital systems to continue pervading our every-day lives, we must strive to build scalable as well as rigorous techniques for ensuring their functional correctness. Formal verification is a rigorous framework for achieving design correctness; it utilizes automata theory and mathematical logic to prove certain properties about digital systems.

Model checking [6, 7] is a predominant technique in formal hardware verification. In model checking, the system to be verified is modeled as a Kripke structure, and the specification is modeled as a formula in some temporal logic. Binary Decision Diagrams (BDDs) [3] are used to symbolically represent sets of states and transition relations in Kripke structures. Bounded Model Checking (BMC) [1, 2] is one branch of model checking that verifies safety and liveness properties of a system using a satisfiability (SAT) solver. In BMC, a propositional formula is generated that is satisfiable if there exists a path ending in a state that violates the property checked. If such a state can not be found at depth i, the search is repeated at depth $i + 1$. One major drawback of BMC, however, is its lack of completeness. In other words, we do not know when to stop incrementing the value of i. It can be shown that to prove the correctness of safety properties, it is sufficient to stop incrementing when $i = ecc(s_0)$, where $ecc(s_0)$ is the eccentricity of the initial state in the corresponding STG.

In this paper, we tackle the above problem: determining the eccentricity of a state in its STG. The contributions of this work are two-fold. First, we formulate eccentricity computation as a logical inference problem for Quantified Boolean Formulas (QBF). Since this formulation results in complex QBFs in terms of the number of variables, we present an eccentricity-preserving graph transformation that drastically reduces the number of variables in constructed formulas. The proposed transformation eliminates the need for evaluating formulas corresponding to particular depths. Second, we study the structure of the resulting simplified QBFs and propose an efficient algorithm to check their satisfiability. The proposed algorithm evaluates the simplified formulas by a sequence of satisfiability (SAT) checks. For these formulas, our algorithm is far more efficient than state-of-the-art QBF solvers.

The paper is organized as follows. In section 2, we review relevant work. Section 3 introduces finite state machines, state transition graphs, and the notion of vertex eccentricity using graph-theoretic concepts. Section 4 formulates eccentricity computation as a logical inference problem for QBFs. In section 5, we describe how to simplify the resulting QBFs by introducing eccentricity-preserving graph modifications. Section 6 is devoted to describing a specialized QBF solver that efficiently checks the simplified formulas. Experimental results on some of the ISCAS 89 benchmarks are discussed in section 7 and the paper is concluded in section 8 with some pointers to future work.

2   Previous Work

The problem of determining when it is safe to stop going into deeper states when checking safety properties has been addressed by Sheeran et al. in [16]. The authors observe that no more iterations should be performed if no new states exist in future iterations. They check for that by searching for a path of length i starting at the initial state (a path, as defined in the next section, should have distinct states.) If no path exists, then there are no new states at depth i and the iteration is stopped. On the other hand, if a path exists, the iteration continues by incrementing the value of i. The authors use a SAT solver to check for the existence of a path at a given depth. A major disadvantage of that approach is the over-approximations that might result. In fact, the existence of a path from the initial state to some state does not guarantee that this state has not been encountered on a different and shorter path. The authors provide a complete formulation in terms of shortest paths but no experimental evaluation is presented. In [2], Biere et al. describe a similar over-approximation of the sequential depth referred to as the recurrence diameter. Computing the recurrence diameter requires solving a sequence of SAT problems instead of QBFs.

In [18], random simulation is utilized to provide an estimate of the sequential depth. A queue is used to keep track of states. The algorithm starts by storing the initial state in the queue and setting its depth to 0. At each iteration, a state is dequeued, and a random input is used to compute its next state. This step is repeated for a number of times determined by a threshold. If the resulting next state has not been encountered before, the state is queued and its depth is set to that of the present state plus one. When the threshold is reached, a new state is dequeued and the above procedure is repeated. When the queue is empty, the algorithm terminates. The sequential depth is the depth of the last state dequeued. The above algorithm samples the state space and consequently might result in under as well as over-approximations of the sequential depth. An improvement of the algorithm that profiles the toggle activity of state variables is also provided.

In [14], Plaisted et al. present an algorithm for solving the satisfiability problem for QBFs and discuss its use in determining fixpoints of repetitive systems. However, no experimental analysis showing the effectiveness of their QBF solver on such systems is presented.

3   Preliminaries

A finite-state machine (FSM) M is defined as a 6-tuple $M = (Q, \Sigma, \Delta, \delta, \lambda, Q^0)$ where Q is a finite set of *states*, Σ is the *input alphabet*, Δ is the *output alphabet*, $\delta: Q \times \Sigma \to Q$ is the *state-transition function*, $\lambda: Q \times \Sigma \to \Delta$ is the *output function*, and Q^0 is the set of initial states.

Synchronous Sequential circuits are modeled using FSMs. The circuit has a finite number m of inputs $(x_1, x_2, ..., x_m)$, a finite number l of outputs $(z_1, z_2, ..., z_l)$, and a finite number n of state or memory elements $(y_1, y_2, ..., y_n)$. The combinational part of the circuit is made up of k internal signals $(w_1, w_2, ..., w_k)$ representing the outputs of combinational gates. A clock signal clk synchronizes the operation of the memory elements. We will refer to $x_1, x_2, ..., x_m$ as the *input variables*, $z_1, z_2, ..., z_l$ as the *output variables*, $y_1, y_2, ..., y_n$ as the *state variables*, and $w_1, w_2, ..., w_k$ as the *internal variables*. Each of these signals takes one of two possible values 0 or 1. An *input* to a sequential circuit is a valuation of its input variables $x_1, x_2, ..., x_m$. An *output* is a valuation of the output variables $z_1, z_2, ..., z_l$. A *state* is a valuation of the state variables $y_1, y_2, ..., y_n$. Usually, the operation of a sequential circuits starts from a given state which is called the *initial state*. The state-transition function $\delta: Q \times \Sigma \to Q$ determines the next state of the machine based on its current state and inputs. The output function $\lambda: Q \times \Sigma \to \Delta$ determines the machine's output based on its current state and inputs. We can write:

$$y^+ = \delta(y, x) \qquad z = \lambda(y, x)$$

where $\quad x \equiv (x_1, ..., x_m), \quad y \equiv (y_1, ..., y_n), \quad y^+ \equiv (y_1^+, ..., y_n^+), \quad z \equiv (z_1, ..., z_l),$ $\delta \equiv (\delta_1, ..., \delta_n)$, and $\lambda \equiv (\lambda_1, ..., \lambda_l)$. We define the circuit's transition relation as:

$$T(y, y^+, x, w) = \bigcap_{i=1}^{k} (w_i \equiv f_i(\tilde{y}, \tilde{x}, \tilde{w})) \cdot \bigcap_{j=1}^{n} (y_j^+ = g_j(\tilde{y}, \tilde{x}, \tilde{w}))$$

where $\tilde{x} \equiv (\tilde{x}_1, ..., \tilde{x}_{\tilde{m}})$ with $\{\tilde{x}_1, ..., \tilde{x}_{\tilde{m}}\} \subseteq \{x_1, ..., x_m\}$. $\tilde{y} \equiv (\tilde{y}_1, ..., \tilde{y}_{\tilde{n}})$ and $\tilde{w} \equiv (\tilde{w}_1, ..., \tilde{w}_k)$ are defined similarly. f_i and g_j are Boolean functions. We can redefine the transition relation without the internal variables by existential quantification:

$$T(y, y^+, x) = (y_1^+ \equiv \delta_1(y, x)) \cdot (y_2^+ \equiv \delta_2(y, x)) \cdot ... \cdot (y_n^+ \equiv \delta_n(y, x)).$$

The state transition graph of an FSM M, $STG(M)$, is a labeled directed graph $\langle V, E \rangle$ where each vertex $v \in V$ corresponds to a state s_i of M (and is labeled with s_i), and each edge $e \in E$ between two vertices s_i and s_j corresponds to a transition from state s_i to state s_j in M. The edge is labeled i_k/o_l where i_k is the input that causes the transition from s_i to s_j and o_l is the output during that transition.

Consider a directed graph $G = \langle V, E \rangle$. A *walk* is a succession of directed edges $v_0 v_1, v_1 v_2, ..., v_{k-1} v_k$; v_0 is the start vertex, v_k the end vertex, and $v_1, ..., v_{k-1}$ the inner vertices of this walk. We denote a walk by $v_0 \to v_1 \to ... \to v_k$. The *length* of a walk is equal to the number of its edges, and is denoted by $|v_0 \to v_1 \to ... \to v_k|$. A walk whose vertices are all different is called a *path*. The distance between two vertices $v, u \in V$, denoted $d(v, u)$ is the length of the shortest path between v and u; if no path exists between v and u, then $d(v, u) = \infty$. The *eccentricity* of v, $ecc(v)$ is the largest of the distances between v and every other vertex in V. We can write $ecc(v) = max_u d(v, u)$ for all $u \in V$. The radius of G is the minimum eccentricity among all its vertices, $radius(G) = min_v(ecc(v))$ for all $v \in V$. The *diameter* of G is the maximum eccentricity among all its vertices, $diameter(G) = max_v(ecc(v))$. Consider two vertices v_0 and v_k. v_k is *reachable* from v_0 if a walk exists from v_0 to v_k. In addition, we assume that each vertex is reachable from itself. We denote by $Reach(v_0)$ the set of vertices reachable from v_0. Note that $Reach(v_0)$ is a subgraph of G.

Consider an FSM M with a single initial state s_0. We define the *sequential depth*[1] of M as the eccentricity of vertex s_0 in the subgraph $Reach(s_0)$ of $STG(M)$, and denote it by $SeqDepth(M)$.

---

1. This is usually referred to as the diameter in the literature

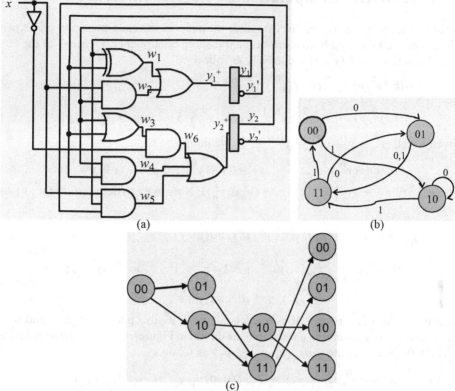

Figure 1   (a) A sequential circuit, (b) its corresponding STG, and (c) its corresponding computation tree

To illustrate the previous definitions, consider the sequential circuit in Figure 1(a). For this circuit, x is the input variable while y_1 and y_2 are the state variables. After eliminating the internal variables w_1, w_2, w_3, w_4, w_5, and w_6, we get:

$$T(y_1, y_2, y_1^+, y_2^+, x) = (y_1^+ \equiv (y_1'x + y_1 \oplus y_2)) \cdot (y_2^+ \equiv (x'(y_1' + y_2) + y_1'y_2 + xy_1y_2'))$$

The corresponding STG is shown in Figure 1(b). The STG is constructed by computing the values of y_1^+ and y_2^+ for all possible values of x, y_1 and y_2. For example, when $(y_1, y_2) = (0, 0)$ and $x = 0$, $(y_1^+, y_2^+) = (0, 1)$. Thus there is an edge from 00 to 01 labeled 0. $00 \to 01 \to 11 \to 01$ is a walk of length 3; it is not a path because the vertex 01 is repeated. The distance between 00 and 11 is computed as $d(00, 11) = min(|00 \to 01 \to 11|, |00 \to 10 \to 11|) = 2$. The eccentricity of 00 is $ecc(00) = max(d(00, 01), d(00, 10), d(00, 11)) = max(1, 1, 2) = 2$.

In this example, we calculated $ecc(00)$ by applying the definitions directly to the STG. However, in practice explicitly building the STG of a sequential circuit can be costly because it has an exponential number of vertices (in terms of the number of state variables.) Consequently, our goal is to compute the eccentricity directly from the sequential circuit without explicitly building its STG.

4   Eccentricity Computation by Logical Inference

Since the definition of eccentricity depends on paths in the state-transition graph, we will define walks and paths in terms of the transition relation $T(y, y^+, x, w)$ of the circuit. We define $walk_i(y^0, y)$ recursively as follows:

$$walk_1(y^0, y) = \exists xw \ T(y^0, y, x, w)$$

$$walk_i(y^0, y) = \exists y^1 y^2 ... y^{i-1} \ walk_1(y^0, y^1) \cdot ... \cdot walk_1(y^{i-1}, y)$$

Given a walk $y^0 \to y^1 \to ... \to y^{i-1} \to y$, we define

$$disend(y^0, y^1, ..., y^{i-1}, y) = (y \oplus y^0)(y \oplus y^1)...(y \oplus y^{i-1})$$

where $(y^k \oplus y^l) = ((y_1^k \oplus y_1^l) + ... + (y_n^k \oplus y_n^l))$. Now, we can define $path_i(y^0, y)$ as follows:

$$path_1(y^0, y) = walk_1(y^0, y) \cdot disend(y^0, y)$$

$$path_i(y^0, y) = \exists y^1 y^2 ... y^{i-1} \ walk_1(y^0, y^1) \cdot ... \cdot walk_1(y^{i-1}, y)$$

$$disend(y^0, y^1) \cdot ... \cdot disend(y^0, y^1, ..., y^{i-1}, y)$$

Intuitively, $path_i(y^0, y) = 1$ if y is reachable from y^0 in a path of length i and is 0 otherwise. Let us consider as an example the circuit in Figure 1(a). From the initial state $y_1^0 y_2^0 = 00$, we compute $walk_1(y_1^0, y_2^0, y_1, y_2)$ as follows:

$$walk_1(y_1^0, y_2^0, y_1, y_2) = \exists x \ T(0, 0, y_1, y_2, x) = y_1' y_2 + y_1 y_2'$$

Thus, there is a walk of length one between the initial state 00 and each of the states 01 and 10. This can be easily verified by inspection of the STG in Figure 1(b). Continuing, we calculate $walk_2(y_1^0, y_2^0, y_1, y_2)$ as follows:

$$walk_2(y_1^0, y_2^0, y_1, y_2) = \exists y_1^1 y_2^1 \ walk_1(y_1^0 y_2^0, y_1^1 y_2^1) \cdot walk_1(y_1^1 y_2^1, y_1 y_2)$$

$$= y_1 y_2' + y_1 y_2$$

which indicates that there is a walk of length 2 with end states 10 and 11.

We have previously defined the eccentricity of the initial state as the largest of all distances between the initial state and every other reachable state. Thus, one way to compute the eccentricity is to find a reachable state that has the largest distance from the initial state. Such a procedure can be performed iteratively. We start at depth 1 and proceed as follows. At depth i, we look for a state whose distance from the initial state is i. Such a state will have the property that no path of length less than i has that state as an end-state; otherwise, the path of length i is not the shortest for that state. A logical formulation of this property is:

$$\exists y \ path_i(y^0, y) \cdot \neg path_{i-1}(y^0, y) \cdot ... \cdot \neg path_1(y^0, y) \tag{1}$$

where y^0 is the initial state.

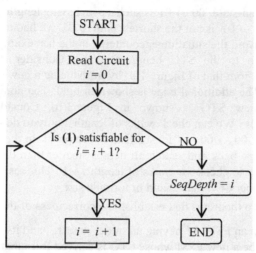

Figure 2   Flowchart of the algorithm for computing vertex eccentricity

If the above formula is satisfiable, the algorithm proceeds to deeper iterations since new states might be found whose distance from the initial state is greater than i. The algorithm terminates when no state satisfying the above property can be found. In other words, the algorithm terminates when the above logical formula is unsatisfiable. A flowchart of this algorithm is illustrated in Figure 2. As an example, consider the STG in Figure 1(a). We set $i = 0$, and check the satisfiability of $\exists y \ path_1(y^0, y)$. This searches for a path of length 1 whose end-state is different from the initial state. $00 \rightarrow 01$ is one such path. As a result, i is incremented to 1. Next, we check the satisfiability of $\exists y \ path_2(y^0, y) \cdot \neg path_1(y^0, y)$; this formula searches for a path of length 2 whose end-state is not an end-state for any path of length 1. $00 \rightarrow 01 \rightarrow 11$ is such a path and i is incremented to 2. In the third iteration, the formula $\exists y \ path_3(y^0, y) \cdot \neg path_2(y^0, y) \cdot \neg path_1(y^0, y)$ searches for a path of length 3 with an end-state that is not an end-state for any path of length 2 or 1. The previous formula is unsatisfiable; no such state can be found. The eccentricity is equal to 2.

Searching for a state s that satisfies the above formula is complicated by the fact that at a given depth i, the search should consider paths of length j for all values of j less than i to ensure that the path to s is the shortest. This can be very expensive rendering the algorithm inapplicable for all but simple circuits. This complexity is a direct consequence of the huge number of variables introduced to model paths of length j ($j < i$.) In the next section we introduce a graph transformation that significantly reduces the complexity of the formula to be checked.

5   Simplified Eccentricity Computation

We would like to simplify (1) so that we do not have to go all the way back to depths of length 1 when searching for a new state. To better understand the problem, consider the computation tree in Figure 1(c). Assume we are checking for new states at depth 3. There is one path of length 3: $00 \rightarrow 10 \rightarrow 11 \rightarrow 01$. Although there is no path of length

2 having 01 as an end-state, $00 \rightarrow 01$ is such a path having length 1. Thus, to conclude that $00 \rightarrow 10 \rightarrow 11 \rightarrow 01$ is not the shortest path to 01 we had to go back to paths of length 1. We can avoid the situation encountered in the last example by introducing a simple modification to the STG being traversed. Consider the STG shown in Figure 3(a) derived from that of Figure 1(a) by introducing a new edge from the initial state 00 to itself. The additional edge is shown in bold. The computation tree corresponding to the new STG is shown in Figure 3(b). Consider again the path $00 \rightarrow 10 \rightarrow 11 \rightarrow 01$. We can check walks of length 2 only to deduce that 01 is not a new state. The walk $00 \rightarrow 00 \rightarrow 01$ establishes such proof. In the STG of Figure 3(a), we do not have to go back and search all depths smaller than the one considered; at depth i, it is enough to check for walks of length $i - 1$. This results in substantial reductions in the number of varialbes and in formula size.

Next, we present two theorems that establish the correctness of the above approach.

Theorem 1 Given an FSM M having an initial state s_0, and its corresponding STG, $STG(M)$. Let M' be a new FSM whose STG is derived from that of M by adding an edge from the initial state to itself. Then $ecc(s_0)$ in $STG(M)$ is equal to $ecc(s_0)$ in $STG(M')$.

Proof   Consider any state s_k of $STG(M)$ and assume that $d(s_0, s_k) = k$. Adding an edge from s_0 to itself doesn't change the distance from s_0 to s_k. Thus, in $STG(M')$, $d(s_0, s_k) = k$. Since all the distances in $STG(M')$ are identical to those in $STG(M)$, and the eccentricity of s_0 is defined as the maximum distance, $ecc(s_0)$ is the same in both graphs.

The above theorem ensures that by adding the transition to the initial state, the eccentricity of the initial state is not altered.

Theorem 2 Given an FSM M having an initial state s_0. Let $STG(M)$ be such that there is a transition from s_0 to itself. If there is a path of length k from s_0 to a state s_k, then there is a walk from s_0 to s_k of length m for every m such that $m \geq k$.

Proof   Since $d(s_0, s_k) = k$, then there is a path $s_0 \rightarrow s_1 \rightarrow \ldots \rightarrow s_{k-1} \rightarrow s_k$ for some inner vertices $s_1, s_2, \ldots\ldots, s_{k-1}$. Since there is an edge from s_0 to itself, we can create a walk of length m for $m > k$ by adding $m - k$ transitions of the form $s_0 \rightarrow s_0$. Thus there is a walk of the form $s_0 \rightarrow s_0 \rightarrow \ldots \rightarrow s_0 \rightarrow s_1 \rightarrow \ldots \rightarrow s_{k-1} \rightarrow s_k$ of length m for all $m \geq k$.

The above theorem ensures that every state reachable through a path of length k from the initial state, will be reachable, as well, through a walk of length m, $m \geq k$.

By setting m in Theorem 2 to $i - 1$ we get:

$$path_1(y^0, y) \Rightarrow walk_{i-1}(y^0, y)$$
$$path_2(y^0, y) \Rightarrow walk_{i-1}(y^0, y)$$
$$\vdots$$
$$path_{i-1}(y^0, y) \Rightarrow walk_{i-1}(y^0, y)$$

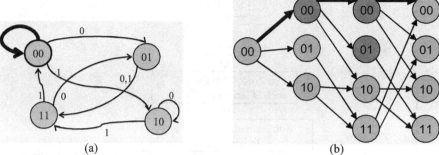

(a)   (b)

Figure 3   (a) The modified STG and (b) its corresponding computation tree

and consequently,

$$(path_{i-1}(y^0, y) + \cdots + path_1(y^0, y)) \Rightarrow walk_{i-1}(y^0, y)$$

In addition, from basic graph theory we have:

$$walk_{i-1}(y^0, y) \Rightarrow (path_{i-1}(y^0, y) + \cdots + path_1(y^0, y))$$

Thus,

$$(path_{i-1}(y^0, y) + \cdots + path_1(y^0, y)) \Leftrightarrow walk_{i-1}(y^0, y)$$

The above theorems enable us to reduce (1) to:

$$\exists y \ path_i(y^0, y) \cdot \neg walk_{i-1}(y^0, y) \tag{2}$$

6   Efficient Evaluation of Eccentricity Formulas

In this section, we present an efficient algorithm for checking the satisfiability of formulas whose structure is similar to (2). First let us take a closer look at (2).We have:

$$path_i(y^0, y) = \exists y^1 \cdots y^{i-1} ((\exists x^0 w^0 \ T(y^0, x^0, w^0, y^1)) \cdots (\exists x^{i-1} w^{i-1} \ T(y^{i-1}, x^{i-1}, w^{i-1}, y)) \cdot$$
$$diff(y^0, \cdots, y))$$
$$= \exists y^1 \cdots y^{i-1} x^0 \cdots x^{i-1} w^0 \cdots w^{i-1} \ (T(y^0, x^0, w^0, y^1) \cdots T(y^{i-1}, x^{i-1}, w^{i-1}, y) \cdot$$
$$diff(y^0, \cdots, y))$$
$$= \exists Y^1 XW \ (T(y^0, x^0, w^0, y^1) \cdots T(y^{i-1}, x^{i-1}, w^{i-1}, y) \cdot diff(y^0, \cdots, y))$$

where $Y^1 = \{y^1, \cdots, y^{i-1}\}$, $X = \{x^0, \cdots, x^{i-1}\}$, $W = \{w^0, \cdots w^{i-1}\}$, and $diff(y^0, \cdots, y)$ represents all constraints of the form $y^i \neq y^j$. We also have:

$$\neg walk_{i-1}(y^0, y) = \neg(\exists r^1 \cdots r^{i-2} (\exists p^0 q^0 \ T(y^0, p^0, q^0, r^1)) \cdots (\exists p^{i-2} q^{i-2} \ T(r^{i-2}, p^{i-2}, q^{i-2}, y)))$$
$$= \neg(\exists r^1 \cdots r^{i-2} p^0 \cdots p^{i-2} q^0 \cdots q^{i-2} \ T(y^0, p^0, q^0, r^1) \cdots T(r^{i-2}, p^{i-2}, q^{i-2}, y))$$
$$= \neg(\exists RPQ \ T(y^0, p^0, q^0, r^1) \cdots T(r^{i-2}, p^{i-2}, q^{i-2}, y))$$
$$= \forall RPQ \ \neg(T(y^0, p^0, q^0, r^1) \cdots T(r^{i-2}, p^{i-2}, q^{i-2}, y))$$

where $R = \{r^1, \cdots, r^{i-2}\}$, $P = \{p^0, \cdots, p^{i-2}\}$, and $Q = \{q^0, \cdots q^{i-2}\}$ .Thus,

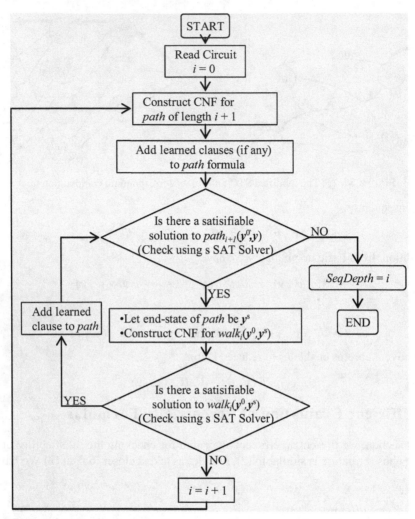

Figure 4 Flowchart of the enhanced algorithm for computing vertex eccentricity

$$\exists y \; path_i(y^0, y) \cdot \neg walk_{i-1}(y^0, y)$$
$$= \exists y \; (\exists YXW \; (T(y^0, x^0, w^0, y^1) \cdots T(y^{i-1}, x^{i-1}, w^{i-1}, y) \cdot diff(y^0, \cdots, y))) \cdot$$
$$(\forall RPQ \; \neg(T(y^0, p^0, q^0, r^1) \cdots T(r^{i-1}, p^{i-1}, q^{i-1}, y)))$$
$$= \exists y YXW \; \forall RPQ \; T(y^0, x^0, w^0, y^1) \cdots \cdots T(y^{i-1}, x^{i-1}, w^{i-1}, y) \cdot diff(y^0, \cdots, y) \cdot$$
$$\neg(T(y^0, p^0, q^0, r^1) \cdots T(r^{i-1}, p^{i-1}, q^{i-1}, y))$$

The above formula is in QBF form and consequently can not be checked immediately using a SAT solver. One possibility is to use a QBF solver. However, experimental results show that even formulas resulting from small circuits are out of the scope of state-of-the-art QBF solvers. In what follows, we describe a SAT-based algorithm to efficiently solve these formulas.

From (2), we notice that $path_i(y^0, y)$ and $walk_{i-1}(y^0, y)$ do not share any intermediate variables. The only variables shared between the two formulas are those corresponding to the initial state y^0 and the final state y. This suggests an efficient eccentricity computation algorithm that is presented in the flowchart of Figure 4.

The algorithms starts by reading the circuit and setting the variable $i = 0$. Let's consider iteration i of the algorithm. At iteration i, the algorithm has to check the satisfiability of (2). Our algorithm decomposes checking the satisfiability of (2) into two simpler checks. The algorithm constructs a CNF formula representing a path of length $i + 1$ starting at the initial state. A SAT solver is used to find a satisfiable assignment for the constructed formula which corresponds to:

$$path_{i+1}(y^0, y) \qquad (3)$$

If (3) is unsatisfiable, we automatically know that (2) is unsatisfiable. Thus, the algorithms terminates and $SeqDepth = i$. If on the other hand, (3) is satisfiable, we construct a CNF formula representing a walk of length i and set its end-state to y^s, the end-state on the path of length $i + 1$ that was returned as a satisfying solution for (3). The formula corresponds to:

$$walk_i(y^0, y^s) \qquad (4)$$

A SAT solver is used to check (4). If (4) is unsatisfiable, then there is no walk of length i starting at the initial state and ending in y^s. Consequently, y^s is a new state at depth $i + 1$ that satisfies (2). Thus, i is incremented and the whole procedure is repeated again. If (4) is satisfiable, then y^s does not satisfy (2) and we have to search for another solution. We do this by adding a constraint to (3) enforcing the SAT solver to find a state different from y^s. If no state can satisfy (3), the algorithm terminate setting the sequential depth to i. Note that all the added constraints to (3) at depth i are used at all the following iterations. This ensures that no time is wasted looking for states that are not on shortest paths.

Let's consider the execution of the algorithm on the circuit of Figure 3(a), whose computation tree is shown in Figure 3(b). The algorithm starts by setting $i = 0$. A CNF formula corresponding to $path_1(00, y)$ is constructed. No learned clauses are present at this time. We check the satisfiability of $path_1(00, y)$. One such solution is $00 \rightarrow 01$. $y = 01$ is a new state so we set $i = 1$.

Next, we construct a formula corresponding to $path_2(00, y)$. A satisfying solution to $path_2(00, y)$ is $00 \rightarrow 01 \rightarrow 11$ where $y = 11$. To check whether 11 is a new state, we construct a formula corresponding to $walk_1(00, 11)$ Such a formula is unsatisfiable (no walk of length 1 exists that terminate in 11.) Thus 11 is a new state; we set $i = 2$.

Again, we construct a formula corresponding to $path_3(00, y)$. A satisfying solution is $00 \rightarrow 10 \rightarrow 11 \rightarrow 01$ where $y = 01$. We check whether $walk_2(00, 01)$ is satisfiable. Since there is a walk $00 \rightarrow 00 \rightarrow 01$, $walk_2(00, 01)$ is satisfiable and 01 is not a new state. We add a constraint to $path_3(00, y)$ forcing y to be different from 01. The resulting formula is $path_3(00, y) \cdot (y \neq 01)$. Checking the new formula

$path_3(00, y) \cdot (y \neq 01)$ yields no satisfying assignment. Thus no new states exist at depth 3. The algorithm terminates; the sequential depth is 2.

Table 1: Runtime of depth formulas on various QBF solvers

Circuit	Depth	SAT/UNSAT	Vars/Clauses	Decide	QuBE-BJ1.0	QuBE-Rel1.3	Quaffle-full
s27	2	SAT	85/142	0.1	0.01	0.07	0.12
s27	3	UNSAT	165/254	20.64	33.79	725.57	T/O
s298	2	SAT	853/1895	T/O	T/O	T/O	T/O
s386	2	SAT	1162/2655	T/O	T/O	T/O	T/O
s641	2	SAT	2539/4494	T/O	T/O	T/O	T/O
s713	2	SAT	2663/4872	T/O	T/O	T/O	T/O

7   Experimental Results

To experimentally evaluate the effectiveness of our algorithm, we implemented a prototype version in C++ and used Chaff [12] as the underlying SAT solver. We report our results on the ISCAS 89 benchmarks. All experiments were conducted on a 2.8 GHz Pentium 4 machine having 1 GB of RAM and running the Linux operating system. To our knowledge, there is no published data on the sequential depth of ISCAS benchmarks computed using SAT-based search only and to which we can compare our results. For that reason, we compare our results to those obtained in [10] where a combination of SAT, BDDs, and partitioning techniques are used to perform reachability analysis. For this comparison, we note the following. First, the techniques in [10] compute the whole set of reachable states using breadth-first search (BFS). In that case, sequential depth is the number of BFS iterations. Our algorithm does not compute the set of reachable states. Second, in [10], the authors set the time limit to 100,000 seconds whereas we set the time limit to 10,000 seconds.

We evaluated some of the formulas for eccentricity computation on state-of-the-art QBF solvers. We used three solvers: Decide [15], QuBE [9], and Quaffle [17]. The results are reported in Table 1. Column 1 lists the name of the circuit. Column 2 shows the value of i in (2) for which the formula was constructed. Column 3 indicates whether the formula is satisfiable (SAT) or unsatisfiable (UNSAT). Columns 4 reports the number of variables and the number of clauses in the formula. Columns 5, 6, 7, and 8 show the runtime (in seconds) when checking the satisfiability of the formula using Decide, QuBE-BJ1.0, QuBE-Rel1.3, and Quaffle-full. T/O indicates that the solver timed out at 10,000 seconds. These solvers (except Quaffle-full) were able to solve the formulas for s27, the smallest of the ISCAS89 benchmarks. However, for larger circuits, all of the solvers timed out at 10,000 seconds. It is instructive to note that the QBF formulas for s298, s386, s641, and s713 correspond to checking the eccentricity at depth 2; the formulas for checking deeper levels are much larger in terms of the number of variables and clauses.

Next, we report experimental results on the ISACS benchmarks when using our specialized QBF solver. The results are shown in Table 2. The name of the circuit appears in column 1. Columns 2 and 3 show the sequential depth and the running time for our algorithm. In case the time limit was reached, we report the maximum depth attained. Column 4 reports the maximum number of reachability steps in [10]; a (c) next to a number indicates that reachability was carried to completion. Empty cells in column 4 indicate non-reported data in [10].

Table 2: Runtime of depth formulas on SAT-based QBF solver

Circuit	Sequential Depth	Time (sec)	Maximum Depth in [10]
s298	18	6.15	-
s386	7	0.06	-
s499	21	0.52	-
s510	46	149.2	-
s641	6	23.8	-
s713	6	35.67	-
s820	10	1.93	-
s953	10	58.86	-
s1196	2	140.35	-
s1269	9	10,000	10(c)
s1423	26	10,000	15
s1488	21	102.47	-
s3271	14	10,000	17(c)
s3330	8	10,000	9(c)
s5378	21	10,000	45(c)
s6669	5	10,000	3

For the small benchmarks, up to s1196, our algorithm is very effective in determining the eccentricity. The least time reported is 0.06 seconds for s386, and the longest time is 140.35 seconds for s1196. Although s386 is deeper than s1196, s386 has 13 reachable states while s1196 has 2616. Consequently, more searching is needed in the case of s1196 before reaching a fixpoint.

For s1269, s1423, s3271, s3330, s5378, and s6669 the time limit was reached. Although reachability analysis completed successfully for these benchmarks in [10], the maximum depth we attained in 10,000 seconds is very close to the sequential depth (except for s5379.) The results for s1423 and s6669 are interesting. Reachability analysis in [10] could not complete: a maximum depth of 15 and 3 were obtained for s1423 and s6669. However, we were able to reach a depth of 26 for s1423 and 5 for s6669.

We also computed the recurrence diameter [2, 16] for some of the ISCAS 89 benchmarks and compared it to the sequential depth. The results are presented in Table 3. Column 2 shows the sequential depth, column 3 shows the recurrence diameter, and column 4 shows the number of reachable states for a given circuit. For s641, s713, and s953 computing the recurrence diameter didn't finish. Thus, we report the depths that were reached. For s499 and s510, the sequential depth and recurrence diameters are equal; these circuits implement some type of a counter design. For s386 and s820, the recurrence diameter approximates the sequential depth within a factor of 2. On the other hand, for s298, s641, s713 and s953, the recurrence diameter is much larger than the sequential depth. We conjecture that the recurrence diameter has a stronger correlation with the set of reachable states that with the sequential depth. As an example, both s820 and s953 has a sequential depth of 10. However, the recurrence diameter of s820 is 17 while that of s953 is greater than 176. The correlation between the recurrence diameter and the number of reachable states for these circuits is obvious.

Table 3: Comparison of sequential depth and recurrence diameter

Circuit	Sequential Depth	Recurrence Diameter	Reachable States
s298	18	100	218
s386	7	11	13
s499	21	21	22
s510	46	46	47
s641	6	> 250	1544
s713	6	> 250	1544
s820	10	17	25
s953	10	> 176	504

8   Conclusion

Determining the eccentricity of a vertex in an exponentially large graph has various practical applications. One such applications is enabling search-based verification techniques such as Bounded Model Checking to prove correctness of designs rather than just find bugs.

To achieve completeness for Bounded Model Checking, we presented a formulation of eccentricity computation as a logical inference problem for Quantified Boolean Formulas. We studied the structure of the resulting formulas and showed that their complexity can be drastically reduced by modifying the corresponding state-transition graphs. We proved that such modifications leave the eccentricity intact and consequently are safe to apply. We also presented an efficient SAT-based algorithm to solve QBFs resulting from our simplification. Our experimental results showed that our SAT-based solution outperforms state-of-the-art QBF solvers.

Since our algorithm learns states that violate some property and adds them to the clause database one at a time, its memory requirements are exponential in the worst case (this situation arises when the number of states is exponential.) To surmount this, we are currently investigating the possibility of adding sets of states at a time. In addition, we are exploring two other directions to improve our approach. We are studying the possibility of applying abstractions to the state-transition graph that preserve the eccentricity of a given state. These abstractions can help simplify the complexity of the QBFs to be checked. We are also investigating better algorithms to solve the formulas that arise during depth computation. One direction we are pursuing is utilizing symmetries in the transition relation and along different paths to simplify the satisfiability check.

Acknowledgement

This work is funded by the DARPA/MARCO Gigascale Silcon Research Center.

References

[1]   A. Biere, A. Cimatti, E. M. Clarke, M. Fujita, and Y. Zhu, "Symbolic Model Checking Using SAT Procedures instead of BDDs," in *Proceedings of the 36th Design Automation Conference*, 1999.

[2]   A. Biere, A. Cimatti, E. M. Clarke, and Y. Zhu, "Symbolic Model Checking without BDDs," in TACAS'99, 1999.

[3]   R. E. Bryant, "Graph-Based Algorithms for Boolean Function Manipulation," in *IEEE Transactions on Computers*, 35(8), pp. 677-691, August 1986.

[4]   B. V. Cherkassky, A. V. Goldberg, and T. Radzik, "Shortest Paths Algorithms: Theory and Experimental Evaluation," in SODA: ACM-SIAM Symposium on Discrete Algorithms, 1993.

[5]   M. Labbe, D. Peeters, and J. -F Thisse, "Location on Networks," in Handbooks in OR & MS, North-Holland, pp. 551-624, 1995.

[6]   J. R. Burch, E. M. Clarke, D. E. Long, K. L. McMillan, and D. L. Dill, "Symbolic Model Checking for Sequential Circuit Verification," in *IEEE Transactions on Computer-Aided Design of Integrated Circuits* 13(4), pp. 401-424, April 1994.

[7]   E. M. Clarke, E. A. Emerson, and A. P. Sistla, "Automatic Verification of Finite-state Concurrent Systems Using Temporal Logic Specifications," in *ACM Transactions on Programming Languages and Systems*, 8(2), pp. 244-263, 1986.

[8]   H. Galperin and A. Wigderson, "Succinct Representations of Graphs," in *Information and Control*, 56, 1983.

[9]   E. Giunchiglia, M. Narizzano, and A. Tacchella, "System Description: QuBE A System for Deciding Quantified Boolean Formulas Satisfiability," in *Proceedings of the International Joint Conference on Automated Reasoning*, 2001.

[10]   A. Gupta, Z. Yang, P. Ashar, L. Zhang, and S. Malik, "Partition-Based Decision Heuristics for Image Computation Using SAT and BDDs," in *Proceedings of the International Conference on Computer-Aided Design*, 2001.

[11]   D. Kroening, and O. Strichman, "Efficient Computation of Recurrence Diameters," in *4th International Conference on Verification, Model Checking, and Abstract Interpretation*, 2003.

[12]   M. Moskewicz, C. Madigan, Y. Zhao, L. Zhang, and S. Malik, "Chaff: Engineering an Efficient SAT solver," in *Proceedings of the Design Automation Conference*, 2001.

[13]   C. Papadimitriou and M. Yannakakis, "A Note on Succinct Representations of Graphs," in *Information and Control*, 71, 1986.

[14]   D. A. Plaisted, A. Biere, and Y. Zhu, "A Satisfiability Procedure for Quantified Boolean Formulae," submitted for publication, Discrete Applied Mathematics.

[15]   J. Rintanen, "Improvements to the Evaluation of Quantified Boolean Formulae," *Proceedings of the 16th International Joint Conference on Artificial Intelligence*, pp. 1192-1197, August, 1999.

[16]   M. Sheeran, S. Singh, and G. Stalmarck, "Checking Safety Properties Using Induction and a SAT-solver," in *Formal Methods in Computer Aided Design*, 2000

[17]   L. Zhang, and S. Malik, "Conflict Driven Learning in a Quantified Boolean Satisfiability Solver," Proceedings of the International Conference on Computer Aided Design (ICCAD2002), 2002.

[18]   C.-C. Yen, K.-C Chen, and J.-Y. Jou, "A Practical Approach to Cycle Bound Estimation for Property Checking," in *11th IEEE/ACM International Workshop on Logic Synthesis*, pp. 149-154, June 2002.

The Combinatorics of Conflicts between Clauses

Oliver Kullmann

Computer Science Department
University of Wales Swansea
Swansea, SA2 8PP, UK
O.Kullmann@Swansea.ac.uk
http://cs-svr1.swan.ac.uk/~csoliver/

Abstract. We study the *symmetric conflict matrix* of a multi-clause-set, where the entry at position (i, j) is the number of clashes between clauses i and j. The conflict matrix of a multi-clause-set, either interpreted as a multi-graph or a distance matrix, yields a new link between investigations on the satisfiability problem (in a wider sense) and investigations on the *biclique partition number* and on the *addressing problem* (introduced by Graham and Pollak) in graph theory and combinatorics. An application is given by the well-known class of what are called 1-*uniform hitting clause-sets* in this article, where each pair of (different) clauses clash in exactly one literal. Endre Boros conjectured at the SAT'98 workshop that each 1-uniform hitting clause-set of deficiency 1 contains a literal occurring only once. Kleine Büning and Zhao showed, that under this assumption every 1-uniform hitting clause-set must have deficiency at most 1. We show that the conjecture is false (there are known star-free biclique decompositions of every complete graph with at least nine vertices), but the conclusion is right (a special case of the Graham-Pollak theorem, attributed to Witsenhausen). A basic notion for investigations on the combinatorics of clause-sets is the *deficiency* of multi-clause-sets (the difference of the number of clauses and the number of variables). We introduce the related notion of *hermitian defect*, based on the notion of the *hermitian rank* of a hermitian matrix introduced by Gregory, Watts and Shader. The notion of hermitian defect makes it possible to combine eigenvalue techniques (especially Cauchy's interlacing inequalities) with matching techniques, and can be seen as the underlying subject of our introduction into the subject.

1 Introduction

Consider a multi-clause-set F and its clause-variable matrix $M(F)$ (a matrix over $\{-1, 0, +1\}$, where each row represents a clause). Using $|M(F)|$ for the matrix obtained from $M(F)$ by taking absolute values for each entry, we see

$$|M(F)| \cdot |M(F)|^t - M(F) \cdot M(F)^t = 2 \cdot \mathrm{C_s}(F),$$

where A^t is the transposition of matrix A, and $\mathrm{C_s}(F)$ is the *symmetric conflict matrix* of F, where the entry at position (i, j) is the number of clashing literals

E. Giunchiglia and A. Tacchella (Eds.): SAT 2003, LNCS 2919, pp. 426–440, 2004.
© Springer-Verlag Berlin Heidelberg 2004

between clauses i and j of F. It follows, that the number of variables in F must be at least the maximum of the number of positive and the number of negative eigenvalues of $C_s(F)$ — the number of variables, which is the number of columns of $M(F)$, is a kind of a bottleneck in the above matrix multiplication; see Lemma 4 together with Lemma 5. For *uniform hitting clause-sets* with m clauses, defined as (multi-)clause-sets F where the entries of $C_s(F)$ are constant except of the diagonal (which is always zero), the symmetric conflict matrix has $m - 1$ negative eigenvalues and one positive eigenvalue, and we conclude that every uniform hitting clause-set must have at most one clause more than it has variables (i.e., its deficiency is at most one). This observation is attributed to Witsenhausen by Graham and Pollak ([9]), and has been the starting point for many investigations in graph theory and combinatorics. We review the basic notions and techniques, but formulated in the language of (multi-)clause-sets, which turns out to be a natural language for this kind of problems, providing a unifying point of view.

The *addressing problem* of Graham and Pollak (see Chapter 9 in [23]) considers a graph G and its *distance matrix* A, which is a square matrix of order $|V(G)|$ with the distance in G between node i and node j at position (i, j). The problem is to attach "addresses" to the nodes of G, which are tuples from $\{0, 1, *\}^n$ for some n to be minimised, such that the distance between nodes i and j equals the "distance" between their addresses, where the distance of two addresses is computed by first removing all positions where at least one of the two addresses has a $*$, and then taking the Hamming distance of the remaining $\{0, 1\}$-vectors. Identifying the positions in the address vectors with variables, and interpreting 0 as "positive", 1 as "negative" and $*$ as "not there" we see, that an addressing of G corresponds to a multi-clause-set $F \in \mathcal{MCLS}$ with $C_s(F) = A$. For example, 1-uniform hitting clause-sets with m clauses correspond to addressings of the complete graph K_m. Interpreting addresses (in our terminology, clauses) as code words, the address problem has been introduced into coding theory in [22].

Given a multigraph G, a *biclique partition* is a list of edge-disjoint sub-graphs of G (not necessarily induced) covering together all edges of G, such that each of these sub-graphs is a complete bipartite graph. Using the adjacency matrix A of G each biclique decomposition corresponds to a multi-clause-set with symmetric conflict matrix A — each node is a clause, each biclique represents a variable, where the "left side" stands for the positive occurrences, while the "right side" stands for the negative occurrences, and each edge represents a conflict. 1-uniform hitting clause-sets correspond to biclique partitions of the complete graph (again). For literature see [15].

This article seeks to present an introduction into a new area of research, and nearly no proofs are given, but all proofs can be found in the underlying report [15]. The organisation is as follows: After presenting some notations, in Section 3 we lay the foundations by *investigating conflict matrices* and their properties, most notable the *hermitian rank* and the classes of conflict matrices with bounded *hermitian defect*, where the hermitian defect is an upper bound for the maximal deficiency. Then in Section 4 we collect observations and results on *hitting clause-sets*, and apply our general results derived from the study of conflict matrices. Finally in Section 5 a *counterexample* to the conjecture of Endre Boros is discussed. Some main results are as follows:

1. In Subsection 3.3 the central (and novel) notion of *hermitian defect* is introduced, the difference of the order of a symmetric matrix and its hermitian rank. Strengthening the Graham-Pollak theorem by using interlacing inequalities, in Theorem 11 we obtain the central result of this section, that for every multi-clause-set F and every strict sub-multi-clause-set F' the deficiency of F' is strictly less than the hermitian defect of F (while the deficiency of F itself is less or equal to the hermitian defect).

2. In Subsection 3.4 *exact multi-clause-sets* are introduced, which realise the symmetric conflict number of its symmetric conflict matrix. While deciding exactness of multi-clause-sets is co-NP complete, the strengthening of this notion by the notion of *eigensharp multi-clause-sets* is decidable in polynomial time, where a clause-set is eigensharp if its hermitian defect coincides with its deficiency. From our central result on the hermitian defect we obtain immediately, that every eigensharp multi-clause-set is matching lean, i.e., has no non-trivial matching autarky (Theorem 14).

3. The class $\mathcal{MCLS}_{\delta_h}(1)$ of multi-clause-sets with hermitian defect at most one is investigated in Subsection 3.5. Theorem 19 completely characterises eigensharp, matching lean and matching satisfiable multi-clause-sets $F \in \mathcal{MCLS}_{\delta_h}(1)$, where it turns out, that the necessary condition $\delta(F) \geq 1$ (i.e., $\delta(F) = 1$) for the first two cases respectively $\delta(F) \leq 0$ for matching satisfiability are also sufficient. In Theorem 26 we show, that unsatisfiable multi-clause-sets in $\mathcal{MCLS}_{\delta_h}(1)$ are exactly the *saturated minimally unsatisfiable clause-sets* of deficiency one.

4. In Subsection 3.6 the classes $\mathcal{MCLS}_{i_+}(k)$ and $\mathcal{MCLS}_{i_-}(k)$ of multi-clause-sets are introduced, where the symmetric conflict matrix has at most k positive respectively negative eigenvalues. Again these classes are stable under formation of sub-multi-clause-sets and under application of partial assignments (Lemma 28). Theorem 29 applies the Perron-Frobenius theorem to show, that the largest eigenvalue of the symmetric conflict matrix of a multi-clause-set is also largest with respect to its absolute value. It follows $\mathcal{MCLS}_{\delta_h}(1) \subset \mathcal{MCLS}_{i_+}(1)$ (Theorem 30).

5. *Hitting clause-sets* are studied in Section 4 (every two clauses clash), and a fair amount of different characterisations and properties are given. Theorem 32 shows, that the class of unsatisfiable hitting clause-sets is exactly the class of all minimally unsatisfiable clause-sets which stay minimally unsatisfiable under applications of any partial assignments. For *uniform hitting clause-sets* (every two clauses clash in the same number of literals) it is shown (Theorem 33) that they are contained in $\mathcal{MCLS}_{\delta_h}(1)$.

6. A *counterexample* to the conjecture of Endre Boros is given in Section 5.

For all proofs see [15], where also a list of open problems is given (see [16] for the status of these (and other) open problems).

2 Some Notations

We use $\mathbb{N} = \{1, 2, \ldots\}$ and $\mathbb{N}_0 = \mathbb{N} \cup \{0\}$. The universe of variables is denoted by $\mathcal{VA}$. Clauses are finite and complement-free sets of literals, the set of all clauses is denoted by $\mathcal{CL}$. The empty clause is denoted by $\bot := \emptyset \in \mathcal{CLS}$. A (finite) multi-clause-set is a map $F : \mathcal{CL} \to \mathbb{N}_0$ such that only for finitely many clauses C we have $F(C) \neq 0$; the set of all multi-clause-sets is denoted by $\mathcal{MCLS}$. The empty multi-clause-set is denoted by $\top := (0)_{C \in \mathcal{CL}} \in \mathcal{MCLS}$. For $F, F' \in \mathcal{MCLS}$ the relation $F \leq F'$ holds iff for all $C \in \mathcal{CL}$ we have $F(C) \leq F'(C)$, while $F \lneq F'$ is equivalent to $F \leq F'$ and $F \neq F'$. The multi-clause-set $F + F'$ is defined via $(F + F')(C) := F(C) + F'(C)$. The number of clauses in F is defined as $c(F) := \sum_{C \in \mathcal{CL}} F(C)$, while the number of occurrences of a literal x in F is denoted by $\#_x(F) := \sum_{C \in \mathcal{CL}, x \in C} F(C)$. The set of variables occurring in F is $\text{var}(F) := \bigcup_{C \in \mathcal{CL}, F(C) \neq 0} \text{var}(C)$, and the number of variables in F is $n(F) := |\text{var}(F)|$. For $F \in \mathcal{MCLS}$ we have $n(F) = 0$ iff for all clauses $C \in \mathcal{CL} \setminus \{\bot\}$ we have $F(C) = 0$. The deficiency of F is $\delta(F) := c(F) - n(F)$, while the maximal deficiency is $\delta^*(F) := \max_{F' \leq F} \delta(F')$. Due to $\delta(\top) = 0$ we have $\delta^*(F) \geq 0$. $\mathcal{PASS}$ denotes the set of all partial assignments, which are maps $\varphi : V \to \{0, 1\}$ for finite set V of variables. The domain of a partial assignment φ is denoted by $\text{var}(\varphi) := V$, while the size of the domain is denoted by $n(\varphi) := |\text{var}(\varphi)|$. The application of partial assignments $\varphi \in \mathcal{PASS}$ to multi-clause-sets $F \in \mathcal{MCLS}$ is denoted by $\varphi * F \in \mathcal{MCLS}$, where for $C \in \mathcal{CL}$ we have

$$(\varphi * F)(C) = \sum_{\substack{C' \in \mathcal{CL} \\ \varphi * \{C'\} = \{C\}}} F(C'),$$

where $\varphi * \{C'\}$ is the empty set in case φ satisfies C', while otherwise $\varphi * \{C'\} = \{C''\}$, where C'' is obtained from C' by removing all literals whose variables are in the domain of φ. Clause-sets, that is, finite sets of clauses, can be identified with multi-clause-sets $F : \mathcal{CL} \to \{0, 1\}$. The set of all clause-sets is denoted by $\mathcal{CLS}$, and the operation of $\mathcal{PASS}$ on $\mathcal{CLS}$ is defined as via $\varphi * F := \hat{t}(\varphi * F)$ for $F \in \mathcal{CLS}$ and $\varphi \in \mathcal{PASS}$, where $\hat{t} : \mathcal{MCLS} \to \mathcal{CLS}$ is the canonical projection morphism, which contracts multiple clause occurrences. For $F \in \mathcal{MCLS}$ and a clause-set $F' \in \mathcal{CLS}$ we define $F \setminus F' \in \mathcal{MCLS}$ by $(F \setminus F')(C) := F(C)$ for $C \in \mathcal{CL} \setminus F'$, while $(F \setminus F')(C) := 0$ for $C \in F'$. By $\mathcal{MUSAT}$ we denote the class of *minimally unsatisfiable clause-sets*, while by $\mathcal{SMUSAT}$ we denote the class of *saturated minimally unsatisfiable clause-set*, which are minimally unsatisfiable clause-sets which become satisfiable when any literal is added to any clause (another notion used for example in [3] instead of "saturated" is "maximal"). For any class $\mathcal{C} \subseteq \mathcal{MCLS}$ of multi-clause-sets and any measure $f : \mathcal{MCLS} \to \mathbb{N}_0$ we define for $k \in \mathbb{N}_0$

$$\mathcal{C}_f(k) := \{F \in \mathcal{C} : f(F) \leq k\}.$$

It follows that for example $\mathcal{MUSAT}_\delta(1)$ is the class of minimally unsatisfiable clause-sets of deficiency 1.

The *clause-variable matrix* $M(F)$ for $F \in \mathcal{MCLS}$ is a $c(F) \times n(F)$-matrix over $\{-1, 0, +1\}$. The order of rows and columns of $M(F)$ is arbitrary, but fixed in any given context. The matrix $E_{m,i,j}$ is the square matrix of order m with entry 1 at position (i, j) and 0 elsewhere. Furthermore we use $J_m := \sum_{i=1}^{m} \sum_{j=1}^{m} E_{m,i,j}$ for the all-one matrix, while the identity matrix is denoted by $I_m := \sum_{i=1}^{m} E_{m,i,i}$. The transposition of a matrix A is denoted by A^t. We consider $m \times n$-matrices for $m, n \in \mathbb{N}_0$; for example the $m \times 0$-matrix is the clause-variable matrix of the multi-clause-set F given by $F(\bot) = m$ and $F(C) = 0$ for $C \in \mathcal{CL} \setminus \{\bot\}$. If we speak in this article of a *graph* respectively of a *directed graph*, we mean a simple undirected graph respectively a simple directed graph (without loops and without parallel edges), while a *(directed) multigraph* is still loopless, but may have parallel edges.

3   Conflict Matrices

A *conflict matrix* is a square matrix of order $m \in \mathbb{N}_0$ with entries in $\mathbb{N}_0$ and with zeros on the main diagonal (in other words, a conflict matrix is the adjacency matrix of some directed multigraph). A *symmetric conflict matrix* is a conflict matrix which is symmetric (in other words, symmetric conflict matrices are the adjacency matrices of multigraphs). Addition of (symmetric) conflict matrices yields again a (symmetric) conflict matrix. For $F \in \mathcal{MCLS}$ let $C_s(F)$ be the *symmetric conflict matrix of* F, which is a square matrix of order $c(F)$ with entries $C_s(F)_{i,j} := |C_i \cap \overline{C_j}|$ (the number of literals $x \in C_i$ with $\overline{x} \in C_j$), where the C_i are the clauses of F (in some order; with multiplicities). And let $C_a(F)$ be the *asymmetric conflict matrix of* F, a square matrix of order $c(F)$ with entries $C_a(F)_{i,j} := |(C_i \cap \mathcal{VA}) \cap \overline{C_j}|$ (the number of positive literals $x \in C_i$ with $\overline{x} \in C_j$). Obviously $C_s(F)$ is a symmetric conflict matrix, while $C_a(F)$ is a conflict matrix, and we have

$$C_s(F) = C_a(F) + C_a(F)^t.$$

Lemma 1. *For every conflict matrix A there is $F \in \mathcal{MCLS}$ with $C_a(F) = A$, and for every symmetric conflict matrix A there is $F \in \mathcal{MCLS}$ with $C_s(F) = A$.*

The underlying (trivial) construction is very wasteful with variables, and it is natural to seek to minimise the number of variables used in F, or, in other words, to maximise the deficiency of F. This problem is the underlying theme of investigations in this context, and will be studied from Subsection 3.1 on. Consider $F, F' \in \mathcal{MCLS}$ with $F' \leq F$. The symmetric conflict matrix $C_s(F')$ is a principal submatrix of $C_s(F)$, and the asymmetric conflict matrix $C_a(F')$ is a principal submatrix of $C_a(F)$ ("principal submatrices" are obtained by deletion of rows and columns with the *same* indices). Since application of partial assignments can be realised by first removing all clauses satisfied by the assignment, and then eliminating all remaining literal occurrences falsified by the assignment, where these occurrences have become pure, for any partial assignment $\varphi \in \mathcal{PASS}$

also $C_s(\varphi * F)$ resp. $C_a(\varphi * F)$ is a principal submatrix of $C_s(F)$ resp. $C_a(F)$. For a class $\mathcal{C}$ of symmetric conflict matrices stable under formation of principal submatrices let

$$\mathcal{MCLS}(\mathcal{C}) := \{F \in \mathcal{MCLS} : C_s(F) \in \mathcal{C}\}$$

be the set of multi-clause-sets with symmetric conflict matrix in $\mathcal{C}$. It follows that $\mathcal{MCLS}(\mathcal{C})$ is stable under sub-multi-clause-set formation and application of partial assignments. Examples for such classes $\mathcal{C}$ are given as the classes of all symmetric conflict matrices A defined by one of the following conditions:

1. $A \geq J_m - I_m$ for arbitrary $m \in \mathbb{N}_0$; the corresponding multi-clause-sets are the *hitting clause-sets* considered in Section 4;
2. $A = k \cdot (J_m - I_m)$ for some fixed $k \in \mathbb{N}$ and arbitrary $m \in \mathbb{N}_0$; the corresponding k-*uniform hitting clause-sets* are considered in Section 4;
3. A has no entry 1; the class of corresponding "resolution-free" clause-sets has been considered in [17];
4. $\delta_h(A) \leq k$ for some fixed $k \in \mathbb{N}_0$ (see Subsection 3.5); the "hermitian defect" $\delta_h(A)$ is introduced in Subsection 3.2, and gives an upper bound on the deficiency of multi-clause-sets F with $C_s(F) = A$; we will see that $\delta_h(k \cdot (J_m - I_m)) = 1$ for $k, m \neq 0$ (cf. example 2);
5. $i_+(A) \leq k$ resp. $i_-(A) \leq k$ for some $k \in \mathbb{N}_0$, where $i_\pm(A)$ is the number of positive resp. negative eigenvalues of A (see Subsection 3.6); we will see that $\delta_h(A) \leq 1$ implies $i_+(A) \leq 1$.

3.1   Symmetric and Asymmetric Conflict Number

For a symmetric conflict matrix A let $\mathbf{n_s}(A)$, the *symmetric conflict number* of A, be the minimal $n(F)$ for $F \in \mathcal{MCLS}$ with $C_s(F) = A$. And for a conflict matrix A let $\mathbf{n_a}(A)$, the *asymmetric conflict number* of A, be the minimal $n(F)$ for $F \in \mathcal{MCLS}$ with $C_a(F) = A$. Obviously we have $n(F) \geq n_a(C_a(F)) \geq n_s(C_s(F))$. For conflict matrices A the asymmetric conflict number $n_a(A)$ is the *biclique partition number* of a directed multigraph with adjacency matrix A, while for symmetric conflict matrices A the symmetric conflict number $n_s(A)$ is the biclique partition number of an undirected multigraph with adjacency matrix A. Thus we can apply [12] and conclude, that the decision problem $n_s(A) \leq k$, where the input size is k plus the size of the graph with adjacency matrix A, is NP-complete.

Lemma 2. *Consider a conflict matrix A.*

1. *The asymmetric conflict number $n_a(A)$ is the minimal $q \in \mathbb{N}_0$, such that there are $\{0,1\}$-matrices $B_1, \ldots, B_q$ of rank 1 with $A = B_1 + \cdots + B_q$.*
2. *If A is symmetric, then $n_s(A)$ is the minimal $q \in \mathbb{N}_0$, such that there are symmetric $\{0,1\}$-matrices $B_1, \ldots, B_q$ of rank 2 with $A = B_1 + \cdots + B_q$.*

Lemma 3. *For every conflict matrix A of order $m \in \mathbb{N}_0$ the asymmetric conflict number $n_a(A)$ is the minimal $q \in \mathbb{N}_0$ such that there is an $m \times q$-matrix X and a $q \times m$-matrix Y over $\{0,1\}$ with $A = X \cdot Y$.*

3.2   The Hermitian Rank

Since every presentation $A = B + B^t$ for a symmetric conflict matrix A and a conflict matrix B corresponds to one possible choice for the polarities of conflicting (clashing) literals, the following lemma holds.

Lemma 4. *For every symmetric conflict matrix A the symmetric conflict number $n_s(A)$ equals the minimal asymmetric conflict number $n_a(B)$ for conflict matrices B with $B + B^t = A$. Thus by Lemma 3 if follows that $n_s(A)$ is the minimal q such that there are $\{0,1\}$-matrices X, Y with $A = XY + Y^t X^t$, where X has q columns and Y has q rows.*

For any symmetric real matrix A let $\boldsymbol{i_+(A)}$ be the number of positive eigenvalues of A, and let $\boldsymbol{i_-(A)}$ be the number of negative eigenvalues of A. Following [10], the *hermitian rank* $\boldsymbol{h(A)}$ of A now is defined as

$$h(A) := \max(i_+(A), i_-(A)).$$

See for example [18], Chapters 10 and 11 for background information on *orthogonal geometries*. The following characterisations of the hermitian rank are direct consequences of the results from [10].

Lemma 5. *Consider a symmetric real matrix A. Now the following three quantities are equal to the hermitian rank $h(A)$:*

1. *the minimal number $k \in \mathbb{N}_0$ such that there are real matrices X, Y with $A = X^t X - Y^t Y$, where X, Y both have k rows;*
2. *the minimal number $k \in \mathbb{N}_0$ such that there are real matrices X, Y with $A = Y^t X + X^t Y$, where X, Y both have k rows;*
3. *the minimal $\mathrm{rank}(B)$ for real matrices B with $A = B + B^t$.*

Corollary 6. *For all symmetric real matrices A, B we have $h(A+B) \le h(A) + h(B)$. It follows, that for $X = A + B$ we have $h(X) \ge h(A) - h(B)$.*

3.3   The Hermitian Defect

Let $\delta_h(A) := m - h(A)$ be the *hermitian defect* of a symmetric real matrix A of order m, and for $F \in \mathcal{MCLS}$ let $\delta_h(F) := \delta_h(C_s(F))$. For rational matrices A the hermitian defect can be computed in polynomial time. Since a conflict matrix has trace zero, and the trace of a symmetric real matrix equals the sum of its eigenvalues (with multiplicities), for every non-empty conflict matrix A we have $\delta_h(A) \ge 1$. By Lemma 4 together with Lemma 5 the following reformulation of the Graham-Pollak Theorem is a direct consequence.

Corollary 7. *For every conflict matrix A we have $n_s(A) \ge h(A)$. It follows that for every multi-clause-set $F \in \mathcal{MCLS}$ the deficiency of F is at most the hermitian defect of F, i.e., $\delta(F) \le \delta_h(F)$.*

For a real symmetric matrix A of order m let $\theta_1(A) \geq \cdots \geq \theta_m(A)$ denote the eigenvalues of A (in descending order). If B is a real symmetric matrix of order $m' \leq m$, then it is said that the eigenvalues of B *interlace* the eigenvalues of A iff we have (see Chapter 9 in [8])

$$\theta_1(A) \geq \theta_1(B), \qquad \theta_2(A) \geq \theta_2(B), \qquad \cdots \qquad \theta_m(A) \geq \theta_m(B)$$
$$\theta_n(A) \leq \theta_m(B), \qquad \theta_{n-1}(A) \leq \theta_{m-1}(B), \qquad \cdots \qquad \theta_{n-m+1}(A) \leq \theta_1(B).$$

Lemma 8. *Consider two real symmetric matrices A, B such that the eigenvalues of B interlace the eigenvalues of A. Then $\delta_h(B) \leq \delta_h(A)$ holds.*

Since the eigenvalues of a principal submatrix of a symmetric real matrix A interlace the eigenvalues of A (see Theorem 9.1.1 in [8]), we obtain the following fundamental property of the hermitian rank.

Corollary 9. *For every real symmetric matrix A and every principal submatrix A' we have $\delta_h(A') \leq \delta_h(A)$. It follows that for every $F, F' \in \mathcal{MCLS}$ with $F' \leq F$ and every partial assignment φ we have $\delta_h(F') \leq \delta_h(F)$ and $\delta_h(\varphi * F) \leq \delta_h(F)$.*

Corollary 10. $\forall F \in \mathcal{MCLS} : \delta^*(F) \leq \delta_h(F)$.

A further strengthening is possible (compare [1]).

Theorem 11. *For all $F, F' \in \mathcal{MCLS}$ with $F' \lneq F$ we have $\delta(F') < \delta_h(F)$.*

Corollary 12. *For multi-clause-sets $F \in \mathcal{MCLS}$ with $\delta(F) \leq \delta_h(F) - 1$ we have $\delta^*(F) \leq \delta_h(F) - 1$.*

3.4   Exact and Eigensharp Multi-clause-Sets

A conflict matrix A is called *eigensharp* if $n_s(A) = h(A)$ holds. Thus a conflict matrix A is eigensharp iff there is $F \in \mathcal{MCLS}$ with $C_s(F) = A$ and $\delta(F) = \delta_h(F) = \delta_h(A)$. Every zero matrix $0 \cdot J_m$ is eigensharp. The notion "eigensharp" was introduced in [12] for biclique partitions of graphs; it seems more natural to put the matrix A into the foreground, while the graph with adjacency matrix A is only one possible representation of A — another possible representation is to consider a graph such that its *distance matrix* is A. To decide whether $F \in \mathcal{MCLS}$ or a symmetric conflict matrix is eigensharp can be done in polynomial time. Call $F \in \mathcal{MCLS}$ *exact* if $n(F) = n_s(C_s(F))$ holds, and call F *eigensharp* if $\delta(F) = \delta_h(F)$ holds. Thus F is eigensharp iff F is exact and $C_s(A)$ is eigensharp. Every $F \in \mathcal{MCLS}$ with $n(F) = 0$ is eigensharp, and if $C_s(F)$ has at least one entry equal to $n(F)$ then F must be exact. Eigensharp multi-clause-sets correspond to *eigensharp addressings* as introduced in [5], while exact multi-clause-sets have been called *optimal addressings* in [5]; the term "exact" is taken from the literature on biclique decompositions. It seems worth to note, that in our context a certain shift of attention takes place: While in [5] an eigensharp addressing is attached to a graph (i.e., distance matrix), in our context the multi-clause-set (addressing) is more dominant, and the distance matrix is derived from it. To decide whether $F \in \mathcal{MCLS}$ is exact is co-NP complete.

Lemma 13. *If a multi-clause-set $F \in \mathcal{MCLS}$ is exact, then F does not contain pure literals, and if F is eigensharp and $F \neq \top$, then we have $\delta(F) \geq 1$.*

For exact multi-clause-sets F in general $\delta(F) \geq 1$ need not hold; e.g. $F = \{\{v_1, \ldots, v_k\}, \{\overline{v_1}, \ldots, \overline{v_k}\}\}$ is exact with $\delta(F) = 2 - k$. However, if F is exact and $C_s(F)$ is the distance matrix of a non-empty connected graph, then Winkler proved $\delta(F) \geq 1$ (Theorem 15). We seek to generalise this result. In [14] the notion of *matching lean clause-sets* has been introduced, and the equivalence of the following conditions for $F \in \mathcal{MCLS}$ have been proven (for us Criterion 3 can be taken as a definition of matching leanness):

1. F is matching lean.
2. F has no non-trivial matching autarky.
3. For all $F' \in \mathcal{MCLS}$ with $F' \lneq F$ we have $\delta(F') < \delta(F)$.

Theorem 11 together with Criterion 3 immediately yields

Theorem 14. *Every eigensharp multi-clause-set is matching lean.*

A matrix is called a distance matrix if it is the distance matrix of some finite metric space. The following result of P. Winkler (see Chapter 9 in [23]) considers special distance matrices, where the underlying metric space is given by the distances of nodes in some (undirected) graph.

Theorem 15. *For every exact multi-clause-set F, where $C_s(F)$ is the distance matrix of some non-empty connected graph, we have $\delta(F) \geq 1$.*

Conjecture 16. Every exact multi-clause-set $F \in \mathcal{MCLS}$, where the symmetric conflict matrix $C_s(F)$ is the distance matrix of some connected graph, is matching lean.

3.5   Multi-clause-Sets with Hermitian Defect One

For $k \in \mathbb{N}_0$ let $\mathcal{HD}(k)$ be the set of all real symmetric $m \times m$-matrices A for $m \in \mathbb{N}_0$ with zero diagonal and hermitian defect $\delta_h(A) \leq k$. Thus we have $\mathcal{MCLS}_{\delta_h}(k) = \mathcal{MCLS}(\mathcal{HD}(k))$. $\mathcal{MCLS}_{\delta_h}(0)$ just contains the empty multi-clause-set $\top$. By Lemma 8 we get immediately the following closure properties.

Lemma 17. *For all $k \in \mathbb{N}_0$ the set $\mathcal{HD}(k)$ is stable under formation of principal submatrices, and so is the subset of $\mathcal{HD}(k)$ consisting of all conflict matrices. Thus $\mathcal{MCLS}_{\delta_h}(k)$ is stable under application of partial assignments and formation of sub-multi-clause-sets.*

In [13] the classes $\mathcal{MCLS}_{\delta^*}(k)$ of multi-clause-sets with bounded maximal deficiency have been introduced, and it has been shown, that several hard problems like satisfiability decision or computation of the maximal autark subset are decidable for these classes in polynomial time; see [6] for an alternative proof of the main result, and see [21] for the fixed parameter tractability of these classes. By Corollary 10 we get

Theorem 18. *For all $k \in \mathbb{N}_0$ we have $\mathcal{MCLS}_{\delta_h}(k) \subset \mathcal{MCLS}_{\delta^*}(k)$.*

Using the set of *matching satisfiable multi-clause-sets* as studied in [14], the following theorem follows immediately from Theorem 14 and Corollary 12; matching satisfiable (multi-)clause-sets have been introduced in [7] under the name "matched formulas", and are characterised by the condition $\delta^*(F) = 0$.

Theorem 19. *Consider $F \in \mathcal{MCLS}_{\delta_h}(1)$.*

1. *The following three conditions are equivalent:*
   (a) F is eigensharp
   (b) F is matching lean
   (c) $\delta(F) = 1 \vee F = \top$.
2. *F is matching satisfiable iff $\delta(F) \leq 0$.*

By Theorem 15 we immediately obtain the following result, which proves Conjecture 16 for all $F \in \mathcal{MCLS}_{\delta_h}(1)$ (as we will see, this contains the two extremal cases of trees and of complete graphs).

Theorem 20. *If the symmetric conflict matrix A is the distance matrix of some graph, and $\delta_h(A) \leq 1$ holds, then A is eigensharp. In terms of multi-clause-sets, consider $F \in \mathcal{MCLS}_{\delta_h}(1)$ such that $C_s(F)$ is the distance matrix of some graph, and F is exact. Then F is eigensharp (and thus F is matching lean, and in case of $F \neq \top$ we have $\delta(F) = 1$).*

Which multi-clause-sets are in $\mathcal{MCLS}_{\delta_h}(1)$? The general question seems to be fairly complicated, but we will completely characterise the *unsatisfiable* elements of $\mathcal{MCLS}_{\delta_h}(1)$ in Theorem 26. Before doing so, we consider a special class of matrices with hermitian defect one. A symmetric real matrix A of order $m \in \mathbb{N}_0$ is of *strictly negative type* if for all $x \in \mathbb{R}^m$, $x \neq 0$ with $\sum_{i=1}^{m} x_i = 0$ we have $x^t A x < 0$, and A has a zero diagonal. By definition we have, that if A is of strictly negative type, then also every principal submatrix of A is of strictly negative type.

Lemma 21. *For every matrix A of strictly negative type we have $\delta_h(A) \leq 1$.*

Lemma 22. *Consider $F \in \mathcal{MCLS}$, such that $C_s(F)$ is the distance matrix of a tree, where each edge length can be an arbitrary positive real number. Then $C_s(F)$ is of strictly negative type (as shown in Corollary 7.2 of [11]), and thus (by Lemma 21) we get $F \in \mathcal{MCLS}_{\delta_h}(1)$.*

It is well known that $J_m - I_m \in \mathcal{HD}(1)$ for $m \in \mathbb{N}_0$.[1] Since for any symmetric matrix A and $\lambda \in \mathbb{R}_{>0}$ we have $h(A) = h(\lambda \cdot A)$, it follows that with $A \in \mathcal{HD}(1)$

. For $m \geq 1$, the matrix $J_m - I_m$ has the eigenvalue -1 with multiplicity $m - 1$ (this follows immediately from $\text{rank}(J_m) = 1$; we remind the reader that for symmetric matrices the algebraic multiplicity of an eigenvalue is equal to the geometric multiplicity), while the eigenvalue $m - 1$ has multiplicity 1 (just note that $(1, \ldots, 1)^t$ is an eigenvector of $J_m - I_m$).

also $\lambda \cdot A \in \mathcal{HD}(1)$ for every $\lambda \in \mathbb{R}_{>0}$ is the case.[2] We can generalise these considerations by the following theorem, admitting one "exceptional position".

Theorem 23. *For every $\lambda \in \mathbb{R}_{>0}$ and $-\lambda < \mu \leq \lambda$ the matrix $A(\lambda, \mu) := \lambda \cdot (J_m - I_m) + \mu \cdot (E_{m,p,q} + E_{m,q,p})$ is of strictly negative type for all $m \in \mathbb{N}$ and $1 \leq p < q \leq m$. In general $A(\lambda, \mu)$ is a distance matrix, and in case of $\mu = 0$ the matrix $A(\lambda, \mu)$ is the distance of some connected graph (namely the complete graph K_m with m nodes).*

Lemma 24. *For any symmetric matrix A with zero diagonal and of order at least 2 in case there are indices $i \neq j$ with $A_{i,j} = 0$ we have $\delta_{\mathrm{h}}(A) \geq 2$.*

Lemma 25. *Consider $F \in \mathcal{MUSAT}_\delta(1)$. Then $\mathrm{C}_{\mathrm{s}}(F)$ is a $\{0,1\}$-matrix (i.e., there are no two clauses with two or more clashing literals), and $F \in \mathcal{SMUSAT}_\delta(1)$ holds if and only if $\mathrm{C}_{\mathrm{s}}(F) = J_{c(F)} - I_{c(F)}$ (i.e., iff every pair of two (different) clauses clash exactly in one literal).*

Theorem 26. *$\mathcal{MCLS}_{\delta_{\mathrm{h}}}(1) \cap \mathcal{USAT} = \mathcal{SMUSAT}_\delta(1)$. In words: The unsatisfiable multi-clause-sets with hermitian defect one are exactly the saturated minimally unsatisfiable clause-sets with defect 1.*

Corollary 27. *Consider $F \in \mathcal{MUSAT}_\delta(1)$. Then F is eigensharp if and only if $F \in \mathcal{SMUSAT}_\delta(1)$ holds.*

Consider $F \in \mathcal{MUSAT}_\delta(1) \setminus \mathcal{SMUSAT}_\delta(1)$. The question is now whether F is not exact or $\mathrm{C}_{\mathrm{s}}(F)$ is not eigensharp. The simplest example for F is given by $F := \{\,\{a\}, \{b\}, \{\bar{a}, \bar{b}\}\,\}$. We see that in fact here $\mathrm{C}_{\mathrm{s}}(F)$ is eigensharp, since we have $\mathrm{C}_{\mathrm{s}}(2 \cdot \{v\} + \{\bar{v}\}) = \mathrm{C}_{\mathrm{s}}(F)$, and thus F is not exact. Is this the case for all elements of $\mathcal{MUSAT}_\delta(1) \setminus \mathcal{SMUSAT}_\delta(1)$?[3]

3.6   Multi-clause-Sets with Only One Positive Eigenvalue

We will see, that $\mathcal{HD}(1)$, the class of symmetric distance matrices A with $\delta_{\mathrm{h}}(A) \leq 1$, consists (exactly) of the empty square matrix together with those A with one non-negative eigenvalue, where all other eigenvalues are negative. To this end let us consider the class of symmetric distance matrices A, which either are zero matrices or have exactly one positive eigenvalue; more succinctly we can describe this class as the class of all conflict matrices A with $i_+(A) \leq$

[.] Perhaps the conceptionally easiest proof is obtained by $h(\lambda I_m - \lambda J_m) \geq h(\lambda I_m) - h(\lambda J_m) = m - 1$, using Corollary 6 (and $h(A) = h(-A)$).

[.] While this paper was in the refereeing process, Xishun Zhao proved that in fact every $F \in \mathcal{MUSAT}_\delta(1) \setminus \mathcal{SMUSAT}_\delta(1)$ is not exact. And the author realised, that furthermore the class of symmetric conflict matrices of $F \in \mathcal{MUSAT}_\delta(1)$ is exactly the class of adjacency matrices of connected graphs (no parallel edges, no loops), and thus there are $F \in \mathcal{MUSAT}_\delta(1)$ with non-eigensharp symmetric conflict matrix.

1, and further generalisation for $k \in \mathbb{N}_0$, considering multi-clause-sets, yields the classes $\mathcal{MCLS}_{i_+}(k)$ and $\mathcal{MCLS}_{i_-}(k)$ where $i_\pm(F) := i_\pm(C_s(F))$. We have $\mathcal{MCLS}_{i_+}(0) = \mathcal{MCLS}_{i_-}(0) = \{ F \in \mathcal{MCLS} : C_s(F)$ is a zero matrix $\}$, and thus $\mathcal{MCLS}_\pm(0)$ is the set of all $F \in \mathcal{MCLS}$ where every literal is a pure literal for F. For every $F \in \mathcal{MCLS}$ we have $h(F) \leq n(F)$, and thus $F \in \mathcal{MCLS}_{i_+}(n(F)) \cap \mathcal{MCLS}_{i_-}(n(F))$. Furthermore

$$F \in \mathcal{MCLS}_{i_+}(1) \Rightarrow h(F) = i_-(F) = \text{rank}(C_s(F)) - i_+(F)$$

holds, where $i_+(F) = 1$ for $F \neq \top$. By the interlacing inequalities from Subsection 3.3 we get

Lemma 28. *If A is a symmetric real matrix, and B is a principal submatrix of A, then $i_+(B) \leq i_+(A)$ and $i_-(B) \leq i_-(A)$ holds. It follows that for every $k \in \mathbb{N}_0$ the classes $\mathcal{MCLS}_{i_+}(k)$ and $\mathcal{MCLS}_{i_-}(k)$ are stable under formation of sub-multi-clause-formation and applications of partial assignments.*

Since $A \in \mathcal{HD}(1)$ of order at least 2 has no zero eigenvalue, the claim that A has exactly one positive eigenvalue, while all other eigenvalues are negative, reduces to the claim $A \in \mathcal{MCLS}_{i_+}(1)$. We use the Perron-Frobenius theorem now to prove this claim, which needs a bit of preparation. A symmetric conflict matrix A is called *irreducible* if the graph with adjacency matrix A is connected (see Section 3.2 in [2]), and we call $F \in \mathcal{MCLS}$ irreducible if $C_s(F)$ is connected. Due to completeness of resolution, every minimally unsatisfiable clause-set is irreducible. If $F \in \mathcal{MCLS}$ does not contain pure literals, then F is not irreducible iff there exist $F_1, F_2 \in \mathcal{MCLS} \setminus \{\top\}$ with $F = F_1 + F_2$ and $\text{var}(F_1) \cap \text{var}(F_2) = \emptyset$, and thus the assumption, that F is irreducible, is in fact a very natural one (otherwise F is the variable-disjoint union of two non-empty multi-clause-sets). The following theorem is a direct consequence of the Perron-Frobenius theorem (see Theorem 31.8 in [23], or Theorem 8.8.1 in [8]). We remind at the notation $\theta_i(A)$ for the eigenvalues of a symmetric real non-empty matrix A in descending order (so that $\theta_1(A)$ is the largest eigenvector), and for $F \in \mathcal{MCLS}$ we use $\theta_i(F) := \theta_i(C_s(F))$.

Theorem 29. *Consider a multi-clause-set $F \in \mathcal{MCLS}$ with $c(F) \neq 0$. For all $1 \leq i \leq c(F)$ we have $|\theta_i(F)| \leq \theta_1(F)$. If $F \neq \{\bot\}$, then in fact $\theta_1(F) > 0$ holds. For any multi-clause-set $\top \lneq F' \lneq F$ we have $\theta_1(F') < \theta_1(F)$. If F is irreducible, then $\theta_1(F) \geq 0$ has multiplicity one ($\theta_1(F)$ is a simple root of the characteristic polynomial of $C_s(F)$), and the eigenspace belonging to $\theta_1(F)$ is generated by a vector whose coordinates are all positive.*

Theorem 30. *We have $\mathcal{MCLS}_{\delta_h}(1) \subset \mathcal{MCLS}_{i_+}(1)$.*

We conclude this section by an important class of matrices with at most one positive eigenvalue. A square real matrix A of order $m \in \mathbb{N}_0$ is of *negative type* if for all $x \in \mathbb{R}^m$ with $\sum_{i=1}^m x_i = 0$ we have $x^t A x \leq 0$, and A has a zero diagonal. By definition we have, that if A is of negative type, then also every principal submatrix of A is of negative type.

Lemma 31. *For every matrix A of negative type we have $i_+(A) \leq 1$.*

4   Hitting Clause-Sets

A *hitting clause-set* is a clause-set where every two (different) clauses clash in at least one literal. We use $\mathcal{HIT}$ for the set of all hitting (multi-)clause-sets; note that in a hitting multi-clause-set in fact no clause can occur more than once. Obviously we have $F \in \mathcal{HIT} \Leftrightarrow C_s(F) \geq J_{c(F)} - I_{c(F)}$ (recall example 1 from Section 3). Another basic characterisation of $\mathcal{HIT}$ is that it is the set of (multi-)clause-sets such that there is no partial assignment falsifying two clauses of F at the same time. Some selected properties of $F \in \mathcal{HIT}$ are:

1. For every $\varphi \in \mathcal{PASS}$ we have $\varphi * F \in \mathcal{HIT}$.
2. For every clause C we have that $F + \{C\} \in \mathcal{HIT}$ holds iff φ_C is a satisfying assignment for F.[4]
3. $\sum_{C \in F} 2^{-|C|} \leq 1$ and $F \in \mathcal{USAT} \Leftrightarrow \sum_{C \in F} 2^{-|C|} = 1$.
4. If F is unsatisfiable, then $F \in \mathcal{SMUSAT}$. It follows $\delta^*(F) = \delta(F)$, and a necessary condition for $F \neq \{\bot\}$ being unsatisfiable is that there must exist a variable $v \in \mathrm{var}(F)$ with $\#_v(F) \leq \delta(F)$ and $\#_{\bar{v}}(F) \leq \delta(F)$ (see Lemma C.2 in [13]); furthermore we must have $\delta(F) \geq 1$.

Every satisfiable hitting clause-set stays satisfiable when flipping the polarities of arbitrary literal occurrences (see [4, 20] for more on such clause-sets). Let $\mathcal{MUSAT}_k$ be the set of minimally unsatisfiable clause-sets F such that for every partial assignment φ with $n(\varphi) \leq k$ (i.e., φ uses at most k variables) also $\varphi * F$ is minimally unsatisfiable; since F is a clause-*set*, the application of partial assignments contracts multiple occurrences of clauses created by the application, so that for example for any assignment φ with $\mathrm{var}(F) \subseteq \mathrm{var}(\varphi)$ the clause-set $\varphi * F$ is minimally unsatisfiable (which for $c(F) \geq 2$ would not be the case for *multi*-clause-sets F). By definition we have $\mathcal{MUSAT}_0 = \mathcal{MUSAT}$, while in [13] it was shown $\mathcal{MUSAT}_1 = \mathcal{SMUSAT}$.

Theorem 32. $\mathcal{HIT} \cap \mathcal{USAT} = \bigcap_{k \in \mathbb{N}_0} \mathcal{MUSAT}_k$.

For $k \in \mathbb{N}$ let $\mathcal{UHIT}_k$, the set of *uniform k-hitting clause-sets*, be the set of all hitting clause-sets where every two (different) clauses have exactly k conflicts. The set of uniform hitting clause-sets is denoted by $\mathcal{UHIT} := \bigcup_{k \in \mathbb{N}} \mathcal{UHIT}_k$. By definition we have $F \in \mathcal{UHIT}_k \Leftrightarrow C_s(F) = k \cdot (J_{c(F)} - I_{c(F)})$ (recall example 2 from Section 3). $\mathcal{UHIT}_1$ can be characterised as the class of all clause-sets such that any two (different) clauses have a (non-tautological) resolvent. It is $\mathcal{UHIT}_k$ stable under application of partial assignments, and for $F \in \mathcal{UHIT}_k$ and any clause C we have $F + \{C\} \in \mathcal{UHIT}_k$ iff φ_C is a satisfying assignment for F which satisfies exactly k literals in every clause of F.

Theorem 33. $\mathcal{UHIT} \subset \mathcal{MCLS}_{\delta_h}(1)$.

---

[*] For a clause C the corresponding partial assignment φ_C falsifies (exactly) all literals in C, while for a partial assignment φ the corresponding clause C_φ consists of all literals falsified by φ (partial assignments are finite).

In [3] the property $\delta(F) \leq 1$ has been concluded for $F \in \mathcal{UHIT}_1$ from a conjecture of Endre Boros; in Section 5 we will give a counterexample for this conjecture, but as we now know, $\delta(F) \leq 1$ in fact must hold (for all $F \in \mathcal{UHIT}$). The following corollary follows also from the result of [3] that every $F \in \mathcal{UHIT}_1$ with $\delta(F) \geq 2$ must be satisfiable (together with known characterisations of minimal unsatisfiable formulas of deficiency 1).

Corollary 34. *Consider $F \in \mathcal{UHIT}_k$ for $k \in \mathbb{N}$. If $k \geq 2$ and $F \neq \{\bot\}$, then $F \in \mathcal{SAT}$. If $k = 1$, then F is unsatisfiable iff F is a saturated minimally unsatisfiable clause-set.*

5   On a Conjecture of Endre Boros

At the SAT'1998 workshop in Paderborn, Endre Boros conjectured that every 1-uniform hitting clause-set F different from $\{\bot\}$ with defect $\delta(F) = 1$ contains a variable $v \in \mathrm{var}(F)$, such that we have $\#_v(F) = 1$ or $\#_{\overline{v}}(F) = 1$. In other words the conjecture states, that every minimum (or exact) biclique decomposition of the complete graph must contain a star (or claw). Now in [19] this question has been discussed in detail, and the smallest counterexample to the conjecture is given by a decomposition of the K_9 into one $K_{2,4}$ and seven $K_{2,2}$ (that is, one variable has a $(2, 4)$-occurrence, while all other variables have $(2, 2)$-occurrence). The clause-variable matrix of a corresponding clause-set is

$$
\begin{pmatrix}
+ & + & & & & + & \\
+ & - & + & & & & \\
- & - & & + & + & & \\
- & - & & & - & - & \\
+ & - & - & - & & & \\
& - & + & & - & - & \\
- & & + & - & & + & \\
- & & + & & + & - & \\
+ & + & & & + & - &
\end{pmatrix}.
$$

In [19] for every K_n with $n \geq 9$ a minimum starfree biclique decomposition has been constructed, using four ad-hoc decompositions of $K_9, \ldots, K_{12}$, and then showing how from a given minimum starfree biclique decomposition of K_n one for K_{n+4} can be obtained. It would be interesting whether a more systematic approach is possible using constructions from the area of (multi-)clause-sets.

A star-free multi-clause-set has $\#_v(F), \#_{\overline{v}}(F) \geq 2$ for all $v \in \mathrm{var}(F)$. Now it is naturally to ask whether for arbitrary $k, \lambda \in \mathbb{N}$ there are eigensharp k-uniform hitting clause-sets with $\#_v(F), \#_{\overline{v}}(F) \geq \lambda$ for all $v \in \mathrm{var}(F)$?!

References

[1] Noga Alon, Richard A. Brualdi, and Bryan L. Shader. Multicolored forests in bipartite decompositions of graphs. *Journal of Combinatorial Theory, Series B*, 53:143–148, 1991.

[2] Richard A. Brualdi and Herbert J. Ryser. *Combinatorial Matrix Theory*. Cambridge University Press, 1991.

[3] Hans Kleine Büning and Xishun Zhao. On the structure of some classes of minimal unsatisfiable formulas. September 2001.

[4] Hans Kleine Büning and Xishun Zhao. Satisfiable formulas closed under replacement. In Henry Kautz and Bart Selman, editors, *LICS 2001 Workshop on Theory and Applications of Satisfiability Testing (SAT 2001)*, volume 9 of *Electronic Notes in Discrete Mathematics (ENDM)*. Elsevier Science, June 2001.

[5] Randall J. Elzinga, David A. Gregory, and Kevin N. Vander Meulen. Addressing the Petersen graph. *Bull. Inst. Math. Acad. Sinicia*, 24(2):87–102, 1996.

[6] Herbert Fleischner, Oliver Kullmann, and Stefan Szeider. Polynomial–time recognition of minimal unsatisfiable formulas with fixed clause–variable difference. *Theoretical Computer Science*, 289(1):503–516, November 2002.

[7] John Franco and Allen Van Gelder. A perspective on certain polynomial-time solvable classes of satisfiability. *Discrete Applied Mathematics*, 125:177–214, 2003.

[8] Chris Godsil and Gordon Royle. *Algebraic Graph Theory*. Springer, 2001.

[9] Ronald L. Graham and H.O. Pollak. On the addressing problem for loop switching. *Bell System Technical Journal*, 50(8):2495–2519, 1971.

[10] David A. Gregory, Valerie L. Watts, and Bryan L. Shader. Biclique decompositions and hermitian rank. *Linear Algebra and its Applications*, 292:267–280, 1999.

[11] Poul Hjorth, Petr Lisoněk, Steen Markvorsen, and Carsten Thomassen. Finite metric spaces of strictly negative type. *Linear Algebra and its Applications*, 270:255–273, 1998.

[12] Thomas Kratzke, Bruce Reznik, and Douglas West. Eigensharp graphs: Decompositions into complete bipartite subgraphs. *Transactions of the American Mathematical Society*, 308(2):637–653, August 1988.

[13] Oliver Kullmann. An application of matroid theory to the SAT problem. In *Fifteenth Annual IEEE Conference on Computational Complexity*, pages 116–124. IEEE Computer Society, July 2000.

[14] Oliver Kullmann. Lean clause-sets: Generalizations of minimally unsatisfiable clause-sets. *Discrete Applied Mathematics*, 2003. To appear.

[15] Oliver Kullmann. On the conflict matrix of clause-sets. Technical Report CSR 7-2003, University of Wales Swansea, Computer Science Report Series, 2003.

[16] Oliver Kullmann. Open problems on the combinatorics of conflicts between clauses. Available at http://cs-svr1.swan.ac.uk/~csoliver/Artikel/OpenProblemsConflicts.html, July 2003.

[17] Gábor Kusper. Solving the resolution-free SAT problem by hyper-unit propagation in linear time. In *Fifth International Symposium on Theory and Applications of Satisfiability Testing*, pages 323–332, 2002.

[18] Steven Roman. *Advanced Linear Algebra*, volume 135 of *Graduate Texts in Mathematics*. Springer-Verlag, New York, 1992. ISBN 0-387-97837-2; QA184.R65 1992.

[19] Allen J. Schwenk and Ping Zhang. Starfree biclique decompositions of complete graphs. *Bulletin of the ICA*, 23:33–62, 1998.

[20] Stefan Szeider. Generalizations of matched CNF formulas. To appear in Annals of Mathematics and Artificial Intelligence, 2002.

[21] Stefan Szeider. Minimal unsatisfiable formulas with bounded clause-variable difference are fixed-parameter tractable. Technical Report TR03-002, revision 1, Electronic Colloquium on Computational Complexity (ECCC), February 2003.

[22] Jacobus H. van Lint. {0, 1, *} distance problems in combinatorics. In Ian Anderson, editor, *Surveys in Combinatorics*, volume 103 of *London Mathematical Society Lecture Note Series*, pages 113–135. Cambridge University Press, Cambridge, 1985.

[23] Jacobus H. van Lint and Richard M. Wilson. *A Course in Combinatorics*. Cambridge University Press, Cambridge, 1992. ISBN 0 521 42260 4.

Conflict-Based Selection of Branching Rules

Marc Herbstritt and Bernd Becker

Institute of Computer Science,
Albert–Ludwigs–University,
79110 Freiburg im Breisgau, Germany
{herbstri,becker}@informatik.uni-freiburg.de

Abstract. We propose an adaptive framework for branching rule selection that is based on a set of branching rules. Each branching rule is attached a preference value that is dynamically adapted with respect to conflict analysis. Thus, our approach brings together two essential features of modern SAT algorithms which were traditionally independent from each other. Experimental results show the feasibility of our approach.

1 Introduction

In modern SAT algorithms two features are very important: the application of a branching rule and the use of non-chronological backtracking. In this paper we propose a technique that is based on a set of branching rules instead of only one. Each branching rule is attached a preference value which reflects the usefulness of the branching rule. Depending on these values we are able to perform a dynamic selection. Moreover, data from non-chronological backtracking is used to customize the dynamic selection of branching rules. Thus, two important parts of modern SAT algorithms are combined which traditionally do not depend on each other. In principal our approach may be integrated into any Davis-Putnam-like SAT algorithm, e. g. GRASP [12], SATO [18], Chaff [15] or BerkMin [5].

In this work we use GRASP as basis for implementation since it already provides a lot of branching rules and also supports non-chronological backtracking.

Experimental results show the feasibility of our approach.

The paper is structured as followed. In the next section we give an overview of popular branching rules. In Section 3 we describe in detail our adaptive framework. Then, in Section 4 experimental results are presented and discussed. Finally the paper is closed with a discussion and a conclusion.

2 Branching Rules

It is widely accepted that the application of a *"good"* branching rule is essential for solving the SAT problem. "Good" means that in case of a satisfiable SAT instance the branching rule selects variables and corresponding assignments in a sequence such that the amount of traversed search space is small compared to

E. Giunchiglia and A. Tacchella (Eds.): SAT 2003, LNCS 2919, pp. 441–451, 2004.

Table 1. DIMACS benchmarks subset.

Class	#Inst.	#SAT	#UNSAT
aim-200	24	16	8
bf	4	0	4
dubois	13	0	13
ii16	10	10	0
ii32	17	17	0
jnh	50	16	34
pret	8	0	8
ssa	8	4	4
$\sum$	134	63	71

Table 2. Application of different branching rules. Results for RDLIS and RAND are average values of 30 experiments.

Branching rule	Time [sec]	Aborts
Böhm	1817.45	8
MOM	1428.04	7
OS-JW	807.82	4
TS-JW	911.28	4
DLCS	746.30	3
DLIS	409.14	1
RDLIS	439.16	1.1
RAND	1431.85	5.7

the size of the whole search space. On the other hand, in case of an unsatisfiable SAT instance the branching rule should select variables and corresponding assignments in a sequence that quickly leads to contradictions and thus minimizes search costs.

It should be noted that the problem of choosing the optimal variable and an assignment to this variable in Davis-Putnam based SAT algorithms has been proven to be NP-hard as well as coNP-hard [10].

In this work we make use of the following branching rules:

☐ Böhm [2]

☐ Maximum Occurences on Clauses of Minimum Size (MOM) [4]

☐ One-Sided and Two-Sided Jeroslaw-Wang (OS-JW and TS-JW) [7]

☐ Dynamic Largest Combined Sum (DLCS) [12]

☐ Dynamic Largest Individual Sum (DLIS) [12]

☐ Random DLIS (RDLIS) [12]

☐ Random selection (RAND)

In [11], a detailed comparison between the branching rules mentioned above is presented. To validate these results we performed experiments for the same set of DIMACS benchmarks (see Table 1) on a AMD Athlon(TM) XP1700+, restricted to 512MB main memory and 180sec of CPU runtime for each instance. For each branching rule we applied GRASP to the set of benchmarks. Note that DLIS is the default branching rule of GRASP. Results are given in Table 2. We have counted aborted instances with 180sec wrt time. Note that the results for RDLIS and RAND are average results from 30 experiments. The standard deviation for time (number of aborts) is 179.04sec (0.94 aborts) for RDLIS, and 198.35sec (1.10 aborts) for RAND, respectively. The application of DLIS produces the best results, in particular only one instance is aborted. These results correspond to the conclusions of [11].

3   Dynamic Selection

The idea of our approach is to not only apply one branching rule during the whole search process, but to give each branching rule the possibility to make a decision assignment from time to time. To do so, we consider a set $B = \{\varrho_1, \ldots, \varrho_t\}$ where each $\varrho \in B$ denotes a branching rule. When a decision assignment must be made, we have to select a branching rule $\varrho \in B$ which selects an unassigned variable and its assignment. What are the criteria for selecting a good branching rule from B?

3.1   Preference Values

We assign a *preference value* $Pref(\varrho)$ to each branching rule $\varrho \in B$ which models the probability of ϱ to be selected. To be consistent with probability theory, we require the following constraints:

$$\forall \varrho \in B : \; 0 < Pref(\varrho) < 1 \tag{1}$$

$$\sum_{\varrho \in B} Pref(\varrho) = 1 \tag{2}$$

3.2   Selecting a Branching Rule

Assuming a valid probability distribution of all $Pref(\varrho)$ we apply selection methods well known from genetic algorithms [14] to the set B at each decision level:

Roulette-Wheel Selection: Each $\varrho \in B$ is selected with probability $Pref(\varrho)$.
Linear Ranking Selection: Consider all $\varrho \in B$ sorted in increasing order by
   their preference values $Pref(\varrho)$. Then, if ϱ has rank r, ϱ is selected with
   probability $r/(\sum_{i=1}^{|B|} i)$.
k-**Tournament Selection:** Select randomly k branching rules $B' = \{\varrho_1, \ldots,$
   $\varrho_k\}$ from B ($k \geq 2$), where each $\varrho \in B$ has equal probability to be selected.
   Now choose from B' the branching rule ϱ_m with maximum preference value
   $Pref(\varrho_m)$, i.e. $\varrho_m = \arg\max_{\varrho \in B'} Pref(\varrho)$. Note that the selected branching
   rule ϱ_m never has the minimum preference value among all branching rules.
   We set $k = 2$ in our experiments.

3.3   Conflict-Triggering Branching Rule

The use of non-chronological backtracking is mandatory to solve unsatisfiable SAT instances in practice. On the other hand in case of a satisfiable SAT instance, it is clear that there exists a conflict-free search path leading to a solution of the instance. This is why triggering a conflict can be positive or negative dependent on the solvability of the underlying instance. Hence we make the following definition.

Definition 31 (Conflict-triggering branching rule)
A branching rule $\varrho \in B$ triggers a conflict iff

1. *A conflict occurs at decision level d.*
2. *Non-chronological backtracking results in backtracking to decision level d'.*
3. *ϱ was applied at decision level $d', (d' \leq d)$.*

3.4   Rewarding or Punishing?

As conflicts are needed for solving unsatisfiable instances, we should give a reward to conflict-triggering branching rules, and in case of satisfiable instances, we should punish them. Therefore, being faced with an unknown instance, we need some estimation on the solvability of the instance. In our approach we use the clauses/variables ratio [13] to estimate the solvability. But these values hold for a *class of instances* only, e. g. 3-SAT. Therefore we introduce some kind of instance specific phase transition approximation.

Definition 32 (Individual averaged #C/#V ratio)
The individual averaged clauses/variables ratio of instance I, $AR(I)$, is set at the beginning of the search according to:

$$AR(I) = no\_of\_clauses(I)/no\_of\_variables(I) \tag{3}$$

During search, each time after a conflict occured, we update $AR(I)$ (after the backtracking process) by setting

$$AR_{new}(I) - \frac{1}{2} \cdot \left(AR_{old}(I) + \frac{no\_unresolved\_clauses(I)}{no\_free\_variables(I)} \right) \tag{4}$$

Note that *no_unresolved_clauses(I)* takes also learned conflict clauses into account. Now we can decide wether to punish or to reward:

1. If
$$\frac{no\_unresolved\_clauses(I)}{no\_free\_variables(I)} < AR_{old}(I) \ , \tag{5}$$
   we reside in a *relatively less constrained* region being more probably satisfiable. Thus conflict-triggering branching rules should be punished. We will denote this by *punishing mode*.
2. If the opposite holds we are in a *relatively more constrained* region being more probably unsatisfiable. Hence conflict-triggering branching rules should be given a reward. This will be denoted by *reward mode*.

In the following let variable *mode* be defined as

$$mode := \begin{cases} 1 & : \quad \text{in punishing mode} \\ 0 & : \quad \text{in reward mode} \end{cases} . \tag{6}$$

Please note that in our implementation we use a 10% deviation for switching *mode* to compensate small "irritations".

To achieve an estimation of the importance of a branching rule, two counters are maintained for each branching rule ϱ:

$Used(\varrho)$: Gives the number of applications of ϱ at some decision levels.
$Trigger(\varrho)$: Gives the number of how often ϱ triggered a conflict according to definition 31.

These counters are updated every time a branching rule is applied at a decision level and every time a branching rule has triggered a conflict, respectively. Based on these counters we can compute an *update factor* wrt to the preference value of branching rule ϱ:

$$Update(\varrho) = 1 + (-1)^{mode} \cdot \frac{Trigger(\varrho)}{Used(\varrho)} \quad . \tag{7}$$

Now, if ϱ triggers a conflict we compute a new preference value of ϱ by setting

$$Pref_{new}(\varrho) = Update(\varrho) \cdot Pref_{old}(\varrho) \quad . \tag{8}$$

Thus, the probability of ϱ to be selected as branching rule at a decision level is decreased in punishing mode and increased in reward mode.

3.5   Difference Distribution

To ensure property (2) we have to distribute the difference $Pref_{diff}(\varrho)$ between the old and the new preference value of ϱ to the other branching rules $B' = B \setminus \{\varrho\}$, where

$$Pref_{diff}(\varrho) = Pref_{old}(\varrho) - Pref_{new}(\varrho) \quad . \tag{9}$$

This distribution can be accomplished in different ways. We have examined the two following distribution mechanisms (t denotes the number of branching rules):

Uniform Distribution: Each branching rule $\psi \in B'$ gets the same portion:

$$Pref_{new}(\psi) \quad = \quad Pref_{old}(\psi) \quad + \quad Pref_{diff}(\varrho) \cdot \frac{1}{(t-1)} \quad . \tag{10}$$

Weighted Distribution: Each branching rule $\psi \in B'$ gets a portion proportional to its own preference value. Using $Pref_{B'} = \sum_{\psi \in B'} Pref(\psi)$, we compute the new preference values by setting

$$Pref_{new}(\psi) \quad = \quad Pref_{old}(\psi) \quad + \quad Pref_{diff}(\varrho) \cdot \frac{Pref_{old}(\psi)}{Pref_{B'}} \quad . \tag{11}$$

Afterwards we have to normalize the preference values in order to fulfill properties (1) and (2).

3.6   Initialization of Preference Values

To complete the framework of our approach suitable initialization values for the preference values should be given. We use the results from single branching rule application given in Table 2. We analyzed the results with respect to runtime, number of aborted instances and both. Applying a linear ranking restricted to this attributes results in initialization values which we will denote by Time-Rank, Abort-Rank, and Time-Abort-Rank, respectively.

It is clear that the initialization of the preference values takes place before starting search.

Table 3. Experimental results by conflict-based selection of branching rules for the benchmarks listed in Table 2.

Select	Dist	Time-Rank				Abort-Rank				Time-Abort-Rank			
		$\oslash$-T	$\sigma(T)$	$\oslash$-A	$\sigma(A)$	$\oslash$-T	$\sigma(T)$	$\oslash$-A	$\sigma(A)$	$\oslash$-T	$\sigma(T)$	$\oslash$-A	$\sigma(A)$
RW	uni	374.48	149.15	1.07	0.89	363.43	181.16	1.00	0.86	337.59	120.87	0.90	0.70
	weight	303.35	147.28	0.83	0.73	394.52	179.05	0.93	0.81	320.60	122.91	0.93	0.77
LR	uni	290.41	168.15	0.87	0.76	207.23	74.93	0.57	0.50	234.18	137.78	0.63	0.66
	weight	329.11	135.78	1.13	0.76	201.07	116.99	0.60	0.49	269.06	118.14	0.83	0.58
2T	uni	387.01	162.21	1.13	0.85	255.16	136.10	0.67	0.75	337.97	148.49	1.07	0.85
	weight	411.52	126.50	1.40	0.76	292.56	150.00	0.83	0.93	316.30	155.51	0.93	0.85

4   Experimental Results

We have integrated our approach into GRASP [12]. Since there are several possibilities for choosing a preference value initialization (3x), branching rule selection method (3x) and difference distribution mechanism (2x) we have conducted experiments with $(3 \cdot 3 \cdot 2) = 18$ configurations.

As our approach is a *randomized* method (namely the application of the proposed selection methods), we handled each instance of the benchmark set (see Table 1) 30 times for each of the 18 parameter settings. All experiments were performed on a AMD Athlon(TM) XP1700+, restricted to 512MB main memory and 180sec of CPU runtime for each instance.

Table 3 gives the results. Column *Select* gives the used branching rule selection method, where RW denotes Roulette-Wheel, LR denotes Linear Ranking, and 2T denotes 2-Tournament. In column *Dist* the applied distribution mechanism is named (*uni* denotes Uniform Distribution and *weight* denotes Weighted Distribution). For each preference value initilization (Time-Rank, Abort-Rank, Time-Abort-Rank) the corresponding column is splitted into a twofold-column wrt time and aborts, respectively. Each column is splitted into two columns for average CPU time in seconds $\oslash(T)$ (average aborts $\oslash(A)$) and standard deviation of time $\sigma(T)$ (standard deviation of aborts $\sigma(A)$). As for the single branching rule experiments, we have counted aborted instances with 180sec.

The application of GRASP using DLIS as default branching rule leads to a total runtime of 409.14 seconds and 1 aborted instance (see Table 2, column DLIS). We refer to this experiment by GRASP-DLIS in the following.

Although the average time used by the different configurations is mostly less than the time of GRASP-DLIS, the average number of aborted instances ranges between good (see column Abort-Rank) and reasonable (see column Time-Rank) compared to GRASP-DLIS. But it is evident that our approach holds the potential to reduce time and aborted instances.

Table 4. Selection of benchmarks from `bejing` [3, 6], BMC [1], `blockworld` [8], `equiv-checking` [9], `pader-hard` [16], and `quasigroup` [18].

Name	#var	#clauses	status
4blocks	758	47820	sat
barrel5	1407	5383	unsat
bw_large.d	6325	131973	sat
c7552_bug	7559	20109	sat
hfo3.002.0	215	920	unsat
hfo3.015.0	215	920	unsat
hfo3.035.0	215	920	unsat
hfo3.039.0	215	920	unsat
hfo5.032.1	55	1163	sat
hfo6.018.1	40	1745	sat
hfo6.020.1	40	1745	sat
qg5-10	1000	43636	unsat
qg6-11	1331	49204	unsat
qg7-11	1331	49534	unsat

Table 5. Results for the benchmarks of Table 4.

Solver	Time		Time	
	$\oslash(T)$	$\sigma(T)$	$\oslash(A)$	$\sigma(A)$
GRASP-DLIS	30288.43	total	7	total
RW+Abort+uni	23648.61	2830.09	5.87	1.26
RW+Abort+weight	22388.31	2494.64	5.37	0.91
LR+Abort+uni	21576.97	1664.68	5.13	0.88
LR+Abort+weight	22156.22	2308.86	4.97	1.02
2T+Abort+uni	23312.63	2486.27	4.90	1.19
2T+Abort+weight	23567.40	2291.61	4.97	0.87

The benchmarks of Table 2 are not challenging for GRASP. Therefore we have performed a second series of experiments using the benchmarks from Table

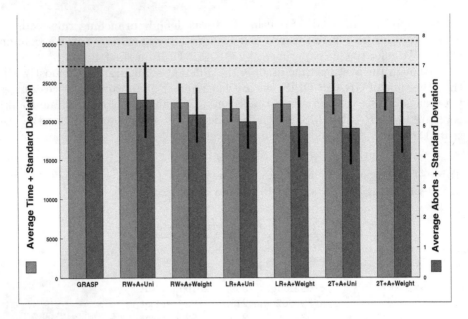

Fig. 1. Results on hard GRASP benchmarks (see Table 4), including satisfiable and unsatisfiable instances.

4. These benchmarks were selected because they are "hard" for GRASP (we refer the reader to [17]). Since the preference initialization Abort-Rank gives the best results in Table 3, we only applied this one for the new experiments.

Again we have applied our approach 30 times per benchmark instance, but now with a time limit of 3000sec (and counting aborted instances with 3000sec)[1]. The results are given in Table 5. Please note that the results for GRASP-DLIS are in the first row. It becomes evident when looking at the chart that our approach is able to significantly outperform GRASP-DLIS. The average time and the average number of aborted instances are always less than the results of GRASP-DLIS. Also, the standard deviations for time ($\sigma(T)$) and the aborted instances ($\sigma(A)$) are acceptable in the sense that we shouldn't have to expect worse results than GRASP-DLIS. Our approach is up to a factor of 1.4x faster compared to GRASP-DLIS (see LR+Abort+uni) and it reduces the average number of aborted instances up to 42% (see 2T+Abort+uni).

Figure 1 illustrates the results from Table 5. The bars on the left-hand side refer to time and the bars on the right-hand side refer to the number of aborts. The bars correspond to the average values wrt time and the number of aborts. The vertical lines within the bars represent the standard deviation at the corresponding averaged value. From Figure 1 we see that all configurations of our

---

[.] The experiments were perfomed on a Intel Xeon(TM) 2Ghz, restricted to 1GB main memory.

approach normally lead to a better result wrt time as well as number of aborts even when we are taking the standard deviation into account (this fact is indicated by the two dashed lines in the top of the figure).

We seperated the results of Table 5 wrt satisfiable and unsatisfiable instances. Figure 2 and Figure 3 illustrate the seperated results, respectively. Clearly, these figures indicate that the overall gain of our approach is mainly achieved on the satisfiable instances. But nevertheless: Regarding only the unsatisfiable instances, all configurations on average can compete with GRASP-DLIS.

5   Discussion and Future Work

The type of branching rules considered in this paper (DLIS, MOM, ...) is not adequate for modern SAT solver like Chaff [15] and BerkMin [5]. But still questions like the following remain valid: Is the branching rule of BerkMin better than that of Chaff? It is conceivable that in the future there will be a lot of Chaff-like branching rules available. The branching rules for Chaff-like SAT solvers are tightly integrated into the search algorithm (e.g. the VSIDS branching rule of Chaff). This however is not the case for the work discussed here and must be taken into account when adapting our framework to Chaff-like SAT-solvers.

There is strong need for a *reliable* and *robust* solvability measure. With reliability we mean that the solvability measure should make the right estimation "most of the time". Put another way, being faced with a huge set of instances the solvability measure should have the correct estimation for the vast majority of the instances. Additionally, robustness means that the solvability measure should work relatively *independent* of the underlying SAT solver such that for the same instance two solvability-measure enhanced SAT solvers report nearly the same estimation about the solvability of the instance.

6   Conclusions

We have presented an approach that combines non-chronological backtracking and a pool of branching rules, resulting in an adaptive framework. In principle, our method can be integrated into every Davis-Putnam-based SAT algorithm. Experimental results point out that our approach results in a faster and more robust behaviour of the SAT algorithm.

Acknowledgements
The authors want to thank Tobias Schubert, Christoph Scholl, and the participants of Dagstuhl Seminar 01051 for helpful discussions, and Matthew Lewis for proof-reading.

References

[1] A. Biere, A. Cimatti, E. Clarke, and Y. Zhu. Symbolic model checking without BDDs. In *Tools and Algorithms for the Constuction and Analysis of Systems*, volume 1579 of *LNCS*. Springer Verlag, 1999.

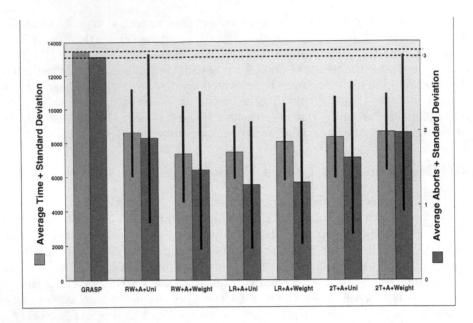

Fig. 2. Results on hard GRASP benchmarks (see Table 4), including only satisfiable instances.

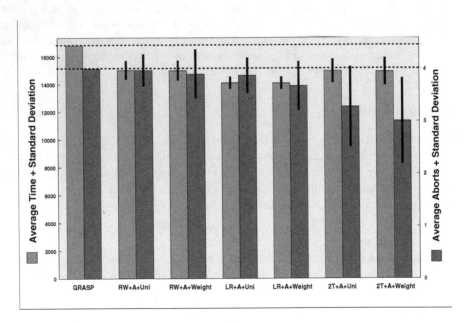

Fig. 3. Results on hard GRASP benchmarks (see Table 4), including only unsatisfiable instances.

[2] M. Buro and H. Kleine-Büning. Report on a SAT competition. Technical report, University of Paderborn, November 1992.

[3] J.M. Crawford. International Competition and Symposium on Satisfiability Testing. International Competition on SAT Testing, Bejing, 1996.

[4] J.W. Freeman. *Improvements to Propositional Satisfiability Search Algorithms.* PhD thesis, University of Pennsylvania, Philadelphia (PA), May 1995.

[5] E. Goldberg and Y. Novikov. BerkMin: A Fast and Robust SAT-Solver. In *Design, Automation, and Test in Europe (DATE '02)*, pages 142–149, March 2002.

[6] H.H. Hoos and T. Stützle. SAT2000: Highlights of Satisfiability Research in the year 2000. In *Frontiers in Artificial Intelligence and Applications*, pages 283–292. Kluwer Academic, 2000.

[7] R.G. Jeroslow and J. Wang. Solving propositional satisfiability problems. *Annals of Mathematics and Artificial Intelligence*, 1:167–187, 1990.

[8] Henry A. Kautz and Bart Selman. Pushing the envelope : Planning, propositional logic, and stochastic search. In *Proceedings of the Twelfth National Conference on Artificial Intelligence (AAAI'96)*, pages 1194–1201, 1996.

[9] W. Kunz and D. Stoffel. *Reasoning in Boolean Networks.* Kluwer Academic Publishers, 1997.

[10] P. Liberatore. On the complexity of choosing the branching literal in DPLL. *Artificial Intelligence*, 116(1-2):315–326, 2000.

[11] J.P. Marques-Silva. The impact of branching heuristics in propositional satisfiability algorithms. In *9th Portuguese Conference on Artificial Intelligence (EPIA)*, 1999.

[12] J.P. Marques-Silva and K.A. Sakallah. GRASP – a new search algorithm for satisfiability. In *Int'l Conf. on CAD*, pages 220–227, 1996.

[13] D. Mitchell, B. Selman, and H. Levesque. Hard and easy distributions of SAT problems. In *National Conference on Artificial Intelligence (AAAI)*, 1992.

[14] M. Mitchell. *An Introduction to Genetic Algorithms.* The MIT Press, 1996.

[15] M.W. Moskewicz, C.F. Madigan, Y. Zhao, L. Zhang, and S. Malik. Chaff: Engeneering an efficient SAT solver. In *Design Automation Conf.*, 2001.

[16] available for downloading at http://sat.inesc.pt/benchmarks/cnf/uni-paderborn.

[17] Laurent Simon and Philippe Chatalic. SATEx: a web-based framework for SAT experimentation. In Henry Kautz and Bart Selman, editors, *Electronic Notes in Discrete Mathematics*, volume 9. Elsevier Science Publishers, June 2001. http://www.lri.fr/~simon/satex/satex.php3.

[18] H. Zhang. SATO: An efficient propositional prover. In *Proceedings of the International Conference on Automated Deduction*, pages 155–160, July 1997.

The Essentials of the SAT 2003 Competition

Daniel Le Berre[1]* and Laurent Simon[2]

* CRIL – CNRS FRE2499,
Université d'Artois, Faculté Jean Perrin
Rue Jean Souvraz SP 18 – F 62307 Lens Cedex, France
leberre@cril.univ-artois.fr
* LRI, Université Paris-Sud
Bâtiment 490, U.M.R. CNRS 8623 – 91405 Orsay Cedex, France
simon@lri.fr

Abstract. The SAT 2003 Competition ran in February – May 2003, in conjunction with SAT'03 (the Sixth Fifth International Symposium on the Theory and Applications of Satisfiability Testing). One year after the SAT 2002 competition, it was not clear that significant progress could be made in the area in such a little time. The competition was a success – 34 solvers and 993 benchmarks, needing 522 CPU days – with a number of brand new solvers. Several 2003 competitors were even able to solve within 15mn benchmarks remained unsolved within 6 hours by 2002 competitors. We report here the essential results of the competition, interpret and statistically analyse them, and at last provide some suggestions for the future competitions.

1 Introduction

Organizing a competition is a large effort for both submitters and organizers. It is thus essential to ensure that this effort is keeping worthwhile for the community. Clearly, such an event keeps the community excited and allows a good visibility of the SAT field outside its own frontiers. For the second edition of the competition, the challenges were to bring up new solvers, to measure progress made since last year and, more importantly, to identify promising techniques and new challenges for SAT solvers.

The aim of this paper is to analyze the good (and bad) points of the competition, to fairly report the performances of solvers and to emphasize what we think was (or was not) a success. Exhaustive results are available via the web, in a dynamic HTML and a raw format, which allows anyone to make its own analysis. The paper is organized as follows: in a first part, after recalling the main rules of the contest, we present the solvers, the benchmarks and the results of the contest identifying the winners. Then, we analyse the first phase results and finally we discuss SAT'03 competition issues and propose some improvements for the next competition. We'll try to demonstrate that the competition is now mature

---

* This work has been supported in part by the IUT de Lens, the CNRS and the Région Nord/Pas-de-Calais under the TACT Program.

E. Giunchiglia and A. Tacchella (Eds.): SAT 2003, LNCS 2919, pp. 452–467, 2004.

enough to evolve in such a way that it can fulfill its three foundation principles: motivate the field, identify new benchmarks and techniques and experimentally study solvers behavior.

2   Presentation of the SAT 2003 Competition

Brief Reminder of the Competition Rules The competitions runs in two phases. The first one is a way to identify the best solvers and the hardest benchmarks, in order to spend more time with them in the second phase. The first phase is thus important to provide exhaustive empirical results for analysis, while the second one provides more limited results, mainly dedicated to award both solvers and benchmarks (see [19] for more details)..

We use the notions of *categories* and *series* for benchmarks. Benchmarks are partitioned into 3 categories (*Industrial, handmade* and *Random*), then in series. The notion of categories is discussed in section 6. The idea of a series is to group together benchmarks that are similar (*e.g.* same generator, ...) to identify solvers that can solve a wide variety of different benchmarks. A series is considered as solved if at least one benchmark in the series is solved. The solvers are ranked after the first phase according to the number of series and then the total number of solved benchmarks. From those anonymized rankings, the competition judges decide which solvers should enter the second phase for each category. Note that only one variant of the same solver is eligible for the second phase. After the second phase, the winner is the solver that solved the greatest number of benchmarks. The competition awards the smallest unsolved SAT (resp. UNSAT) benchmark and 6 solvers: 3 categories of benchmarks in which we award the best complete solver (answers SAT+UNSAT) and the best solver on satisfiable benchmarks (Complete + Incomplete solvers).

The competition ran this year on two clusters of Linux boxes. One, from the "Laboratoire de Recherche en Informatique" (LRI, Orsay, France), was composed of 15 Athlon 1800+ with 1Gb memory. The second one, from "Dipartimento di Informatica Sistemica e Telematica" (DIST Genoa, Italy), was composed of 8 PIV 2.4GHz with 1Gb memory.

2.1   Submitted Solvers

Solver submission closed on February 14th but late submissions were accepted *hors-concours*[3] during the I/O compliance testing.

We had 28 **complete non randomized** solvers: berkmin (5.61 and 6.2, awarded last year) from E. Goldberg and Y. Novikov [7]; bmsat (hors concours) from X. Xu; compsat (0.5) from A. Biere. Because a bug was discovered, we report the fixed version, compsat-fixed, as a hors-concours solver; farseer (1.00) from M. Girard, forklift from E. Goldberg and Y. Novikov ; funex (0.2)

[.] Those solvers entered the first phase, which allowed them to compete with current state-of-the-art SAT solvers, without being awardable from A. Biere; jerusat (1.1 *a, b, c*) from N. Alexander [2]; jquest2 from I. Lynce and J. Marques-Silva [13]; kcnfs from G. Dequen and O. Dubois [5, 4]; limmat (1.3), awarded last year, from A. Biere; lsat (1.1) from R. Ostrowski, B. Mazure and L. Sais [15]; march sp/tt/xq from M. Dufour, M. Heule, J. van Zwieten and H. van Maaren. Because a bug was found in marchxq, we report here the fixed version, named marchxq-fixed, as a hors-concours solver; oepir (0.2) from J. Alfredsson; oksolver from O. Kullmann [9] as an *invited* hors-concours solver (oksolver was awarded twice last year); opensat (0.44) from G. Audemard, D. Le Berre and O. Roussel [3]; satnik from N. Sörensson [6]; sato (3.4) from H. Zhang [21]; satzilla v1/v2/v2-fixed (0.9) (v2s are hors-concours) from K. Leyton-Brown, E. Nudelman, G. Andrew, C. Gomes, J. McFadden, B. Selman and Y. Shoham [11, 10]. Authors discovered during the contest that satzilla2 had bugs in its learning phase that might lead to wrong choices of solvers. The corrected version is referred to as satzilla2-fixed; satzoo v0/v1 (0.98) from N. Eén [6]; tts (1.0), hors-concours, from I. Spence; and zchaff from Y. Yu and L. Zhang [14, 22], awarded last year.

We had 2 **complete randomized** solvers: siege, hors-concours, from L. Ryan and xqinting (0.1) from X. Yu Li, M. Stallmann and F. Brglez.

And, at last, we had 4 **incomplete** (one-sided SAT) solvers: amvo from S. vogt and A. Machen; qingting (1.0) from X. Yu Li, M. Stallmann and F. Brglez [12]; saturn (2) from S. Prestwich [16]; and unitwalk (0.981) from E. Hirsch and A. Kojevnikov [8].

2.2 Submitted Benchmarks

We had 17 submitters, but a lot of benchmarks were submitted by the same person (BMC from IBM, generators for random/structured formulae). We also reused all SAT'02 challenges (except some from SAT'02 random challenges, because there was too many of them). We tried to equilibrate the CPU effort on categories: Industrial with 323 benchmarks in 45 series, Handmade with 353 benchmarks in 34 series, Random with 317 benchmarks in 30 series. As we did last year, all the benchmarks were shuffled to avoid easy syntactical recognition of benchmarks by solvers. Regarding random and handmade benchmarks, we only considered 3 benchmarks per point. That means that once we fixed all the parameters for the generator, we generated 3 (and only 3) benchmarks with 3 different seeds.

The **Industrial** category is composed of a total number of 323 benchmarks with, on average, 42703 variables, 163947 clauses and 452283 literals: hard_eq_check (16 benchmarks in 1 series), from E. Goldberg; addm (6 benchs, 1 series) from J. kukula (equivalence check of two randomly structured adders with multiple addends); li-exam and test (12 benchs, 2 series) from R. Li (unit covering problem of logic minimization of logic circuit benchmark test4); ferry, gripper and hanoi (24 benchs, 3 series) from F. Maris (based on TSP (tunable satplan), a generator for planing problems in PDDL format); l2s (11 benchmarks, 1 series) from V. Schuppan (Bounded model checking of Bwolen Yangs collection of benchmarks for SMV in conjunction with a method to convert live-ness to safety); zarpas (223 benchmarks in 25 series) from E. Zarpas (a lot of (large) BMC formulae).

The **Handmade** category is composed of a total number of 353 bench-marks having, on average 3233 variables, 37355 clauses and 154485 literals: sgi (120 benchmarks, 8 series) from C. Anton (subgraph isomorphisms); graphs (56 benchmarks, 7 series) from R. Bevan (based upon various graphs as first described by Tseitin in 1968); qgh (19 benchmarks, 2 series) from C. Gomes (quasigroups with holes); chesscolor (5 benchs, 1 series) from F. Lardeux (en-code an original chessboard coloring problem); multLin and mm (17 benchmarks, 1 series) from K. Markstrom (various instances of fast matrix multiplication); genurq (10 benchmarks, 1 series) from R. Ostrowski (modified Urquhart prob-lems, all satisfiable); clusgen (9 benchmarks, 3 series) from A. Slater (randomly generated clustered 3SAT problems; hwb (21 benchmarks, 1 series) from T. Stan-ion (equivalence checking problem of two different circuits computing the hidden weighted bit function).

Finally, the **Random** category is made of 317 benchmarks that *look like (cf section 6)* uniform random benchmarks with on average 442 variables, 4164 clauses and 25401 literals: balanced and hidden (111 benchmarks, 8 series) from C. Moore (1 or 2 hidden assignments and balanced literals); hgen* (57 bench-marks, 4 series) from E. Hirsch (generators based on hgen2, last year winner); gencnf (63 benchmarks, 7 series) from G. Dequen (uniform generator at thresh-old for $k \in [4, 10]$); uniform (45 benchmarks, 5 series) from L. Simon (uniform generator for k=3, one series for each ratio $r \in \{3.0, 4.0, 4.25, 4.5, 5.5\}$).

3   First Phase Results

The first phase began on March 19th and ended on April 14th. During the whole process of the competition, solver submitters were able to check virtually in real time the progress of their own solvers. All the traces were available, and suspicious launches (due to hardware/network error, ...) were ran again. The two solvers compsat and marchxq were found incorrect during the first phase due to minor bugs. Let us recall that we present here the result of their fixed version, compsat-fixed and marchxq-fixed. We will not discuss here the results of the satisfiable subcategory in both the industrial and handmade categories, since there are not enough benchmarks to draw any conclusions.

First Phase on Industrials The figure 1 illustrates the results of the complete solvers on all the benchmarks of the industrial category. This representation, used in the previous competition report [19], allows to check the choice of the CPU timeout and to have clues about solvers behaviors on all the benchmarks in a given category. Last year winners in the category are zchaff and limmat. Those solvers close a group of zchaff-like solvers leaded by forklift. If those solvers are taken as the references for state-of-the-art SAT solvers in 2002 then some significant progress was made in the industrial category. However, one can

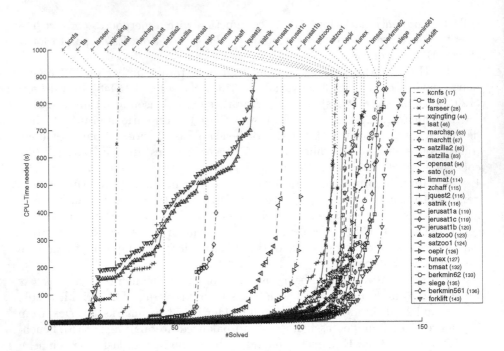

Fig. 1. # of solved instances vs. CPU time for complete solvers on all industrial benchs

note that few solvers are stronger than `berkmin62`, which entered the SAT'02 competition but had a bug that prevented it to solve more than one hundred of benchmarks [19]. The current version of `berkmin62` is of course fixed, so it is reasonable to consider it as a progress witness for the year 2002. In that case, few improvements were made: only three solvers performed better that `berkmin62` and two of them are from the same authors, one solver being a variant of one of the engine of `berkmin62`, while the other one is a new solver that can be seen as an extension of `berkmin62` with binary clause reasoning. At last, one can notice that there were many zchaff-like solvers this year (13 solvers, 17 variants, against 4 solvers last year). This can be explained by the huge interest in the practical resolution of SAT in the EDA community and by the fact that zchaff was a breakthrough in the community in the last two years. So, the good point is that a lot of new solvers were proposed and showed relatively good performances for their first public evaluation. But the pernicious effect of the competition is that, in the run for good performances, many solvers are very similar (zchaff-like).

First Phase on Handmade The figure 2 illustrates the results of the complete solvers on handmade benchmarks. One solver, `lsat`, clearly shows a particular behavior. This solver does not implement only resolution, but can handle (even long) chains of equivalency clauses and can more generally retrieve from a CNF

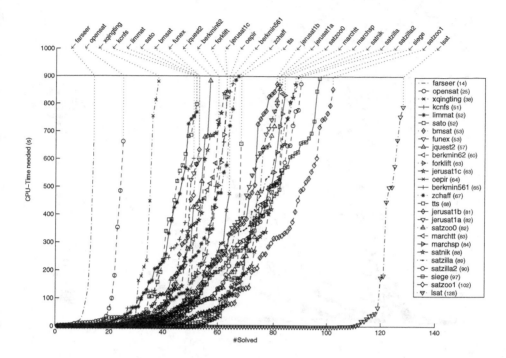

Fig. 2. # of instances solved vs. CPU time for complete solvers on all handmade benchs

some boolean function on which it applies some simplifications [15]. That point gives it the ability to quickly solve tricky formulas, commonly used in this category. The figure also shows relatively efficient solvers between `marchxq-fixed` and `satzoo1` on this category.

First Phase on Random Results for the two random categories are given on figures 3 and 4. On the first figure, one can easily identify all complete solvers suited for random benchmarks (between `oksolver` and `satzilla2-fixed`). The curves have a relatively slow growth (certainly due to the fact that random instances scales smoothly), especially for `kcnfs`, which seems to scale well. This is not the same picture for `satzilla`, which present the same kind of curve than `lsat` on handmade: a lot of benchmarks are solved quickly, and then the curve have an important growth. On SAT instances, `unitwalk` and `satzilla2` present the same kind of curves.

The Case of Satzilla The solver takes an unusual, "portfolio" approach to the SAT problem [11, 10]. It uses machine learning models to choose the best solver among a set of existing SAT algorithms. For `satzilla` 0.9, these include: `2clseq`, `limmat`, `jerusat`, `oksolver`, `relsat`, `sato`, `satzrand`, and `zchaff`

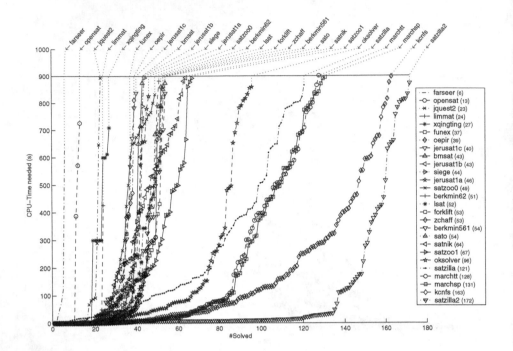

Fig. 3. # of instances solved vs. CPU time for complete solvers on all random benchs

(Hors Concours solver `satzilla2` uses two additional complete solvers, `eqsatz` and `heerhugo`). It also executes the `autoSat` algorithm for a short time before starting any other computation, and is thus able to filter out easy satisfiable problems (this point can explain the kind of curves we observed). We conducted a few analysis on `satzilla2-fixed` on the random category and report here how many times it launched a given solver and how many times the solver succeeded (number in parenthesis): `jerusat` 1 (1), `zchaff` 12 (2), `eqsatz` 18 (18), `2clseq` 13 (1), `satzrand` 124 (37) and `oksolver` 35 (1). One may notice that `zchaff` was chosen 12 times on random instances (and solved 2 benchmarks). Note that `eqsatz` was launched 18 times on random problems and succeeded each time. This is because the `hardnm` benchmarks from C. Moore contain equivalency chains that can be solved in linear time by specific solvers such as `eqsatz` of `lsat`. For such a reason, the behavior of `satzilla2` on those "random" benchmarks must be taken with care.

Solvers Entering the Second Phase This year, the three judges decided, from a complete anonymized version of the rankings presented table 1 which solvers should enter the second phase of the competition. Table 1 also emphasizes the solvers that were alone to solve at least one benchmark (so called State-Of-The-Art Contributor, SOTAC, in [20]) during the first phase. In addition to

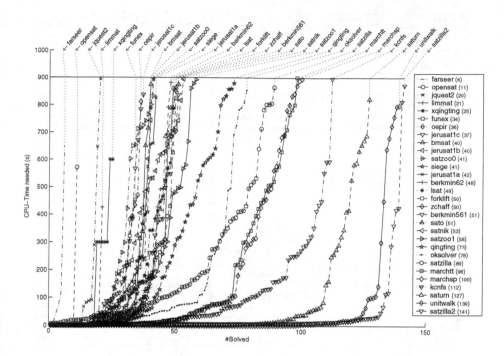

Fig. 4. # of instances solved vs. CPU time for all solvers on SAT random benchs

the one reported in the tables, `marchxq-fixed` (Industrial) and `compsat-fixed` (Industrial, HandMade) are also SOTAC.

The Hardest Benchmarks The pool of benchmark remained unsolved during the first phase is available on SATLIB[4]. We encourage developers to focus on this set of hard benchmarks. The industrial category contains 174 benchmarks (from 16 series, mean (min–max) sizes: #clauses=222811 (7814–3113540), #variables=61322 (2269–985042), #literals=627908 (21206–7976612)), the handmade one 165 benchmarks (from 18 series, #clauses=54422 (832–630000), #variables=4518 (75–270000), #literals=263992 (2818–1350000)) and the Random category 91 benchmarks (from 17 series, #clauses=7838 (391–36000), #variables=286 (38–700), #literals=61515 (888–360000)). Among the unsolved benchmarks, we often find large industrial formulas and, in the random category, k-cnf with large k.

4 The Second Phase: the Winners

"Best" Solvers The table 2 summarizes the whole second phase of the contest. Here, the CPU timeout was set to 2 hours (instead of 15mn). The 4 winners are

---
* http://www.satlib.org

Solvers	#series	#benchs (comments)	Solvers	#series	#benchs (comments)
Industrial					
Complete solvers on all benchs			All solvers on SAT benchs		
forklift	34	143 (SOTAC)	forklift	9	33
berkmin561	32	136	berkmin62	8	32
satzoo1	31	124	zchaff	8	31
oepir	30	126	oepir	8	31
funex	29	127 (SOTAC)	jerusat1b	8	31
satnik	29	116	funex	8	31
jquest2	29	116	satnik	8	30
zchaff	29	115 (SAT'02 winner)	limmat	8	30 (SAT'02 winner)
jerusat1b	28	120	satzoo1	7	30
limmat	28	114	jquest2	7	29
HandMade					
Complete solvers on all benchs			All solvers on SAT benchs		
jerusat1b	22	81	satzoo1	11	37 (SOTAC)
satzoo1	21	102	jerusat1b	11	35
satzilla	21	89	satzilla	11	25
satnik	20	88	forklift	10	30 (SOTAC)
marchsp	19	84	oepir	9	31 (SOTAC)
lsat	15	128 (SOTAC)	satnik	9	30
			berkmin561	9	30 (v62 was SAT'02 winner)
			saturn	7	29 (SOTAC)
Random					
Complete solvers on all benchs			All solvers on SAT benchs		
kcnfs	25	163 (SOTAC)	kcnfs	18	112
satzilla	19	121	unitwalk	15	139 (SOTAC)
marchsp	18	131	saturn	14	127
oksolver	16	96	satzilla	14	89
satnik	13	64	marchsp	13	100

Table 1. Ranking of solvers entering the second phase.

in bold face in each category. Briefly, **forklift** won the two (SAT+UNSAT and SAT) Industrial categories, **satzoo1** won the two handmade categories. **kcnfs** won the Random SAT+UNSAT category and **unitwalk** the Random SAT category. Note that all awarded solvers showed good performances during the first stage, and that this year, an incomplete solver was awarded in the random SAT category. At last, we have to point out that winners are often determined by very small difference. This is not really surprising since the shapes of curves are almost vertical on figures 1–4 for these solvers, but may be disappointing since after one month of contest, the winner was determined on its performances on a very few benchmarks.

Hardest Benchmarks In each categories, we selected the smallest benchmarks to use during the second phase, according to their number of literals, in order to award the smallest unsolved benchmarks in both SAT and UNSAT categories.

The winner for the smallest unsat benchmark is `random/hirsch/hgen8/hgen8--n260-01-S1597732451`, a benchmark submitted by E. Hirsch. This unsatisfiable benchmark contains 391 clauses, 260 variables and only 888 literals.

On the SAT category, we did not award any benchmark. The winner was `random/simon/sat02-random/hgen2-v400-s161064952` (renamed `sat03-1681` this year), a benchmark submitted by E. Hirsch last year. It contains 1400 clauses, 400 variables and 4200 literals. That benchmark was solved in the second phase

Solver	#benchs	Solver	#benchs	Solver	#benchs
Industrial Complete		Industrial SAT		HandMade Complete	
forklift	**12**	**forklift**	**4**	**satzoo1**	**9**
berkmin561	11	jerusat1b	3	lsat	7
satzoo1	5	satnik	3	satzilla	6
jerusat1b	5	funex	1	satnik	3
satnik	4	zchaff	1	jerusat1b	1
zchaff	4	oepir	1	marchsp	0
funex	3	berkmin62	1		
oepir	1	satzoo1	1		
jquest2	0	limmat	0		
limmat	0	jquest2	0		
Random Complete		Random SAT		HandMade SAT	
kcnfs	**12**	**unitwalk**	**12**	**satzoo1**	**5**
satzilla	6	kcnfs	4	satzilla	3
oksolver	4	saturn	2	forklift	3
marchsp	3	satzilla	2	berkmin561	1
satnik	0	marchsp	0	saturn	0
				oepir	0
				satnik	0
				jerusat1b	0

Table 2. Ranking of solvers after the second phase

Solver	Total Time (s)	#solved benchs	Solver	Total Time (s)	#solved benchs
Complete on Industrials			Complete on handmade		
forklift	171240	143	lsat	207371	128
siege	175168	135	satzoo1	241664	102
berkmin561	176088	136	satnik	249446	88
bmsat	178445	132	siege	250043	97
berkmin62	179121	133	marchtt	250711	83
Complete on Random			All On SAT Random		
satzilla2	147208	172	satzilla2	35402	141
kcnfs	165655	163	unitwalk	38572	139
unitwalk	166372	139	saturn	52136	127
saturn	179936	127	kcnfs	67493	112
marchsp	192688	131	marchsp	83030	100

Table 3. Top-5 ranking of solvers à la satex

last year, but not this year. Such a phenomenon may be due to bad luck this year for randomized instances, or simply to the re-shuffling of the instance. We'll discuss this kind of problem in the section 5.

5   Discussion

To check the strength of the competition final results or to find other ways to analyse the results, one may want to try other methods to rank solvers and to compare the rankings. For instance, in the contest, we tried to forget about the cputime, by counting only the number of series and benchmarks solved. Of course the CPU time had a direct impact on the competition due to the timeout, but it is important to try to characterize efficient solvers, in term of CPU time.

Ranking à la SatEx In SatEx [18], rankings are done by counting the total CPU time needed for one solver to solve the whole category of benchmarks, using the timeout as the default penalty. Note that the timeout was 15 mn, which is not a big penalty, so a fast solver can easily be ranked above a solver that solved more benchmarks. Table 3 summarizes the top-5 solvers in each category that we studied in this paper. forklift is still a clear winner (siege was hors-concours) on Industrial benchmarks. On handmade benchmarks, such a ranking would have awarded lsat instead of satzoo1. lsat quickly solved a large amount of benchmarks without search, but when the DLL engine of lsat was needed (on hard instances like in the second phase), lsat was slower than satzoo1. On random benchmarks, satzilla2 (which is hors-concours) would be ranked first, which is in part due to the 18 hardnm benchmarks outlined before.

Relative Efficiency of Solvers One of the hard things to handle for ranking solvers is that only partial informations are available, because of the timeout. One possible way is to compare only pairs of solvers (X,Y) on the subset of benchmarks they both solves (let use write $s(X)$ as the set of benchmarks solved by X), then we can compare X and Y on their respective performances on the set $s(X) \cap s(Y)$). When doing this, we have a strong way of comparing the *relative efficiency* (RE) of X with Y : $re(X, Y) = s(X) \cap s(Y)/s(Y)$ gives the percentage of instances of Y that X solves too. Let us write now $cpu(X, b)$ the cpu time needed for X to solve all the benchmarks in b, without any timeout. Because there was a timeout in the competition, only $cpu(X, s')$, with $s' \subseteq s(X)$ are defined here for the solver X. Now, we can compare the relative efficiency of X with Y by computing $crr(X, Y) = cpu(X, s(X) \cap s(Y))/cpu(Y, s(X) \cap s(Y))$. This last measure is called here the *cputime reduction ratio* (CRR), and means that, on their common subset of solved benchmarks, X needs only $crr(X, Y)$ percent of the time needed by Y. Finally, we average these two measures over all the possible values for Y, while keeping X fixed, and we thus defined $re(X)$ and $crr(X)$, two new measures to compare solvers performances.

In the Industrial (complete) category, we have the following values for re : forklift (98.9%), berkmin561 (98.7%), siege (98.6%), berkmin62 (98.4%), oepir (97.6%), funex (97.4%), bmsat (97.2%), compsat-fixed (96.8%), satzoo0 (96.4%) and satzoo1 (96.3%). The crr values are forklift (23%), siege (28%), berkmin561 (46%). funex and oepir needs only 55% of the time of the other solvers. The crr values of solvers then jumps to more than 70% for the other solvers. It is interesting to remark that both forklift and siege are very efficient solvers ($crr < 25\%$ and an impressive value for re). satzoo0 is also interesting because it has a high re and a crr at 164%. This is the same kind of remark for jquest2, which has a re of 94.5% and a crr at only 388% (certainly due to the java penalty). In a sense, we think that such a measure may emphasize a robust solver (if $s(X)$ is large enough): zchaff has a re of 91.1% and a crr of 87%, and limmat a re of 93.9% and a crr of 138%. Thus, jquest2 and limmat are slower than zchaff, but can on average solve more benchmarks.

Interestingly, in the Random (SAT) category, we have the following re: satzilla2 (92.8%), satzilla2-fixed (92.8%), marchxq-fixed (87.0%), marchsp (82.2%), marchtt (80.1%) and kcnfs (78.9%). The best crr are dif-

ferent here: unitwalk, have only a *re* of 71.4% but a very remarkable *crr* of 6%!
saturn has a *re* of 71.1% and a *crr* of 28%. That means that unitwalk and saturn
solve problems different from other solvers, and solve the common problems very
quickly. In order, after these two incomplete solvers, the *crr* are sazilla2-fixed
and satzilla2 (29%) and kcnfs (81%). The *crr* then falls to more than 100%.
In some sense, unitwalk presents the same kind of results than lsat: a low *re*
value that indicates here an original method. If *crr* < 1 then this method is fast,
and, at last, the method is *interesting* if $s(x)$ is large enough. This is clearly the
case for lsat in the handmade category and unitwalk in the random one.

The *Lisa Syndrom* It was brought to the fore in [12] that solvers may have
drastically different behaviors on a range of problems generated from one original
benchmark by randomly shuffling[5] it. This year, the same benchmark (a SAT'02
challenge called *lisa*) was placed in two different categories by error. Before we
detected the problem, it was shuffled twice and solvers ran on both instances.
Here are some examples of solvers that showed different behaviors on those two
benchmarks ("–" means *unsolved*): jerusat1a (–, 40.71) ; jerusat1b (236.85, –
) ; oepir (–, 12.62) ; siege (67.43, –), satzoo1 (–, 287.12) and satzilla (196.8,
727.52). If such a result can question the validity of the competition, it may be
smoothed by the large number of benchmarks used. This is not the case for the
second phase. Considering the small number of benchmarks solved, being lucky
can make the difference.

6 Proposals for SAT'04 Competitions: Back to the Roots

Some improvements were made concerning the way solvers were evaluated this
year: only the best variant of a given solver was eligible for the second phase,
solver submitters were able to follow almost in real time their solver during the
first phase of the competition, the judges chose the solvers that should enter
the second phase, which was more flexible and fair than using an arbitrary rule
(top 5 solvers..), etc. But nothing really changed regarding the way we handled
benchmarks. We need to focus on this issue for the next year competition. The
idea of creating three categories of benchmarks, namely industrial, handmade
and random, was to cover the different interests of researchers in the practical
resolution of SAT.

Industrial Category The industrial category was aimed to address the practical resolution of SAT problems for SAT-based reasoning engines. Ideally, that
category should reflect the strengths or weaknesses of the solvers as a SAT
component, in a "real/industrial" setting. Unfortunately, since we shuffled the
benchmarks, we did not really evaluate the solvers in "real conditions". The
order of the clauses, the identifiers of the variables may provide some structural information about the initial problem and some solvers built to solve those

---

* The shuffler used for the competition generates members of the *PC-class* of [12].

kind of problems were designed to handle that information. By shuffling the benchmarks, we hide that information. Thus a solver can have a very different behaviour during the competition and "in situ", which is in contradiction with one of the aims of the contest. *we propose to launch the solvers on the original benchmark and X variants of the same benchmark class. The results of the solvers on the original benchmarks should then reflect the real strength of the solvers in an application setting while the results on each class of benchmark would emphasize the robustness of the solver.*

Random Category The random category was designed to evaluate the progress made on a class of theoretically well known and hard (NP-complete) problems: random k-sat. There are not that many ways to generate random k-SAT problems, so most of the benchmarks used for that category where based on some uniform k-SAT generators. However, we received some generators able to build problems looking syntactically like random uniform problems, but are forced SAT or UNSAT (like all the generators from Edward Hirsch). The question is: were those generators in the right category? Even worst is the case of the `hardnm` benchmarks related previously. *We propose to rename the random category as "uniform random" category and all the non uniform random generators should move to another category. Whether that category should be a new one or should be the handmade category is not yet decided.*

Handmade The handmade (crafted) category was aimed to point out new techniques not based on resolution, which is one of the ten propositional challenges proposed in [17]. That category was also a reminder that if some solvers can solve huge benchmarks in the industrial category, some small benchmarks are still out of reach of those solvers. The solvers awarded so far in that category (zChaff, Berkmin and satzoo) are using resolution. So the category is awarding very efficient solvers (typically with the greatest number of decisions per second) rather than clever ones. Note that this year, an hybrid solver recovering potential structural knowledge and using it to simplify the initial problem solved the greatest number of problems in that category, but was not awarded. Awarding the most efficient solver is of interest, pointing out "clever" solvers would be a plus for the competition. *We propose that the judges can award "remarkable" solvers, based on an analysis of the first phase results.*

At last, we observed in both industrial and handmade category very few satisfiable benchmarks. *One way to solve this problem is to filter satisfiable benchmarks from a pool of benchmarks submitted to the competition. In that case the benchmarks are needed long before the solvers, which means to change the way the benchmarks are submitted. Another solution is to ask benchmarks submitters a balanced number of SAT and UNSAT benchmarks.*

Briefly, a lot of other points may be discussed before next year competition:

- To begin the competition earlier, just discard solvers not complying with the input/output requirements.
- Do not run randomized solvers 3 times since deterministic solvers can be lucky too (cf the lisa syndrom).
- Collect benchmarks in a different way: we need more control over benchmark submission in order to ensure a meaningful competition (variety of benchmarks, series of equivalent size, etc.).
- Allow benchmarks submitters to follow "live" the solvers on their benchmarks.

7 Conclusion

The SAT 2003 competition was exciting and worthwhile. We presented in this paper a number of different views and analysis based on the data gathered during the first phase. Many new solvers were submitted this year, a majority of them were zchaff-like solvers. There was no outstanding improvements compared to last year, but a reasonable set of strong solvers for "industrial" benchmarks is now available. Forklift, which inherits from berkmin62, leads those solvers. In the handmade category, lsat demonstrated an interesting behavior (a former version participated to SAT'02 but was incorrect). Satzoo showed better performances in terms of diversity of benchmarks and scalability. In the random category, the awarded solvers are not a surprise: the solver kcnfs in the complete category, the incomplete solver unitwalk in the satisfiable category. The most interesting result of the competition is the behavior of the portfolio algorithm satzilla, which compared favourably with the awarded solvers in the random category. Whether this result can be replicated in the other categories is an interesting challenge. Next year, a new competition will be organized in conjunction with SAT2004, in Vancouver. We showed in this paper that the competition is now mature enough to fulfill three main goals: (1) motivate the field, (2) promote new solvers and hard benchmarks and (3) learn from the observation of solvers behaviors.

Acknowledgements

We especially want to thanks our three judges, John Franco, Hans van Maaren and Toby Walsh for their involvement in every phase of the competition. Authors would also like to thank the "Laboratoire de Recherche en Informatique" (LRI, Orsay, France) and the "Dipartimento di Informatica Sistemica e Telematica" (DIST Genoa, Italy) for providing us with clusters of machines and Paul Purdom for presenting the phylogenetic classification of SAT solvers during the conference (not included here for lack of space). At last, we thank all the authors of solvers and benchmarks for their participation and their effort to make their solver I/O compliant with our format. Without them there would be no competition at all!

References

[1] *Sixth International Conference on Theory and Applications of Satisfiability Testing*, S. Margherita Ligure - Portofino (Italy), May 2003.

[2] N. Alexander. Backtrack search algorithms for propositional satisfiability: Review and innovations. Master's thesis, Hebrew University of Jerusalem, November 2002.

[3] G. Audemard, D. Le Berre, O. Roussel, I. Lynce, and J.P. Marques Silva. OpenSAT: An Open Source SAT Software Project. In *Proc. of SAT2003* [1].

[4] G. Dequen and O. Dubois. Renormalization as a function of clause lengths for solving random k-sat formulae. In *Proceedings of Fifth International Symposium on Theory and Applications of Satisfiability Testing*, pages 130–132, 2002.

[5] O. Dubois and G. Dequen. A backbone-search heuristic for efficient solving of hard 3-sat formulae. In *Proceedings of the Seventeenth International Joint Conference on Artificial Intelligence (IJCAI'01)*, Seattle, Washington, USA, August 4th-10th 2001.

[6] Niklas Eén and Niklas Sörensson. An extensible SAT solver. In *In this issue*, 2003.

[7] E. Goldberg and Y. Novikov. BerkMin: A fast and robust SAT-solver. In *Design, Automation, and Test in Europe (DATE '02)*, pages 142–149, March 2002.

[8] E. A. Hirsch and A. Kojevnikov. UnitWalk: A new SAT solver that uses local search guided by unit clause elimination. PDMI preprint 9/2001, Steklov Institute of Mathematics at St.Petersburg, 2001.

[9] O. Kullmann. Heuristics for SAT algorithms: Searching for some foundations, September 1998.

[10] K. Leyton-Brown, E. Nudelman, G. Andrew, J. McFadden, and Y. Shoham. A portfolio approach to algorithm selection. In *Proceedings of IJCAI'03*, 2003.

[11] K. Leyton-Brown, E. Nudelman, and Y.Shoham. Learning the empirical hardness of optimization problems: The case of combinatorial auctions. In *Proceedings of CP'02*, 2002.

[12] X. Y. Li, M.F. Stallmann, and F. Brglez. QingTing: A Fast SAT Solver Using Local Search and Efficient Unit Propagation. In *Proc. of SAT2003* [1].

[13] Inês Lynce and Jo ao Marques Silva. On implementing more efficient data structures. In *Proc. of SAT2003* [1].

[14] M. W. Moskewicz, C. F. Madigan, Y. Zhao, L. Zhang, and S. Malik. Chaff: Engineering an efficient SAT solver. In *Proceedings of the 38th Design Automation Conference (DAC'01)*, pages 530–535, June 2001.

[15] R. Ostrowski, E. Grégoire, B. Mazure, and L. Sais. Recovering and exploiting structural knowledge from cnf formulas. In *Proc. of the Eighth International Conference on Principles and Practice of Constraint Programming (CP'2002)*, LNCS, pages 185–199, Ithaca (N.Y.), September 2002. Springer.

[16] S. D. Prestwich. Randomised backtracking for linear pseudo-boolean constraint problems. In *Proc. of 4th Int. Workshop on Integration of AI and OR techniques in CP for Combinatorial Optimisation Problems*, 2002.

[17] Bart Selman, Henry A. Kautz, and David A. McAllester. Ten challenges in propositional reasoning and search. In *Proceedings of the Fifteenth International Joint Conference on Artificial Intelligence (IJCAI'97)*, pages 50–54, 1997.

[18] Laurent Simon and Philippe Chatalic. SATEx: a web-based framework for SAT experimentation. In Henry Kautz and Bart Selman, editors, *Electronic Notes in Discrete Mathematics*, volume 9. Elsevier Science Publishers, June 2001.

[19] Laurent Simon, Daniel Le Berre, and Edward E. Hirsch. The sat2002 competition report. *Annals of Mathematics and Artificial Intelligence*, 2003. Special issue for SAT2002, to appear.

[20] Geoff Sutcliff and Christian Suttner. Evaluating general purpose automated theorem proving systems. *Artificial Intelligence*, 131:39–54, 2001.

[21] Hantao Zhang. SATO: an efficient propositional prover. In *Proceedings of the International Conference on Automated Deduction (CADE'97), volume 1249 of LNAI*, pages 272–275, 1997.

[22] L. Zhang, C. F. Madigan, M. W. Moskewicz, and S. Malik. Efficient conflict driven learning in a Boolean satisfiability solver. In *International Conference on Computer-Aided Design (ICCAD'01)*, pages 279–285, November 2001.

Challenges in the QBF Arena:
the SAT'03 Evaluation of QBF Solvers[*]

Daniel Le Berre[1], Laurent Simon[2], and Armando Tacchella[3]

[·] CRIL, Université d'Artois,
Rue Jean Souvraz SP 18 – F 62307 Lens Cedex, France
leberre@cril.univ-artois.fr
[·] LRI, Université Paris-Sud
Bâtiment 490, U.M.R. CNRS 8623 – 91405 Orsay Cedex, France
simon@lri.fr
[·] DIST, Università di Genova,
Viale Causa, 13 – 16145 Genova, Italy
tac@mrg.dist.unige.it

Abstract. The implementation of effective reasoning tools for deciding the satisfiability of Quantified Boolean Formulas (QBFs) is an important issue in several research fields such as Formal Verification, Planning, and Reasoning about Knowledge. Several QBF solvers have been implemented in the last few years, most of them extending the well-known Davis, Putnam, Logemann, Loveland procedure (DPLL) for propositional satisfiability (SAT). At the same time, a substantial breed of QBF benchmarks emerged, both in the form of statistical models for the generation of random formulas, and in the form of real-world instances. In this paper we report about the – first ever – evaluation of QBF solvers that was run as a joint event to SAT'03 Conference on Theory and Applications of Satisfiability Testing. Owing to the relative youngness of QBF tools and applications, we decided to run the comparison on a non-competitive basis, using the same technology that powered SAT'02 and SAT'03 competitions of SAT solvers. Running the evaluation enabled us to collect all sorts of data regarding the relative strength of different solvers and methods, the quality of the benchmarks, and to understand some of the current challenges for researchers involved in the QBF arena.

1 Introduction

The implementation of effective reasoning tools for deciding the satisfiability of Quantified Boolean Formulas (QBFs) is an important issue in several research fields such as Formal Verification [1, 2], Planning [3, 4], and Reasoning about

---

[*] The work of the first two authors is partially supported by the "Action Spécifique 83 du STIC/CNRS"; the work of the third author is partially supported by MIUR, ASI and by a grant from the Intel Corporation. The authors would like to thank all the participants to the QBF evaluation for submitting benchmarks and solvers and the University of Genoa for providing the computers to run the evaluation.

E. Giunchiglia and A. Tacchella (Eds.): SAT 2003, LNCS 2919, pp. 468–485, 2004.
© Springer-Verlag Berlin Heidelberg 2004

Knowledge [5]. Several QBF solvers have been implemented in the last few years, most of them extending the well-known Davis, Putnam, Logemann, Loveland procedure (DPLL) [6, 7] for propositional satisfiability (SAT). Starting from the seminal papers of Kleine-Büning, Karpinski and Flögel about the system QKN and, later on, of Cadoli, Giovanardi and Schaerf [8] about the system EVALU-ATE, we have witnessed the development of several complete QBF solvers, such as DECIDE [3] (and its new version QSAT [9]), OPENQBF, QBFL, QSOLVE [10] (and its new version SSOLVE), QUAFFLE [11], QuBE [12], SEMPROP [13], and WALKQSAT [14], the first QBF solver leveraging incomplete stochastic search.

Accompanying the evolution of QBF solvers, a substantial breed of QBF benchmarks emerged, both in the form of statistical models for the generation of random formulas [15, 16], and in the form of real-world instances [3, 2, 1, 4, 5]. In particular, Cadoli et al. in [15] proposed an extension of the fixed clause length generation method for SAT [17] to QBF, which was later improved by Gent and Walsh in [16] to eliminate trivial instances. The generation model named "A" in [16] has been used in other papers (e.g., [10, 18]) to experimentally evaluate heuristics and optimization techniques. Since the problem of deciding the satis-fiability of a QBF is paradigmatic for the class of PSPACE-complete problems, there is a substantial number of real-world reasoning tasks that can be recast into the task of solving a QBF. Examples come from the planning community (see, e.g., [3, 4]), where planning domains described with complex action lan-guages have been converted and successfully solved using QBF-based reasoning, and from the formal verification community (see, e.g., [2, 1]), where QBF-based encodings started to show up as an alternative to SAT-based encodings for hard-ware verification problems. Even more recently Vardi and Pan showed how it is possible to convert problems expressed in the modal logic K into equi-satisfiable instances expressed in QBF in [5]. As reported in [2, 1] and in [5], QBF solvers are still lagging behind special purpose SAT and modal-K solvers, respectively. Therefore the effective application of QBF tools on real-world reasoning tasks is still a challenging issue.

In this paper we report about the – first ever – evaluation of QBF solvers that was run as a joint event to SAT'03 Conference on Theory and Applications of Satisfiability Testing [19]. Owing to the relative youngness of QBF tools and applications, we decided to run the comparison on a non-competitive basis, using SatEx [20], the same technology that powered SAT'02 and SAT'03 competitions of SAT solvers. The evaluation was not intended to award one particular solver, but instead to draw a picture of the current state of the art in QBF solvers and benchmarks. Indeed, running the evaluation enabled us to collect all sorts of data regarding the relative strength of different solvers and methods, the quality of the benchmarks, and to understand some of the current challenges for researchers involved in the QBF arena.

The paper is structured as follows. In Section 2 we describe all the QBF solvers that participated in the evaluation (Sub. 2.1) and the benchmarks that we used to compare the solvers (Sub. 2.2). In Section 3 we present the results of the evaluation arranged by solver (Sub. 3.2), and by benchmark category

(Sub. 3.1); in Sub. 3.3 we complete the picture by highlighting the relative merits of different solvers, as well as providing some details about the difficulty of the currently available benchmarks. Section 4 proposes some challenges for the QBF community at large. We conclude the paper in Section 5 with a balance about the evaluation.

2   State of the Art in QBF

2.1   QBF Solvers

Eleven solvers participated to the evaluation. The requirements where similar to the ones of the SAT competition: reading the problem in a common format (the Q-DIMACS format [21]) and providing the answer in a given output format. Among the 11 solvers, many did not directly input problems in the Q-DIMACS format. Preprocessing tools to convert the benchmarks from the Q-DIMACS format to each solver format were available, but it was necessary to convert each benchmark into 4 different formats in order to run the evaluation. Furthermore, many of the solvers did not respect the output requirements, so we needed to either go through the source code to adjust it, or to modify SatEx scripts.

A short description of the solvers submitted to the evaluation follows.

OPENQBF by Gilles Audemard, Daniel Le Berre and Olivier Roussel. The solver is a simple extension to QBF of the standard DPLL algorithm. It uses a branching rule based on the quantifier order in the prefix, performs unit propagation and pure literal propagation as described in [15]. The solver, written in Java, is based on the OpenSAT framework [22] and it uses the watched literals data structure [23].

QBFL by Florian Letombe. A simple and evolutionary C solver that uses the SAT-solver Limmat (version 1.3) as a sub-procedure. It is simple, because it is a standard DPLL that implements Cadoli, et al. [15] trivial falsity and truth. It is evolutionary, because frequency for testing triviality changes with the success rate of the tests. It choses variables based on Böhm's and Speckenmeyer's lexicographic order [24]. Unit and pure literal propagation for QBF [15] are implemented too.

QSAT by Jussi Rintanen [3, 9]. An extension of the DPLL procedure featuring a lookahead heuristic with failed literal rule, sampling, partial unfolding and quantifier inversion (see [9] for a description of these techniques).

SSOLVE by Rainer Feldmann and Stefan Schamberger [10]. The SSOLVE algorithm is a DPLL-based search algorithm. It includes the test for trivial truth of Cadoli et al. [15] and a modified version of Rintanen's method of inverting quantifiers [3]. The data structures used are extensions of the data structures of Max Böhm's SAT-solver [25].

QUAFFLE by Lintao Zhang and Sharad Malik [11, 26], featuring non-chronological backtracking based on conflict/solution learning [11].

QUANTOR by Armin Biere. This solver features elimination of innermost universal quantifiers by copying. It is a very preliminary version, which was

Series		Satisfiability		Formulas Data					
name	count	T%	F%	variables	clauses	sets	exist	forall
k_branch	42	50%	50%	5679.05	28927.43	27.00	5614.67	64.38
k_d4	42	50%	50%	847.76	2796.17	30.90	804.36	43.40
k_dum	42	50%	50%	531.19	1373.24	59.04	504.22	26.98
k_grz	42	50%	50%	529.93	1940.69	17.00	510.31	19.62
k_lin	42	50%	50%	1431.22	15117.00	7.91	1418.95	12.27
k_path	42	50%	50%	769.41	2138.72	26.00	733.93	35.48
k_ph	42	50%	50%	3364.34	75153.57	4.88	3356.00	8.34
k_poly	42	50%	50%	910.5	2054.50	72.00	841.5	69.00
k_t4p	42	50%	50%	1256.29	3922.38	45.00	1196.31	59.98
C432	8	37.5%	37.5%	594.50	1525.00	6.50	588.00	6.50
C499	8	37.5%	12.5%	881.50	2528.50	7.00	875.00	6.50
C5315	8	25%	12.5%	5509.00	14818.25	10.00	5459.50	49.00
C6288	8	12.5%	0%	4702.25	13798.75	6.00	4651.25	50.50
C880	8	25%	0%	1009.00	2571.50	7.50	994.50	14.00
comp	8	50%	50%	299.75	815.25	4.00	297.25	2.50
term1	8	50%	25%	1090.75	3760.25	8.00	1084.75	5.75
z4ml	8	50%	50%	63.75	193.00	3.50	62.25	1.50

Table 1. Data of submitted benchmarks.

assembled in less than 2 person weeks. A lot of optimizations are reported to be missing [27].

QuBE-Bj by Enrico Giunchiglia, Massimo Narizzano, and Armando Tacchella [28]. The basic architecture of QuBE is a standard counter-based structure, featuring unit propagation, pure literal propagation, and trivial truth (in the spirit of [15]); the heuristic is and adaptation to QBF of the Böhm's and Speckenmeyer's heuristic for SAT. QuBE-Bj adds to this conflict and solution backjumping, while

QuBE-Rel by Enrico Giunchiglia, Massimo Narizzano, and Armando Tacchella [29] also features good (solution) and no-good (conflict) learning, plus unit clause propagation on universal literals (see [29] for details).

semprop from Reinhold Letz [30]. Semantic tree (DPLL) with dependency directed backtracking and lemma/model caching for false/true subproblems.

watchedCSBj Andrew G. D. Rowley and Ian Gent features lazy data structures similar to those of Chaff (see [31]), constraint and solution backjumping in the spirit of QuBE-Bj.

walkQsat by Andrew G. D. Rowley, Ian Gent, Holger Hoos and Kevin Smyth [14] is the first incomplete QBF solver based on stochastic search methods. It is a version of watchedCSBj with Walksat [32] as a SAT oracle and guidance heuristic. walkQsat is reported to be "an inherently incomplete QBF solver, which still solves many unsatisfiable instances as well as satisfiable ones" [14].

2.2   QBF Benchmarks

In this Section we present the benchmarks used in the evaluation. A total of 1720 benchmarks divided into 134 series have been run during the evaluation. Of these, 856 are generated according to some probabilistic model, and 864 are non-random (i.e., real-world and hand-made) instances. 442 instances have been

Series		Satisfiability		Formulas Data				
name	count	T%	F%	variables	clauses	sets	exist	forall
Adder-s1	8	100%	0%	2775.00	3742.50	4.00	1216.50	667.50
Adder-s2	8	100%	0%	9071.00	9927.50	6.00	7623.50	556.50
Adder-u1	8	0%	100%	2791.00	3750.50	3.00	1556.50	343.50
Adder-u2	8	0%	100%	9071.00	9919.50	7.00	7725.50	454.50
DFlipFlop	10	0%	100%	62919.50	82655.30	3.00	62058.50	37.50
MutexP	7	100%	0%	8890.43	4172.71	2.00	3078.86	145.14
SzymanskiP	12	100%	0%	82461.75	90270.75	3.00	72526.08	1786.67
VonNeumann	11	0%	100%	441489.00	643479.00	3.00	438327.00	240.00
ToiletA-s	21	100%	0%	304.00	4341.81	3.00	297.43	6.57
ToiletA	56	0%	100%	219.25	5246.98	3.00	211.61	7.64
ToiletC-s	29	100%	0%	406.03	2997.93	3.00	402.97	3.07
ToiletC	56	0%	100%	214.79	986.48	3.00	211.61	3.18
ToiletG	7	100%	0%	49.57	204.29	3.00	46.43	3.14
Tree	14	35%	65%	52.86	34.14	39.29	26.43	26.43
Robots_D2	30	70%	0%	6422.00	21359.00	2.00	5172.50	27.50
Robots_D3	30	57%	30%	6422.00	21360.00	2.00	5172.50	27.50
Robots_D4	30	47%	53%	6422.00	21361.00	2.00	5172.50	27.50
Robots_D5	30	47%	53%	6422.00	21362.00	2.00	5172.50	27.50
Blocks-s	4	100%	0%	435.75	5098.00	3.00	431.00	4.75
Blocks	9	0%	100%	507.11	7570.44	3.00	501.78	5.33
Chain	12	100%	0%	1997.50	11173.75	3.00	1980.00	17.50
Impl	10	100%	0%	46.00	90.00	23.00	35.00	11.00
Logn	4	0%	100%	1318.00	61599.00	2.00	1317.50	0.50
R3cnf-s	13	100%	0%	150.00	383.08	5.15	135.00	15.00
R3cnf	7	0%	100%	150.00	381.43	4.71	135.00	15.00
Toilet-s	5	100%	0%	754.80	3434.60	3.00	751.80	3.00
Toilet	3	0%	100%	240.33	869.00	3.00	238.00	2.33

Table 2. Data of benchmarks from QBFLIB.

submitted for the evaluation, and 1278 have been selected from the instances available on QBFLIB [21]. At the time of this report, all the instances used for the evaluation (and more) are available on QBFLIB.

In more details, the submitted benchmarks are:

Scholl/Becker 64 benchmarks in 8 series, encode equivalence checking for partial implementations problems (see [1]).

Pan 378 benchmarks in 18 series, encode the satisfiability of modal K formulas into QBF (see [5]). The original benchmarks have been proposed during the TANCS'98 comparison of theorem provers for modal logics [33].

The benchmarks that have been selected from QBFLIB are:

Ayari 72 benchmarks in 8 series. A family of problems related to the formal equivalence checking of partial implementations of circuits (see [2]).

Castellini 169 benchmarks in 5 series. Various QBF-based encodings of the bomb-in-the-toilet planning problem (see [4]).

Letz 14 benchmarks. A family of formulas proposed in [30] generated according to the pattern $\forall x_1 x_3 \ldots x_{n-1} \exists x_2 x_4 \ldots x_n (c_1 \wedge c_n)$ where $c_1 = x_1 \wedge x_2$, $c_2 = \neg x_1 \wedge \neg x_2$, $c_3 = x_3 \wedge x_4$, $c_4 = \neg x_3 \wedge \neg x_4$, and so on. The instances consists of simple variable-independent subproblems but they should be hard for standard (i.e., without non-chronological backtracking) QBF solvers.

Narizzano/Random 842 benchmarks in 82 series. QBFs randomly generated with model A proposed by Gent and Walsh in [16].

Narizzano/Robot 120 benchmarks in 4 series. QBF-based encoding of the robot navigation problems presented in [4].

Rintanen 67 benchmarks (44 true, 23 false) in 9 series consisting of planning, hand-made and random problems, some of which have been presented in [3].

To run the submitted solvers on the above benchmarks we used a farm of eight identical PCs equipped with a Pentium IV 2.4 Ghz processor, and 1024MB of RAM (DDR 333Mhz), running Linux RedHat 7.2 (kernel version 2.4.7-10). For practical reasons, a timeout of 15 CPU minutes has been placed on the run time of the solvers on each instance.

3   Results of the Evaluation

3.1   Benchmark-Centric View

In Tables 1 and 2 we summarize the data about the benchmarks for the submitted instances (Table 1) and for the the instances from QBFLIB (Table 2). In the tables each row represents different families of instances (series) for which we report the number of benchmarks in the family (count), the percentage of true (T%) and false (F%) instances, the average number of variables (variables), clauses (clauses) and alternations in the prefix (sets); we also report the average number of existentially quantified variables (exists) and of universally quantified variables (forall). Notice that:

- when the sum of T% and F% is less than 100%, it means that the status of the problems cannot be known in advance, and not all the benchmarks could be solved, so the numbers shown are the best current estimate of the satisfiability status of the formulas;
- for the sake of compactness, Pan instances (18 families) have been grouped into 9 entries, each entry consisting of the data of two families differing only for the satisfiability status, e.g., the two families k_branch_n (21 problems all true) and k_branch_p (21 instances all false) are grouped in the entry k_branch;
- finally, we do not report about the Random benchmarks generated according to model A in detail. Indeed, these are a selection of fixed clause length QBFs (clause length is 5), with 25, 50, 100 and 150 variables per quantifier set, having 2,3,4,5 quantifier sets and different number of clauses. The number of clauses was empirically chosen to cover the 100%-0% satisfiability curve.

3.2   Solver-Centric View

Table 3 summarizes the results of the evaluation for all the solvers on all the benchmarks[4] In the Table we report for each solver the total number of series

---

- We count an instance as solved by a solver iff the solver (*i*) does not crash, and (*ii*) answers within the time limit. Therefore, in the results presented, we do not distinguish between timeouts and other kind of failures, e.g., memory outs.

solver	Overall		Random			Non-random		
	#series	#benchs	#series	#benchs	Remark	#series	#benchs	Remark
openqbf	110	725	65	404		45	321	
qbfl	118	709	75	415		43	294	
qsat	130	1152	82	701		48	451	SOTAC(22)
quaffle	99	812	60	389		39	423	SOTAC(9)
quantor	64	456	19	115		45	341	
qubebj	131	1383	82	842	SOTAC(4)	49	541	SOTAC(8)
quberel	133	1324	82	803		51	521	SOTAC(12)
semprop	133	1389	82	810		51	579	SOTAC(58)
ssolve	131	1293	82	840	SOTAC(5)	49	453	
walkqsat	117	823	80	600		37	223	
watchedcbsj	125	1004	80	602		45	402	SOTAC(2)

Table 3. Overall results of the evaluation.

solved (Overall - #series), the total number of benchmarks solved (Overall - #benchs), and then we present the same data separating random (columns 4-6) and non-random (columns 7-9) instances. A series is considered *solved* iff at least one of its benchmarks is solved, and a solver is considered a *state-of-the-art contributor* (SOTAC) iff it was the only one able to solve a given problem (as in [34]). The columns "Remark" indicate whether a given solver is SOTAC and, if so, how many problems it uniquely conquered. Looking at Table 3 we immediately see that two QBF solvers, SEMPROP and QuBE-REL, were able to solve all but one series (133 out of 134 series) and that SEMPROP, was able to solve the highest number of benchmarks with 1389 instances solved out of 1720. Also, with the exception of QUANTOR, most solvers did quite well on the random benchmarks, while non-random ones proved more challenging for all the solvers. Notice that all series where solved by at least one solver. In the following, we discuss the results of the evaluation considering the results on random and non-random instances separately.

Random benchmarks Figure 1 presents a pictorial view of the evaluation results considering random instances only. The plot in Figure 1 is obtained as follows: the x-axis represents the number of benchmarks solved and the y-axis the time taken to solve a particular instance; the curve representing the performances of the solver is obtained by sorting its CPU times in ascending order. Therefore, the curve represents the distribution of run times for the given solver: the first point is the 0% percentile, the intermediate point is (approximately) the 50% percentile (median), and so on. The distributions have all the same qualitative behavior, where a solver can easily conquer a few problems and then inevitably reaches its "level of incompetence" (similar phenomena are noticed in [34]).

Figure 1 shows clearly that solvers are clustered in groups of similar ability as far as solving random benchmarks is concerned. Starting to analyze the plot in Figure 1 from left (slower) to right (faster), we see that QUANTOR does not perform very well on this category (only 115 benchmarks solved), so it is likely that the approach used in QUANTOR is not adapted to random problems; then we see a cluster made of QUAFFLE, QBFL and OPENQBF (around 400 benchmarks solved). The reason why QUAFFLE is lagging behind OPENQBF and QBFL is likely

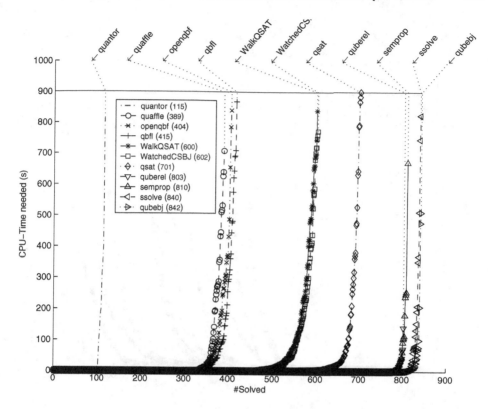

Fig. 1. Number of instances solved vs. CPU time for random benchmarks

to be twofold: the heuristic used by QUAFFLE (an adaptation of VSIDS [23]) is probably not adapted to random problems, and the pure literal rule is missing from QUAFFLE. Since both QBFL and OPENQBF perform pure literal detection, we conjecture that this rule might be quite important when solving random benchmarks. Still going right in Figure 1, we encounter WATCHEDCSBJ and WALKQSAT (about 600 benchmarks solved). In a way, this result is not surprising, since WALKQSAT shares its core architecture with WATCHEDCSBJ. On the other hand, the performance of WALKQSAT is very interesting if we consider that it is incomplete and, still, it can rival with its complete sibling on *mixed* instances, i.e., satisfiable and unsatisfiable ones. Back to Figure 1, QSAT stands alone with about 700 benchmarks solved. Finally, a group made of SEMPROP and QUBE-REL (resp. 802 and 810 benchmarks solved) and leaded by SSOLVE and QUBE-BJ (resp. 840 and 842 benchmarks solved) closes the listing.

Considering the results for random benchmarks, it seems that the features needed to solve effectively this kind of instances are those of QUBE-BJ and SSOLVE: both solvers feature traditional (counter-based) data structures, unit clause propagation, pure literal detection, trivial truth and an adaptation to QBF of Böhm's and Speckenmeyer's lexicographic heuristic as search strategy;

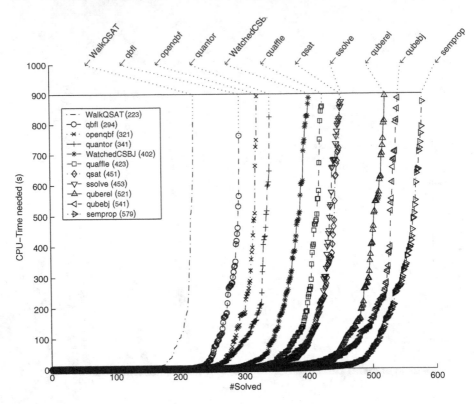

Fig. 2. Number of instances solved vs. CPU time for non-random benchmarks

SSOLVE adds to this trivial falsity, while QuBE-BJ features constraint and solution backjumping. From the evaluation data, we can conclude that the common features of QuBE-BJ and SSOLVE are playing a major role in solving random benchmarks effectively. Moreover, given that random problems are generated according to model A, from the literature (see [18]), we also know that trivial truth is particularly effective on 2- and 4-QBFs, while backjumping is particularly effective on 3- and 5- QBFs. We do not know of a similar analysis for trivial falsity, but looking at SSOLVE performances, also this technique seems quite effective for random benchmarks. Furthermore, if we contrast the performances of SEMPROP and QuBE-REL with those of SSOLVE and QuBE-BJ, we can conjecture that learning does not seem effective on random problems. This is evident if we restrict the comparison to QuBE-REL and QuBE-BJ which differ only for the look-back engine. Finally, considering the performances of QSAT, the gap of 150 benchmarks w.r.t. SSOLVE induces the suspect that, besides learning, also partial unfolding, quantifier inversion and sampling are not particularly effective on the class of random benchmarks.

Non-random benchmarks Figure 2 presents the results on non-random benchmarks arranged in the same way as Figure 1. Going from left (slower) to right (faster), we first encounter WALKQSAT with 223 benchmarks solved. Since its sibling WATCHEDCSBJ obtained a good standing also on non-random benchmarks, we conjecture that the problem with WALKQSAT on this kind of problems might lie in the Walksat oracle not informing the search in an effective way. Still going right in Figure 2, we see a cluster comprised of QBFL, OPENQBF and QUANTOR (resp. 294, 321 and 341 problems solved). It is interesting to notice that QUANTOR works much better on non-random benchmarks, possibly because these problems have much less entropy than random ones [27]. On non-random benchmarks, a basic implementation of the universal quantifiers elimination approach, such as QUANTOR, is competitive with basic implementations of the search approach, such as QBFL and OPENQBF. Another cluster of solvers is made up by WATCHEDCSBJ, QUAFFLE, QSAT and SSOLVE (resp. 402, 423, 451, and 453 benchmarks solved) which can be considered as average performers for non-random problems. These solvers, with the exception of SSOLVE, are also SOTAC for the category of non-random problems, with QSAT being more effective than QUAFFLE and WATCHEDCSBJ (resp. 22, 9 and 2 instances uniquely conquered). Since QSAT features inversion of quantifiers, sampling and partial expansion, we can conjecture that these techniques are particularly well suited for some of the benchmarks that we used for the evaluation, while the look-back optimizations featured by solvers of similar strength like QUAFFLE and WATCHEDCSBJ are less effective on these specific benchmarks. Finally, QUBE-REL, QUBE-BJ, and SEMPROP (resp. 521, 541 and 579 instances solved) close the listing with the best performances on non-random benchmarks. In particular, SEMPROP stands out for being able to solve the highest number of benchmarks and for being SOTAC on non-random problems with 58 uniquely solved instances.

A few comments on these results are in order. First of all, the three solvers that conquered more than 500 instances implement both conflict and solution backjumping. SEMPROP and QUBE-REL add to this conflict and solution learning. While backjumping seems to be definitely useful for non-random benchmarks, the bad performances of QUBE-REL seem to point out that learning might be ineffective on these benchmarks. This conjecture seems to be reinforced by the not-so-positive performances of QUAFFLE which features conflict and solution learning as well. Nevertheless, the evidence on QUAFFLE cannot be considered as conclusive, since this solver does not feature some of the optimizations (e.g., pure literal, trivial truth) that are found in QUBE-BJ, QUBE-REL and SEMPROP. Moreover, SEMPROP itself implements learning, while QUBE-REL learning engine does not feature fancy data structures and carefully engineered optimizations. In conclusion, non-random problems seem to favor look-ahead (sampling, inversion of quantifiers, expansion) techniques and lightweight look-back techniques (backjumping). However, the effectiveness of learning, and the trade-offs between look-ahead and look-back remain largely to be explored in this context.

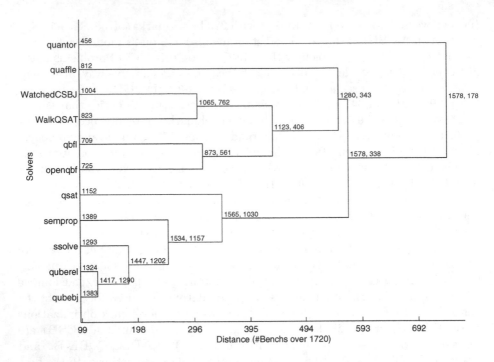

Fig. 3. Clustering of the solvers according to their ability to solve the benchmarks.

3.3   Overall Analysis

Strength of solvers In Figure 3 we present yet another view of the competition. In the Figure, for each solver we draw a line which represents the distance of the solver from the ideal one, i.e., one that solves all the benchmarks used for the evaluation: the longer the line, the worse the performances of the given solver. Distance is on the x-axis, solvers are listed on the y-axis; the listing is done by grouping together solvers that conquered analogous sets of instances (e.g., not surprisingly, QUBE-BJ and QUBE-REL solve a lot of instances in common). The number of benchmarks solved is showed on the line representing the solver, and the clustering of solvers is represented by two lines joining. For each cluster we indicate the union of the instances solved by the members of the cluster, and the intersection thereof (e.g., by running QUBE-BJ and QUBE-REL in parallel we could solve 1417 benchmarks, but the benchmarks solved by both of them are only 1290).

Looking at Figure 3 we can divide the solvers that participated to the evaluation into 3 groups. QUANTOR is the only member of the first group. Then we have a group of *medium strength* solvers comprised of OPENQBF, QBFL, WALKQSAT, WATCHEDCSBJ (SOTAC) and QUAFFLE (SOTAC). We grouped together OPENQBF with QBFL, since they both represent very basic solvers (about 700 benchmarks solved), and WATCHEDCSBJ with WALKQSAT, since they share a common architecture (overall, WALKQSAT is slightly worse than WATCHED-CSBJ). The cluster is completed by QUAFFLE, which brings the total number of benchmarks solved by this group to 1280. Finally, there is the group of *strong* solvers, composed by QSAT, SEMPROP, SSOLVE, QuBE-REL and QuBE-BJ. All of these solvers are SOTAC: clustered together, they can solve 1565 problems, and around 1000 problems can be solved by every solver in this group.

Summing up, the ideal solver comprised of all the solvers running in parallel (SOTA solver in [34]) would solve 1578 benchmarks out of the 1720 used for the evaluation (a success rate of about 90%). QuBE-BJ and SEMPROP turn out to be the strongest solvers of the evaluation (resp. 1383 and 1389 benchmarks solved), where QuBE-BJ turns out to be better suited for random benchmarks and SEMPROP for non-random ones, although their performance on both categories is respectable. Conflict and solution backjumping, pure literal rule, trivial truth and an informed heuristic seem to be important features to build a strong QBF solver. All the other features (learning, quantifier inversion, sampling, expansion) are definitely useful since the solvers implementing them are SOTAC, but a deeper analysis is needed to understand their effectiveness as general purpose techniques when used alone or in combination with other optimizations.

Hardness of benchmarks Considering the numbers of Figure 3, all the solvers were able to conquer 178 *easy* benchmarks, while 142 *hard* instances remain still unsolved. Among the easy instances, 5 come from Scholl/Becker, 19 from Pan instances, and the remaining from QBFLIB; among the hard instances, 25 come from Scholl/Becker instances, 72 from Pan instances, and the rest from QBFLIB instances: 25, 7 and 12 instances from Ayari, Narizzano/Random and Narizzano/Robot, respectively. Since 1030 problems can be conquered by all the strong solvers, this leaves 862 (1030 instances minus the 178 easy ones) *medium* instances. The bulk of the medium instances comes from Narizzano/Random instances (663); among the submitted benchmarks, 14 from Scholl/Becker instances and 61 from Pan instances are of medium difficulty. The smallest hard instance is `blif_0.10_0.20_0_0_out_exact` in the C432 family of benchmarks submitted by Scholl and Becker [1], having 568 variables, 3 alternations, and 1439 clauses. The biggest problem that has been solved is the `ripple-carry-15-c` instance in the VonNeumann family of benchmarks submitted by A. Ayari [2], having 1311754 variables, 3 alternations and 1923754 clauses. Indeed, this "monster" QBF could be solved by QUANTOR, and by the medium strength solvers OPENQBF and QBFL. Among the strong solvers, only QuBE-REL and SEMPROP conquered the instance.

4   Challenges in the QBF Arena

In this Section we briefly summarize some of the open issues that occurred to us while evaluating the solvers and analyzing the results. We believe these issues will challenge developers and researchers in the QBF arena for the next few years.

Challenge 1 *Solve the 142 hard QBF benchmarks remained unsolved during the evaluation.*

The first, and main, contribution of the evaluation was to isolate the instances that are too hard to be solved given the alloted resources and the current state of the art. These instances will be made available as a single collection on QBFLIB so that developers can test and tune their solvers to conquer these problems.

Challenge 2 *Random generation model that respects the polynomial hierarchy.*

Currently the best random generators, i.e., shown experimentally to produce non-trivial formulas with controllable properties, are those based on models A and B proposed by Gent and Walsh in [16]. As noticed in previous experiments (see, e.g., [18]) and confirmed during the evaluation, formulas generated according to model A have the property of being reasonably easy in their 3,5-QBF incarnations and quite hard in their 2,4-QBF incarnations. The phenomenon is evident when using non-chronological backtracking, and it clashes with theoretical complexity results that require that 3-QBFs (resp. 5-QBFs) are harder to solve than 2-QBFs (resp. 4-QBFs), unless the polynomial hierarchy collapses. The open issue is whether a generator for random instances which is well behaved in this respect can be produced.

Challenge 3 *Solution learning as enabling technologies for new simplification techniques and heuristics.*

Solution learning in QBF has been introduced somewhat contemporaneously in [29, 30, 11]. Implementing solution learning forces us to think of the formula under test as a mixed CNF/DNF instance, the DNF part coming from the accumulation of learned solutions. This conceptual shift enables the introduction of new pruning techniques, e.g., unit clauses on universal literals (see [29]) and, possibly, new and more effective heuristics. The open point here is the exploration of the expanded possibilities that solution learning gives to developers, also considering that learning did not come out of this evaluation as an all-time winner.

Challenge 4 *Conceive and experiment new simplification and intelligent backtracking techniques.*

Although several simplification and backtracking techniques have been proposed and experimented for QBF, and that the effectiveness of solution backjumping – without a counterpart in SAT – was demonstrated during the evaluation, we still see a lot of room for improvement. Since QBF is deductively more challenging than SAT, more reasoning forward (simplification) and backward (intelligent backtracking) has a good chance of paying off. The trade-off between search and reasoning must still be explored for QBF.

Challenge 5 *Conceive and experiment new effective heuristics for QBF.*

Most heuristics that have been proposed for QBF are simple adaptations of their SAT counterpart. Currently, to the extent of our knowledge, there are no comparative studies assessing the relative effectiveness of heuristics for QBF, while, judging from the evaluation results, effective heuristics might prove crucial in conquering some of the problems.

Challenge 6 *Investigate the combination of technologies such as inversion of quantifiers, random sampling, partial expansion together with simplification and backtracking techniques.*

Solvers like QSAT and QUANTOR feature aggressive look-ahead techniques like quantifier expansion (resp. partial and complete), sampling and inversion of quantifiers. Others, like SEMPROP and QUBE-BJ, rely more on look-back techniques. The combination of such different philosophies, which might lead to more powerful solvers, remains largely to be investigated.

Challenge 7 *Proving in the field that QBF-based reasoning can beat SAT-based reasoning.*

One of the reasons why SAT research flourished so much in the last decade has been the application of SAT-based encoding in several fields, e.g., planning [35] and formal verification [36] to cite only some of the most significant. Indeed, there are problems that can be solved by means of a SAT-based encoding and that admit an exponentially more compact QBF encoding. This is the case, e.g., of symbolic reachability which underlies the applications considered in [35] and [36]. Therefore the question is: is it worth using the more compact QBF encoding at the price of a more complex solver, or expanding into an equivalent (and possibly huge) SAT encoding is paying more dividends? This is a crucial question that QBF developers must answer in the next few years, one that can establish whether QBF technology is an enabling technology for real-world applications.

Challenge 8 *Build an efficient QBF solver not based on AND/OR search.*

Ten out of eleven solvers participating to the evaluation are extensions to QBF of the DPLL method for SAT. The only exception is QUANTOR, featuring an approach based on quantifier expansion. The evaluation showed that QUANTOR is competitive with basic implementations of the DPLL method for QBF, which makes its alternative paradigm appealing for further research.

Challenge 9 *Propose new techniques for randomized/incomplete QBF solvers.*

WALKQSAT scored quite well on the evaluation, classifying among medium strength *complete* solvers. Although it did not turn out to be SOTAC for some category, we see the research line of WALKQSAT as interesting and promising.

Challenge 10 *Build special purpose solvers, i.e., solvers tuned specifically for k-QBFs with k ∈ {2, 3}*

Although the average number of alternations on all the benchmarks is about 9, most of the benchmarks that we used for the evaluation have a limited number of alternations (2,3) and some of the hard problems as well. Therefore, it might be interesting to investigate methods which are specifically tuned to solve these kind of formulas, rather than formulas with arbitrary alternation depth.

5   Conclusions

The final balance of this first comparative evaluation of QBF solvers is definitely positive:

- 11 solvers participated, 10 complete and 1 incomplete; this is almost twice the number of solvers that participated in the first SAT solvers competition [25], showing that research on QBF tools reached a decent level of maturity. Since most of the solvers have been proposed in recent years, the field shows a good rate of fertility too.
- 442 benchmarks were submitted coming from such diverse fields as Formal Verification [2], and Knowledge About Reasoning [5]; 1278 benchmarks were selected from more than 4000 instances available on QBFLIB [21]. These numbers witness that also the field of applications of QBF reasoning is growing steadily.
- State of the art solvers, both for random and non-random benchmarks, have been identified; also, a total of 142 challenging benchmarks that cannot be solved by none of the current QBF solvers have been identified to set the reference point for future developments in the field.
- Gathering and analyzing the data of the evaluation allowed us to compile a tentative list of future challenges in the QBF arena.

The evaluation also evidenced some critical points:

- The Q-DIMACS standard input format for QBFs in prenex clausal normal form has been around since 3 years (see [21] for a technical report about it). Nevertheless, many solvers submitted to the evaluation still used non-standard – although very close to Q-DIMACS – formats; this fact considerably complicated the work of the evaluators, by requiring the use of translation scripts;
- The numbers of the QBF evaluation are still dwarfed by the number of the SAT competition: 34 SAT solvers and 993 benchmarks from 17 different sources were submitted this year to the SAT competition.
- QBF encoding of real-world applications (e.g., Ayari's hardware verification problems, Scholl-Becker's circuit equivalence problems, etc.) contributed a lot to the pool of 142 challenging benchmarks. This shows that QBF developers must improve the performance of their solvers before these can be practical for industrial-sized benchmarks (as it is already true for SAT solvers).

- The question of how to check the answer of the QBF solvers in an effective way is still unanswered. Specifically, the questions of what is a good certificate of satisfiability/unsatisfiability for QBF, and, if this proves too huge to be practical, what is a good approximation of such certificate, remain open.

Overall, the evaluation showed the vitality of QBF as a research area. Despite some technological limitations and some maturity issues, it is our opinion that the development of effective QBF solvers and the use of QBF-based automation techniques can be regarded as promising research directions.

References

[1] Cristoph Scholl and Bernd Becker. Checking equivalence for partial implementations. Technical report, Institute of Computer Science, Albert-Ludwigs University, October 2000.

[2] Abdelwaheb Ayari and David Basin. Bounded model construction for monadic second-order logics. In *12th International Conference on Computer-Aided Verification (CAV'00)*, number 1855 in LNCS, pages 99–113. Springer-Verlag, 2000.

[3] Jussi Rintanen. Improvements to the evaluation of quantified boolean formulae. In *Proceedings of the Sixteenth International Joint Conferences on Artificial Intelligence (IJCAI'99)*, pages 1192–1197, Stockholm, Sweden, July 31-August 6 1999. Morgan Kaufmann.

[4] C. Castellini, E. Giunchiglia, and A. Tacchella. Sat-based planning in complex domains: Concurrency, constraints and nondeterminism. *Artificial Intelligence*, 147(1):85–117, 2003.

[5] Guoqiang Pan and Moshe Y. Vardi. Optimizing a BDD-based modal solver. In *Proceedings of the Nineteenth Internations Conference on Automated Deduction*, 2003. to appear.

[6] Martin Davis and Hilary Putnam. A computing procedure for quantification theory. *Journal of the ACM*, 7(3):201–215, July 1960.

[7] Martin Davis, George Logemann, and Donald Loveland. A machine program for theorem proving. *Communications of the ACM*, 5(7):394–397, July 1962.

[8] Marco Cadoli, Andrea Giovanardi, and Marco Schaerf. An algorithm to evaluate quantified boolean formulae. In *Proceedings of the Fifteenth National Conference on Artificial Intelligence (AAAI'98)*, pages 262–267, Madison (Wisconsin - USA), 1998.

[9] Jussi Rintanen. Partial implicit unfolding in the Davis-Putnam procedure for quantified Boolean formulae. In *Logic for Programming, Artificial Intelligence and Reasoning. 8th International Conference*, number 2250 in LNAI, pages 362–376. Springer, 2001.

[10] R. Feldmann, B. Monien, and S. Schamberger. A distributed algorithm to evaluate quantified boolean formula. In *Proceedings of the Seventeenth National Conference in Artificial Intelligence (AAAI'00)*, pages 285–290, 2000.

[11] Lintao Zhang and Sharad Malik. Towards a symmetric treatment of satisfaction and conflicts in quantified boolean formula evaluation. In *Proceedings of the Eighth International Conference on Principles and Practice of Constraint Programming*, pages 200–215, 2002.

[12] Enrico Giunchiglia, Massimo Narizzano, and Armando Tacchella. QuBE : A system for deciding Quantified Boolean Formulas Satisfiability. In *Proceedings of*

the International Joint Conference on Automated Reasoning (IJCAR'01), number 2083 in LNCS, pages 364–369. Springer-Verlag, 2001.

[13] Reinhold Letz. Advances in decision procedures for quantified boolean formulas. In *Proceedings of the First International Workshop on Quantified Boolean Formulae (QBF'01)*, pages 55–64, 2001.

[14] Andrew G D Rowley Ian P Gent, Holger H Hoos and Kevin Smyth. Using stochastic local search to solve quantified boolean formulae. Technical Report APES-58-2003, APES Research Group, January 2003.

[15] Marco Cadoli, Marco Schaerf, Andrea Giovanardi, and Massimo Giovanardi. An algorithm to evaluate quantified boolean formulae and its experimental evaluation. *Journal of Automated Reasoning*, 28(2):101–142, 2002.

[16] Ian P. Gent. Beyond NP: the QSAT phase transition. In *Proceedings of the Sixteenth National Conference on Artificial Intelligence (AAAI'99)*, pages 648–653, 1999.

[17] David G. Mitchell, Bart Selman, and Hector J. Levesque. Hard and easy distributions for SAT problems. In *Proceedings of the Tenth National Conference on Artificial Intelligence*, pages 459–465. AAAI Press, 1992.

[18] Enrico Giunchiglia, Massimo Narizzano, and Armando Tacchella. On the effectiveness of backjumping and trivial truth in quantified boolean formulas satisfiability. In *Proceedings of the First International Workshop on Quantified Boolean Formulas (QBF'01)*, pages 40–54, 2001.

[19] *Sixth International Conference on Theory and Applications of Satisfiability Testing*, S. Margherita Ligure - Portofino (Italy), May 2003.

[20] Laurent Simon and Philippe Chatalic. SATEx: a web-based framework for SAT experimentation. In Henry Kautz and Bart Selman, editors, *Electronic Notes in Discrete Mathematics*, volume 9. Elsevier Science Publishers, June 2001. http://www.lri.fr/ simon/satex/satex.php3.

[21] M. Narizzano. QBFLIB - The Quantified Boolean Formulas Satisfiability Library. http://www.qbflib.org.

[22] Gilles Audemard, Daniel Le Berre, Olivier Roussel, Inês Lynce, and Jo ao Marques Silva. OpenSAT: An Open Source SAT Software Project. In *Sixth International Conference on Theory and Applications of Satisfiability Testing* [19].

[23] M. W. Moskewicz, C. F. Madigan, Y. Zhao, L. Zhang, and S. Malik. Chaff: Engineering an efficient SAT solver. In *Proceedings of the 38th Design Automation Conference (DAC'01)*, pages 530–535, June 2001.

[24] M. Bohm and E. Speckenmeyer. A fast parallel sat–solver — efficient workload balancing. *Annals of Mathematics and Artificial Intelligence*, 17:381– 400, 1996.

[25] M. Buro and H. Kleine Büning. Report on a SAT competition. *Bulletin of the European Association for Theoretical Computer Science*, 49:143–151, 1993.

[26] Lintao Zhang and Sharad Malik. Conflict driven learning in a quantified boolean satisfiability solver. In *Proceedings of International Conference on Computer Aided Design (ICCAD'02)*, San Jose, CA, USA, Nov. 2002.

[27] Armin Biere, 2003. Personal communications.

[28] Enrico Giunchiglia, Massimo Narizzano, and Armando Tacchella. Backjumping for quantified boolean logic satisfiability. In *Proceedings of the Seventeenth International Joint Conferences on Artificial Intelligence (IJCAI'01)*, Seattle, Washington, USA, August 4th-10th 2001. to appear.

[29] Enrico Giunchiglia, Massimo Narizzano, and Armando Tacchella. Learning for quantified boolean logic satisfiability. In *The Eighteenth National Conference on Artificial Intelligence (AAAI'02)*, Edmonton, Alberta, Canada, July 28–August 1 2002.

[30] Reinhold Letz. Lemma and model caching in decision procedures for quantified boolean formulas. In *Proceedings of Tableaux 2002*, LNAI 2381, pages 160–175, Copenhagen, Denmark, 2002. Springer.

[31] Andrew G. D. Rowley, Ian Gent, Enrico Giunchiglia, Massimo Narizzano, and Armando Tachella. Watched data structures for QBF solvers. In *Sixth International Conference on Theory and Applications of Satisfiability Testing (SAT'03)*, pages 348–355, 2003. Extended Abstract.

[32] Bart Selman, Henry A. Kautz, and Bram Cohen. Noise strategies for improving local search. In *Proceedings of the Twelfth National Conference on Artificial Intelligence (AAAI'94)*, pages 337–343, Seattle, 1994.

[33] Peter Balsiger, Alain Heuerding, and Stefan Schwendimann. A benchmark method for the propositional modal logics k, kt, s4. *Journal of Automated Reasoning*, 24(3):297–317, 2000.

[34] Geoff Sutcliff and Christian Suttner. Evaluating general purpose automated theorem proving systems. *Artificial Intelligence*, 131:39–54, 2001.

[35] Henry A. Kautz and Bart Selman. Pushing the envelope : Planning, propositional logic, and stochastic search. In *Proceedings of the Twelfth National Conference on Artificial Intelligence (AAAI'96)*, pages 1194–1201, 1996.

[36] A. Biere, A. Cimatti, E. Clarke, and Y. Zhu. Symbolic Model Checking without BDDs. In *International Conference on Tools and Algorithms for Construction and Analysis of Systems*, volume 1579 of *LNCS*, pages 193–207. Springer-Verlag, 1999.

kcnfs: An Efficient Solver for Random *k*-SAT Formulae

Gilles Dequen[1][*] and Olivier Dubois[2]

[*] LaRIA, Université de Picardie, Jules Verne, CURI, 5 rue du moulin neuf, 80000 Amiens, France,
dequen@laria.u-picardie.fr
[*] LIP6, CNRS-Université Paris 6, 4 place Jussieu, 75252 Paris cedex 05, France,
Olivier.Dubois@lip6.fr

Abstract. In this paper we generalize a heuristic that we have introduced previously for solving efficiently random 3-SAT formulae. Our heuristic is based on the notion of backbone, searching variables belonging to local backbones of a formula. This heuristic was limited to 3-SAT formulae. In this paper we generalize this heuristic by introducing a sub-heuristic called a re-normalization heuristic in order to handle formulae with various clause lengths and particularly hard random *k*-SAT formulae with $k \geq 3$. We implemented this new general heuristic in our previous program *cnfs*, a classical DPLL algorithm, renamed *kcnfs*. We give experimental results which show that *kcnfs* outperforms by far the best current complete solvers on any random *k*-SAT formula for $k \geq 3$.

1 Introduction

In [1] we presented a new branching variable selection heuristic for solving random 3-SAT formulae using a DPLL-type procedure. This heuristic relied on the notion of *local backbones* of a SAT formula, derived from a notion introduced in a statistical physics study [2]. The heuristic described in [1] was designed and works efficiently only on 3-SAT formulae. It cannot work on formulae with different clause lengths. In this paper we present a generalization of our previous heuristic covering formulae with clauses of any length and in particular for solving the classical random *k*-SAT formulae with $k \geq 3$. The generalization consists in inserting in our previous "backbone-search heuristic" (BSH for short), a sub-heuristic, which we call a *renormalization* heuristic, taking account of the various lengths of clauses. We describe in this paper such a renormalization heuristic and integrating it into the backbone-search heuristic, we get a general heuristic. We show that this general heuristic quite markedly improves the best current performances of solving of random *k*-SAT formulae as far as values of *k* can be handled in reasonable time.

The paper is organized as follows. In section 2 we first recall the principle of the backbone-search heuristic, we illustrate how it works differently from the

---

[*] Contract grant sponsor: The Région Picardie under HTSC project

E. Giunchiglia and A. Tacchella (Eds.): SAT 2003, LNCS 2919, pp. 486–501, 2004.

classical heuristics and we resume and update the experimental results of performances of this heuristic implemented in the program *cnfs* [1], a DPLL-type procedure. Then, in section 3 we describe our renormalization heuristic. Having implemented it in a general program which we renamed *kcnfs*, we give the performance results obtained on solving hard 3-SAT, 4-SAT, 5-SAT ... random formulae as compared with those obtained with the best known solvers.

2 The Backbone-Search Heuristic for Solving Random 3-SAT Formulae

2.1 The Principle of the Computation of the Backbone-Search Heuristic (BSH)

Let $\mathcal{F}$ be a 3-SAT formula and consider a DPLL-type algorithm ([3], [4]), for solving $\mathcal{F}$. Such an algorithm develops a solving tree until all clauses of $\mathcal{F}$ are satisfied by values 0 or 1 assigned to the variables of $\mathcal{F}$ at nodes of the tree or until no satisfying truth assignment has been found. This type of algorithm is called a complete algorithm since it gives an answer for unsatisfiable as well as satisfiable formulae. To solve efficiently satisfiable formulae, right truth values to be assigned to the variables are to be found. To solve efficiently unsatisfiable formulae, an implicit solving tree with a size as small as possible is to be developed. It appears that these two goals are hard and even somewhat incompatible to reach by a same heuristic. However developing a solving tree leads automatically to a solution of a formula if it is satisfiable. Trying to find right truth values for satisfying a formula does not lead automatically to be able to answer that the formula is unsatisfiable if it cannot be satisfied. Moreover developing a solving tree cannot be less efficient in the worst case (and in fact in any case for random formulae) for a satisfiable formula than for an unsatisfiable formula with the same size. For these reasons, designing an efficient heuristic for a complete DPLL-type algorithm amounts to supposing that one tries to solve only unsatisfiable formulae. This is exactly the way we designed the heuristic BSH which we describe below.

Let x be one of variables of the 3-SAT formula $\mathcal{F}$. The idea behind BSH is to measure the correlations of the literal x on the one hand and the literal $\bar{x}$ on the other hand with all the other variables of the formula through the clauses where x and $\bar{x}$ appear. More precisely let us take for example a clause where the literal x appears : $(x \lor u \lor v)$. If both literals u and v can be TRUE in most truth assignments satisfying many (possibly all) clauses, then x is a little correlated with the other variables. If one of the literals u and v must be FALSE if the other one is TRUE in most truth assignments satisfying many clauses, then x is a little more correlated with the other variables than in the previous case. Finally if both literals u and v are FALSE in most truth assignments satisfying many clauses, then x is strongly correlated to the other variables. Our heuristic BSH is intended to identified the variables for which both associated literals are strongly correlated with the others, by means of the function $h(t)$ we gave in [1]

Set $i \leftarrow 0$, let t be a literal and let MAX be an integer.

Integer $h(t)$

begin

    $i \leftarrow i + 1$

    compute $\mathcal{I}(t)$

    IF $i < MAX$

        return $\displaystyle\sum_{\cdot u \vee v \cdot \in \mathcal{I} \cdot t \cdot} h(\bar{u}) \times h(\bar{v})$

    ELSE

        return $\displaystyle\sum_{\cdot u \vee v \cdot \in \mathcal{I} \cdot t \cdot} (2p \cdot (\bar{u}) + p \cdot (\bar{u})) \times (2p \cdot (\bar{v}) + p \cdot (\bar{v}))$ (1)

    END IF

end

Fig. 1. Function $h(t)$ of the 3-SAT backbone search heuristic BSH

and described in Fig. 1, where t represents the literal x or $\bar{x}$. BSH chooses as a branching variable to be assigned successively 0 and 1 at a current node of the solving tree, one of the variables x having the highest score $h(x) \times h(\bar{x})$.

The computation carried out by $h(t)$ is as follows. t is any literal not assigned a value at a current node of the solving tree and $h(t)$ is the score of t. The evaluation $h(t)$ is based on the set $\mathcal{I}(t)$ of binary clauses such that for each of them if it is FALSE then it forces t to be TRUE unless a contradiction occurs. There are two cases to be considered. The first case is one of the occurrences of t appears in a ternary clause such as $(t \vee u \vee v)$. Then, if the binary clause $(u \vee v)$ is FALSE, as a result t is necessary TRUE so that $(t \vee u \vee v)$ is satisfied. The binary clause $(u \vee v)$ is therefore put in the set $\mathcal{I}(t)$. The second case is one of the occurrences of t appears in a binary clause such as $(t \vee u)$. Then, one searches recursively for the binary clauses which can force $\bar{u}$ to be TRUE. Such binary clauses force consequently t to be TRUE through the unary clause u or a chain of unary clauses. These binary clauses are also put in the set $\mathcal{I}(t)$. In this way, $\mathcal{I}(t)$ is the union of all binary clauses which can force t to be TRUE directly or indirectly (through unary clauses) if they were FALSE.

$h(t)$ is said to be evaluated at the level 1, 2, 3 ... if the integer MAX is set to 1, 2, 3, ... At the level 1, $h(t)$ is given directly by the sum of products (1) in Fig. 1. $p_2(\bar{u})$ and $p_3(\bar{u})$ are the numbers of binary and ternary clauses, respectively, where $\bar{u}$ appears in the sub-formula at the current node where $h(t)$ is evaluated. It is the same for $p_2(\bar{v})$ and $p_3(\bar{v})$. Each term of the sum (1) represents the weighted number of possibilities that a binary clause $(u \vee v)$ of $\mathcal{I}(t)$ is FALSE if one assumes that every occurrence of $\bar{u}$ and every occurrence of $\bar{v}$ can be forced to be TRUE in the clauses where $\bar{u}$ and $\bar{v}$ appear. The weighting of p_2 by 2 is due to the fact that as a rough estimate (for a uniform distribution of assignments) $\bar{u}$ has a double probability to be forced to be TRUE in a binary clause with respect to a ternary clause. On the whole, $h(t)$ can be interpreted as a measure of some correlation between t and all the other variables which can force t to be true, in the sub-formula at the current node where $h(t)$ is evaluated. This measure can be computed at level 1, 2, 3 ... recursively by setting MAX at 1, 2, 3 .... The measure is considered all the more accurate as the level of recursive computation is high. However it must be noticed that in the computation of $h(t)$, it is not taken into account that the same variable can be met several times with the same sign or opposite signs. One can imagine that this event occurs all the more often as the evaluation level is high. That is the reason why the level of evaluation must be limited in order that $h(t)$ as a measure of some correlation is not distorted (see next section). Finally if t should be TRUE in all assignments satisfying the maximum number of clauses (possibly all) of the sub-formula at the current node where $h(t)$ is evaluated, the correlation that $h(t)$ is intended to measure must be the strongest and then $h(t)$ should have one of the highest values among all values which it can take at the considered node. In this sense $h(t)$ tends to find out the variables belonging to the local backbone of the sub-formula at the considered node.

2.2 Limitations in the Implementation of the Heuristic BSH

There are two limitations which must restrict the use of the heuristic BSH for an efficient implementation. First it can be noticed that if at a current node of the solving tree the sub-formula associated with the considered node has many binary clauses, the computation of $h(t)$ can be too time-consuming with respect to simpler heuristics essentially based on the binary clauses. In *kcnfs* a limitation has been defined empirically as a function of the proportion of binary clauses in a sub-formula. The second limitation concerns the level of evaluation of $h(t)$ defined by the constant MAX. It is due both to the formal argument mentioned in the previous section and again to the cost of the computation. A good limitation must be a compromise between the accuracy of the evaluation which must remain significant so that it is not distorted by variables taken into account more than one time and the cost of the computation. In *kcnfs* this limitation has been defined empirically as a function of the number of variables and the maximum length of the clauses in the formula.

2.3 Effect of BSH Compared with Basic Heuristics

One can give a small example in order to illustrate the quite different effect of the backbone-search heuristic BSH compared with usual rules of choosing of a branching variable which heuristics implemented in many SAT solvers use. In Table 1, an unsatisfiable 3-SAT formula with 7 variables and 13 clauses is given.

Consider the choice of the first variable at the root of the tree to be developed by a DPLL-type algorithm for solving this formula. A basic heuristic consists in choosing a variable having the maximum number of occurrences in minimum size clauses (MOMS for short). All clauses of the formula being of the same length at the root, all variables are regarded equally. In Table 2, the first four highest numbers of occurrences per variable and the corresponding variables are given. An

Table 1. Unsatisfiable 3-SAT formula with 7 variables and 13 clauses

$$(\bar{a} \vee b \vee \bar{f}) \wedge (\bar{a} \vee \bar{c} \vee \bar{d}) \wedge (\bar{a} \vee \bar{b} \vee \bar{g}) \wedge (\bar{a} \vee \bar{c} \vee d) \wedge$$
$$(a \vee \bar{c} \vee d) \wedge (a \vee \bar{c} \vee \bar{d}) \wedge (a \vee b \vee f) \wedge (a \vee b \vee g) \wedge$$
$$(b \vee d \vee \bar{e}) \wedge (b \vee d \vee e) \wedge (b \vee c \vee \bar{d}) \wedge$$
$$(\bar{b} \vee c \vee e) \wedge (\bar{b} \vee c \vee \bar{e})$$

additional criterion, with respect to the previous one for the choice of a branching variable, is also quite commonly used in the heuristics. It is the respective proportions of the positive (uncomplemented) and negative (complemented) occurrences of each variable. A classical heuristic choice tends to maximize both the number of occurrences and the balance of the proportions of opposite literals of the chosen variable. A simple function which tries to combine these two criterions is for example the product of the numbers of occurrences of opposite literals of the variables. The Table 2 lists, according to this product heuristic, the highest four scores which correspond to the same variables as the first four ones according to the MOMS heuristic. Finally as comparison, the highest four scores obtained with BSH applied to all variables of the formula are also listed in Table 2. These highest four scores correspond again to the same variables.

Table 2. comparison of the highest four scores obtained for the variables in the formula of Table 1, between a basic heuristic and BSH

variable	#occ (*pos ; neg*)	Product	BSH
a	8 (*4+ ; 4−*)	16	810
b	9 (*6+ ; 3−*)	18	800
c	7 (*3+ ; 4−*)	12	2016
d	7 (*4+ ; 3−*)	12	1296

As noted, on the whole, for the three heuristics the highest four scores correspond to the same variables. However with the scores of BSH the variables are put in a different order from the one corresponding to the two other basic heuristics and above all, BSH is able to differentiate strongly the four variables between them, while they look alike for the two other heuristics. The latter point is the most important effect of BSH which helps to distinguish better a good branching variable with respect to the size of the solving tree. In our example this is particularly clear because as usual at the root of a solving tree, many variables look very similar for heuristics and this is the case for the four variables in Table 2. However BSH distinguishes the variable c as a much better variable than the three other ones. This is essentially due to the fact that the nature of the evaluation of the variables by BSH is quite different from the classical heuristics. BSH evaluates the correlation of a variable with the other variables, in other words the message sent by the variables of a formula, through the clauses, to a specific variable to be set at some value.

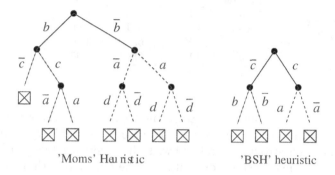

Fig. 2. Solving trees of a DPLL-type procedure developed with MOMS and BSH heuristics on the formula given in Table 1.

Practically the benefit brought by BSH can be seen for our example on the solving trees Fig. 2. At the left hand, one of the possible solving trees developed with the MOMS heuristic is drawn and at the right hand, the root of the tree is chosen as being the variable *c* selected with BSH and then for the subsequent nodes, each branching variable is chosen also with the MOMS heuristic. As it can be seen for the examples drawn on the Fig. 2, one can check that the size of the tree with the root variable *c*, chosen with BSH, is smaller than any other tree developed with MOMS. The more sophisticated heuristics implemented in the best current solvers tend to improve the evaluation of the possible branching variables with respect to MOMS but not as BSH can do it. Let us take for example the heuristic of the solver *satz* which is one of the best current solvers for random formulae [5]. The evaluation by this heuristic of the variables of our example Table 1, is given Table 3 in comparison with the Product heuristic and BSH. With the *satz* heuristic the variable *c* has the best score as with BSH. But the differences of score with the other variables are not as definite as with BSH. For this reason one can easily suppose that on large formulae and not a toy as our illustration example, the *satz* heuristic distinguishes less accurately than BSH the best variables to choose as branching variables. This is clearly confirmed by the comparison of solving performances of hard random formulae related in Section 3. Thus what is seen on the Fig. 2, illustrates what happens on large size formulae even with heuristics more sophisticated than MOMS.

2.4   Local Treatments: Picking and Pickback, in Addition to BSH

As said above, the heuristic BSH is intended to measure in some way the correlation of a variables with the other variables through clauses of a formula. But

Table 3. Comparison of the highest four scores obtained for the variables in the formula of Table 1, between the Product heuristic, *satz* heuristic and BSH

variable	Product	*satz* heuristic	BSH
a	16	504	810
b	18	561	800
c	12	690	2016
d	12	256	1296

this measurement does not take into account two things, the size of the neighborhood containing the correlated variables and the variables for which only one associated literal is strongly correlated with the other variables. If the size of the neighborhood is small a contradiction might be detected with a small tree and then to lead to a backtrack. If one literal associated with a variable is strongly correlated in a small neighborhood then the variable might be set to a fixed truth value by developing also a small tree. Picking and pickback are local treatments to detect such contradictions or such variables to be fixed. These local treatments are only applied on variables having the highest scores of correlation. They consist for the former in a look ahead and for the latter in a look back.

2.5   Performance Comparison Results on Random 3-SAT Formulae

The Tables 4 and 5 list performances comparative results of *cnfs* against several of the best current solvers : *posit* [6], *csat* [7], *salz* [8]and *OkSolver* [9] which was the winner of the SAT 2002 competition[3] in the category of random formulae. For information, we give also the performances of known solvers as *zchaff (v. 2003.7.1)* [10] and *sato* [11] not being specifically devoted to solving random formulae. The performance results are given in terms of mean sizes of trees (Table 4) and mean solving times (Table 5) on sample of 1000 random 3-SAT formulae with 200, 300 and 400 variables at a ratio, clauses to variables, equal to 4.25 which corresponds to the SAT-UNSAT transition area where the formulae are on average the hardest. These results show that *cnfs* outperforms the other solvers. Moreover it is interesting to note that the detailed results showed that *cnfs* was systematically faster than the other solvers on any solved unsatisfiable formula. On hard formulae larger than 400 variables the gap of performance between *cnfs* and the other solvers increases as a function of size of the formula. Finally *cnfs* was able to solve formulae up to 700 variables which were until now out of reach with any other solver.

[.] http://www.satlive.org/SATCompetition/2002/

Table 4. Mean tree sizes on hard random 3-SAT formulae having 200, 300 and 400 variables with *sato, csat, posit, satz, OkSolver* and *cnfs* solvers

SAT Solvers	unsat (N) sat (Y)	200 V 850 C (482 N) (518 Y) mean #nodes (*std dev.*)	300 V 1275 C (456 N) (534 Y) mean #nodes (*std dev.*)	400 V 1700 C (245 N) (255 Y) mean #nodes in millions (*std dev.*)
zchaff	unsat	57666 (*23708*)	1.7 10ˈ (*0.9 10⁶*)	-
	sat	9365 (*11435*)	512687 (*405756*)	-
sato	unsat	14076 (*5225*)	443313 (*109694*)	-
	sat	4229 (*4754*)	151728 (*157102*)	-
csat	unsat	2553 (*997*)	90616 (*37729*)	3.6 (*1.7*)
	sat	733 (*763*)	26439 (*31224*)	1.8 (*2.4*)
OkSolver	unsat	1346 (*430*)	36137 (*12181*)	1.1 (*0.5*)
	sat	314 (*381*)	8454 (*8594*)	0.3 (*0.4*)
posit	unsat	1992 (*754*)	82572 (*35364*)	3.4 (*1.7*)
	sat	789 (*694*)	34016 (*31669*)	1.5 (*1.4*)
satz	unsat	623 (*206*)	18480 (*7050*)	0.5 (*0.2*)
	sat	237 (*216*)	6304 (*6541*)	0.2 (*0.2*)
cnfs	unsat	470 (*156*)	12739 (*4753*)	0.3 (*0.1*)
	sat	149 (*154*)	3607 (*4089*)	0.1 (*0.1*)

Table 5. Mean solving times on hard random 3-SAT formulae having 200, 300, 400 variables with *sato, csat, posit, satz* and *cnfs* solvers

SAT Solvers	unsat (N) sat (Y)	200 V 850 C (482 N) (518 Y) mean time (*std dev.*)	300 V 1275 C (456 N) (534 Y) mean time (*std dev.*)	400 V 1700 C (245 N) (255 Y) mean time (*std dev.*)
zchaff	unsat	18.9s (*21.5s*)	2h 54m (*2h*)	-
	sat	1.4s (*2.1s*)	20m (*22m*)	-
sato	unsat	89.2s (*84.6s*)	18h 27m (*7h*)	-
	sat	13.2s (*38.0s*)	4h (*6h 40m*)	-
csat	unsat	0.5s (*0.2s*)	29.6s (*13.3s*)	29m 01s (*14m 36s*)
	sat	0.1s (*0.1s*)	9.0s (*10.7s*)	7m 47s (*10m 01s*)
OkSolver	unsat	0.6s (*0.2s*)	32.1s (*11.2s*)	26m 03s (*12m 14s*)
	sat	0.1s (*0.2s*)	7.6s (*7.7s*)	5m 58s (*9m 23s*)
posit	unsat	0.3s (*0.1s*)	15.1s (*6.6s*)	13m 07s (*6m 35s*)
	sat	0.1s (*0.1s*)	6.3s (*5.9s*)	5m 56s (*5m 36s*)
satz	unsat	0.2s (*0.1s*)	4.9s (*1.8s*)	2m 44s (*1m 09s*)
	sat	0.1s (*0.1s*)	1.7s (*1.7s*)	1m 05s (*1m 03s*)
cnfs	unsat	0.1s (*0.0s*)	3.7s (*1.3s*)	1m 50s (*0m 43s*)
	sat	0.0s (*0.0s*)	1.1s (*1.2s*)	0m 37s (*0m 40s*)

3 Renormalization in the Search-Backbone Heuristic BSH for Formulae with Clauses Having Various Lengths

3.1 The Principle of the Renormalization Computation

Recall that the function $h(t)$ of BSH described in Section 2.1 is intended to measure some correlation of the literal t with the other variables in the considered sub-formula. This correlation of one occurrence of t depends intuitively on the length of the clause where the considered t appears and on the correlation of the other literals themselves associated with t in this clause. For example in the following clause of length 6 : $(t \lor u_1 \lor u_2 \lor u_3 \lor u_4 \lor u_5)$, t is likely (in a random formula) much less correlated through $(u_1 \lor u_2 \lor u_3 \lor u_4 \lor u_5)$ with the rest of the variables than in the following clause of length 3 : $(t \lor v_1 \lor v_2)$ through $(v_1 \lor v_2)$. In contrast, if in this latter clause v_1 and v_2 are weakly correlated with the other variables of the formula, for example (which is extreme) if they are pure literals, then through a clause of any length, as large as it can be, where t occurs, t is more correlated with the rest of the variables than through $(v_1 \lor v_2)$. We have just identified two factors of correlation of an occurrence of t, the length of the clause where t appears and the correlation of the literals associated with t. This latter factor is naturally taken into account by the level of evaluation defined for $h(t)$ in section 2.1, whose principle extends directly to clauses with various lengths. The first factor is linked to the computation of $h(t)$ itself. We have seen in section 2.1 that this computation is based on the set $\mathcal{I}(t)$ only consisting in clauses of length 2. The number of possibilities that each of the clauses of $\mathcal{I}(t)$ is FALSE, is estimated as the product of two terms associated with each literal of a binary clause. Considering a formula with clauses of various lengths and applying the same principle described in Section 2.1, the set $\mathcal{I}(t)$ contains also clauses of various lengths. This can brings about disparities in estimating the correlations of different occurrences of t with the rest of the variables. For example if two occurrences of t can be forced to be TRUE through two clauses of different lengths, say z_1 and z_2, with $z_2 \geq z_1$, a direct generalization of the computation of $h(t)$ in Section 2.1 could yield a larger number of possibilities that t is forced to be TRUE by the clause having the largest length z_2 than by the clause with the length z_1. This seeming incoherence does not mean that the principle of computation described in Section 2.1 is not sound. The reason is merely that the correlations are not estimated at the same scale and therefore must be normalized. First, we point out that the reason the lengths of clauses in $\mathcal{I}(t)$ are homogeneous and equal to 2 for 3-SAT formulae is that if an occurrence of t occurs in a binary clause associated with only one literal, then binary clauses which could forced t to be TRUE are searched recursively through only a chain of unary clauses preserving thereby the number of possibilities that t is forced to be TRUE. In presence of clauses with various lengths this property no longer holds. More sophisticate structures could be considered such as trees, but the computation of $h(t)$ would turn out probably prohibitive. To overcome the difficulty, we introduce a "fictitious" literal with which we fill the clauses in the set $\mathcal{I}(t)$ such that they have all the same length equal to the largest possible length of clauses in the sub-formula with which $h(t)$ is evaluated. Then it must be associated a reasonable value with this fictitious literal so that the evaluation of the correlation of t makes sense. We choose to assign to it the mean value of all values computed for the literals in the clauses where t appears. Finally since the literal is fictitious we can choose adequately its truth value as FALSE. Hence there is no uncertainty about its truth value, the probability of being FALSE is 1, whereas the probability for "real" literals to be FALSE or TRUE can be roughly estimated (in a random assignment) at $1/2$. Consequently each time the fictitious literal is used its value is weighted by a factor 2. Thus we obtain a complete normalization in the computation of the correlation of t with the rest of variables. The normalization computation is to be done at every node of the solving tree and is specific for each sub-formula associated with the considered node. That is the reason why we say that we have to renormalize in the course of the development of the solving tree.

3.2 The Backbone-Search Renormalized Heuristic BSRH

From a practical viewpoint the backbone-search renormalized heuristic is described by the function $hr(t)$ in Fig. 3.

This presentation of BSRH should not be considered as an implementation (it would be very inefficient), it is just a formal description for the sake of clarity. The computation carried out by $hr(t)$ is as follows. t is any literal not assigned a value at a current node of the solving tree and $hr(t)$ is the score of t. k is the maximum length of the clauses in the sub-formula associated with the considered node.

The evaluation $hr(t)$ is based on the sets $\mathcal{I}_j(t)$ of clauses of length j such that for each of them if it is FALSE then it forces t to be TRUE unless a contradiction occurs. j varies from 2 (using the principle of the chains of unary clauses described in section 2.1) up to at most $k-1$ (the clauses containing at least t). As previously for $h(t)$, $hr(t)$ can be evaluated at the level 1, 2, 3 . . . according to a recursive computation, by setting the integer MAX to 1, 2, 3, ... The measure is considered all the more accurate as the level of recursive computation is high. $hr(t)$ is the result of a computation based on every set $\mathcal{I}_j(t)$. At level 1, this computation is given directly by the sum of products (1) in Fig. 3. The literals $\bar{l}_i$ are evaluated by the function $eval(t)$, described in Fig. 4, which is a generalization of the computation in the corresponding sum (1) in Fig. 1. $p_j(t)$ is the number of occurrences of t in the clauses of length j. Each term of the sum (1) represents the weighted number of possibilities that a clause of length j of $\mathcal{I}(t)$ is FALSE if one assumes that every occurrence of its negated literals can be forced to be TRUE in the clauses where they appear. The sum (2) adds all the values associated with every evaluated literal of the clauses in all the sets $\mathcal{I}_j(t)$ and (3) compute the renormalization coefficient. Finally (4) gives the evaluation of $hr(t)$ according to the renormalization computation principle described in the preceding section. it must be noticed that, as for $h(t)$, in the computation of $hr(t)$, it is not taken into account that a same variable can be met several times with the same sign or opposite signs. As said previously in section 2.1, one can

Set $i \leftarrow 0$, let t be a literal, let k be the maximum clause length of the considered formula and let MAX, Sum_{eval} and $Card_{eval}$ be integers.

Integer $hr(t)$

begin

$\quad i \leftarrow i + 1$

$\quad Sum_{eval} \leftarrow 0,\ Card_{eval} \leftarrow 0$

$\quad$ FOR $j \in \{2,\ \ldots\ , k-1\}$

$\quad\quad$ compute $\mathcal{I}_j(t)$

$\quad\quad$ IF $\quad i < MAX$

$$S_j \leftarrow \sum_{\bullet l_1 \vee \ldots \vee l_j \bullet \in \mathcal{I}_j \bullet t \bullet} hr(\bar{l}_\bullet) \times \ldots \times hr(\bar{l}_j)$$

$$Sum_{eval} \leftarrow Sum_{eval} + \sum_{\bullet l_1 \vee \ldots \vee l_j \bullet \in \mathcal{I}_j \bullet t \bullet} hr(\bar{l}_\bullet) + \ldots + hr(\bar{l}_j)$$

$\quad\quad$ ELSE

$$S_j \leftarrow \sum_{\bullet l_1 \vee \ldots \vee l_j \bullet \in \mathcal{I}_j \bullet t \bullet} eval(\bar{l}_\bullet) \times \ldots \times eval(\bar{l}_j) \tag{1}$$

$$Sum_{eval} \leftarrow Sum_{eval} + \sum_{\bullet l_1 \vee \ldots \vee l_j \bullet \in \mathcal{I}_j \bullet t \bullet} eval(\bar{l}_\bullet) + \ldots + eval(\bar{l}_j) \tag{2}$$

$\quad\quad$ END IF

$$Card_{eval} \leftarrow Card_{eval} + j \times |\mathcal{I}_j(t)|$$

$\quad$ END FOR

$$\mathcal{E}(l_v) \leftarrow \frac{Sum_{eval}}{Card_{eval}} \tag{3}$$

$$\text{return} \sum_{j \in \{\bullet,\ \ldots\ , k-\bullet\}} S_j \times \mathcal{E}(l_v)^{k-j-\bullet} \times 2^{k-j} \tag{4}$$

end

Fig. 3. the function $hr(t)$ of the backbone-search renormalized heuristic

imagine that this event occurs all the more often as the evaluation level is high and therefore the level must be limited in order that $hr(t)$, as a measure of some correlation, is not distorted. For that as well as for efficiency reasons, we have applied in the implementation of the heuristic BSRH in *kcnfs* the two types of limitations described in section 2.2.

3.3   Performance Comparison Results on Random 4-SAT, 5-SAT, ...Formulae

In the same line of the comparative experiments carried out for random 3-SAT formulae, we compared the performances of *kcnfs* against most of the solvers used for 3-SAT comparisons, which are also the best known solvers on random k-SAT formulae for $k > 3$. Tables 6, 7 and Tables 8, 9 give the mean tree sizes and the mean computing times respectively on samples of 1000 random 4-SAT and 5-SAT formulae with a ratio clauses to variables around the SAT-UNSAT transition. It appears that apart from *kcnfs*, the other solvers are not in the same order according to their performances relating to the tree sizes or to the

Let t be a literal, let k be the maximum clause length of the considered formula and s be an integer.

Integer $eval(t)$
begin
    $s \leftarrow 0$
    FOR $j \in \{2, \ldots, k\}$
        $s \leftarrow s + (p_j(t) \times 2^{k-j})$
    END FOR
    return s
end

Fig. 4. Final evaluation function of a literal t

computing times. In contrast *kcnfs* has both the best performances relating to the two criterions. Moreover as for 3-SAT formulae, the detailed results showed that *kcnfs* was systematically faster than the others on any solved unsatisfiable formula. Further experiments were carried out on formulae up to 8-SAT against *posit* which appeared to be the best among the other solvers. The Fig. 6 show the curves representing the ratio of the computing times of *cnfs* to those of *posit* as a function of the numbers of variables of formulae from 4-SAT up to 8-SAT with a ratio again around the SAT-UNSAT transition. It can be observed that the steepness of the curves increases as the length of clauses becomes larger. This confirms that the gap of performances of *kcnfs* with the other known solvers enlarges as the length of clauses in random formulae grows.

Table 6. Mean tree sizes on hard random 4-SAT formulae having 100, 130 and 150 variables with the solvers *csat*, *posit*, *satz*, *OkSolver* and *kcnfs* solvers

Solvers	SAT unsat (N) sat (Y)	100 V 988 C (106 N) (94 Y) mean #nodes (*std dev.*)	130 V 1285 C (80 N) (120 Y) mean #nodes in millions (*std dev.*)	150 V 1482 C (85 N) (115 Y) mean #nodes in millions (*std dev.*)
csat	unsat	52080 (*9298*)	0.966 (*0.151*)	6.062 (*0.967*)
	sat	15976 (*16867*)	0.342 (*0.286*)	2.361 (*2.012*)
posit	unsat	38016 (*6443*)	0.744 (*0.124*)	5.496 (*0.906*)
	sat	17396 (*9585*)	0.385 (*0.255*)	2.778 (*1.811*)
satz	unsat	30610 (*5144*)	0.484 (*0.081*)	3.173 (*0.531*)
	sat	9049 (*7076*)	0.159 (*0.128*)	1.249 (*1.048*)
OkSolver	unsat	24989 (*4543*)	0.375 (*0.064*)	2.381 (*0.386*)
	sat	7040 (*7781*)	0.119 (*0.119*)	0.913 (*0.751*)
kcnfs	unsat	10450 (*1740*)	0.174 (*0.026*)	1.111 (*1.903*)
	sat	2616 (*2957*)	0.051 (*0.047*)	0.410 (*0.381*)

Table 7. Mean tree sizes on hard random 5-SAT formulae having 60, 80, 95 variables with *csat*, *posit*, *satz*, *OkSolver* and *kcnfs* solvers

SAT unsat (N) Solvers sat (Y)		100 V 988 C (106 N) (94 Y) mean #nodes (*std dev.*)	130 V 1285 C (80 N) (120 Y) mean #nodes in millions (*std dev.*)	150 V 1482 C (85 N) (115 Y) mean #nodes in millions (*std dev.*)
csat	unsat	52080 (*9298*)	0.966 (*0.151*)	6.062 (*0.967*)
	sat	15976 (*16867*)	0.342 (*0.286*)	2.361 (*2.012*)
posit	unsat	38016 (*6443*)	0.744 (*0.124*)	5.496 (*0.906*)
	sat	17396 (*9585*)	0.385 (*0.255*)	2.778 (*1.811*)
satz	unsat	30610 (*5144*)	0.484 (*0.081*)	3.173 (*0.531*)
	sat	9049 (*7076*)	0.159 (*0.128*)	1.249 (*1.048*)
OkSolver	unsat	24989 (*4543*)	0.375 (*0.064*)	2.381 (*0.386*)
	sat	7040 (*7781*)	0.119 (*0.119*)	0.913 (*0.751*)
kcnfs	unsat	10450 (*1740*)	0.174 (*0.026*)	1.111 (*1.903*)
	sat	2616 (*2957*)	0.051 (*0.047*)	0.410 (*0.381*)

Table 8. Mean computation time on hard random 4-SAT formulae having 100, 130, 150 variables with *csat*, *posit*, *satz*, *OkSolver* and *kcnfs* solvers

SAT unsat (N) Solvers sat (Y)		60 V 1290 C (102 N) (98 Y) mean #nodes (*std dev.*)	80 V 1720 C (123 N) (77 Y) mean #nodes in millions (*std dev.*)	95 V 2042 C (131 N) (69 Y) mean #nodes in millions (*std dev.*)
csat	unsat	33511 (*2210*)	0.721 (*0.054*)	7.246 (*0.602*)
	sat	15629 (*8843*)	0.316 (*0.191*)	2.960 (*1.981*)
posit	unsat	20729 (*1419*)	0.473 (*0.035*)	4.925 (*0.437*)
	sat	11478 (*6652*)	0.236 (*0.148*)	2.989 (*1.673*)
satz	unsat	33645 (*2954*)	0.697 (*0.075*)	6.887 (*0.668*)
	sat	16132 (*9933*)	0.314 (*0.188*)	2.53 (*1.721*)
OkSolver	unsat	19537 (*1397*)	0.364 (*0.023*)	3.310 (*0.286*)
	sat	9046 (*5656*)	0.144 (*0.110*)	1.233 (*1.042*)
kcnfs	unsat	7902 (*543*)	0.147 (*0.011*)	1.456 (*0.125*)
	sat	3096 (*2263*)	0.059 (*0.044*)	0.593 (*0.414*)

3.4   SAT 2003 Competition

kcnfs has been submitted to the SAT 2003 competition in the category of random formulae. In this category, the performances of 34 solvers have been appreciated on 316 benchmarks. 222 out of the 316's have been solved by at least one solver and 30 by only one. As stressed already in the experiments given before in this paper, all solved unsatisfiable benchmarks have been solved by *kcnfs*. Moreover *kcnfs* has solved the greatest number of benchmarks with respect to its competitors. In terms of computing time *kcnfs* has been on average 35% faster than its

Table 9. Mean computation time on hard random 5-SAT formulae having 60, 80, 95 variables with *csat, posit, satz, OkSolver* and *kcnfs* solvers

SAT unsat (N) Solvers sat (Y)		100 V 988 C (106 N) (94 Y) mean time (*std dev.*)	130 V 1285 C (80 N) (120 Y) mean time (*std dev.*)	150 V 1482 C (85 N) (115 Y) mean time (*std dev.*)
satz	unsat	8.8s (*1.5s*)	188.1s (*119.3s*)	23m 16s (*3m 50s*)
	sat	2.6s (*2.1s*)	62.1s (*49.9s*)	9m 05s (*7m 32s*)
OkSolver	unsat	7.7s (*1.4s*)	166.4s (*26.1s*)	21m 54s (*3m 30s*)
	sat	0.1s (*0.2s*)	7.6s (*7.7s*)	5m 58s (*9m 23s*)
csat	unsat	5.3s (*1.0s*)	119.3s (*18.9s*)	15m 22s (*2m 28s*)
	sat	1.6s (*1.7s*)	42.2s (*35.5s*)	5m 57s (*5m 13s*)
posit	unsat	3.9s (*0.6s*)	92.2s (*15.5s*)	12m 30s (*2m 07s*)
	sat	1.9s (*1.1s*)	43.2s (*30.1s*)	6m 27s (*4m 12s*)
kcnfs	unsat	2.6s (*0.4s*)	57.7s (*8.5s*)	7m 12s (*1m 12s*)
	sat	0.7s (*0.7s*)	16.7s (*15.5s*)	2m 33s (*2m 23s*)

best competitors (respectively for each formula). It has been declared the winner in the random category as a complete solver.

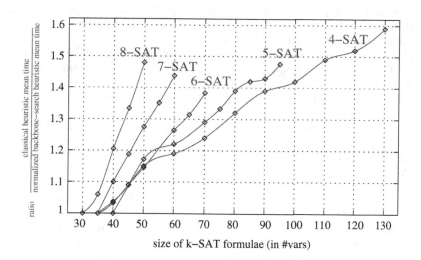

Fig. 5. ratio of mean time of *kcnfs* to mean time of *posit* on hard random *k*-SAT formulae from *k* = 4 up to *k* = 8

4   Conclusion

The new heuristic presented in this paper has yielded improvements in the performances of solving hard random formulae with any length of clauses, on a scale which had not been observed since a long time ago. This heuristic, as mentioned in Section 2.1, has been designed with respect of the solving of unsatisfiable formulae. Very efficient algorithms processing specifically satisfiable formulae as WALKSAT, [12], or SP, [13], for very large size of formulae, could be combined with our heuristic in a DPLL-type algorithm without damaging significantly the performances on unsatisfiable formulae reported in this paper. It would be a worthwhile operation to enhance on the whole the solving of random formulae. In a more long-term view our feeling is that, in contrast with some pessimistic conclusion having given previously in the literature as in [14], one can expect much more considerable improvements than those reported in this paper, for random as well as for structured formulae. This should come from a much more understanding both of the structure of the space of solutions and the combinatorial structure of formulae. Contributions of various communities could be helpful as progresses already achieved regarding the structure of space of solutions by statistical physics studies in [15, 16].

References

[1] Dubois, O., Dequen, G.: A Backbone Search Heuristic for Efficient Solving of Hard 3-SAT Formulae. In: Proc. of the 17th Internat. Joint Conf. on Artificial Intelligence, Seattle (2001) 248–253

[2] Monasson, R., Zecchina, R., Kirkpatrick, S., Selman, B., Troyansky, L.: Determining Computational Complexity from Characteristic 'Phase Transitions'. Nature 400 (1999) 133–137

[3] Davis, M., Putnam, H.: A Computing Procedure for Quantification Theory. Journal Association for Computing Machine (1960) 201–215

[4] Davis, M., Logemann, G., Loveland, D.: A Machine Program for Theorem-Proving. Journal Association for Computing Machine (1962) 394–397

[5] Li, C.M., Anbulagan: Heuristics Based on Unit Propagation for Satisfiability Problems. In: Proc. of the 15th Internat. Joint Conf. on Artificial Intelligence, Nagoya (Japan), IJCAI (1997) 366–371

[6] Freeman, J.W.: Hard Random 3-SAT Problems and the Davis-Putnam Procedure. Artificial Intelligence 81 (1996) 183–198

[7] Dubois, O., Andre, P., Boufkhad, Y., Carlier, J.: SAT versus UNSAT. DIMACS Series in Discr. Math. and Theor. Computer Science (1993) 415–436

[8] Li, C.M.: A Constraint-Based Approach to Narrow Search Trees for Satisfiability. Information Processing Letters 71 (1999) 75–80

[9] Kullman, O.: Heuristics for SAT algorithms: A systematic study. In: Extended abstract for the Second workshop on the satisfiability problem (SAT'98). (1998)

[10] Moskewicz, M.W., Madigan, C.F., Zhao, Y., Zhang, L., Malik, S.: Chaff: Engineering an Efficient SAT Solver. In: Proceedings of the 38th Design Automation Conference (DAC'01). (2001)

[11] Zhang, H.: SATO. An Efficient Propositional Prover. In: Proc. of the 14th Internat. Conf. on Automated Deduction, CADE-97, LNCS (1997) 272–275

[12] Selman, B., Kautz, H.A., Cohen, B.: Noise Strategies for Improving Local Search. In: Proc. of the 12th National Conf. on Artificial Intelligence. Volume 1, Menlo Park, CA, USA, AAAI Press (1994) 337–343

[13] A. Braunstein, M. Mezard, R.Z.: Survey propagation: an algorithm for satisfiability. arXiv - cond-mat/0207194 (2002)

[14] Li, C.M., Gerard, S.: On the Limit of Branching Rules for Hard Random Unsatisfiable 3-SAT. In: Proc. of European Conf. on Artificial Intelligence, ECAI (2000) 98–102

[15] G. Biroli, R. Monasson, M.W.: A variational description of the ground state structure in random satisfiability problems. Eur. Phys. J. B 14 551 (2000)

[16] M. Mezard, G. Parisi, R.Z.: Analytic and algorithmic solutions of random satisfiability problems. Science 297 (2002)

An Extensible SAT-solver

Niklas Eén and Niklas Sörensson

Chalmers University of Technology, Sweden
{een,nik}@cs.chalmers.se

Abstract. In this article[*], we present a small, complete, and efficient SAT-solver in the style of conflict-driven learning, as exemplified by CHAFF. We aim to give sufficient details about implementation to enable the reader to construct his or her own solver in a very short time. This will allow *users* of SAT-solvers to make domain specific extensions or adaptions of current state-of-the-art SAT-techniques, to meet the needs of a particular application area. The presented solver is designed with this in mind, and includes among other things a mechanism for adding arbitrary boolean constraints. It also supports solving a series of related SAT-problems efficiently by an incremental SAT-interface.

1 Introduction

The use of SAT-solvers in various applications is on the march. As insight on how to efficiently encode problems into SAT is increasing, a growing number of problem domains are successfully being tackled by SAT-solvers. This is particularly true for the *electronic design automation* (EDA) industry [BC+99, Lar92]. The success is further magnified by current state-of-the-art solvers being adapted to meet the specific characteristics of these problem domains [AR+02, ES03].

However, modifying an existing solver, even with thorough knowledge of both the problem domain and of modern SAT-techniques, can be a time consuming journey into the inner workings of a ten-thousand-line software package. Likewise, writing a solver from scratch often means spending much time rediscovering the intricate details of a correct and efficient solver. The problem is that although the *techniques* used in a modern SAT-solver are well documented, the details necessary for an *implementation* have not been adequately presented before.

In the fall of 2002, the authors implemented the solvers SATZOO and SATNIK. In order to sufficiently understand the implementation tricks needed for a modern SAT-solver, it was necessary to consult the source-code of previous implementations.[2] We find that the material contained therein can be made more accessible, which is desirable for the SAT-community. Thus, the principal goal of this article is to bridge the gap between existing descriptions of SAT-techniques and their actual implementation.

---

[*] Extended version available at http://www.cs.chalmers.se/~een

[*] LIMMAT at http://www.inf.ethz.ch/personal/biere/projects/limmat
ZCHAFF at http://www.ee.princeton.edu/~chaff/zchaff

E. Giunchiglia and A. Tacchella (Eds.): SAT 2003, LNCS 2919, pp. 502–518, 2004.
© Springer-Verlag Berlin Heidelberg 2004

We will do this by presenting the code of a minimal SAT-solver MINISAT, based on the ideas for conflict-driven backtracking [MS96], together with watched literals and dynamic variable ordering [MZ01]. The original C++ source code (downloadable from `http://www.cs.chalmers.se/~een`) for MINISAT is under 600 lines (not counting comments), and is the result of rethinking and simplifying the designs of SATZOO and SATNIK without sacrificing efficiency. We will present all the relevant parts of the code in a manner that should be accessible to anyone acquainted with either C++ or Java.

The presented code includes an incremental SAT-interface, which allows for a series of related problems to be solved with potentially huge efficiency gains [ES03]. We also generalize the expressiveness of the SAT-problem formulation by providing a mechanism for defining arbitrary *constraints* over boolean variables.

From the documentation in this paper we hope it is possible for *you* to implement a fresh SAT-solver in your favorite language, or to grab the C++ version of MINISAT from the net and start modifying it to include new and interesting ideas.

2   Application Programming Interface

We start by presenting MINISAT's external interface, with which a user application can specify and solve SAT-problems. A basic knowledge about SAT is assumed (see for instance [MS96]). The types *var*, *lit*, and *Vec* for variables, literals, and vectors respectively are explained in more detail in section 4.

class *Solver* – *Public interface*		
var	*newVar*	()
bool	*addClause*	(*Vec⟨lit⟩* literals)
bool	*add...*	(...)
bool	*simplifyDB*	()
bool	*solve*	(*Vec⟨lit⟩* assumptions)
Vec⟨bool⟩ model		– *If found, this vector has the model.*

The "*add...*" method should be understood as a place-holder for additional constraints implemented in an extension of MINISAT.

For a standard SAT-problem, the interface is used in the following way: Variables are introduced by calling *newVar()*. From these variables, clauses are built and added by *addClause()*. Trivial conflicts, such as two unit clauses $\{x\}$ and $\{\overline{x}\}$ being added, can be detected by *addClause()*, in which case it returns FALSE. From this point on, the solver state is undefined and must not be used further. If no such trivial conflict is detected during the clause insertion phase, *solve()* is called with an empty list of assumptions. It returns FALSE if the problem is *unsatisfiable*, and TRUE if it is *satisfiable*, in which case the model can be read from the public vector "model".

The *simplifyDB()* method can be used before calling *solve()* to simplify the set of problem constraints (often called the *constraint database*). In our implementation, *simplifyDB()* will first propagate all unit information, then remove all satisfied constraints. As for *addClause()*, the simplifier can sometimes detect a conflict, in which case FALSE is returned and the solver state is, again, undefined and must not be used further.

If the solver returns *satisfiable*, new constraints can be added repeatedly to the existing database and *solve()* run again. However, more interesting sequences of SAT-problems can be solved by the use of *unit assumptions*. When passing a non-empty list of assumptions to *solve()*, the solver temporarily assumes the literals to be true. After finding a model or a contradiction, these assumptions are undone, and the solver is returned to a usable state, even when *solve()* return FALSE, which now should be interpreted as *unsatisfiable under assumptions*.

For this to work, calling *simplifyDB()* before *solve()* is no longer optional. It is the mechanism for detecting conflicts independent of the assumptions – referred to as a *top-level* conflict from now on – which puts the solver in an undefined state. For an example of the use if unit assumptions, see [ES03].

An alternative interface would be for *solve()* to return one of three values: *satisfiable, unsatisfiable,* or *unsatisfiable under assumptions*. This is indeed a less error-prone interface as there is no longer a pre-condition on the use of *solve()*. The current interface, however, represents the smallest modification of a non-incremental SAT-solver.

3   Overview of the SAT-solver

This article will treat the popular style of SAT-solvers based on the DPLL algorithm [DLL62], backtracking by conflict analysis and clause recording (also referred to as *learning*) [MS96], and boolean constraint propagation (BCP) using *watched literals* [MZ01]. We will refer to this style of solver as a *conflict-driven SAT-solver*. The components of such a solver, and indeed a more general constraint solver, can be conceptually divided into three categories:

- **Representation.** Somehow the SAT-instance must be represented by internal data structures, as must any derived information.

- **Inference.** Brute force search is seldom good enough on its own. A solver also needs some mechanism for computing and propagating the direct implications of the current state of information.

- **Search.** Inference is almost always combined with search to make the solver complete. The search can be viewed as another way of deriving information.

A standard conflict-driven SAT-solver can represent *clauses* (with two literals or more) and *assignments*. Although the assignments can be viewed as unit-clauses, they are treated specially, and are best viewed as a separate type of information.

The only inference mechanism used by a standard solver is *unit propagation*. As soon as a clause becomes *unit* under the current assignment (all literals except one are false), the remaining unbound literal is asserted, possibly making more clauses unit. The process continues until no more information can be propagated.

The search procedure of a modern solver is the most complex part. Heuristically, variables are picked and assigned values (*assumptions* are made), until the propagation detects a *conflict* (all literals of a clause have become false). At that point, a so called *conflict clause* is constructed and added to the SAT problem. Assumptions are then canceled by backtracking until the conflict clause becomes unit, at which point it is propagated and the search process continues.

MINISAT is extensible with arbitrary boolean constraints. This will affect the *representation*, which must be able to store these constraints; the *inference*, which must be able to derive unit information from these constraints; and the *search*, which must be able to analyze and generate conflict clauses from the constraints. The mechanism we suggest for managing general constraints is very lightweight, and by making the dependencies between the SAT-algorithm and the constraints implementation explicit, it adds to the clarity of the solver.

Propagation. The propagation procedure of MINISAT is largely inspired by that of CHAFF [MZ01]. For each literal, a list of constraints is kept. These are the constraints that *may* propagate unit information (variable assignments) if the literal becomes TRUE. For clauses, no unit information can be propagated until all literals except one have become FALSE. Two unbound literals p and q of the clause are therefore selected, and references to the clause are added to the lists of $\bar{p}$ and $\bar{q}$ respectively. The literals are said to be *watched* and the lists of constraints are referred to as *watcher lists*. As soon as a watched literal becomes TRUE, the constraint is invoked to see if information may be propagated, or to select new unbound literals to be watched.

An effect of using watches for clauses is that on backtracking, no adjustment to the watcher lists need to be done. Backtracking is therefore cheap. However, for other constraint types, this is not necessarily a good approach. MINISAT therefore supports the optional use of *undo lists* for those constraints; storing what constraints need to be updated when backtracking unbinds a variable.

Learning. The learning procedure of MINISAT follows the ideas of Marques-Silva and Sakallah in [MS96]. The process starts when a constraint becomes conflicting (impossible to satisfy) under the current assignment. The conflicting constraint is then asked for a set of variable assignments that make it contradictory. For a clause, this would be all the literals of the clause (which are FALSE under a conflict). Each of the variable assignments returned must be either an *assumption* of the search procedure, or the result of some *propagation* of a constraint. The propagating constraints are in turn asked for the set of variable assignments that made the propagation occur, continuing the analysis backwards. The procedure is repeated until some termination condition is met, resulting in a set of variable assignments that implies the conflict. A clause prohibiting that particular assignment is added to the clause database. This *learnt* (conflict) clause will always be implied by the original problem constraints.

Learnt clauses serve two purposes: they drive the backtracking and they speed up future conflicts by "caching" the reason for the conflict. Each clause will prevent only a constant number of inferences, but as the recorded clauses start to build on each other and participate in the unit propagation, the accumulated effect of learning can be massive. However, as the set of learnt clauses increase, propagation is slowed down. Therefore, the number of learnt clauses is periodically reduced, keeping only the clauses that seem useful by some heuristic.

Search. The search procedure of a conflict-driven SAT-solver is somewhat implicit. Although a recursive definition of the procedure might be more elegant, it is typically described (and implemented) iteratively. The procedure will start by selecting an unassigned variable x (called the *decision variable*) and assume a value for it, say TRUE. The consequences of $x =$ TRUE will then be propagated, possibly resulting in more variable assignments. All variables assigned as a consequence of x is said to be from the same *decision level*, counting from 1 for the first assumption made and so forth. Assignments made before the first assumption (decision level 0) are called *top-level*.

All assignments will be stored on a stack in the order they were made; from now on referred to as the *trail*. The trail is divided into decision levels and is used to undo information during backtracking. The decision phase will continue until either all variables have been assigned, in which case we have a model, or a conflict has occurred. On conflicts, the learning procedure will be invoked and a conflict clause produced. The trail will be used to undo decisions, one level at a time, until precisely one of the literals of the learnt clause becomes unbound (they are all FALSE at the point of conflict). By construction, the conflict clause cannot go directly from conflicting to a clause with two or more unbound literals. If the clause is unit for several decision levels, it is advantageous to chose the lowest level (referred to as *backjumping* or *non-chronological backtracking* [MS96]).

```
loop
    propagate()    – propagate unit clauses
    if not conflict then
        if all variables assigned then
            return SATISFIABLE
        else
            decide()    – pick a new variable and assign it
    else
        analyze()    – analyze conflict and add a conflict clause
        if top-level conflict found then
            return UNSATISFIABLE
        else
            backtrack()    – undo assignments until conflict clause is unit
```

Activity heuristics. One important technique introduced by CHAFF [MZ01] is a dynamic variable ordering based on activity (referred to as the VSIDS heuristic). The original heuristic imposes an order on *literals*, but borrowing from SATZOO, we make no distinction between p and $\bar{p}$ in MINISAT.

Each variable has an *activity* attached to it. Every time a variable occurs in a recorded conflict clause, its activity is increased. We refer to this as *bumping*. After the conflict, the activity of all the variables in the system are multiplied by a constant less than 1, thus *decaying* the activity of variables over time.

Activity is also used for clauses. When a learnt clause takes part in the conflict analysis, its activity is bumped. Inactive clauses are periodically removed.

Constraint removal. The constraint database is divided into two parts: the *problem constraints* and the *learnt clauses*. The set of learnt clauses is periodically reduced to increase the performance of propagation. This may result in a larger search space, as learnt clauses are used to crop future branches of the search tree. The balance between the two forces is delicate, and there are SAT-instances for which a big learnt clause set is advantageous, and others where a small set is better. MINISAT's default heuristic starts with a small set and gradually increases the size.

Problem constraints can also be removed if they are satisfied at the top-level. The API method *simplifyDB()* is responsible for this. The procedure is particularly important for incremental SAT-problems.

Top-level solver. The pseudo-code for the search procedure presented above suffices for a simple conflict-driven SAT-solver, but a solver *strategy* can improve the performance. A typical strategy applied by modern conflict-driven SAT-solvers is the use of *restarts* to prevent from getting stuck in a futile part of the search tree. In MINISAT we also vary the number of learnt clauses kept at a given time. Furthermore, the *solve()* method of the API supports incremental assumptions, not handled by the above pseudo-code.

4   Implementation

The following conventions are used in the code. Atomic types start with a lower-case letter and are passed by value. Composite types start with a capital letter and are passed by reference. Blocks are marked by indentation level. The bottom symbol $\perp$ always mean *undefined*; FALSE is used to denote the boolean false.

We will use, but not specify an implementation of, the following abstract data types: $Vec\langle T \rangle$ an extensible vector of type T; *lit* the type of literals containing a special literal $\perp_{lit}$; *lbool* for the lifted boolean domain containing elements TRUE$_\perp$, FALSE$_\perp$, and $\perp$; $Queue\langle T \rangle$ a queue of type T. We also use *var* as a type synonym for *int* (for implicit documentation) with the special constant $\perp_{var}$. The literal data type has an *index()* method which converts a literal to a "small" integer suitable for array indexing.

4.1   The Solver State

A number of things need to be stored in the solver state. *Figure 2* shows the complete set of member variables of the solver type of MINISAT. A number of trivial, one-line functions will be assumed to exist, such as *nVars()* for the number of variables etc. The interface of *VarOrder* is given in *Figure 1*, and is further explained in section 4.6. Note that the state does *not* contain a boolean "conflict" to remember if a top-level conflict has been reached. Instead we impose as an invariant that the solver must never be in a conflicting state.

4.2   Constraints

MINISAT can handle arbitrary constraints over boolean variables through the abstraction presented in *Figure 3*. Each constraint type needs to implement methods for constructing, removing, propagating and calculating reasons. In addition, methods for simplifying the constraint and updating the constraint on backtrack can be specified. The contracts of these methods are as follows:

Constructor. The constructor may only be called at the top-level. It must create and add the constraint to appropriate watcher lists after enqueuing any unit information derivable under the current top-level assignment. Should a conflict arise, this must be communicated to the caller.

Remove. The remove method supplants the destructor by receiving the solver state as a parameter. It should dispose the constraint and remove it from the watcher lists.

Propagate. The propagate method is called if the constraint is found in a watcher list during propagation of unit information *p*. The constraint is removed from the list and is required to insert itself into a new or the same watcher list. Any unit information derivable as a consequence of *p* should be enqueued. If successful, TRUE is returned; if a conflict is detected, FALSE is returned. The constraint may add itself to the undo list of *var(p)* if it needs to be updated when *p* becomes unbound.

Simplify. At the top-level, a constraint may be given the opportunity to simplify its representation (returns TRUE) or state that the constraint is satisfied under the current assignment (returns FALSE). A constraint must *not* be simplifiable to produce unit information or to be conflicting; in that case the propagation has not been correctly defined.

---

class **VarOrder** – *Public interface*
    V arOder (**Vec**⟨*lbool*⟩ ref_to_assigns, **Vec**⟨*double*⟩ ref_to_activity)

void *newVar*()	– *Called when a new variable is created.*
void *update*(**var** x)	– *Called when a variable has increased in activity.*
void *updateAll*()	– *Called when all variables have been assigned new activities.*
void *undo*(**var** x)	– *Called when variable is unbound (and may be selected again).*
var *select*()	– *Called to select a new, unassigned variable.*

Fig. 1. Assisting ADT for the dynamic variable ordering of the solver. The constructor takes references to the assignment vector and the activity vector of the solver. The method *select()* will return the unassigned variable with the highest activity.

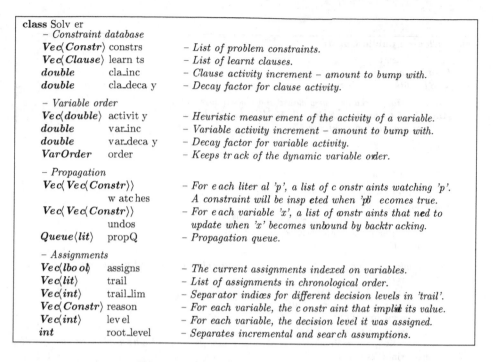

class Solv er
- *Constraint database*

Vec⟨*Constr*⟩	constrs	– *List of problem constraints.*
Vec⟨*Clause*⟩	learn ts	– *List of learnt clauses.*
double	cla_inc	– *Clause activity increment – amount to bump with.*
double	cla_deca y	– *Decay factor for clause activity.*

- *Variable order*

Vec⟨**double**⟩	activit y	– *Heuristic measur ement of the activity of a variable.*
double	var_inc	– *Variable activity increment – amount to bump with.*
double	var_deca y	– *Decay factor for variable activity.*
VarOrder	order	– *Keeps tr ack of the dynamic variable order.*

- *Propagation*

Vec⟨**Ve**c⟨*Constr*⟩⟩	w atc hes	– *For e ach liter al 'p', a list of c onstr aints watching 'p'. A constraint will be insp eted when 'p' b ecomes true.*
Vec⟨**Ve**c⟨*Constr*⟩⟩	undos	– *For e ach variable 'x', a list of constr aints that n ed to update when 'x' becomes unbound by backtr acking.*
Queue⟨*lit*⟩	propQ	– *Propagation queue.*

- *Assignments*

Vec⟨*lbo ol*⟩	assigns	– *The current assignments indexed on variables.*
Vec⟨*lit*⟩	trail	– *List of assignments in chronological order.*
Vec⟨*int*⟩	trail_lim	– *Separ ator indices for different decision levels in 'trail'.*
Vec⟨*Constr*⟩	reason	– *For each variable, the c onstr aint that implid its value.*
Vec⟨*int*⟩	lev el	– *For each variable, the decision level it was assigned.*
int	root_level	– *Separates incremental and sear ch assumptions.*

Fig. 2. Internal state of the solver.

class *Constr*

virtual *void* remove	(*Solver* S)	– *must be defined*
virtual *b ool* propagate	(*Solver* S, *lit* p)	– *must be defined*
virtual *b ool* simplify	(*Solver* S)	– *defaults to return false*
virtual *void* undo	(*Solver* S, *lit* p)	– *defaults to do nothing*
virtual *void* calcReason	(*Solver* S, *lit* p, **Ve**c⟨*lit*⟩ out_reason)	– *must be defined*

Fig. 3. Abstract base class for constraints.

Undo. During backtracking, this method is called if the constraint added itself to the undo list of *var*(p) in *propagate()*. The current variable assignments are guaranteed to be identical to that of the moment before *propagate()* was called.

Calculate Reason. This method is given a literal p and an empty vector. The constraint is the *reason* for p being true, that is, during propagation, the current constraint enqueued p. The received vector is extended to include a set of assignments (represented as literals) implying p. The current variable assignments are guaranteed to be identical to that of the moment before the constraint propagated p. The literal p is also allowed to be the special constant $\perp_{lit}$ in which case the reason for the clause being *conflicting* should be returned through the vector.

The code for the *Clause* constraint is presented in *Figure 4*. It is also used for learnt clauses, which are unique in that they can be added to the clause database

```
class Clause : public Constr
    bool      learn t
    float     activit y
    Vec⟨lit⟩ lits

    – Constructor – creates a new clause and adds it to watcher lists:
    static bool Clause_new(Solver S, Vec⟨lit⟩ ps, bool learn t, Clause out_clause)
        "Implementation in Figure 5"

    – Learnt clauses only:
    bool locked(Solver S)
        return S.reason[var(lits[0])] == this

    – Constraint interface:
    void remove(Solver S)
        remov eElem(this, S.watches[index(¬lits[0])])
        remov eElem(this, S.watches[index(¬lits[1])])
        delete this

    bool simplify(Solver S)                    – only called at top-level with empty prop. queue
        int j = 0
        for (int i = 0; i < lits.size(); i++)
            if (S.value(lits[i]) == TRUE⊥)
                return TRUE
            else if (S.value(lits[i]) == ⊥)
                lits[j++] = lits[i]            – false liter als ar e not copied (only occur for i ≥ 2)
        lits.shrink(lits.size() − j)
        return F ALSE

    bool propagate(Solver S, lit p)
        – Make sure the false literal is lits[1]:
        if (lits[0] == ¬p)
            lits[0] = lits[1], lits[1] = ¬p

        – If 0th watch is true, then clause is already satisfied.
        if (S.value(lits[0]) == TRUE⊥)
            S.watc hes[index(p)].push(this)    – re-insert clause into watcher list
            return TRUE

        – Look for a new literal to watch:
        for (int i = 2; i < size(); i++)
            if (S.value(lits[i]) != FALSE⊥)
                lits[1] = lits[i], lits[i] = ¬p
                S.watches[index(¬lits[1])].push(this)   – insert clause into watcher list
                return TRUE

        – Clause is unit under assignment:
        S.watches[index(p)].push(this)
        return S.enqueue(lits[0], this)        – enqueue for prop agation

    void calcReason(Solver S, lit p, vec⟨lit⟩ out_reason)
        – invariant: (p == ⊥) or (p == lits[0])
        for (int i = ((p == ⊥) ? 0 : 1); i < size(); i++)
            out_reason.push(¬lits[i])          – invariant: S.value(lits[i]) == F ALSE⊥
        if (learnt) S.claBumpActivity(this)
```

Fig. 4. Implementation of the *Clause* constraint.

```
bool Clause_new(Solver S, Vec⟨lit⟩ ps, bool learnt, Clause out_clause)
    out_clause = NULL

    – Normalize clause:
    if (!learnt)
        if ("any literal in ps is true")      return TRUE
        if ("both p and ¬p occurs in ps")  return TRUE
        "remove all false literals from ps"
        "remove all duplicates from ps"

    if (ps.size() == 0)
        return FALSE
    else if (ps.size() == 1)
        return S.enqueue(ps[0])                          – unit facts are enqueued
    else
        – Allocate clause:
        Clause c = new Clause
        ps.moveTo(c.lits)
        c.learn t  = learnt
        c.activity = 0                                   – only relevant for learnt clauses

        if (learn t)
            – Pick a second literal to watch:
            "Let max_i be the index of the literal with highest decision level"
            c.lits[1] = ps[max_i], c.lits[max_i] = ps[1]

            – Bumping:
            S.claBumpActivity(c)              – newly learnt clauses should be considered active
            for (int i = 0; i < ps.size(); i++)
                S.varBumpActivity(ps[i])          – variables in conflict clauses are bumped

        – Add clause to watcher lists:
        S.watches[index(¬c.lits[0])].push(c)
        S.watches[index(¬c.lits[1])].push(c)
        out_clause = c

    return TRUE
```

Fig. 5. Constructor function for clauses. Returns FALSE if top-level conflict is detected. 'out_clause' may be set to NULL if the new clause is already satisfied under the current top-level assignment. **Post-condition:** 'ps' is cleared. For learnt clauses, all literals will be false except 'lits[0]' (by design of *analyze()*). For the propagation to work, the second watch must be put on the literal which will first be unbound by backtracking.

while the solver is not at top-level. This makes the constructor code a bit more complicated than it would be for a normal constraint.

Implementing the *addClause()* method of the solver API is just a matter of calling *Clause_new()* and pushing the new constraint on the "constrs" vector, storing the list of problem constraints. The *newVar()* method of the API has a trivial implementation.

```
Constr Solver.propagate()
    while (propQ.size() > 0)
        lit p = propQ.dequeue()
        - 'p' is the enqueued fact to propagate
        Vec⟨Constr⟩ tmp
        watches[index(p)].moveTo(tmp)
        - 'tmp' contains the watcher list for 'p'

        for (int i = 0; i < tmp.size(); i++)
            if (!tmp[i].propagate(this, p))
                - Constraint is conflicting;
                - copy remaining watches to
                - 'watches[p]' and return
                - constraint:
                int j = i+1
                for (; j < tmp.size(); j++)
                    watches[index(p)].push(tmp[j])
                return tmp[i]
    return NULL
```

```
bool Solver.enqueue(lit p, Constr from)
    if (value(p) != ⊥)
        if (value(p) == FALSE)
            - Conflicting enqueued assignment:
            propQ.clear()
            return FALSE
        else
            - Existing consistent assignment;
            -   don't enqueue:
            return TRUE
    else
        - New fact, store it:
        assigns [var(p)] = lbool(!sign(p))
        level  [var(p)] = decisionLevel()
        reason [var(p)] = from
        trail.push(p)
        propQ.insert(p)
        return TRUE
```

Fig. 6. *propagate():* Propagates all enqueued facts. If a conflict arises, the *conflicting* clause is returned, otherwise NULL. *enqueue():* Puts a new fact on the propagation queue, and immediately updates the variable's value in the assignment vector. If a conflict arises, FALSE is returned and the propagation queue is cleared. 'from' contains a reference to the constraint from which 'p' was propagated (defaults to NULL if omitted).

4.3   Propagation

Given the mechanism for adding constraints, we now move on to describe the propagation of unit information on these constraints.

The propagation routine keeps a set of literals (unit information) that is to be propagated. We call this the *propagation queue*. When a literal is inserted into the queue, the corresponding variable is immediately assigned. For each literal in the queue, the watcher list of that literal determines the constraints that may be affected by the assignment. Through the interface described in the previous section, each constraint is asked by a call to its *propagate()* method if more unit information can be inferred, which will then be enqueued. The process continues until either the queue is empty or a conflict is found.

An implementation of this procedure is displayed in *Figure 6*. It starts by dequeuing a literal and clearing the watcher list for that literal by moving it to "tmp". The propagate method is then called for each constraint of "tmp". This will re-insert watches into new lists. Should a conflict be detected during the traversal, the remaining watches will be copied back to the original watcher list.

The method for enqueuing unit information is relatively straightforward. Note that the same fact can be enqueued several times, as it may be propagated from different constraints, but it will only be put on the queue once.

4.4   Learning

We describe the basic conflict-analysis algorithm by an example. Assume the database contains the clause $\{x, y, z\}$ which just became unsatisfied during propagation. This is our conflict. We call $\overline{x} \wedge \overline{y} \wedge \overline{z}$ the reason set of the conflict. Now x is false because $\overline{x}$ was propagated from some constraint. We ask that constraint to give us the reason for propagating $\overline{x}$ (the *calcReason()* method). It will respond with another conjunction of literals, say $u \wedge v$. These were the variable assignment that implied $\overline{x}$. The constraint may in fact have been the clause $\{\overline{u}, \overline{v}, \overline{x}\}$. From this little analysis we know that $u \wedge v \wedge \overline{y} \wedge \overline{z}$ must also lead to a conflict. We may prohibit this conflict by adding the clause $\{\overline{u}, \overline{v}, y, z\}$ to the clause database. This would be an example of a *learnt* conflict clause.

In the example, we picked only one literal and analyzed it one step. The process of expanding literals with their reason sets can be continued, in the extreme case until all the literals of the conflict set are decision variables. Different learning schemes based on this process have been proposed. Experimentally the *First UIP* heuristic has been shown effective [ZM01]. We will not give the definition here, but just state the algorithm: In a breadth-first manner, continue to expand literals of the current decision level, until there is just one left.

In the code for *analyze()*, displayed in *Figure 7*, we make use of the fact that a breadth-first traversal can be achieved by inspecting the trail backwards. Particularly, the variables of the reason set of p is always before p in the trail. In the algorithm we initialize p to $\perp_{lit}$, which makes *calcReason()* return the reason for the conflict. Besides returning a conflict clause, *analyze()* sets the backtracking level, which is the lowest decision level the conflict clause is unit.

4.5   Search

The search method in *Figure 8* works basically as the pseudo-code presented in section 3 but with the following additions:

Restarts. The first argument of the search method is "nof_conflicts". The search for a model or a contradiction will only be conducted for this many conflicts. If failing to solve the SAT-problem within the bound, all assumptions will be canceled and $\perp$ returned. The surrounding solver strategy will then restart the search.

Reduce. The second argument, "nof_learnts", sets an upper limit on the number of learnt clauses that are kept. Once this number is reached, *reduceDB()* is called. Clauses that are currently the reason for a variable assignment are said to be *locked* and cannot be removed by *reduceDB()*. For this reason, the limit is extended by the number of assigned variables, which approximates the number of locked clauses.

Parameters. The third argument groups some tuning constants. In the current version of MiniSat, it only contains the decay factors for variables and clauses.

Root-level. To support incremental SAT, the concept of a *root-level* is introduced. The root-level acts a bit as a new top-level. Above the root-level are the incremental assumptions passed to *solve()* (if any). The search procedure is not allowed to backtrack above the root-level, as this would change the incremental assumptions. If we reach a conflict at root-level, the search will return FALSE.

```
void Solver.analyze(Constr confl, Vec⟨lit⟩ out_learn t,Int out_btlev el)

    Vec⟨bool⟩ seen(nVars(), FALSE)
    int        counter = 0
    lit        p       = ⊥_lit
    Vec⟨lit⟩   p_reason

    out_learnt.push()                                   – leave room for the asserting literal
    out_btlev el = 0
    do
        p_reason.clear()
        confl.calcReason(this, p, p_reason)             – invariant here: confl != NULL

        – TRACE REASON FOR P:
        for (int j = 0; j < p_reason.size(); j++)
            lit q = p_reason[j]
            if (!seen[var(q)])
                seen[var(q)] = TRUE
                if (level[var(q)] == decisionLevel())
                    counter++
                else if (level[var(q)] > 0)             – exclude variables from decision level 0
                    out_learn t push(¬q)
                    out_btlev el = max(out_btlev el, level[var(q)])

        – SELECT NEXT LITERAL TO LOOK AT:
        do
            p    = trail.last()
            confl = reason[var(p)]            ┌──────────────────────────────────────────
            undoOne()                         │ void Solver.record( Vec⟨lit⟩ clause)
        while (!seen[var(p)])                 │
        counter−−                             │   Clause c
    while (counter > 0)                       │   Clause_new(this, clause, TRUE, c)
    out_learn t[0] = ¬p                       │   enqueue(clause[0], c) – cannot fail
                                              │   if (c != NULL) learn ts push(c)
```

Fig. 7. Analyze a conflict and produce a reason clause. **Pre-conditions:** (1) 'out_learnt' is assumed to be cleared. (2) Current decision level must be greater than root level. **Post-conditions:** (1) 'out_learnt[0]' is the asserting literal at level 'out_btlevel'. **Effect:** Will undo part of the trail, but not beyond last decision level. *record():* records a clause and drives backtracking; 'clause[0]' must contain the asserting literal.

A problem with the approach presented here is conflict clauses that are unit. For these, *analyze()* will always return a backtrack level of 0 (top-level). As unit clauses are treated specially, they are never added to the clause database. Instead they are enqueued as facts to be propagated (see the code of *Clause_new()*). There would be no problem if this was done at top-level. However, the search procedure will only undo until root-level, which means that the unit fact will be enqueued there. Once *search()* has solved the current SAT-problem, the surrounding solver strategy will undo any incremental assumption and put the solver back at the top-level. By this the unit clause will be forgotten, and the next incremental SAT problem will have to infer it again. There are many possible solutions to this problem. In practice we have not seen any performance difference in our applications [ES03, CS03].

Simplify. Provided the root-level is 0 (no assumptions were passed to *solve()*) the search will return to the top-level every time a unit clause is learnt. At that point it is legal to call *simplifyDB()* to simplify the problem constraints according to the top-level assignment. If a stronger simplifier than presented here is implemented, a contradiction may be found, in which case the search should be aborted. As our simplifier is not stronger than normal propagation, it can never reach a contradiction, so we ignore the return value of *simplify()*.

4.6   Activity Heuristics and Constraint Removal

In the *VarOrder* data type of MiniSat, the list of variables is kept sorted on activity at all time. The search will always accurately choose the most active variable. The original suggestion for the VSIDS dynamic variable ordering was to sort periodically. MiniSat implements variable decay by bumping with larger and larger numbers. Only when the limit of what is representable by a floating point number is reached need activities be scaled down.

Activity for conflict clauses are also maintained. The method for reducing the set of learnt clauses based on this activity, as well as the top-level simplification procedure can be found in *Figure 9*.

4.7   Top-Level Solver

The method implementing MiniSat's top-level strategy can be found in *Figure 8*. It is responsible for making the incremental assumptions and setting the root level. Furthermore, it completes the simple backtracking search with restarts, which are performed less and less frequently. After each restart, the number of allowed learnt clauses is increased.

5   Conclusions and Related Work

By this paper, we have provided a minimal reference implementation of a modern conflict-driven SAT-solver. We have tested MiniSat against zChaff and Berkmin 5.61 on 177 SAT-instances. These instances were used to tune Satzoo for the *SAT 2003 Competition*. As Satzoo solved more instances and series of problems, ranging over all three categories (*industrial, handmade,* and *random*), than any other solver in the competition, we feel that this is a representative test-set. No extra tuning was done in MiniSat; it was just run once with the constants presented in the code. At a time-out of 10 minutes, MiniSat solved 158 instances, while zChaff solved 147 instances and Berkmin 157 instances.

Another approach to incremental SAT and non-clausal constraints was presented by Aloul, Ramani, Markov, and Sakallah in their work on Satire and PBS [WKS01, AR+02]. Our implementation differs in that it has a simpler notion of incrementality, and that it contains an interface for non-clausal constraints.

Finally, a set of reference implementations of modern SAT-techniques is present in the OpenSAT project. However, the project aim for completeness rather than minimal exposition, as we have chosen in this paper.

```
lbool Solver.search(int nof_conflicts,
          int nof_learn ts, Se ar chParms params)
    int conflictC = 0
    var_deca y = 1 / params.var_deca y
    cla_decay = 1 / params.cla_decay
    model.clear()

    loop
        Constr confl = propagate()
        if (confl != NULL)
            - CONFLICT

            conflictC++
            if (decisionLevel() == root_level)
                return F ALSE
            Vec(lit) learn tclause
            int       bac ktrlevel
            analyze(confl, learnt_clause, backtr_level)
            cancelUntil(max(bac ktrlevel,root_level))
            record(learnt_clause)
            de cayActivities()
        else
            - NO CONFLICT

            if (decisionLevel() == 0)
                - Simplify the set of problem clauses:
                simplifyDB()  - our simplifier cannot
                                      return false here
            if (learn tssize()−nAssigns()≥nof_learnts)
                - Reduc e the set of learnt clauses:
                reduc eDB()
            if (nAssigns() == nVars())
                - Model found:
                model.growTo(nVars())
                for (int i = 0; i < nVars(); i++)
                    model[i] = (value(i) == TRUE)
                canc elUntil(root_level)
                return TRUE
            else if (conflictC ≥ nof_conflicts)
                - Reached bound on num. of conflicts:
                canc elUntil(root_level) - force restart
                return ⊥
            else
                - New variable decision:
                lit p = lit(order.select())
                - may have heuristic for polarity here
                assume(p)      - cannot return false
```

```
bool Solver.solve(Vec(lit) assumps)
    Se ar chParms params(0.95, 0.999)
    double nof_conflicts = 100
    double nof_learnts  = nConstraints()/3
    lbool   status      = ⊥

    - PUSH INCREMENTAL ASSUMPTIONS:
    for (int i = 0; i < assumps.size(); i++)
        if (!assume(assumps[i])
            || prop agate() != NULL)
            canc elUntil(0)
            return F ALSE
    root_level = de cisionLevel()

    - SOLVE:
    while (status == ⊥)
        status = search((int)nof_conflicts,
            (int)nof_learn ts, params)
        nof_conflicts *= 1.5
        nof_learnts  *= 1.1

    cancelUntil(0)
    return status == TRUE
```

```
void Solver.undoOne()
    lit     p = trail.last()
    var     x = var(p)
    assigns [x] = ⊥
    reason  [x] = NULL
    level   [x] = -1
    order.undo(x)
    trail.pop()

    while (undos[x].size() > 0)
        undos[x].last().undo(this, p)
        undos[x].pop()
```

```
bool Solver.assume(lit p)
    trail_lim.push(trail.size())
    return enqueue(p)
```

```
void Solver.cancel()
    int c = trail.size() − trail_lim.last()
    for (; c != 0; c−−)
        undoOne()
    trail_lim.pop()
```

```
void Solver.cancelUntil(int level)
    while (decisionLevel() > level) canc el()
```

Fig. 8. *search():* assumes and propagates until a conflict is found, from which a conflict clause is learnt and backtracking performed until search can continue. **Pre-condition:** root_level == *decisionLevel()*. *solve():* **Pre-condition:** If assumptions are used, *simplifyDB()* must be called right before using this method. If not, a top-level conflict (resulting in a non-usable internal state) cannot be distinguished from a conflict under assumptions. *assume():* returns FALSE if immediate conflict. **Pre-condition:** propagation queue is empty. *undoOne():* unbinds the last variable on the trail. *cancel():* reverts to the state before last *push()*. **Pre-condition:** propagation queue is empty. *cancelUntil():* cancels several levels of assumptions.

```
void Solver.reduceDB()
    int    i, j
    double lim = cla_inc / learnts.size()

    sortOnActivity(learn ts)
    for (i=j=0; i < learnts.size()/2; i++)
        if (!learn ts[i].locked(this))
            learn ts[i].remove(this)
        else
            learn ts[j++] = learnts[i]
    for (; i < learn ts.size(); i++)
        if (!learn ts[i].locked(this)
            && learnts[i].activity() < lim)
            learn ts[i].remove(this)
        else
            learn ts[j++] = learnts[i]
    learnts.shrink(i − j)
```

```
bool Solver.simplifyDB()
    if (propagate() != NULL)
        return FALSE

    for (int type = 0; type < 2; type++)
        Vec⟨Constr⟩ cs = type ?
            (Vec⟨Constr⟩)learnts : constrs
        int j = 0
        for (int i = 0; i < cs.size(); i++)
            if (cs[i].simplify(this))
                cs[i].remove(this)
            else
                cs[j++] = cs[i]
        cs.shrink(cs.size()−j)
    return TRUE
```

Fig. 9. *reduceDB():* Remove half of the learnt clauses minus some locked clauses. A locked clause is a clauses that is reason to a current assignment. Clauses below a threshold activity are also be removed. *simplifyDB():* Top-level simplify of constraint database. Will remove any satisfied constraint and simplify remaining constraints under current (partial) assignment. If a top-level conflict is found, FALSE is returned. **Pre-condition:** Decision level must be zero. **Post-condition:** Propagation queue is empty.

References

[AR* 02]   F. Aloul, A. Ramani, I. Markov, K. Sakallah. **"Generic ILP vs. Specialized 0-1 ILP: an Update"** in *International Conference on Computer Aided Design (ICCAD)*, 2002.

[BC* 99]   A. Biere, A. Cimatti, E.M. Clarke, M. Fujita, Y. Zhu. **"Symbolic Model Checking using SAT procedures instead of BDDs"** in *Proceedings of Design Automation Conference (DAC'99)*, 1999.

[CS03]   K. Claessen, N. Sörensson. **"New Techniques that Improve MACE-style Finite Model Finding"** in *CADE-19, Workshop W4. Model Computation – Principles, Algorithms, Applications*, 2003.

[DLL62]   M. Davis, M. Logman, D. Loveland. **"A machine program for theorem proving"** in *Communications of the ACM*, vol 5, 1962.

[ES03]   N. Eén, N. Sörensson. **"Temporal Induction by Incremental SAT Solving"** in *Proc. of First International Workshop on Bounded Model Checking*, 2003.

[Lar92]   T. Larrabee. **"Test Pattern Generation Using Boolean Satisfiability"** in *IEEE Transactions on Computer-Aided Design*, vol. 11-1, 1992.

[MS96]   J.P. Marques-Silva, K.A. Sakallah. **"GRASP – A New Search Algorithm for Satisfiability"** in *ICCAD. IEEE Computer Society Press*, 1996

[MZ01]   M.W. Moskewicz, C.F. Madigan, Y. Zhao, L. Zhang, S. Malik. **"Chaff: Engineering an Efficient SAT Solver"** in *Proc. of the 38th Design Automation Conference*, 2001.

[ZM01]   L. Zhang, C.F. Madigan, M.W. Moskewicz, S. Malik. **"Efficient Conflict Driven Learning in Boolean Satisfiability Solver"** in *Proc. of the International Conference on Computer Aided Design (ICCAD)*, 2001.
[WKS01]   J. Whittemore, J. Kim, K. Sakallah. **"SATIRE: A New Incremental Satisfiability Engine"** in *Proc. 38th Conf. on Design Automation*, ACM Press 2001.

Survey and Belief Propagation on Random K-SAT

Alfredo Braunstein[1,2] and Riccardo Zecchina[2]

$\cdot$ SISSA, Via Beirut 9, 34100 Trieste, Italy,
$\cdot$ ICTP, Str. Costiera 11, 34100 Trieste, Italy

Abstract. Survey Propagation (SP) is a message passing algorithm that can be viewed as a generalization of the so called Belief Propagation (BP) algorithm used in statistical inference and error correcting codes. In this work we discuss the connections between BP and SP.

1 Introduction

The Survey Propagation (SP) algorithm is an iterative procedure which combines the evaluation of the probability distribution function (pdf) over the probability space of SAT assignments of n variables appearing in a given Boolean formula with a reduction process in which the most biased variables are fixed. The SP algorithm was proposed in [1] and further developed in [2] and has allowed to find satisfying assignments for n sufficiently large [3] of hard random K-SAT instances with a computational cost roughly scaling as $n \log n$ [2]. The general version of SP given in [1] can also be used to find assignments which minimize the number of violated clauses in the UNSAT regime.

The iterative equations providing the pdfs can be viewed as the algorithmic version of the statistical physics analysis (the so called cavity method [8,10]) which has been used in refs. [1,4,5] to compute the SAT/UNSAT thresholds for random K-SAT. Recent results [11,12] show that such statistical physics methods provide thresholds that are rigorous upper bounds. In what follows, we shall review the formalism underlying the SP equations and make a comparison with a similar and simpler algorithm, the Belief Propagation (BP) algorithm [13,14], used in statistical inference problems.

2 The SAT Problem and Its Factor Graph Representation

Given a vector of $\{0, 1\}$ Boolean variables $\mathbf{x} = \{x_i\}_{i \in I}$ where $I = \{1, \ldots, n\}$, consider a SAT formula defined by

$$\mathcal{F}(\mathbf{x}) = \bigwedge_{a \in A} C_a(\mathbf{x})$$

where A is an arbitrary finite set (disjoint with I); $C_a(\mathbf{x}) = \bigvee_{i \in I(a)} J_{a,i}(x_i)$; any *literal* $J_{a,i}(x_i)$ is either x_i or $\neq x_i$; and finally, $I(a) \subset I$ for every $a \in A$. Similarly to $I(a)$ we can define the set $A(i) \subset A$ as $A(i) = \{a : i \in I(a)\}$.

E. Giunchiglia and A. Tacchella (Eds.): SAT 2003, LNCS 2919, pp. 519–528, 2004.
© Springer-Verlag Berlin Heidelberg 2004

Given a formula $\mathcal{F}$, the problem of finding a variable assignment $\mathbf{s}$ such that $\mathcal{F}(\mathbf{s}) = 1$ if it exists, or the problem of deciding if one such $\mathbf{s}$ exists at all is known to be NP-complete. We will define $S_{\mathcal{F}} = \mathcal{F}^{-1}(\{1\}) \subset \{0,1\}^n$ as the set of truth assignments of $\mathcal{F}$, and the goal of this work is to describe an algorithm that is has experimentally shown to be able to find an $s \in S_{\mathcal{F}}$ for typical random K-SAT formulas $\mathcal{F}$ on the so-called *hard* region of the parameters.

We will use the "factor graph" representation: given a formula $\mathcal{F}$, we will define its associated *factor graph* as a bipartite undirected graph $G = (V; E)$, having two types of nodes, and edges only between different types of nodes:

- Variable nodes, each one labeled by a variable index in $I = \{1, \ldots, n\}$ and
- Function or factor nodes, each one labeled by a clause index $a \in A$.
- An edge (a, i) will belong to the graph if and only if $a \in A(i)$ or equivalently $i \in I(a)$.

In other words, $V = A \cup I$ and $E = \{(i, a) : i \in I, a \in A(i)\}$ or equivalently $E = \{(i, a) : a \in A, i \in I(a)\}$.

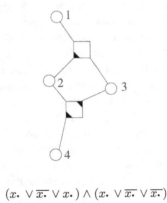

$$(x_\bullet \vee \overline{x_\bullet} \vee x_\bullet) \wedge (x_\bullet \vee \overline{x_\bullet} \vee \overline{x_\bullet})$$

Fig. 1. A simple formula with its corresponding factor graph. A triangle-shaped decoration indicates here that the corresponding literal appears negated in the clause. The factor graph of a formula is also known as the incidence graph

2.1   Random K-SAT

A K-SAT formula is a SAT formula having $|I(a)| = k$ for each $a \in A$. The *random* $K, m-$SAT formula distribution is obtained by picking a formula with m clauses, each one chosen independently and uniformly between all K-tuples of variables (no variable repetition), in which negations are distributed afterward randomly with a fair coin.

We will be interested in solving formulas from this distribution for large m, n but $m = n\alpha$ for some constant α. Asymptotic analysis [1,4] and numerical experiments indicate that *hard* sat formulas exist in a region $[\alpha_d, \alpha_c]$ below the SAT/UNSAT threshold α_c [6,7] (e.g., for random 3-SAT $\alpha_d \in [3.92, \alpha_c]$ and $\alpha_c = 4.2667$ [1,4]). In such *hard-sat* region the geometrical structure of solutions changes dramatically: while at low α solutions belong to a single giant cluster, in the hard-sat region solutions become divided into an exponential number of smaller clusters [9] each one having a finite fraction of backbone (also referred to as constrained or frozen) variables [15].

3   An Enumerative Graph-Growing Algorithm

Define $I_k = \{1, \ldots, k\}$, $A_k = \{a \in A : I\{a\} \in I_k\}$ and $\mathcal{F}_k, G_k$ its associated formula and factor graph. Consider the following inductive procedure to compute $S_k = S_{\mathcal{F}_k}$. Clearly $S_k \subset S_{k-1} \times \{0, 1\}$, and

$$S_0 = \emptyset$$
$$S_k = \{(\mathbf{s}, s_k) : \mathbf{s} \in S_{k-1}, s_k \in \{0, 1\} / C_a(\mathbf{s}, s_k) = 1 \text{ for } a \in A_k \setminus A_{k-1}\} \quad (1)$$

Observe that $A_k \setminus A_{k-1} \subset I(k)$, so typically only a small number of checks have to be done for every s to take it from S_{k-1} to S_k. This procedure can help us compute $S_n = S$ iteratively: at each step we add a variable to the graph, and then "filter out" contradictory configurations (i.e. configurations which are UNSAT once the last variable was added).

3.1   Using Probabilities

Of course that this procedure is typically exponential (in time and space) in n, partially because the set S can be (and typically is) exponentially large.

A more efficient idea would be to carry on less information instead. Consider

$$P_S(s_i = v)$$

where P_S is the uniform probability measure over the set S and $v = 0, 1$. If we can compute these quantities, then we can certainly use this information to get a single $s \in S$ (we will see exactly how in Section 6).

Trying to mimic Eq. 1, we see that if we get to know the joint probability distribution of variables $\{s_j\}_{j \in I(A_i(i)) \setminus \{i\}}$ (that is, the neighboring variables variables of i) on the space S_{i-1} then clearly we can trivially rebuild the state of these variables in any S_{i-1} configuration, and then compute the statistics for s_i on the extended space including this variable, i.e. $P_{S_i}(s_i)$.

The basic assumption behind the algorithm we are describing is that these variables $\{s_j\}_{j \in I(A_i(i)) \setminus \{i\}}$ are weakly correlated, that is that their joint probability distribution approximately factorizes. This factorization hypothesis is believed to be true in the limit $n \to \infty$ with high probability thanks to the so called tree-like structure of the typical underlying factor graph [16].

In this case, we only need to know $P_{S_{i-1}}(s_j = 0, 1)$ for $j \in I(A(i)) \setminus \{i\}$ to compute $P_{S_i}(s_i = 0, 1)$ and so we can write an explicit equation relating these quantities.

Let's assume that we have computed all $P_{S_{i-1}}$ (we will drop the S_{i-1} index for notational simplicity), and we want to compute the statistics for s_i in S_i.

3.2   u Messages

In order to do so explicitly, we will define the following quantities: given a configuration for $\{s_j\}_{j \in I(a) \setminus \{i\}}$ on S_{i-1} we will denote the set-valued variable $u_{a \to i} \subset \{0, 1\}$ as the subset of available (satisfying) configurations for the last variable s_i, i.e.:

$$u_{a \to i} = \left\{ s_i \in \{0, 1\} : C_a \left(\{s_j\}_{j \in I(a)} \right) = 1 \right\} \tag{2}$$

Note that given the particular structure of the SAT problem, any clause can be satisfied by a given participating variable (if chosen to an appropriate value), disregarding all other ones in that clause, i.e. $J_{a,i}(1) \in u_{a \to i}$ always, and that eliminates $\emptyset$ and $\{\neg J_{a,i}(1)\}$ from possible u outputs (meaning that effectively each u is a 2-valued variable).

The available states for variable s_i will be given by the set

$$h_i = \bigcap_{a \in A_i(i)} u_{a \to i} \tag{3}$$

3.3   Statistics for $u_{a \to i}$ and h_i

We will have that $u_{a \to i} = \{J_{a,i}(1)\}$ when the clause is not already satisfied by the remaining participating variables, and $u_{a \to i} = \{0, 1\}$ otherwise. Then

$$P(u_{a \to i} = \{J_{a,i}(1)\}) = \prod_{j \in I(a) \setminus \{i\}} P(s_j = \neg J_{a,j}(1))$$

$$P(u_{a \to i} = \{\neg J_{a,i}(1)\}) = 0$$

$$P(u_{a \to i} = \emptyset) = 0$$

$$P(u_{a \to i} = \{0, 1\}) = 1 - \prod_{j \in I(a) \setminus \{i\}} P(s_j = \neg J_{a,j}(1)) \tag{4}$$

Now we have to compute the statistics for h_i. Following Eq. 3 we have that

$$P(h_i = \{0, 1\}) = \prod_{a \in A_i(i)} P(u_{a \to i} = \{0, 1\})$$

$$P(h_i = \{v\}) = \prod_{a \in A(i)} \left(P(u_{a \to i} = \{0, 1\}) + P(u_{a \to i} = \{v\}) \right)$$

$$- \prod_{a \in A(i)} P\left(u_{a \to i} = \{0, 1\}\right)$$

$$P\left(h_i = \emptyset\right) = 1 - P\left(h_i = \{0, 1\}\right) - P\left(h_i = \{0\}\right) - P\left(h_i = \{1\}\right) \quad (5)$$

4   Belief Propagation

Once we have computed the statistics for h_i on S_{i-1}, it is straightforward to compute the statistics for s_i on S_i:

- each S_{i-1} configuration leading to $h_i = \{0, 1\}$ will be split on two new S_i configurations with respectively $s_i = 0$ and 1;
- each S_{i-1} configuration leading to $h_i = \{v\}$ will correspond to a single S_i configuration with $s_i = v$, and finally
- each configuration leading to $h_i = \emptyset$ will be eliminated or ignored.

Explicitly:

$$P_i\left(s_i = v\right) = \frac{P\left(h_i = \{v\}\right) + P\left(h_i = \{0, 1\}\right)}{P\left(h_i = \{0\}\right) + P\left(h_i = \{1\}\right) + 2P\left(h_i = \{0, 1\}\right)} \quad (6)$$

Note that the normalization is needed here because each configuration of the input variables can produce from zero up to two output configurations in the extended space. [13,14].

4.1   Cavities

The problem to use Eq. 6 to build up a recursion is that we have successfully computed $P_{S_i}(s_i)$ but we need also all other $P_{S_i}(s_j)$ for $j < i$. This problem is due to the fact that the quantities we can put together in an equation, namely $P_{S_{i-1}}(s_j)$ and $P_{S_i}(s_i)$ are rather different in nature when we observe them on the same set S_i: while the former are quantities "ignoring" variable i, the latter is a "complete" quantity.

This problem can be simply avoided by switching everywhere to "incomplete" quantities. We can define the *cavity* objects $I^{(i)} = \{1, \ldots, n\} \setminus \{i\}$, $A^{(i)} = \{a \in A : I(a) \in I^{(i)}\}$, $I^{(i)}(a) = I(a) \cap I^{(i)}$ and $\mathcal{F}^{(i)}$, $S^{(i)}$, $i \in I$ the associated formula and solution space, and try to define equations relating statistics over $S^{(i)}$ (see Figure 2).

Now the "output" computed statistics for s_i would need to be also with respect to some $S^{(j)}$ for obtaining "closed" equations, and so we will have to single out another neighboring variable $j \in I(b)$ for $b \in A(i)$ (note that as all others $k \in I(b), k \neq i, j$ will become disconnected from i, they won't enter in the equations. See the variable labeled k in the figure). Then eqs. 4 become

$$P^{(i)}\left(u_{a \to i} = \{J_{a,i}(1)\}\right) = \prod_{h \in I(a) \setminus \{i\}} P^{(i)}\left(s_h = \neg J_{a,h}(1)\right)$$

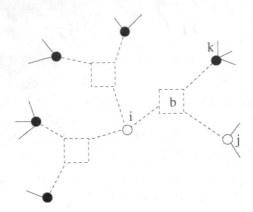

Fig. 2. We use as inputs $P^{(i)}(\bullet)$, corresponding to cavity probabilities ignoring var i . The computed probabilities $P^{(j)}(i)$ will be the one computed ignoring var j

$$P^{(i)}\left(u_{a\to i} = \{\neg J_{a,i}\,(1)\}\right) = 0$$
$$P^{(i)}\left(u_{a\to i} = \emptyset\right) = 0$$
$$P^{(i)}\left(u_{a\to i} = \{0,1\}\right) = 1 - \prod_{h\in I(a)\backslash\{i\}} P^{(i)}\left(s_h = \neg J_{a,h}(1)\right) \qquad (7)$$

and eqs. 5 become:

$$P^{(j)}\left(h_i = \{0,1\}\right) = \prod_{a\in A(i)\backslash\{b\}} P^{(i)}\left(u_{a\to i} = \{0,1\}\right)$$

$$P^{(j)}\left(h_i = \{v\}\right) = \prod_{a\in A(i)\backslash\{b\}} P^{(i)}\left(u_{a\to i} = \{0,1\}\right) + P^{(i)}\left(u_{a\to i} = \{v\}\right)$$

$$- \prod_{a\in A(i)\backslash\{b\}} P^{(i)}\left(u_{a\to i} = \{0,1\}\right) \qquad (8)$$

$P^{(j)}\left(h_i = \emptyset\right)$ can be computed by the normalization condition.

It is worth noticing that for n small enough the random K-SAT formulas will have a considerable relative number of pairs of clauses sharing more than one variable. (e.g. for 3-SAT one finds on average for large n, $9\alpha^2$ pairs of clauses sharing two literals). The presence of such pairs invalidates the cavity procedure which assumes independence of input pdfs. Better equations can be obtained by considering effective function nodes in which the dangerous pairs are substituted by single function nodes connected to four variable each. Extended numerical experiments for small n ($n = 100, 500$) at $\alpha = 4.2$ show that in this way performance can be improved by 30%. We expect similar modification of the SP equations based on other types of local structure to further improve the algorithm performance at small n.

In order to actually use the above BP equations to find the required probability distributions we will parametrize $\eta_{i,j} = P^{(j)}(s_i)$ and use Eq. 8 together with 6 to obtain an operator $\Lambda : \{\eta_{i,j}\}_{i\in I, j\in I(A(i))} \mapsto \{\eta'_{i,j}\}_{i\in I, j\in I(A(i))}$. We will look for a fixed point for this operator, that can be achieved by

$$\left\{\eta_{i,j}^{fix}\right\} = \lim_{t\to\infty} \overbrace{\Lambda \circ \cdots \circ \Lambda}^{t\,times} \left(\{\eta_{i,j}^0\}\right)$$

for some random initial $\{\eta^0\}$.

These equations can be easily implemented, and extensive experimentations and stability analysis show that they indeed converge in the low α region and can be used to efficiently find satisfying assignments by the decimation procedure described in the inset below. However, in the hard-sat region the use of BP equations is limited. In 3-SAT, they stop to converge already at $\alpha = 3.87$ while for $K > 3$ they always converge below $\alpha_c(K)$ and yet the pdfs one obtains seem to be less efficient for obtaining SAT assignments with respect to the pdfs provided by SP. The reason for such limitation seems to reside in the constraintness property of variables which is ignored by the BP procedure.

5   SP Equations

We refer to ref. [1] for the original statistical physics derivation of the SP equations, valid in $n \to \infty$ limit. The chief difference with respect to the BP formalism of the previous section consists in taking care of clusters of solutions: within one cluster a variable can be either "frozen" to some value (that is, the variable takes always the same value for all sat assignments within the cluster) or it may be "unfrozen" (that is it fluctuates from solution to solution within the cluster). Scope of the SP equations will be to properly describe the cluster to cluster fluctuations.

Going back to Eq. 6 on the previous section, we want to try to propagate also the "unfrozen" state of variables.

Then s_i will be a 3-valued variable: it can take values $0, 1$ or a new "joker" or "don't care" state $*$, corresponding $h_i = \{0, 1\}$. In fact s_i will become nothing but a renormalized h_i (renormalization still needed because we ignore configurations for which $h_i = \emptyset$, i.e. configurations which are incompatible with the addition of the new variable).

In order to take care of the new joker state, the eqs. 6 are modified to

$$P^{(j)}(s_i = v) = \frac{P^{(j)}(h_i = \{v\})}{P^{(j)}(h_i = \{0\}) + P^{(j)}(h_i = \{1\}) + P^{(j)}(h_i = \{0, 1\})}$$

$$P^{(j)}(s_i = *) = \frac{P^{(j)}(h_i = \{0, 1\})}{P^{(j)}(h_i = \{0\}) + P^{(j)}(h_i = \{1\}) + P^{(j)}(h_i = \{0, 1\})} \quad (9)$$

In these equations, every configuration in the restricted graph can be extended to: zero (canceled) or just one configuration. There is no longer any *splitting* into two configurations like it happens in BP and this is most likely the cause of the convergence improvement of SP with respect to BP.

6   Complete Probabilities and Decimation

Once we have found a fixed point for Eqs. 8 and 9, we need to get back "complete" probabilities in order to be able to solve the original sat problem. Of course, this is simply a matter of *not* ignoring the additional variable j in equation 8 so it becomes

$$P(h_i = \{0,1\}) = \prod_{a \in A(i)} P^{(i)}(u_{a \to i} = \{0,1\})$$

$$P(h_i = \{v\}) = \prod_{a \in A(i)} P^{(i)}(u_{a \to i} = \{0,1\}) + P^{(i)}(u_{a \to i} = \{v\}) \qquad (10)$$

$$- \prod_{a \in A(i)} P^{(i)}(u_{a \to i} = \{0,1\})$$

And $P(h_i = \emptyset)$ is obtained by normalization. Equations 9, 9 become

$$P(s_i = v) = \frac{P(h_i = \{v\})}{P(h_i = \{0\}) + P(h_i = \{1\}) + P(h_i = \{0,1\})}$$

$$P(s_i = *) = \frac{P(h_i = \{0,1\})}{P(h_i = \{0\}) + P(h_i = \{1\}) + P(h_i = \{0,1\})} \qquad (11)$$

With these quantities on hand, the following simple decimation procedure has been implemented:

1. $\{\eta\} \leftarrow$ random
2. SP

   (a) Compute $\{\eta'\}$ from $\{\eta\}$ following eqs. 7, 8 and 9, 9.
   (b) If $|\eta' - \eta| > \epsilon$ SET $\{\eta\} \leftarrow \{\eta'\}$ and GOTO 2a
3. Compute $P(s_i = 0)$, $P(s_i = *)$, $P(s_i = 1)$ following eqs. 10-11.
4. For $B_i = P(s_i = 1) - P(s_i = 0)$, Choose i such that $|B_i|$ is maximum.
5. IF $|B_i| < \epsilon$ STOP and output sub-formula.
6. FIX $s_i \leftarrow 1$ if $B_i > 0$, $s_i \leftarrow 0$ otherwise.
7. GOTO 2

The behavior of the algorithm on sufficiently large ($n > 10^3$) random 3-SAT instances is the following:

 - for low α ($\alpha < \alpha_d$), the variables turn out to be all unfrozen
 - in the hard-sat region the output probabilities are non-trivial and the decimation procedure leads to sub-formulas which are under-constrained and easily solved by standard algorithms. Very close to α_c the decimation procedure may fail in finding solutions in the first run.

For generic K-SAT ($K > 3$) the behavior is similar and the performace should be compared with other efficient incomplete solvers such as Walksat [18] or Record-to-Record Travel local search [19]. Some further results on the behaviour of SP on random 3-SAT can be found in [20]

For small n the structural "rare events" of the random SAT formulas, like clauses sharing more than one variable or other types of short loops, require an appropriate (in principle simple) modification of the SP iterations[21]. More in general, the presence of loops of different length scales may introduce correlations which may require further non-trivial generalization of the whole SP procedure in the framework of the so called Cluster Variation Method [22]. Work is in progress along this line.

References

1. M. Mézard and R. Zecchina, *Random K-satisfiability: from an analytic solution to a new efficient algorithm*, Phys.Rev. E **66** 056126 (2002)
2. A. Braunstein, M. Mezard, R. Zecchina, *Survey Propagation: an Algorithm for Satisfiability*, preprint URL: `http://lanl.arXiv.org/cs.CC/0212002`
3. Code and benchmarks available http://www.ictp.trieste.it/~zecchina/SP/.
4. M. Mézard, G. Parisi and R. Zecchina, *Analytic and Algorithmic solutions to Random Satisfiability Problems*, Science 297, 812 (2002) (Sciencexpress published on-line 27-June-2002; 10.1126/science.1073287)
5. S. Mertens, M. Mézard, R. Zecchina, *High precision values for SAT/UNSAT thresholds in random K-SAT*, in preparation, (2003)
6. Friedgut E., Necessary and Sufficient conditions for sharp thresholds of graph properties, and the k-SAT problem, J. Amer. Math. Soc. **12**, 1017-1054 (1999).
7. Dubois O. , Monasson R., Selman B. & Zecchina R. (Eds.), Phase Transitions in Combinatorial Problems, Theoret. Comp. Sci. **265** (2001)
8. M. Mezard, G. Parisi, M.A. Virasoro, *SK model: replica solution without replicas*, Europhys. Lett. **1**, 77 (1986); M. Mezard, G.Parisi, *The cavity method revisited*, Eur. Phys. J. B 20, 217 (2001)
9. M. Talagrand, Rigorous low temperature results for the p-spin mean field spin glass model, *Prob. Theory and Related Fields* **117**, 303–360 (2000).
10. The cavity and the replica methods deal with average quantities over some probability distribution of problem instances (e.g. average fraction of violated clauses in random K-SAT problems). See ref. *Complex Systems: a Physicist's View*, G. Parisi (2002), cond-mat/0205297 for a recent review.
    In ref. [1] it was realized that quite in general the cavity approach could be brought down to the level of single problem instances thus revealing the algorithmic potentialities of the formalism. As discussed in this paper, a simple version of survey propagation was already known to Gallager since 1963 and used in practice as decoding algorithm in Low Density Parity Check codes and Turbo Codes.
11. F. Guerra, F.L. Toninelli, *The infinite volume limit in generalized mean field disordered models*, Commun. Math. Phys.**230**:1, 71 (2002)
12. S. Franz, M. Leone, *Replica bounds for optimization problems and diluted spin systems*, J. Stat. Phys. 111 (2003) 535

13. *Low-Density Parity-Check Codes*, R. G. Gallager, Cambridge, MA: MIT Pres,1963.
14. *Probabilistic Reasoning in Intelligent Systems*, J. Pearl, 2nd ed. (San Francisco, MorganKaufmann,1988)
15. Over the n-dimensional hypercube, we say that two vertices are connected if they represent solutions at Hamming distances one. We define a cluster as a connected component over the whole hypercube. SP looks at clusters with a finite fraction of backbone variables [17,1]. Exact enumerations on small random formulas indeed confirm the onset of clustering below α_c already for rather small values of n (A. Braunstein, V. Napolano, R. Zecchina, in preparation, 2003). Rigorous results about clustering phenomenon taking place in random K-XOR-SAT can be found in:
    S. Cocco, O. Dubois,J. Mandler, R. Monasson, *Rigorous decimation-based construction of ground pure states for spin glass models on random lattices*, Phys. Rev. Lett. 90, 047205 (2003)
    M. Mezard, F. Ricci-Tersenghi, R. Zecchina *Two solutions to diluted p-spin models and XORSAT problems* J. Stat. Phys. 111, 505 (2003)
    We mention that for random 3-SAT, there exists another clustering transition at $\alpha \simeq 3.87$ [5] where clustering appears and yet no variable is constrained. Such a transition is absent for any $K > 3$.
16. In the large n limit, properties of typical random K-SAT instances are:
    – $P(|A(i)| = k) = Poisson(i, k)$ so $|A(i)| = o(1)$ for each i with high probability.
    – The most loops are of length $O(ln(n))$
    Such properties are usually referred to as locally tree-like. Cavity variables will be at typical distance of order $\log n$ and hence conjectured to have marginal probability distributions approximatively uncorrelated.
17. R. Monasson, R. Zecchina, S. Kirkpatrick, B. Selman and L. Troyansky, *Computational complexity from "characteristic" phase transitions*, Nature (London) **400**, 133 (1999).
18. B. Selman, H. Kautz, B. Cohen, *Local search strategies for satisfiability testing*, Proceedings of DIMACS, p. 661 (1993).
19. S. Seitz, P. Orponen, *An efficient local search method for random 3-satisfiability*, LICS'03 Workshop on Typical Case Complexity and Phase Transitions (Ottawa, Canada, June 2003).
20. G. Parisi, *On the survey-propagation equations for the random K-satisfiability problem*, cs.CC/0212009, preprint
21. J.S. Yedidia, W.T. Freeman and Y. Weiss, Generalized Belief Propagation, in *Advances in Neural Information Processing Systems 13* eds. T.K. Leen, T.G. Dietterich, and V. Tresp, MIT Press 2001, pp. 689-695.
22. R. Kikuchi, Phys. Rev. **81**, 988 (1951)

Author Index